Archaeological and Paleontological Research in Lagoa Santa

Pedro Da-Gloria • Walter A. Neves
Mark Hubbe
Editors

Archaeological and Paleontological Research in Lagoa Santa

The Quest for the First Americans

Springer

Editors
Pedro Da-Gloria
Laboratory for Human Evolutionary and Ecological Studies
Department of Evolutionary Biology and Genetics in the Biosciences Institute
University of São Paulo
São Paulo, SP, Brazil

Walter A. Neves
Laboratory for Human Evolutionary and Ecological Studies
Department of Evolutionary Biology and Genetics in the Biosciences Institute
University of São Paulo
São Paulo, SP, Brazil

Mark Hubbe
Department of Anthropology
The Ohio State University
Columbus, Ohio, USA

Instituto de Arqueologia y Antropología
Universdiad Católica del Norte, Chile

Translation from the Portuguese language edition: "Lagoa Santa: História das pesquisas arqueológicas e paleontológicas" by Pedro Da-Gloria, Walter A. Neves, Mark Hubbe © Annablume Editora, 2016. All rights reserved. Translated by Martin Charles Nicholl with the support of the grant 2016/07279-9 from the São Paulo Research Foundation (Fundação de Amparo à Pesquisa do Estado de São Paulo – FAPESP). The opinions, hypotheses and conclusions expressed in this material are the authors' responsibility and don't necessarily reflect FAPESP's vision.

ISBN 978-3-319-57465-3 ISBN 978-3-319-57466-0 (eBook)
DOI 10.1007/978-3-319-57466-0

Library of Congress Control Number: 2017943688

Printed on acid-free paper

This Springer imprint is published by Springer Nature
The registered company is Springer International Publishing AG
The registered company address is: Gewerbestrasse 11, 6330 Cham, Switzerland

Foreword

Is Lagoa Santa a Tropical “Classic”?

What are the characteristics of a scientific contribution that make it become a “classic” in a given field of knowledge? How and why does an intellectual production (article, chapter, book) become elevated to a central position, acquire fundamental importance for a given field of knowledge, and become a “classic”? Is every work considered to be a “classic” necessarily old, dating back decades or even centuries, or can it be something more recent, almost as recent as yesterday?

A few years ago, I had some long and pleasant discussions with my friend Susan Lindee, a science historian at the University of Pennsylvania in the USA, regarding what it is that makes a work become a “classic.” Memories of those conversations came to my mind on several occasions during my reading of this important and stimulating book, organized by Pedro Da-Gloria, Walter A. Neves, and Mark Hubbe, on the past and present of the archaeological and paleontological research in Lagoa Santa.

On that occasion, Susan and I were writing the introductory text to a thematic issue for the journal *Current Anthropology* with articles originally presented at a Wenner Gren Foundation seminar on the history of physical/biological anthropology on the global scale (Lindee and Santos 2012). The question arose because, when thinking about the trajectory of a given intellectual field, immediately questions come to mind associated to certain authors and works that have been singled out in the course of time. The departure point for my comment was that anthropology, understood in its broadest sense as being composed of the subareas of social anthropology, archaeology, linguistics, and biological anthropology, does not currently possess a specific set of “classics” that are obligatory reading for students of its various specialties. Furthermore, there are important differences of temporalities to consider. For example, unlike social anthropology, in which the training process generally involves contact with authors from earlier centuries, especially the nineteenth and beginning of the twentieth century, like Emile Durkheim, Bronislaw Malinowski, or

Margaret Mead, the training of researchers in the area of biological anthropology rarely emphasizes the reading of works with a more distant temporality.

So why was that debate important for our argument? The question revolved around the fact that, quite often, researchers who work with human diversity, whether in the case of genetic studies or studies on plasticity (e.g., physical growth), do not read authors from the nineteenth or the early twentieth century who have written on the subject.

Susan asked me what I thought would be a "classic" for biological anthropology. Without hesitation, but not before having sipped a little of my wine (the conversations went on into the evening), I mentioned Franz Boas's studies of the physical characteristics of the children of immigrants to the USA in the 1910s, as well as the influential work of population geneticist Richard Lewontin entitled "The apportionment of human diversity," published in 1972. Those two texts, separated by half a century and based on very different theoretical and methodological perspectives, broadly redefined our views of the factors involved in the expression of human biological diversity. We spent hours reflecting on how important it could be for the training of future generations of biological anthropologists not only to include contact with more recent scholarship but also to become familiar with authors who, in their respective social, cultural, and historical contexts, had reflected on issues of human biological differences in the past.

From those exchanges, I came to the conclusion that whereas all works considered to be "classics" have also been historically significant for a certain field, the reverse is not always applicable. A given book may be of fundamental importance in the history of the constitution of a given study discipline, but those regarded as "classics," for the most varied theoretical and/or methodological reasons, tend to constantly reappear in the rounds of discussions or even be reborn from the ashes. They may be less remembered at one time or another, depending on the intellectual fashion of the day, but they insist on reappearing, re-nurturing, and re-signifying the debates. Another point that was reasonably apparent to me was the wide variability of the temporality of the "classics" in science as a whole. For example, if philosophy has many of its seminal works way back in the times of the Greeks, thousands of years ago, areas such as robotics or molecular biology may have their "classics" emerging the mere blink of an eye ago. In the case of anthropology, the temporality patterns of the "classics" are extremely varied. I would not go so far as to talk about archaeology and linguistics, areas in which I have great friendships with colleagues but at the same time only shallow theoretical/methodological knowledge, but in a similar way to social anthropology and biological anthropology, as mentioned above, they show marked differences in the temporality of their "classics."

Well, here I am in the seventh paragraph of this foreword and readers (at least those that have read this far) must be wondering whether the editor has not made a mistake and printed that of another book altogether. No, please do not give up (as yet). Everything that I have said above has been to emphasize that this volume, which involves almost two dozen Brazilian and foreign authors and addresses a vast set of themes, exemplifies exactly how "classic" (and contemporary) the discussions around the Lagoa Santa question are.

Allow me to introduce another concept (I promise to be brief) that I came across some time ago and which I think may be useful for a reflection on anthropology and the history of bioanthropological research in Brazil (and beyond, obviously). It is the concept of "significant sites of cognition and critical reflection" proposed by Australian science historian Warwick Anderson (2012). One of Anderson's main interests is the history of twentieth-century research in human biological diversity in a postcolonial perspective. He argues that there are certain places in the world which, in a singular and intense way, have been (and continue to be) differentiated focuses of research activities, many of which have been conducted for years or even decades. From such contexts, he argues, interpretations (and continual reinterpretations) have been generated that have strongly influenced the direction of scientific knowledge on a scale that goes well beyond the local or regional sphere.

There may be a few others, but in what situations in the Brazil of today, other than Lagoa Santa, is scientific knowledge in archaeology and bioarchaeology being produced which has such a vast spectrum of implications (time of occupation of the Americas by human populations, relations of that presence with environmental transformations, etc.) in a dialogue with ideas and perspectives emanating from investigations conducted more than 150 years ago? As can be concluded from this collection of chapters, many examples could be cited, but we can rate just one of them as being the most "classic" among them (also in the sense of perhaps being the most well-known): the so-called hypothesis of the contemporaneity of humans and the megafauna in Lagoa Santa. That theme, originally proposed by Lund in the first half of the nineteenth century, persists as a central one in the archaeological and paleontological debates around Lagoa Santa.

To make use of a more modern metaphor, perhaps we could say that, just like stem cells, Lagoa Santa has certain aspects of "totipotency," an incredible capacity for differentiation into many other types of contents. As the authors of the chapters that follow show us, ever since the time of Lund's pioneering studies, Lagoa Santa has been visited and revisited innumerable times over the years by scientific researchers and missions of the most varied countries of origin such as Denmark, France, the USA, and Germany, as well as by Brazilians, obviously.

While each new cycle of research has introduced new questions of interest and new investigatory techniques (including, e.g., new interpretations based on use of isotope dating procedures, beginning in the 1950s), themes and materials collected by Lund have commonly been reworked and debated in successive waves of investigation. On many occasions, going beyond the scientific interpretations, questions surrounding Lagoa Santa have influenced the relations among researchers and among institutions. An example is the interesting debates of amateur naturalists versus professional scientists about the region's multiple archaeological and paleontological aspects like those involving the Minas Gerais Academy of Sciences and the National Museum in the 1930s and 1940s.

Lagoa Santa's almost magical powers of generation and multiplication are present in the very constitution of the scientific collections of many of the more important institutions involved in research in that region in the course of the last century and a half. That is exemplified by the group led by Walter A. Neves at the University of Sao Paulo and the recent "Origins" project which has been in course since the year 2000 and is closely related to the propositions that Lund launched. The project has been discussing new hypotheses for the human occupation of the Americas and at the same time has been training a whole generation of researchers by involving them in the studies of the Lagoa Santa caves.

Like me, I hope that those who delve into the following pages will similarly be captivated by the complex mix of past and present that is so evident in the scholarship on Lagoa Santa, which this volume convincingly portrays as a remarkable example of a "significant site of cognition and critical reflection."

Ricardo Ventura Santos
Senior Researcher at the National
School of Public Health
Oswaldo Cruz Foundation, and Associate
Professor in the Department of Anthropology
National Museum, Federal University of Rio de Janeiro
Rio de Janeiro, Brazil

References

Anderson, W. (2012). Hybridity, race, and science: the voyage of the Zaca, 1934–35. *Isis*, *103*(2):229–253.

Lindee, S., & Santos, R.V. (orgs.). (2012). The biological anthropology of living human populations: World histories, national styles, and international networks. *Current Anthropology*, *53*(S5):S1–S266.

Contents

Part II Research Topics

About the Authors

Adriana T.A. Martins Keuller holds a degree in history from the Catholic University of Rio de Janeiro, a master's degree in the social history of culture from the same university, and a Ph.D. in the history of science from the University of São Paulo. She is the author of an e-book that received an award from the Faculty of Philosophy and Human Sciences of the University of São Paulo entitled *Os estudos físicos de antropologia do Museu Nacional do Rio de Janeiro: cientistas, ideias, objetos e instrumentos científicos (1889–1936)* (*Physical Anthropology Studies of the National Museum of Rio de Janeiro: Scientists and Scientific Ideas, Objects, and Instruments*) published by Humanitas. Currently, she is a collaborating researcher of the Astronomy and Related Sciences Museum (*Museu de Astronomia e Ciências Afins,* MAST) in Rio de Janeiro conducting research in the field of the history of anthropology.

E-mail: adriana.keuller@gmail.com

Alenice Baeta is an archaeologist, historian, and collaborating researcher of the Archaeology Sector of the Museum of Natural History and Botanical Gardens of the Federal University of Minas Gerais (MHNJB/UFMG). Baeta holds a master's in education majoring in ethno-archaeology and heritage education from the Faculty of Education of the Federal University of Minas Gerais, a Ph.D. in archaeology from the Museum of Archaeology and Ethnology at the University of São Paulo (MAE/USP), and a postdoctoral qualification in archaeology from the MHNJB--FAFICH/UFMG.

E-mail: alenicebaeta@yahoo.com.br

Andersen Liryo holds a degree in archaeology and a master's degree and a Ph.D. in the field of public health. He is assistant professor in the Anthropology Department of the National Museum/Federal University of Rio de Janeiro (UFRJ). Specialized in the field of biological anthropology, he has vast experience in human skeletal anatomy and is responsible for the subjects of paleopathology and forensic anthropology in the undergraduate and graduate courses at the UFRJ.

E-mail: liryo@hotmail.com

Andrei Isnardis is assistant professor in the Anthropology and Archaeology Department and researcher of the Archaeology Sector at the Museum of Natural History and Botanical Gardens of the Federal University of Minas Gerais (UFMG). He holds a Ph.D. in archaeology from the University of São Paulo and a postdoctoral qualification from the same university. His current research approaches Late Holocene settlements and precolonial rupestrian graphisms in the center and north of the state of Minas Gerais and the peopling of Brazil in the Pleistocene/Holocene transition and the Early Holocene.
E-mail: isnardis@gmail.com

André Prous is former professor at the Universidade de Angers (France), at the University of São Paulo, and at the Federal University of Minas Gerais, where he created the Archaeology Sector of the Natural History Museum in 1975. He took part in the Franco-Brazilian Mission to Lagoa Santa (1971/1976) and following that was responsible for the Minas Gerais Brazilian Rupestrian Art and Lithic Technology Mission. He is scholar of the prehistory of the meridional coastlands, archaeology of Minas Gerais, and Brazilian rupestrian art and lithic technology. He is currently working on Tupi-Guarani themes and preparing the publication of the results of research conducted in the upper Trombetas River.
E-mail: aprous80@gmail.com

André Strauss holds a social sciences degree from the Catholic University of São Paulo, a bachelors degree in geology from the Geosciences Institute of the University of São Paulo, an M.Sc. from the Genetics and Evolutionary Biology Department of the University of São Paulo and a Ph.D. in archaeology from the Eberhards Karls Universität Tübingen. He was a fellow of the Max Planck Institute for Evolutionary Anthropology and of the Konrad Lorenz Institute. His main lines of research include first Americans, human evolution, the archaeology of death, geometric morphometrics, and geoarchaeology. Since 2011, he is the archaeologist in charge of the excavations at the Lapa do Santo site.
E-mail: andre.strauss@uni-tuebingen.de

Astolfo Gomes de Mello Araujo is a member of the Institute of Advanced Studies (IEA), University of São Paulo; Professor at the Museum of Archaeology and Ethnology (MAE), University of São Paulo; Honorary Senior Lecturer at the Department of Archaeology, University of Exeter, England. He holds a BSc. in geology from the Institute of Geosciences of the University of São Paulo, with M.Sc. and Ph.D. in archaeology at the Museum of Archeology and Ethnology at the University of São Paulo. He is a coordinator of the Laboratory for Interdisciplinary Studies in Evolution, Culture, and Environment (LEVOC) and associate researcher at the Laboratory for Human Evolutionary and Ecological Studies (LEEEH) at the University of São Paulo. He has experience in the field of archaeology, with emphasis in geoarchaeology and geomorphology, working mainly in the following subjects: archaeology, quaternary studies, geomorphology, paleoenvironments, and archaeological methods.
E-mail: astwolfo@usp.br

Augusto S. Auler holds a B.Sc. in geology from the Federal University of Minas Gerais; M.Sc. in karst hydrology from the Western Kentucky University, USA; and Ph.D. in karst geochronology and paleoclimatology from the University of Bristol (UK). A professional consultant specialized in karst terrains, he is actively engaged in karst research throughout Brazil and abroad. He has produced around 270 karst and cave-related publications.
E-mail: aauler@gmail.com

Birgitte Holten is a Danish historian, Ph.D., living near Lagoa Santa in the mid-- 1980s, she became aware of the differences between the narratives on Lund in Denmark and Brazil. In collaboration with Michael Sterll, she has published the books: *P.W. Lund og Knokkelhulerne i Lagoa Santa* (Copenhagen 2010), Port.: *P.W. Lund e as grutas com ossos em Lagoa Santa* (Belo Horizonte 2011), (P.W. Lund and the Bone-caves in Lagoa Santa); and *Den forsvundne maler – P.W. Lund og P.A. Brandt i Brasilien*, (Copenhagen 2004), Port.: *O artista desaparecido – P.W. Lund e P.A. Brandt no Brasil* (Belo Horizonte 2012), (The Missing Artist – P.W. Lund and P.A. Brandt in Brazil). Currently, she is preparing an edition of the correspondence of Lund with the Danish zoologists J.C.H. Reinhardt and J.Th. Reinhardt.
E-mail: holtenbirgitte@gmail.com

Castor Cartelle holds teaching degrees in classic language and literature, philosophy, and biology. He has a master's degree in geosciences at the Federal University of Rio Grande do Sul and a Ph.D. in morphology from the Federal University of Minas Gerais (UFMG). He is a retired full professor of the UFMG. He was a permanent member of the Environment Policy Council of the State of Minas Gerais and president of the Chambers of the State Environmental Foundation and of the Biodiversitas Foundation. Currently, he is curator of the Paleontology Collection and professor at the Museum of Natural Sciences of the Catholic University of Minas Gerais.
E-mail: cartelle@pucminas.br

Claudia Rodrigues-Carvalho is an archaeologist (Estácio de Sá University) and holds a Ph.D. in public health from the National School of Public Health/Oswaldo Cruz Foundation (ENSP/FIOCRUZ), with a specialization in paleopathology and a master's degree in paleopathology from the same institution. She is assistant professor in the Biological Anthropology Sector of the Anthropology Department of the National Museum at the Federal University of Rio de Janeiro (UFRJ). She undertakes studies in the areas of biological anthropology with an emphasis on bioarchaeology and paleopathology. She is also involved in questions associated to human evolution and forensic anthropology. She is responsible for research topics in the master's degree course in archaeology at the National Museum and the course of specialization in Quaternary geology at the same institution. She also participates in the undergraduate course in biomedical sciences at the Federal University of Rio de Janeiro (UFRJ) and is currently director of the National Museum/UFRJ.
E-mail: claudia@mn.ufrj.br

Danilo Vicensotto Bernardo holds a degree in biological sciences from the Biosciences Institute of the São Paulo State University, Botucatu Campus. He holds a master's and a Ph.D. in biological sciences (biology/genetics) from the University of São Paulo. Currently, he is assistant professor in the archaeology course at the Human Sciences and Information Institute of the Federal University of Rio Grande. He has experience in biological anthropology, evolutionary biology, and prehistoric archaeology. His lines of research involve human evolution and microevolution, bioarchaeology, and the peopling of the New World.

E-mail: danilobernardo@furg.br

Fernando Walter da Silva Costa is a native of Belo Horizonte and is currently living in Espírito Santo state. He holds a history degree from the Faculty of Philosophy and Human Sciences from the Federal University of Minas Gerais and a master's degree and Ph.D. in archaeology from the Museum of Archaeology and Ethnology at the University of São Paulo.

E-mail: fwscosta65@gmail.com

Lucas de Melo Reis Bueno is assistant professor of the History Department and graduate program in history at the Federal University of Santa Catarina. He holds a Ph.D. in prehistoric archaeology from the University of São Paulo (2005), with a postdoctoral qualification in archaeology from the Muséum National d'Histoire Naturelle (2011), in the Laboratory for Human Evolutionary and Ecological Studies at the University of São Paulo (2008–2011), and at the Natural History Museum of the Federal University of Minas Gerais (2007–2008). He is a CNPq researcher developing projects in Central Brazil, Amazon, and Southern Brazil regarding the settlement of the Americas with an emphasis on technology.

E-mail: lucasreisbueno@gmail.com

Luís Beethoven Piló is a geographer with a degree from the Federal University of Minas Gerais. He holds a Ph.D. in karstic geomorphology from the University of São Paulo. He undertook postdoctoral studies at the Laboratory for Human Evolutionary and Ecological Studies in the Biosciences Institute at the University of São Paulo, where he has worked with geoarchaeology and chronology of the Quaternary megafauna. Currently, he holds a research fellowship from the Chico Mendes Institute for Biodiversity Conservation/National Center for Cave Conservation and Research (ICMBio/CECAV), in addition to working as a consultant in the environmental field. He has published various articles and chapters of books on karsts, speleology, and related topics.

E-mail: lbpilo7@gmail.com

Mark Hubbe received his Ph.D. in biological anthropology from the University of São Paulo. He is currently an assistant professor in the Department of Anthropology at The Ohio State University (USA). His main lines of research are concentrated on the dispersion of human groups at the beginning of the Holocene in South America and the emergence of social inequality in prehistoric societies of the Atacama Oasis in Northern Chile.

E-mail: hubbe.1@osu.edu

Michael Sterll is a Danish science historian, (Magister). He dedicated during many years to the history of science, with special emphasis on the natural sciences. He became involved in the studies of P.W. Lund via the collections in the Zoological Museum in Denmark. He is a co-author with Birgitte Holten of books on P.W. Lund and P.A. Brandt (see Birgitte Holten profile).

Email: sterllmichael@gmail.com

Pedro Da-Gloria holds a degree and a master's degree in biology from the Biosciences Institute at the University of São Paulo (IB-USP). He received his Ph.D. in biological anthropology from The Ohio State University, USA, and is currently undertaking postdoctoral studies at the Laboratory for Human Evolutionary and Ecological at the University of São Paulo (IB-USP). The areas of concentration of his research are the health and lifestyle of prehistoric populations in Lagoa Santa and of present-day populations in the Amazon. His other areas of interest include bioarchaeology, bioanthropology, and human health and evolution.

E-mail: da-gloria@ib.usp.br

Renato Kipnis holds a history degree from the University of São Paulo (USP), a Ph.D. in anthropology/archaeology from the University of Michigan–Ann Arbor, and a postdoctoral qualification from the Biosciences Institute of the USP. Currently, he is carrying out research in Central Brazil and in the Amazon into human evolution, human ecology, archaeology, prehistory, hunter-gatherers, zooarchaeology, and paleoecology, as well as in the area of environmental sciences with an emphasis on the assessment, mitigation, and compensation for impacts on the archaeological, historical, and paleontological heritage.

E-mail: rkipnis@scientiaconsultoria.com.br

Rodrigo Elias Oliveira holds a degree and a master's degree from the Faculty of Dentistry at the University of São Paulo (FOUSP) and a Ph.D. from the Biosciences Institute of the USP (IB-USP). He is a postdoctoral researcher in periodontology at the FOUSP. Since 2006, he has been a collaborating researcher at the Laboratory for Human Evolutionary and Ecological Studies (LEEH, IB-USP), and his main study areas are bioarchaeology and bone and teeth pathologies in archaeological remains.

E-mail: eliaso@usp.br

Sheila Mendonça de Souza is a doctor, with an M.Sc. in anatomy and a Ph.D. in public health. She has dedicated years of study to bioarchaeology and paleopathology and has published many works, nationally and internationally. Currently, she is an active researcher attached to the National Public Health School of the Oswaldo Cruz

Foundation, where she participates in a research group investigating paleoparasitology and paleoepidemiology.

E-mail: sferraz@ensp.fiocruz.br

Verlan Valle Gaspar Neto holds a social sciences degree from the Federal University of Juiz de Fora (UFJF) and a master's degree and a Ph.D. in sociocultural anthropology from the postgraduate program in anthropology of the Fluminense Federal University (PPGA-UFF). He is author of *Na Pegação: Encontros Homoeróticos Masculinos em Juiz de Fora (Making Out: Male Homoerotic Encounters in Juiz de Fora)* (UFF publisher, 2014). His fields of interest and scientific production are: science and society, nature and culture, the history of anthropology, and the interfaces between biology and anthropology.

E-mail: verlan02@yahoo.com.br

Walter A. Neves is an evolutionary anthropologist. He implanted and coordinated the Laboratory for Human Evolutionary and Ecological in the Department of Evolutionary Biology and Genetics of the Biosciences Institute at the University of São Paulo (IB-USP). He has considerable scientific production in the areas of evolutionary biology, archaeology, ecological anthropology, and paleontology of megafauna. He holds a Ph.D. in evolutionary biology from the IB-USP and one postdoctoral qualification from the Northwestern University, USA, and another from the Department of Anthropology of the Faculty of Philosophy, Letters, and Human Sciences of the USP. He is full professor in the Department of Evolutionary Biology and Genetics at the IB-USP. He has published around 130 articles in Brazil and other countries. In the last two decades, he has been noted for his studies of the origins of man on the Americas. Currently, he is conducting paleoanthropological research in Jordan and archaeological research in the north of Peru. From the beginning of his professional career, he has been intensely active in extension and outreach to the general public, dedicating articles and books to that end as well as organizing various museological exhibitions.

E-mail: waneves@ib.usp.br

Chapter 1
Introduction

Pedro Da-Gloria, Walter A. Neves, and Mark Hubbe

Abstract In the past two centuries, the Lagoa Santa Karst has been intensely explored by paleontologists and archaeologists, given the richness of the fossiliferous and archaeological record on the local caves and rockshelters. This chapter introduces and briefly describes the chapters that compose the volume and highlights how the multidisciplinary nature of the research in the region contributed to make Lagoa Santa one of the most important archaeological regions in Brazil and South America in general. The book is divided into two parts. The first consists of chapters referring to each one of the great research interventions that took place in the region, beginning with the work of Peter Lund and culminating with the most recent intervention in the region. The second part of the book consists of synthetic reviews of important topics for research in the region such as migration, health, mortuary rituals, paleontology, rock paintings, lithic technology, and geoarchaeology. Together, the 17 chapters of this volume bring for the first time a comprehensive review of the archaeological work in Lagoa Santa to the international public.

In the course of more than 180 years of research undertaken in the region, the Lagoa Santa Karst in Minas Gerais has received, and continues to receive, special attention in the Brazilian scientific scenario for having been one of the first places in our Brazilian territory to be the object of systematic archaeological and paleontological investigation. The findings in that region have constantly gone beyond national borders and have had direct impacts on the discussions and concepts surrounding the question of the antiquity and the mode of human occupation of the Americas. Nevertheless, gaining an understanding of the long history of the research

P. Da-Gloria (✉) • W.A. Neves
Laboratory for Human Evolutionary and Ecological Studies, Department of Evolutionary Biology and Genetics in the Biosciences Institute, University of São Paulo, R. do Matão 277, sala 218, São Paulo, SP, Brazil
e-mail: da-gloria@ib.usp.br; waneves@ib.usp.br

M. Hubbe
Department of Anthropology, The Ohio State University, 174 West 18th Avenue, Columbus, OH, USA
e-mail: hubbe.1@osu.edu

P. Da-Gloria et al. (eds.), *Archaeological and Paleontological Research in Lagoa Santa*, DOI 10.1007/978-3-319-57466-0_1

conducted in Lagoa Santa has always required a huge effort on the part of researchers interested in local prehistory. The primary sources regarding the excavations are scattered in various archives in Belo Horizonte, Rio de Janeiro, São Paulo, and Copenhagen, and many of them are little more than handwritten manuscripts or internal reports. In that light, the idea emerged of bringing together, in a single work, researchers involved in the excavations, curatorship, and research in Lagoa Santa that would be capable of synthesizing all that dispersed information. Many of such authors have participated directly in the research trajectory described in this volume, and their reports and accounts are rich in details of how knowledge of the region was constructed. This book offers a space to representatives of multiple institutions that have contributed to research in Lagoa Santa, among which the National Museum of Rio de Janeiro, the Federal University of Minas Gerais, and the University of São Paulo are outstanding examples.

The concept of this book aims to reach out to a public of Brazilian and international researchers in the fields of archaeology, paleontology, cultural anthropology, and human sciences in general, and we have made every effort to adopt a precise scientific language well referenced in the specialized literature. On the other hand, the multidisciplinary nature of the research undertaken in Lagoa Santa requires a clear language that allows for efficient discussion among all those readers interested in details of the region's past. In that regard, we have orientated the authors of each chapter to avoid using excessively technical jargon. As the chapters of this volume clearly reveal, the studies in Lagoa Santa go beyond the boundaries of individual study disciplines and embrace dialogues among biology, archaeology, and anthropology, and we are very pleased to have been able to show that aspect here. Actually, the research efforts in Lagoa Santa have been and continue to be driven by very broad long-term scientific questions rather than being circumscribed by any single traditional knowledge field. As an example, many of the questions that have motivated research in Lagoa Santa, such as the man's antiquity in the Americas and his contemporary existence with the extinct mega-mammals, are the legacy of Peter Lund, a Danish naturalist who worked in the region in the 1830s and 1840s at a time when researchers could still transit freely among very different scientific fields.

The historical depth of the research in Lagoa Santa has made the region an area of great historical, cultural, and scientific value for Brazil. Indeed, the studies in Lagoa Santa are especially relevant and important for Brazil's historical and archaeological heritage. For decades, the region has been the target of mining activities exploiting the limestone rock and, therefore, threatening the preservation of the caves and archaeological sites. Besides the damage caused by mining, the destruction of archaeological sites by depredation and uncontrolled excavations still threaten the regional heritage. The preservation of the area has been the object of public-private initiatives such as the *Rota das Grutas Peter Lund* (Peter Lund Caves Circuit) endeavoring to harmonize preservation and the economic return generated by tourism. Within that context of preserving the Brazilian heritage, nothing could be more valuable than to publish a work that describes in detail all the rich array of knowledge produced in Lagoa Santa, with collections deposited in national and international public museums and universities. Indeed, it is our hope that the book will attract attention to Lagoa Santa's scientific importance not only of academics of the respective fields but also of the relevant authorities in the national and international spheres.

The book has been organized in two parts. The first consists of chapters referring to each one of the great research interventions that took place in the region, beginning with the work of Peter Lund and culminating with the most recent intervention led by one of us (WAN). Those chapters provide the reader with the institutional context of the period, the identities of the main researchers involved, and the debates and scientific production generated by each of those individual researchers or research groups. The second part of the book consists of synthetic reviews of important topics for research in the region such as migration, health, mortuary rituals, paleontology, rock paintings, lithic technology, and geoarchaeology. The chapters describe the construction of a body of knowledge on each topic, including recent contributions and changes in concepts over the course of history. In addition, these chapters delineate prospects for future research, identifying possible gaps in knowledge of the region and the pathways that could be traced to fill them.

The first part begins with Chap. 2 "Peter Wilhelm Lund – life and work," which tells the history of Lund's life. The chapter describes in detail moments of transition in the naturalist's life, such as when he decided to come to Brazil for the first time in 1826 and again when he met Peter Claussen on his second trip to Brazil. At the same time, as the authors describe those events, they also report on the discoveries Lund made in the caves of the Lagoa Santa region. The number of excavations carried out is extraordinary and so is the diversity of topics he addressed, such as the formation of the caves and an accurate description of the region's extinct and living animal species.

Chapter 3, "Peter Wilhelm Lund's Scientific Contributions," summarizes the scientific discoveries that were made in the period from 1835 to 1844, when Lund was carrying out excavations in the Lagoa Santa region. Written by researchers trained in the geomorphology of caves, the chapter describes Lund's solid contribution to paleontology, geology, Karst geomorphology, stratigraphy of sediments, and cave taphonomy. The chapter shows, once more, the importance of Lund's work as the initial starting point for scientific investigation in Lagoa Santa.

Chapter 4 describes "the anthropological studies of Lagoa Santa in the National Museum of Rio de Janeiro: insertion, debates, and scientific controversies at the turn of the 19th to the 20th century." Written by a historian, the chapter brings a detailed account of the insertion of the Lagoa Santa studies in the institutional context of the National Museum of Rio de Janeiro. At that time, the Museum played an outstanding role in Brazilian science, not only in the natural sciences but also in anthropology. It is possible to identify the organization of important museological exhibitions such as the Brazilian Anthropological Exhibition of 1882 and field expeditions undertaken to collect material to enable an understanding of the origin of American man. In this chapter, the author delineates a panorama of the scientific dialogues that took place between the national museum and other national and international institutions.

Chapter 5, "The Physical Anthropology Archives of the National Museum of Rio de Janeiro: Lagoa Santa in the First Half of the 20th Century," explores the documents in the national museum's physical anthropology archives, mostly from the work of Jorge Augusto Padberg-Drenkpohl and José Bastos de Ávila about the Lagoa Santa region. The former researcher found human bones in five sites and incorporated a large number of human skeletons to the national museum's collection, while the second worked in the Carrancas caves, where he also came across very ancient human skeletons. It is worth noting in this chapter how the question formulated by Peter Lund served as the driving force for much of the subsequent research in Lagoa Santa. The debate between the national museum and the Minas Gerais Academy of Sciences on the contemporaneity of humans and the extinct megafauna in the Lagoa Santa region is a typical example of the kind of debate that was occurring at that time.

Chapter 6 is entitled "The Minas Gerais Academy of Sciences – Lund's inheritors." It is written by an archaeologist, who seeks to redeem an important contribution the Academy made to the studies in the Lagoa Santa region. Even though they enjoyed no kind of public financing, the group undertook various archaeological and paleontological interventions in the region, publishing books and articles in Portuguese and English and organizing museological exhibitions. The author underscores how prolific were the academy publication record and its important role played in keeping alive the scientific questions formulated by Lund.

Chapter 7, "Archaeological Missions to the Lagoa Santa Region in the Second Half of the 20th Century," is a protagonist report on the American-Brazilian missions, led by Wesley Hurt in 1955 and 1956, and the Franco-Brazilian mission, led by Anette Laming-Emperaire in the 1970s. That period marked the beginning of expeditions of professional archaeologists in the Lagoa Santa region, involving international partnerships and large research teams. In addition to the archaeological data generated, the author of the chapter shows the long-term legacy that stemmed from those exchanges. In the case of the Franco-Brazilian mission, the author provides details of the excavations that only someone who participated in them directly could be in a position to share.

Chapter 8 is entitled "The National Museum's Contributions to Lagoa Santa Research in the Second Half of the 20th Century." It has been written by two professionals attached to the National Museum of Rio de Janeiro, representing an institution that played an outstanding role in the academic discussion of Lagoa Santa, as well as being the repository of a significant part of the collection generated by those research activities. The authors make a review of the collections in the museum's custody, totaling 1,500 archaeological and 41 paleontological items, in addition to 2,300 entries of human bones from Lagoa Santa. Furthermore, the chapter describes the museum's scientific missions and projects in the second half of the twentieth century, undertaken by researchers like Marília de Mello e Alvim, Maria Beltrão, Luiz de Castro Faria, Carlos de Paula Couto, Fausto de Souza Cunha, and Martha Locks Guimarães.

Chapter 9 is called "The Origins Project and the First Americans' Controversy." It has been written by members of a team of the large-scale project that unfolded in the Lagoa Santa region between 2000 and 2009. Led by one of us (WAC), the project "Origins and Microevolution of Man in America: A Paleoanthropological Approach" was structured around four research objectives referring to the fields of archaeology, biological anthropology, paleontology, and paleobotany. It is worth noting the large volume of academic products the project generated, with more than 200 written contributions to the four areas of research. The project involved fieldwork and curatorship of the archaeological and paleontological material, attempting to respond to important scientific questions outstanding since Lund's day.

Chapter 10, "The Repercussions of the Human Skeletons from Lagoa Santa in the International Scenario," inaugurates the second part of the book dedicated to thematic reviews of the research history in Lagoa Santa. In this chapter, the authors review the impact on the international scenario of the studies of human crania found in Lagoa Santa, from the time of Lund's publications up until the most recent articles discussing the first occupations of the Americas. The chapter demonstrates how important the Lagoa Santa material is for understanding the first human settlement in the Americas, for the morphological and chronological contextualization of the first wave of migration, and for the question of New World morphological diversity.

Chapter 11, "The Lagoa Santa Skeletons and the Cranial Morphology of the First Americans," explores the implications for Brazilian and international physical anthropology of the craniometrics studies conducted in Lagoa Santa. In it, we find a description of the changes in scientific approaches along the twentieth century, such as the introduction of multivariate statistical analysis and the use of big databases to gain an understanding of current and past morphological variability. Finally, the chapter addresses the recent discussions on the *Two Main Biological Components Model*, which is a genuinely Brazilian contribution toward understanding the arrival of humans in the Americas.

Chapter 12 is entitled "History of the Research into Health and Lifestyle in Lagoa Santa." In this chapter, the authors make a review of texts published on the health of the prehistoric population in Lagoa Santa, observing that it is a line of research that has been marginalized compared with studies on migration and craniometrics. In the second part of the text, the authors propose new ways of investigating health and lifestyle in the region, including studies focusing on human skeletons and others on living populations. Those new approaches need the support of a solid archaeological context, which has only recently been incorporated to the studies in Lagoa Santa.

Chapter 13, "Burial Practices in the Lagoa Santa Region," offers a review of another topic that has received little attention in the history of Lagoa Santa research. Up until recently, the descriptions of mortuary patterns in the region suggested a scenario that was at once simple and homogenous. However, recent excavations at the sites of Lapa das Boleiras and Lapa do Santo have made it clear that burial patterns are much more varied and elaborated than was formerly thought. The author describes three funeral patterns observed at the Lapa do Santo site, including dismembering, cremation, decapitation, multiple burials, and removal of flesh. The author interprets such manipulation of the body after death as a local, symbolic manifestation, possibly reifying cosmological principles.

Chapter 14, "Constructing the Past. A look at Lagoa Santa Paleontology," presents a review of the paleontology studies of Lagoa Santa. This chapter describes the importance of the legacy bequeathed by Peter Lund for posterior work in the region. Among such work is that of Herluf Winge and his monumental treatise *E Museo Lundii* and of Carlos de Paulo Couto, whose writings extended over a period of 30 years. At the end of the chapter, the author delineates two projects that have yet to be undertaken with the paleontological material: an investigation of a list of recent mammals that Lund made and the examination of the collection of micro-mammals deposited in the Museum of Natural History in Copenhagen.

Chapter 15 is entitled "The History of the Studies of Prehistoric Rock Paintings in the Lagoa Santa Karst." The authors describe the history of the study of the graphisms in the region, ranging from the illustrations of Peter Brandt in the nineteenth century to the chrono-stylistic review conducted by Alenice Baeta in the twenty-first century. In the second part of the chapter, the authors present some of the fundamental elements of the work with rock painting material, that is, the chronology, the characterization of the supports, the placement and relations among the figures, and their insertion in the landscape. Subsequently, the chapter includes the large stylistic traditions found in the region, such as the Planalto tradition and the Ballet figures. Finally, the authors describe the occurrence of overlapping, additions, and retouching of older paintings, showing how dynamic such registrations were.

Chapter 16 "Lithic Technology in Lagoa Santa in the Early Holocene" presents a complete discussion of the lithic material excavated in Lagoa Santa. The main lithic collections analyzed are those to be found in the National Museum of Rio de Janeiro and in the Institute of Biosciences at the University of São Paulo. The description of the lithic material comes with graphs showing the distribution of the pieces by raw material, by archaeological level, and by category of remains. That descriptive part is then used to underscore the challenging of the concept of homogeneous and simple industry in the region, insofar as Lagoa Santa shows a quite unique industry, made up of small artifacts with the occurrence of polished material and composite instruments. Another aspect underscored in the text is the exogenous nature of the raw materials, which may lead to an understanding of mobility and the complex interregional interactions. Once more, it becomes apparent that the archaeological scenario at Lagoa Santa has proved to be far more complex than its early descriptions would suggest.

Chapter 17 is called "Towards the Development of a Tropical Geoarchaeology: Lagoa Santa as an Emblematic Case Study." This chapter approaches the developments of a tropical geoarchaeology in the Lagoa Santa region. Although previous researchers approached the archaeological matrix of the sites, only recently, in the Origins Project, a robust effort was made to understand the tropical soils of the region. The authors show the importance of the anthropic component to explain the deep archaeological packets in the rock-shelters of the region. In addition, geoarchaeology studies has shown its relevance for understanding paleoclimate, contributing to the discussion of the middle Holocene "Archaic Gap" and to the understanding of the sediment deposition in open-air sites. Finally, the authors highlight the key role of cave geology and dating techniques to solve the long-held problem of the coexistence between man and extinct megafauna.

In short, we have brought together in this volume a great diversity of topics and theoretical approaches to the material in Lagoa Santa, emphasizing the temporal profundity of studies in the region. We hope the reader will enjoy the products of this richness of information and come to value and recognize the importance of the scientific heritage that the Lagoa Santa region represents. Good reading!

Part I
History of Research in Lagoa Santa

Chapter 2
Peter Wilhelm Lund: Life and Work

Birgitte Holten and Michael Sterll

Abstract Peter Wilhelm Lund (1801–1880) was a most remarkable nineteenth century Danish naturalist. During 10 years of intense work in the Brazilian limestone caves in the 1830s and 1840s, he generated new explanations for the evolution of the Earth, its fauna and flora, and human beings. His scientific methods were characterized by his unprecedented focus on complete systematic registration. Lund discovered and described an enormous range of extinct animals. He demonstrated that the extinction of prehistoric animals had by no means been complete, that actual animals in fact had lived side by side with animals that are now extinct and even that human beings had been contemporaneous with the extinct megafauna. Lund's discoveries and analyses contributed toward the lasting and persistent reformulation of the history of the Earth and of humanity that eventually paved the way for a more general acceptance of Darwin's revolutionary ideas. In fact, he offered to Darwin a long-term view of animal evolution. In addition, Lund performed a pioneering attempt of determining an absolute dating of the contents of a cave. This chapter gives an introductory survey of Lund's work in the caves of the Lagoa Santa region, as well as a summary of his most important studies on cave fauna.

Peter Wilhelm Lund (1801–1880) was among the most remarkable Danish naturalists of the nineteenth century (Fig. 2.1). During 10 years of intense work in Brazilian limestone caves, he generated new explanations for the origin and evolution of the Earth, its fauna and flora, and human beings. His scientific methods were characterized by his unprecedented focus on complete systematic registration – whether it involved the contents of an individual cave, the taxonomy of the fauna of the Brazilian central plateau, or the vegetation around Lagoa Santa.

Lund's discoveries and analyses contributed toward the lasting and persistent reformulation of the history of the Earth and of humanity that eventually paved the way for a more general acceptance of Darwin's revolutionary ideas. The overriding aim of P.W. Lund's excavation work was not merely to collect as many bones as possible but, above all, to create a comprehensive view of the "Brazilian animal

B. Holten (✉) • M. Sterll
Lector emeritus, the Botanical Garden, University of Copenhagen,
Kurvemagerstien 67, DK 2300 Copenhagen, Denmark
e-mail: holtenbirgitte@gmail.com; sterllmichael@gmail.com

P. Da-Gloria et al. (eds.), *Archaeological and Paleontological Research in Lagoa Santa*, DOI 10.1007/978-3-319-57466-0_2

Fig. 2.1 Photographic portrait of Lund aged 64 taken by Eugen Warming (Natural History Museum, University of Copenhagen)

world." He did that by making detailed descriptions of the species, whether extinct or not, and their kinship relations.

Peter Wilhelm Lund was born into a wealthy family of cloth merchants in the Danish capital, Copenhagen. His father had left him and his two brothers a considerable fortune at his death in 1820. That inheritance, competently administered by his brothers, made Lund economically independent for the rest of his days.

In 1818, Lund entered the University of Copenhagen as a medical student, as the natural sciences were not yet included in the curriculum. Even so, Lund spent most of his time at the Museum of Natural History, presided over by Professor J.C. Reinhardt,[1] and at the botanical gardens with Professor J.F. Schouw.[2] At the age of 23, Lund graduated with the presentation of two dissertations, both of which were awarded the university gold medal. One of them, a treatise on vivisection (Lund 1825) became widely disseminated as a textbook at the universities of Europe. The subject of the other treatise was crustaceans, one of Lund's principal interests under his first stay in Rio (Lund 1826).

Lund set off for Brazil shortly after his graduation. Unlike most of his colleagues, he did not need a travel allowance from the state; he was instead endowed with a monetary grant for the express purpose of making a collection of zoological material for the Museum of Natural History.

We have no explicit indications of why Lund chose to go to Brazil. We know, however, that the country's recent independence had opened a new and unexplored field for naturalists. The resulting ample possibilities for making important discoveries and the warm climate is a valid explanation for the choice made by an ambitious and independent – as well as adventurous – young man.

---

[1] Johannes Christopher Hagemann Reinhardt (1776–1845), zoologist and Lund's professor at the University of Copenhagen. Father of Johan Theodor Reinhardt (1816–1882), zoologist and curator of the Lund collection at the Museum of Natural History in Copenhagen.

[2] Joakim Frederik Schouw (1789–1852), professor of Botany and an important political figure in Denmark.

Lund spent most of his first journey in Rio de Janeiro, with some incursions in Niteroi and a more prolonged stay near Nova Friburgo. Apart from the material he collected for the museum, the journey rendered good results for Lund's scientific endeavor and yielded important studies on tanagers (which became the object of Lund's doctoral thesis; Lund 1829),[3] mollusks (Lund 1834), vultures (Lund 1832), and ants. Lund returned to Europe in 1828 full of new experiences and with a profound and long-lasting love for the Brazilian Nature. This first trip was decisive for the course of his life.

Back in Denmark in 1829, Lund soon set off on his grand tour across Europe, visiting museums and making contacts with colleagues in the big cities. He passed though Berlin and Vienna, and on arrival in Italy, he took off to Sicily to study botany and marine zoology. While there, he received the news that his mother had passed away. Even devastated by the loss, he welcomed his freedom to travel and stay abroad. First, he decided to spend the winter of 1830–1831 in Paris where he attended George Cuvier's lectures on comparative anatomy and made acquaintance with leading figures in the field of natural history like Alexander von Humboldt, Henri Milne-Edwards, and Jean Victoire Audouin.[4] In the spring of 1831, he informed his family of his wish to make another trip to Brazil to complete the material already collected (Holten and Sterll 2010:69–85).

At the end of 1832, Lund set off once more for Brazil, arriving in Rio de Janeiro in January 1833. After a stay in Rio, he arranged to make an expedition to the interior with the botanist Luiz Riedel, a veteran of the ill-fated Langsdorff expedition.[5] They set off for São Paulo in 1833, spent the rainy season in Campinas, and then headed north, intending to pass through the province of Goiás. However, repeated bouts of fever delayed them so much that they opted for a shorter route across the northern part of Minas Gerais province (Holten and Sterll 2010:106–28). The immediate result of this expedition was an important botanical paper, which like Lund's other work was published in Copenhagen (Lund 1837a). Herein, Lund discusses the effects of fires on the *cerrado* (Brazilian savannah vegetation), stating his conviction that they occurred before the arrival of the Europeans in Brazil. This understanding would later serve his interpretation about the extinct megafauna way of life (Lund 1841b, 1846).

In the village of Curvelo, in the north of Minas Gerais province, Lund had an accidental and decisive encounter with his compatriot Peter Claussen.[6] A common industry of this region was the extraction of saltpeter from caves for the manufacture

[3] Small passerine bird inhabiting forests in Brazil, Argentina, and Paraguay (Family: Thraupidae). Lund studied the *Euphone* genus which today is attributed to the Fringillidae family, subfamily Euphoninae.

[4] Georges Cuvier (1769–1832), French zoologist, instrumental in establishing the fields of comparative anatomy and paleontology. Alexander von Humboldt (1769–1859), German geographer, explorer and diplomat, known for his expeditions in Latin America. Henri Milne-Edwards (1800–1885), French zoologist, publisher of *Annales des sciences naturelles*. Jean Victoire Audouin (1797–1841) French entomologist and ornithologist.

[5] Luiz (Ludwig) Riedel (1790–1861), German botanist, became director for the department of botany and the botanical garden of the Natural History Museum of Rio de Janeiro.

[6] Peter Claussen (1804–1855), also known as Pedro Claudio Dinamarquez and Chevalier Claussen. Danish natural history collector, business man, and adventurer.

of gunpowder. Claussen was heavily involved in this industry, as he possessed several caves at his nearby farm. At the end of the extraction process, huge bones were frequently found, which the local people believed were the remains of "giants." Claussen, however, was better informed. He had participated in the expeditions of Friedrich Sellow in the 1820s and had at that time established a lucrative trade, selling fossilized bones to European museums.[7]

Lund immediately realized the significance of Claussen's findings and changed his travel plans. After accompanying Riedel to Ouro Preto, Lund returned to Curvelo to explore the caves with Claussen. Two important incidents marked Lund's return to Curvelo. First, at Claussen's home, he met the Norwegian Peter Andreas Brandt, who for almost 30 years went to work with him as his illustrator and assistant.[8] Second, Lund realized that he could not stand Claussen's company. Indeed, they crossed paths many times in the years that followed (for further details, see Holten and Sterll 2010:131–34, 164–172).

In October 1835, Lund and Brandt traveled from Curvelo to Lagoa Santa where they intended to spend the rainy season. Lund, however, was captivated by the little town by the lake, and after 2 year of residence, he purchased a house on a large plot of land that went right down to the lake's edge. This house became his home for the next 43 years and the center of his scientific activity. The grounds were large enough for an ever-increasing number of sheds to store and study the findings from the caves. The garden was planted with specimens of the regional vegetation, including orchids, and gardening became Lund's main leisure activity. It also housed several animals – armadillos (Fig. 2.2), bush dogs, sloths, monkeys – that the boys from the town used to bring to him. In addition to company, these animals were utilized as study objects and references for the past megafauna behavior.

Fig. 2.2 Armadillo. Drawing by Lund (Museum of Natural History, University of Copenhagen)

[7] Friedrich Sellow (1789–1831), German botanist and naturalist. One of the earliest scientific explorers of Brazil.

[8] Peter Andreas Brandt (1792–1862), Norwegian artist and publisher; Lund's illustrator and assistant.

Excavations

During 10 years of work in the Lagoa Santo region, Lund excavated and researched innumerable caves. What follows is a summary of his enterprise.

In 1836, Lund undertook three journeys. Two of them were to the Fazenda do Mocambo and the Cerca Grande cave complex (Lund 1837c), and one was to the more distant Sete Lagoas region.

In 1837, Lund made two trips, the first returning to the Fazendo do Mocambo where Lund visited the Cerca Grande cave and the Lapa do Baú. On his second trip, Lund made a brief and unfruitful visit to the Sumidouro cave before returning to Cerca Grande and Baú.

In the years that followed, Lund cut down his trips to one every year, as he had already collected an enormous quantity of bones. In 1838, he visited the Lapa do Baú cave and made another try in the region of Sumidouro. In 1839, he began explorations at Sumidouro and visited the Escrivânia cave complex for the first time, before completing the excavations at Cerca Grande.

The decisive year of 1840 was marked by the first discovery of fossilized human remains. It was made at Lapa de Sumidouro, which is usually flooded. Lund wrote:

> Sunday, (July) 26. Visit to two caves in the rock at the eastern extremity of the Sumidouro Lake one of which I had been inside the year before but I had not got very far because it had been full of water. In a corridor of that cave, which had been under water the year before and probably becomes submerged periodically during the rainy season, two extraordinarily old, completely petrified human skeletons were found, in addition to some animal remains.[9]

In 1841, Lund concentrated his efforts on the Baú and the Sumidouro caves, as both continued to yield good results. The following year, he visited Baú and Escrivânia until the work was interrupted by a local insurrection, which stopped any further travel.

The year of 1843 became Lund's most productive one – he undertook no fewer than six expeditions, simultaneously excavating various caves and continuing work until the height of the rainy season. The largest project was the emptying of the Sumidouro cave to such an extent that it became possible to obtain a complete view of its geology and evaluate the deposits of human bones found together with those of extinct animals. The project was destined to yield one of his most important papers (Holten and Sterll 1998; Lund 1845a, b).

The last year of excavations, 1844, was dedicated to the complete removal and registration of all the material from the Lapa da Escrivânia V (Lund 1846). This was a daring experiment attempting to obtain an absolute dating of the contents of the cave. The cave's entrance was a vertical shaft stretching downward from the rock surface. A scaffold was mounted over the entrance, and during three and a half months, a dozen men extracted 6552 barrels of earth from within. The mandibles of small mammals contained in one barrel - chosen at random - were counted. Then followed a series of complicated calculations that considered the reduced numbers

[9] Lund: Travel Diaries (manuscript); Royal Library, Copenhagen, Add. 1128 4°.

of animals in the deeper layers of sediment. From that inventory, Lund estimated that the cave had contained 7,590,650 animals, most of them mammals and representing a total of 56 species.

The next step was the result of many years of ornithological observations. Lund was quite familiar with the white owls (*Glaucidium perlatum*) living in the caves, one couple to each cave, and he knew that each couple captures and devours about four mice a day, afterward regurgitating their bones. If the cave had been occupied without interruptions, the owl's devouring of 7.5 million mice would have required no less than 5137 years.

This estimate was in concordance with the high age of the Brazilian central plateau that Lund had emphasized in his scientific memoir on the excavations at Sumidouro (Lund 1845c). Lund's investigation of the owl pellets is remarkable as one of the first qualified attempts ever made to obtain an age estimate of paleontological material based on empirical observations.

Working Methods

Work in the caves was often carried out in complete dark, only sparsely illuminated by torches or candles. To reach the bone breccia, the layers of stalagmite covering most of the sedimentary deposits had to be removed using hammer and chisel. When the breccias were brought to the surface, the next step in the process was to evaluate their contents, cutting free the more interesting parts for supplementary investigation. In addition to thousands of isolated bones, Lund's collection included a great number of bone breccia to help the interpretation of how the bones were deposited. Lund himself described the organization of this work in these words:

> First the breccias are separated and divided into two classes: those that should be preserved just as they are, and those that should be sacrificed to extract the bones. The former, according to the circumstances, may undergo a finishing process, partly to eliminate any useless parts, and partly to obtain characteristic surfaces showing not only the bones but also the material surrounding them. The second class calls for more work and the use of different instruments, according to the hardness of the mass and the condition of the bones. For three months, two people have been busy at this work under my constant supervision. Similarly, the bones are separated into two groups: one comprising the complete bones and those with ancient fractures; the other comprising those with recent fractures and all the broken parts with the same characteristics. In spite of all the care taken, many bones are broken during the excavation or in the separation process and it is not always possible to localize and adjust the parts that should fit together. Piles of broken bones and bone fragments are always accumulating, and the cleaning up this veritable Augean stable calls for great patience and implies a great loss of time. It is like a puzzle, putting each of those pieces in its proper place and afterwards gluing and reconstructing the complete bones. Naturally, only I can carry out that work and it has taken me several months. Afterwards, each bone has to be examined, determined, numbered and introduced in the catalogue and, as you, Sir, are well aware, my own house does not have space enough for all those things, so I have packed up a large part of them and sent them to Rio. Only after all those preparations have

Fig. 2.3 Bone breccias (Museum of Natural History of the University of Copenhagen)

> I been able to form an approximate vision of the contents and, even then, only in relation to the larger animals because, in regard to the smaller mammal species and the bird and reptile classes, that is absolutely unthinkable in view of the stunning quantity of pieces, which is so great that, even if I were to use thousands as my unit, there would be great difficulty to count them, much less to examine them appropriately.[10]

During 10 years of excavation, Lund collected more than 1000 bone breccias (Fig. 2.3), thousands of fossilized and recent bones of animals, many of which were type specimens, fossilized human bones, and a reference collection of contemporary animal skeletons. In addition, he used the animals in his garden for behavioral studies.

The results of this assiduous activity were published in Danish in the Journal of the Danish Scientific Society (1837–1846). Excerpts of these publications were translated and published in journals in England and in France. At the end of the nineteenth century, four scientific memoirs were translated into Portuguese via the French versions, and in 1950, these Portuguese translations were reedited and supplemented by Carlos Couto (1950). The memoirs reveal quite clearly how Lund's thinking was transformed during his work in the caves; they show his growing understanding of the geological evolution of the Brazilian highland and bring evidence of the evolution of his view regarding the taxonomy of species from the caves. We shall proceed with a brief introduction to Lund's more important writings.

[10] P.W. Lund to J.Ch. Reinhardt, Letter dated April 26, 1844, Royal Library, Copenhagen, NKS 2838 4°.

Lapa Nova de Maquiné (Lund 1837b)

Maquiné was the first cave Lund ever visited, and his ideas and even his destiny were strongly marked by the singular beauty and scientific potential he found in this cave. The memoir on Maquiné is imbued with an almost lyrical tone, and Peter Andreas Brandt provided a detailed mapping of the cave as well as his first scientific illustrations to Lund's work. The memoir supplies a precise geological description and a full report of the remains of animals found below the stalagmite (Fig. 2.4). Lund and Brandt visited the cave together with Claussen and spent a week in the excavation.

Lund based his analysis of the Maquiné cave on Cuvier's catastrophism, proposing that a flood had swept the animals into the cave. The British geologist William Buckland (1784–1856) had proposed in 1823 a radical catastrophism derived from the biblical diluvium (Buckland 1823). Lund, however, clearly dissociates his views from any biblical understanding, accepting the term "diluvian" proposed by Buckland "in spite of the hypothetical identification of that event with the Mosaic diluvium that the name could, and indeed, did lead to" (Lund 1837b:230).

Fig. 2.4 Articulated skeleton underneath stalagmite crust in the Lapa Nova de Maquiné cave. Pencil drawing by Brandt (Natural History Museum of the University of Copenhagen)

More interesting is that Lund, already in this first memoire, introduces one very important question, which was to occupy his thoughts for the following years: how the animals have entered – and remained – in the caves. Throughout his scientific memoirs, Lund discusses this question repeatedly, proposing a total of five different explanations. In this first attempt, he considers three of them (numbers 1, 2, and 5):

1. Sick or old animals may have sought shelter in the caves and died inside.
2. Animals may have been carried inside by predators.
3. Animals may have entered the caves to lick the salty earth, getting lost in the dark corridors.
4. Animals may have fallen into the caves through vertical fissures in the roof that would apply both to extinct animals like the giant sloths and to actual ones like cattle.
5. Animals may have been swept inside by currents of water.

At Cerca Grande, Lund witnessed a demonstration of his first two explanations, while his observations of cattle and other actual animals made him include explanations 3 and 4.

Cerca Grande (Lund 1837b)

The Cerca Grande cave is situated on the Mocambo farm near Lagoa Santa, and Lund visited the many caves on this farm several times (Fig. 2.5). In Cerca Grande, Lund found his first fossil carnivore: a large, robust canine species that Lund called cave wolf (*Canis troglodytes*). In view of the great quantity of bones found in the cave, both wolfs and their prey, it was possible to establish that this was the first

Fig. 2.5 The Indians' Rock near to the Mocambo farm. Painting in Indian ink by Brandt (Royal Library, Copenhagen)

animal found that had indeed lived in the caves. Furthermore, the finds at Cerca Grande gave Lund his first clues for his description of the process of petrification.

The Cerca Grande canines lived in the caves and carried their prey, mostly pacas (*Coleogenys laticeps*), inside. Both the fossil canines and the fossil pacas were bigger than the contemporary ones. The establishment of this observation as a rule was a major inspiration for Darwin, giving him a concrete example of a very long-lasting evolution process and earning Lund and Claussen a reference in *On the Origin of Species*.

The petrification process had transformed many of the fossil bones into molds, where only the form remained while the bony tissue itself had been transformed into stone. Lund explained this process as being result of the continuous immersion of the bones in the calcareous waters of the cave.

> As regards the internal constitution of those bones and their chemical composition, they exhibited yet another phenomenon; one which I had not previously had the opportunity of observing in bones from that period. With the loss of their animal constituents, which these bones generally have lost to a greater or lesser extent, they normally become brittle and present a weak, earthy fracture in which the organic structure is clearly recognizable. That was not the case here. Every trace of bony substance and organic structure had disappeared, being replaced with calcite. Consequently, those bones were very heavy and stronger than fresh bones, and dropped on the floor they made the same sound as pieces of stalactite. Only a few bones avoided this transformation but in some of them the process was more or less complete making it apparent that the transformation had begun on the surface and advanced towards the inside.

The Cave Fauna Memoirs (Lund 1841a, b, c, d, 1842a, b, c, 1845c, 1846)

"A view of the animal world in Brazil before the last cataclysm of the Earth" (Blik paa Brasiliens Dyreverden før den sidste Jordomvæltning) is the collective title of a series of six scientific memoirs with various addenda that Lund published in the Journal of the Danish Scientific Society from 1841 to 1846. The memoirs consist of approximately 450 written pages and 56 illustrations produced by Brandt.

The Earth's geological evolution was the object of intense debate at the time. Had the Earth's history proceeded as a smooth evolution or as a sequence of sudden catastrophes? Was the extinction of a species a conceivable idea? Could one species be transformed into another?

The title of the memoirs contains a clear reference to Cuvier's catastrophism, according to which one or several "terrestrial convulsions" had wiped out the animals in a given area, later to be replaced by a new fauna. Cuvier did not speculate as to where that new fauna might come from, much less about the origins of human beings. The same is true of Lund, who always warned against any attempt to subordinate empirical observations to a system of thought. Lund, however, let himself become gradually convinced of the inadequacy of catastrophism and veer toward the uniformitarianism proposed by Charles Lyell.[11] He bought Lyell's book in 1840 and in later years regretted the Cuvier-inspired title chosen for his major work.[12]

The first memoir was written the year after the first memoirs on the Maquiné and Cerca Grande caves. The investigation and excavation of an additional 59 caves led him to alter his explanation for the cave formation and development. When he wrote the memoir on Maquiné, he imagined that the cave had been formed by the infiltration of water from above. Later, he abandoned that idea, demonstrating that far more violent currents of water would have been necessary.

> In these caves you feel as if wandering at a rocky coastline seeing how the naked rock walls are eroded and abraded by the waves. The origin is the same, and one is forced to change the period of creation of the caverns to those remote times when either great terrestrial lakes covered all of the now dry land, or the times when they still rested at the bottom of the sea. What is evident, is that infiltration of water from above through clefts in the limestone is in no way sufficient to explain those phenomena, especially in the case of deep, although blind, erosions of the ceilings. (Lund 1841a:32)

The first step in the process, the degradation of the rock by infiltrated water, is therefore attenuated and completed by a much more intense and more complicated process. First, it is necessary to bear in mind the slow rising up from the sea bed of the whole South America in the course of thousands of years. That was a subject of constant discussion among the geologists of the day, and Lund could read the regularly published notes by Charles Darwin, among others, about the measurement of the rising of the landmass on the coast of Chile, in the journal *Neues Jahrbuch für Mineralogie, Geologie und Paläontologie.*

Following the introductory notes on the region's geology, the formation of caves, and the contemporary animal species, the main contents of the memoirs consist in a progressive presentation of extinct animal species from the caves. It starts from the description of individual bones, and, as knowledge cumulated and new bone deposits were discovered, the complete image of the animal and its taxonomic relations is gradually delineated.

Lund's most valuable tool for his work with the extinct animals was the discipline of comparative anatomy, which is Cuvier's most lasting contribution to natural history. Lund had studied his methodology in depth and had become highly skilled at extrapolating conclusions from the detail to the whole, from the tiny bone to the complete animal. This method was especially useful, as many of the fossil bones were separated from their connections and often more or less broken.

Lund's basic premise for his work was that "… the set of animals which, prior to the Earth's last modification, populated the Brazilian tropical central plateau, was, in terms of basic types, the same as that which inhabits it now" (Lund 1841b:114) – on the other hand, however, "the last set of mammals to disappear from the South American continent, in terms of species, was completely different from those which

---

[11] Charles Lyell (1797–1875), British geologist. His book *Principles of Geology* proposed the principle of uniformitarianism, that the Earth's evolution has involved the same processes from its origin until today (Lyell 1830–1833).

[12] P.W. Lund a J.C. Reinhardt, Letter dated April 26, 1844, Royal Library, Copenhagen, NKS 2838 4°.

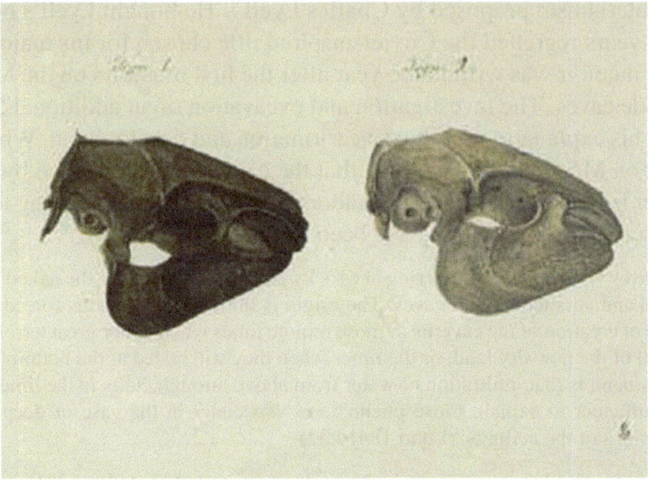

Fig. 2.6 Paca crania: extinct species *Coleogenys laticeps* on the left and living species *Coleogenys paca* on the right. Gouache by Brandt (Royal Library, Copenhagen)

today populate the same continent" (Lund 1841b: 118). In other words, the families have many characteristics in common, but the species have changed.

Despite that underlying certainty, Lund made a deliberate effort to discover if any species were identical in both periods, thus having succeeded to cross the "frontier" separating the extinct from the contemporary. This was a most difficult question to solve, as proof of being identical would require complete skeletons, whereas only a few bones were needed to prove the diversity of species. For a while, Lund believed to have established the identity between fossil and fresh bones of the paca (Cuniculidae family), but, in the end, he had to conclude the existence of three species, two extinct and one contemporary (Fig. 2.6). These species were only distinguishable from one another by a slight cranial variation (Lund 1841b:102).

The following year, a similar investigation, however, led to the opposite result. Lund was able to conclude that the fossil and fresh bones of the spiny rat (*Loncheres elegans*) belonged to the same species: "... a result which, if confirmed, will represent an exception to the rule whereby the species are different in every case for the two periods mentioned" (Lund 1841c:245). Finally, in the fourth memoir written in 1841, he put forward a thesis diametrically opposed to the one proposed in 1837:

> ... it is, nevertheless, very strange how the number of such forms from the past, which show a more or less notable affinity with present day forms, continually increases so that the probability increases day by day that the complete set of animals living today existed in those former times in a model that was more or less similar but that, at the same time, alongside those similar present-day animal forms, others existed, which to a greater degree were distant from them and that these last mentioned, generally speaking, are distinguished by their more considerable body masses. (Lund 1842b:141)

An important part of the "view of the animal world in Brazil before the last cataclysm of the Earth" is the increasingly diversified and embracing list of the currently established living and fossil mammal species attached to each memoir. In the later memoirs, these lists demonstrate clearly when a close affinity occurs between a fossil and its similar living species. When Lund, for instance, names the spiny rat *Loncheres elegans* and its fossil equivalent *Loncheres affinis eleganti* (i.e., "in affinity with "elegans"), he is killing two birds with one stone, indicating both the affinity and the slight differences between the two species. In the list attached to the fourth memoir, Lund enumerates 88 living and 111 extinct species. Among them, the term "affinis" is attributed to 38 fossil species, marking their representation among both the contemporary and the fossil species.

Already in his 1796 description of the *Megatherium*, Cuvier had noted the close relation between this huge animal and the tiny, modern day sloth. Lund continued in the same direction with his indication of the close relations between extinct and living species, despite the remarkable differences in size. In this way, he provided a demonstration of a long-term development which was highlighted by Darwin in *The Origin of Species* as a decisive inspiration:

> That relationship is even more clearly seen in the wonderful collection of fossil bones made by MM. Lund and Clausen in the caves of Brazil. I was so much impressed with these facts that I strongly insisted, in 1839 and 1845, on this 'law of succession of types' – on 'this wonderful relationship in the same continent between the dead and the living.' (Darwin 1859, p. 338).

Lund's discovery of human remains in the same layers as those of extinct animals posed the crucial question of contemporaneity of humans and extinct species. In 1843, Lund was able to demonstrate that humans had been contemporary with of at least five extinct animal species, and, in his report on the human findings at Sumidouro cave, he made a strong argumentation in support of this thesis (Lund 1845a, b).

Lund's Personal Evolution

In his student days, Lund had been influenced by the Romantic natural philosophy in vogue at the beginning of the nineteenth century. Faced with the reality of the Brazilian caves, however, he abandoned this line of thought and became adept of an implacable empiricism, which led him to sacrifice any theory that could be contested by observed facts. He was highly precocious in his readiness to admit that the question of the evolution of species was one that must be left totally open.

Lund was a product of his time but advanced beyond it in various aspects, becoming a true pioneer in various areas. One example is the historicity applied in his writings on the vegetation of the Brazilian central plateau (Lund 1837a), where he emphasized the human role in the evolution of nature. Another example is his investigation of owl pellets (1846) establishing for the first time an empirical calculation of the time necessary to fill a cave, making the first steps of establishing a methodology of absolute dating.

Lund continues to surprise us. How was it possible for a lone researcher, far removed from the European scientific milieu, to go so far with his scientific methods and to anticipate scientific development in so many areas? Our belief is that he was able of transforming his isolation from a weakness to a force and that his intense and prolonged contact with the nature he was studying opened the way for his novel approaches. It enabled him to make complete records of the local fauna during an immense time span by the Herculean task of emptying various caves of their entire contents. When he stumbled on a question of special interest, he was able to intensify his searches in this particular area. This enabled him to make one of his most important discoveries, namely, the close similarities between the living and extinct species. In addition to using that revolutionary approach in his own work, he passed it on as a legacy to two of his students who visited him in Lagoa Santa. He gave J.T. Reinhardt the assignment of making a complete collection of the fish fauna of the Das Velhas River and asked Eugen Warming for and an equally complete registration of the Lagoa Santa flora.[13]

References

Buckland, W. (1823). *Reliquiæ Diluvianæ, or, observations on the organic remains attesting the action of a universal deluge*. London: John Murray.

Couto, C. P. (1950). *Memórias sobre a paleontologia brasileira*. Rio de Janeiro: Instituto Nacional do Livro.

Darwin, C. (1859). *On the origin of species*. London: Penguin Classics. 1985.

Holten, B., & Sterll, M.. (1998). Et genfundet brev – naturforskeren P.W. Lunds afsluttende rapport fra udgravningerne ved Lagoa Santa, *Fund og Forskning*, Copenhagen, p. 129–218. (1999 Uma carta reencontrada – Relatório conclusivo das escavações feitas em Lagoa Santa pelo naturalista P.W. Lund. *Revista do Instituto Histórico e Geográfico Brasileiro* 160:371–398).

Holten, B., & Sterll, M. (2010). *P.W, Lund og knokkelhulerne i Lagoa Santa*, Natural History Museum of Denmark, Copenhagen. (2011 *P.W. Lund e as grutas com ossos em Lagoa Santa*. Editora UFMG, Belo Horizonte).

Lund, P.W. (1825) *Physiologische Resultate der Vivisectionen neuerer Zeit*. Copenhagen.

Lund, P. W. (1826). Om Krebsenes Aaresystem. Udtog af en af Kjøbenhavns Universitet kronet Priisafhandling. *Tidsskrift for Naturvidenskaberne, 11*, 233–248.

Lund, P.W. (1829). *De genere Euphones Præsertim de singulari canalis intestinalis structura in hocce avium genere*. Copenhagen: F. Brummer.

[13] Johannes Theodor Reinhardt (1816–1882), Danish zoologist, son of Johannes Christopher Hagemann Reinhardt. Curator of the Lund-collection at the Zoological Museum in Copenhagen, visited Lund several times in Lagoa Santa. His complete record of the fishes of the Rio das Velhas (Lütken 1875) became an important source for the Manuelzão Project initiated in 1997 for revitalizing this river. Eugene Warming (1841–1924), Danish botanist, Lund's assistant in Lagoa Santa (1861–1863), later director of the Botanical Garden in Copenhagen. His thesis *Lagoa Santa, et Bidrag til den biologiske Plantegeografi* (Warming 1892) and later works laid the ground for defining the notion of ecology.

Lund, P. W. (1832). Sur la confirmation particulière du Jabot chez l'Urubu (Percnopterus jota). *Annales des Sciences Naturelles, 5*, 333–336.

Lund, P. W. (1834). Recherches sur les enveloppes d'oeufs des Mollusques gasteropodes pectinibranches avec des observations physiologiques sur les embryons qui y sont contenus. *Annales des Sciences Naturelles Zoologie, 1*, 84–115.

Lund, P. W. (1837a). Bemærkninger over Vegetationen paa de indre Höisletter af Brasilien, især i plantehistorisk Henseende. *Det Kongelige danske Videnskabernes Selskabs Skrifter (DKDVS), 6*, 146–188.

Lund, P.W. (1837b). Om Huler i Kalksteen i det Indre af Brasilien, der tildeels indeholde fossile Knokler. Første Afhandling. *DKDVS* 6: 207–48. (Cavernas existentes no interior do Brasil, contendo algumas delas ossadas fósseis. Primeira memória. In: Couto, C.P. 1950 *Memórias sobre a paleontologia brasileira* (Lund, P.W.). Instituto Nacional do Livro, Rio de Janeiro, pp. 67–93).

Lund, P.W. (1837c). Om Huler i Kalksteen i det Indre af Brasilien, der tildeels indeholde fossile Knokler. Anden Afhandling. *DKDVS* 6: 309–30. (Cavernas existentes no interior do Brasil, contendo algumas delas ossadas fósseis. Segunda memória. In: Couto, C.P. *Op.cit.*, pp. 93–106).

Lund, P.W. (1841a). Blik paa Brasiliens Dyreverden før sidste Jordomvæltning. Første Afhandling: Indledning. *DKDVS* 8:27–60. (Primeira memória sobre a fauna das cavernas. Introdução. In: Couto, C.P. *Op. Cit.*, pp.107–30).

Lund, P.W. (1841b). Blik paa Brasiliens Dyreverden før sidste Jordomvæltning. Anden Afhandling: Pattedyrene. *DKDVS* 8:61–144. (Segunda memória sobre a fauna das cavernas. Mamíferos. In: Couto, C.P. *Op. Cit.*, pp.131–206).

Lund, P.W. (1841c). Blik paa Brasiliens Dyreverden før sidste Jordomvæltning. Tredje Afhandling: Fortsættelse af Pattedyrene. *DKDVS* 8:217–72. (Terceira memória sobre a fauna das cavernas. Continuação dos mamíferos. In: Couto, C.P. *Op. Cit.*, pp.207–50).

Lund, P.W. (1841d). Tillæg til de to sidste Afhandlinger over Brasiliens Dyreverden før sidste Jordomvæltning. *DKDVS* 8:272–96. (Suplemento às duas últimas memórias sobre o reino animal no Brasil, antes da última revolução do globo. In: Couto, C.P. *Op. Cit.*, pp. 273–92).

Lund, P.W. (1842a). Fortsatte Bemærkninger over Brasiliens Uddøde Dyreskabning. *DKDVS* 9:121–36. (Apéndice às observações sobre os animais fósseis do Brasil. In: Couto, C.P. *Op. Cit.*, pp. 293–305).

Lund, P.W. (1842b). Blik paa Brasiliens Dyreverden før sidste Jordomvæltning. Fjerde Afhandling: Fortsættelse af Pattedyrene. *DKDVS* 9:137–208. (Quarta memória sobre a fauna das cavernas. Continuação dos mamíferos. In: Couto, C.P. *Op. Cit.*, pp. 317–377).

Lund, P.W. (1842c). Tillæg til Dr. P.W. Lunds Blik paa Brasiliens Dyreverden før sidste Jordomvæltning. Fjerde Afhandling. *DKDVS 9361–63.* (Suplemento a 4a memória sobre a fauna das cavernas. In: Couto, C.P. *Op. Cit.*, pp. 379–380).

Lund, P.W. (1845a). Bemærkninger om forstende Menneskeknokler, fundne i Huler i Brasilien. *Antiquarisk Tidsskrift.* (Notícia sobre ossadas humanas fósseis achadas numa caverna no Brasil. In: Couto, C.P. *Op. Cit.*, pp. 465–484).

Lund, P. W. (1845b). Notice sur des Ossements Humains fossiles, trouvés dans une Caverne du Brésil. *Memoires de la Société royale des Antiquaires du Nord, 1845–49*, 49–77.

Lund, P.W. (1845c). Blik paa Brasiliens Dyreverden før sidste Jordomvæltning. Femte Afhandling: Fortsættelse af Pattedyrene. Om de nulevende og uddøde Arter af Rovdyrenes Familie. *DKDVS* 11:1–83. (Quinta memória. As espécies de carnívoros atuais e fósseis nos planaltos centrais do Brasil tropical. In: Couto, C.P. *Op. Cit.*, pp. 381–463).

Lund, P.W. (1846). Meddelelse af det Udbytte de i 1844 undersøgte Knoglehuler have afgivet til Kundskaben om Brasiliens Dyreverden før sidste Jordomvæltning. *DKDVS* 12:1–94. (Comunicação sobre o material das cavernas de ossadas exploradas em 1844 e sobre sua contribuição para o conhecimento da vida animal no Brasil antes da última revolução do globo. In: Couto, C.P. *Op. Cit.*, pp. 505–535).

Lütken, C. F. (1875). *Velhas-flodens Fiske: et Bidrag til Brasiliens ichtyologi efter J Reinhardts Indsamlinger og optegnelser*. Copenhagen: Bianco Luno.
Lyell, C. (1830–1833). *Principles of Geology: Being an attempt to explain the former changes of the Earth's surface, by reference to causes now in operation*. London: John Murray.
Warming, E. (1892). *Lagoa Santa, et Bidrag til den biologiske Plantegeografi (Lagoa Santa, a contribution to the biological plant geography)*. Copenhagen: Dreyer.

Chapter 3
Peter Wilhelm Lund's Scientific Contributions

Augusto S. Auler and Luís Beethoven Piló

> *"Future explorers of the caves in those regions will find, in these pages, a reliable guide that will orientate them in the research of their treasures and in determining their age".*
>
> Lund (1837) discoursing on the complex stratigraphy of the sediments in the caves.

Abstract Following Peter Wilhelm Lund's fortuitous meeting with compatriot Peter Claussen and his first visit to the Maquiné cave in 1834, Lund converted himself into a highly specialized scientist, formulating original concepts not only regarding the fossil bones but also the geology, speleology, and the age and modes of deposition of bones and cave sediments. Most of those topics were addressed in order to provide a robust framework in which the description of the fossil bones, his karst main interest, could be properly understood. In particular, Lund dedicated considerable efforts to understanding the depositional context of the cave bones. In his first forays into caves, Lund attempted to match his observations with the cataclysmic theory of his master, George Cuvier, which stated that a universal deluge would have suddenly killed all animals. However, Lund soon realized that living species were also found in caves, sometimes mixed with older ones. Lund cautiously distanced himself from Cuvier's ideas, proposing five mechanisms of fossil emplacement in caves: (i) carried in by predators, (ii) the fall of animals in vertical passages, (iii) animals that wandered into caves and got lost, (iv) the death of species that live inside caves, and (v) washed by runoff into caves. The discovery of human fossil bones at the Sumidouro cave allowed Lund to propose the great antiquity of these bones and their contemporaneity with the extinct species; both ideas have been proved correct by modern studies. Lund's contribution to various disciplines in the Lagoa Santa karst has retained its scientific value nearly 200 years later.

A.S. Auler (✉)
Instituto do Carste, Rua Barcelona 240/302, 30360-260 Belo Horizonte, MG, Brazil
e-mail: aauler@gmail.com

L.B. Piló
Grupo Bambuí de Pesquisas Espeleológicas, Belo Horizonte, MG, Brazil
e-mail: lbpilo7@gmail.com

P. Da-Gloria et al. (eds.), *Archaeological and Paleontological Research in Lagoa Santa*, DOI 10.1007/978-3-319-57466-0_3

Introduction

Peter Wilhelm Lund's relations with Lagoa Santa began with his fortuitous meeting with fellow countryman Peter Claussen (1801–1872) in Curvelo, toward the end of 1834. Claussen was the owner of a farm on which he resided, the Fazenda Porteirinhas, located near to the present-day municipality of Cordisburgo, in the state of Minas Gerais. It was in the first cave that Claussen showed him, the Lapa Nova de Maquiné, today known as the Gruta de Maquiné, that Lund, overwhelmed by the beauty and scientific richness of the cave, decided to set his life on a new course and, consequently, changed the scientific history of Lagoa Santa and of Brazilian paleontology as well. Even though the Gruta de Maquiné and all the other caves visited by Lund in his early years are not part of what is today referred to as the Lagoa Santa Karst, those pioneering studies must be included in any study embracing Lund or Lagoa Santa because they marked Lund's initial trajectory in the caves so strongly.

Lund's work distinguishes itself from the production of the various other naturalists that explored Brazil in the nineteenth century because of its notable degree of specialization. Lund's separation from the generalism that typified the naturalists of the time occurred immediately after his first visit to the caves. He withdrew from the general zoology and general botany that were his previous target areas, and for about 10 years, his productive activity was completely dedicated to the analysis of fossil material from the caves. His incursions into geology and the generalization of themes such as archaeology were focused on producing a backcloth that would serve to contextualize the voluminous material extracted from the caves. The depth and detailing of Lund's work are among the reasons that make his legacy so persistently relevant, even after almost 200 years have passed.

During his second stay in Brazil, which started in 1833, Lund met the German botanist Ludwig Riedel (1790–1861) (Fig. 3.1). Together with Riedel, Lund planned a study trip that was supposed to pass through São Paulo and Goiás and return via Minas Gerais to Rio de Janeiro. That route would take him to the region of the limestone caves that are found to the north of Belo Horizonte. Lund had previously visited caves in Germany (Lund 1836), and he often cited the works of Buckland, such as "*Reliquiae Diluvianae,*" which was based on bone finds in caves (Buckland 1823). Furthermore, Lund was well aware that other naturalists who had preceded him in the region, such as Johann von Spix and Carl von Martius, had excavated fossils in caves (Spix and Martius 1828). However, it seems to us improbable that Lund would have dedicated himself to his speleological incursions if he had not received the stimulus represented by his meeting with Claussen. Furthermore, his equally casual encounter with Peter Andreas Brandt (1792–1862) (Fig. 3.1), during his second stay at Claussen's farm in 1835, must be considered crucial for Lund's career, because it provided him with the means of mapping the caves and making scientific illustrations of the bone finds. The incredible coincidence of there being a "Peter" who was familiar with the caves and valued them and their scientific content (Claussen, even if only in commercial terms), a "Peter" with the artistic skills and technical capacity for mapping the caves (Brandt), and a third "Peter" with scientific qualification that enabled him to interpret all those finds (Lund) is really notable,

Fig. 3.1 The German Ludwig Riedel (*left*) and the Norwegian Peter Andreas Brandt (*right*) were Lund's two main companions. The portrait of Riedel is by Hercule Florence and was made a few years before his expedition with Lund. The portrait of Brandt is by an unknown artist

especially in the hinterland of the state of Minas Gerais in the years 1834–1835. It was that highly improbable combination of talents that made it possible for Lund to transform himself, in the very short space of a few months, from being a naturalist focusing on botany and zoology into a specialist in caves.

This chapter endeavors to synthesize Peter Wilhelm Lund's contribution to the body of scientific knowledge of the Lagoa Santa caves and their fossil contents. It consists of a complemented and completely revised synthesis of earlier articles on the subject published in the speleological magazine *O Carste* (Auler 2002; Piló 2002; Rubbioli and Auler 2002).

Lund and the Cave World

From the end of the eighteenth century on, the deposits of saltpeter (potassium nitrate) frequently found in the soil of the caves became the object of intense exploitation because they were the raw material for the manufacture of gunpowder (Gomes and Piló 1992). The economic value of the caves meant that the locations of many of them became well known in the region. Lund was greatly benefited by that knowledge, and it enabled him to visit various caves in a short period of time. For example, in the period from September 2 to October 17, 1835, Lund's team investigated 19 caves (Reinhardt 1888b) in the municipalities of Santo Hipólito and Lagoa Santa, in a trajectory of 200 km. Figure 3.2 shows the area researched by Peter Lund with the location of some of the most important caves.

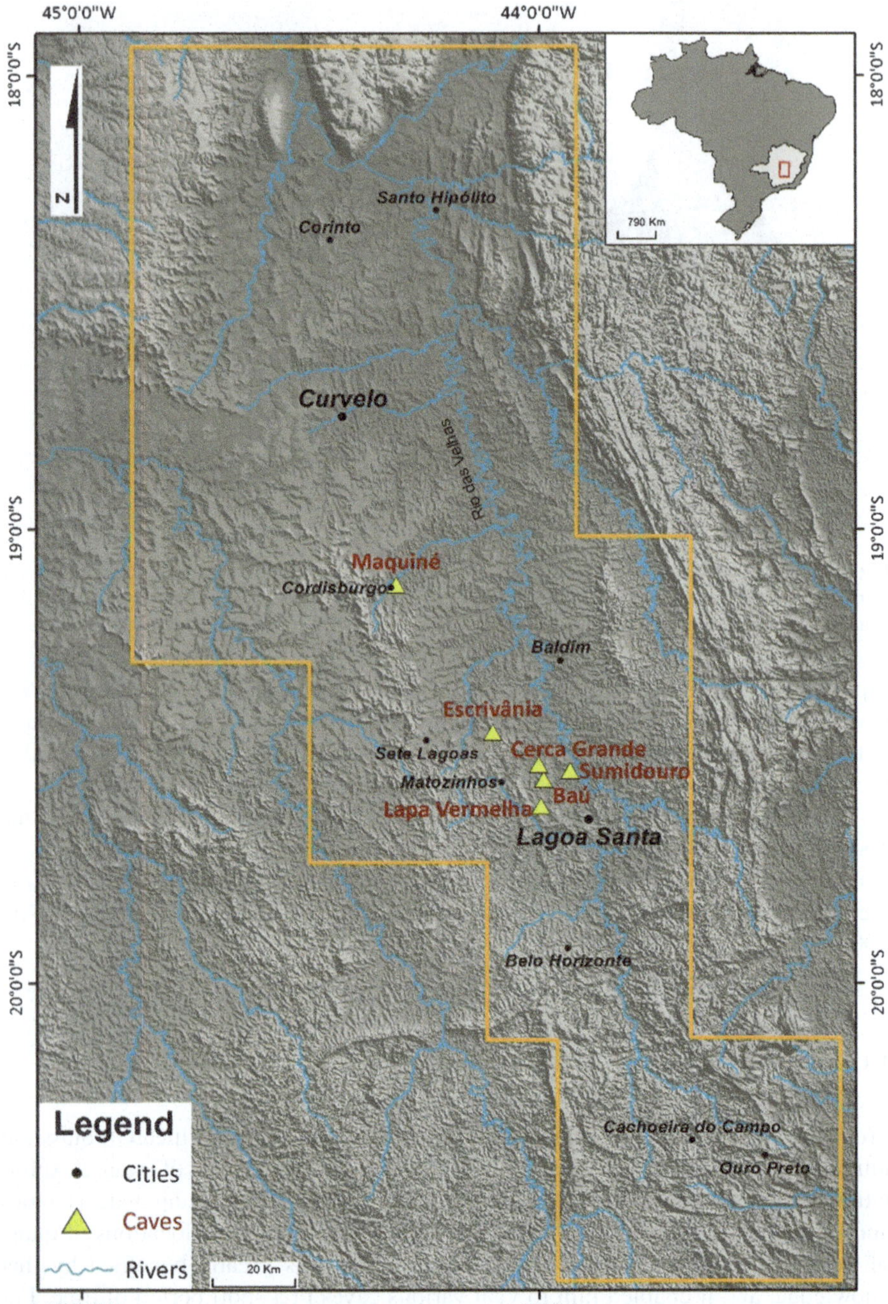

Fig. 3.2 Region explored by Peter Lund between 1834 and 1844, showing the locations of some of the more important caves

Over the years the number of caves investigated gradually increased. By February 1837, Lund recorded that he had visited 88 caves (Lund 1838). In September 1838, the number was up to 106 (Lund 1840), and the number at the end of his work in March 1844 (Lund 1845b) was over 800. That number is practically equal to the total number of caves officially registered for the region in the present day, which only goes to show the detailed approach of the work undertaken by him. In addition to those caves that were already known because of the extraction of saltpeter, Lund was in the habit of employing men to search out new caves, while the greater part of the team carried on with excavations (Lund 1846). Among the main caves Lund explored, there are some that were outstanding for the number and importance of the fossils they yielded such as the Maquiné, Cerca Grande, and Baú caves, the Lapa Vermelha of Lagoa Santa, and the Escrivânia and Sumidouro caves. The equipment used, as can be seen from the drawings produced by Brandt, included candles and torches for illumination as well as ladders to get up and down the steeper parts.

Apart from the excavating activities, it was the mapping, which was Brandt's responsibility, that required the highest level of technical knowledge. Although the earliest known cave maps preceded Brandt's work by almost a century (Shaw 1992), the maps produced in Minas Gerais represent collections of international importance. We have knowledge of maps of 27 caves, most of them elaborated during Lund's first trip between Curvelo and Lagoa Santa in 1835 (Table 3.1). Those maps are presented in plan and profile views (Fig. 3.3) and contain information on the types of soil, the location of the speleothems, and various other details. The accuracy of Brandt's maps was the object of comparison with maps produced using modern equipment whereby recognizable points in the Maquiné cave and the Lapa do Saco Comprido were compared in terms of their azimuth values (horizontal angles) and distances (Rubbioli and Auler 2002). It was found that there is an error of around 10° associated to the azimuths. The distances between points tend to have been overestimated by Brandt, in some cases by as much as 100%. That observation led Rubbioli and Auler (2002) to conclude that Brandt must have used a compass, but he did not have any apparatus that would enable him to measure the distances. The drawings show a good proportion among the passages and considerable artistic richness. Considering the period and the conditions in which they were created (Brandt probably conducted the mapping alone – work which today would require a team of at least three members), the maps produced are situated on the threshold separating work of good technical accuracy from a work of art (Rubbioli and Auler 2002).

Formulating Concepts at the Maquiné Cave

Lund was most certainly not expecting what he saw on his first visit to a Brazilian cave, the Gruta de Maquiné. The beauty of the cavern with its vast chambers, profusion of fossils, and the innumerable scientific questions it raised made the normally circumspect Lund wax emotional in his description of it (Lund 1836). It is important to stress that Lund was among the discoverers of most of the great internal

Table 3.1 Caves mapped by Peter Andreas Brandt listed in the order they appear in Brandt's sketchbook

Cave	Municipality	Representation, all with indicated north and graphic scale unless otherwise specified	Year and reference
Gentios	Curvelo (?)	Plan/profile/drawing of entrance	1835 Brandt (undated)
Onça	Cordisburgo	Plan/profile	1835 Brandt (undated)
Velha do Maquiné	Cordisburgo	Plan/profile	1835 Brandt (undated)
Santo Amaro	Cordisburgo	Plan/profile	1835 Brandt (undated)
Lagoa da Pedra	Cordisburgo	Plan/profile	1835 Brandt (undated)
Capim Branco	Corinto (?)	Plan/profile	1835 Brandt (undated)
Velha de Mocambo	?	Plan with heights	1835 Brandt (undated)
Quatro Bocas	?	Plan/profile	1835 Brandt (undated)
Mosquito	Curvelo	Plan/profile	1835 Brandt (undated)
Saco Comprido	Curvelo	Plan/profile	1835 Brandt (undated)
Dona Ana Felícia	?	Plan/profile	1835 Brandt (undated)
Soares	?	Plan/profile	1835 Brandt (undated)
Santa Rita	?	Plan/profile	1835 Brandt (undated)
Morcegos	?	Plan/profile/no scale	1835 Brandt (undated)
Três Bocas	Corinto (?)	Plan with heights/no scale	1835 Brandt (undated)
Cagaiteira	?	Plan/profile/no scale	1835 Brandt (undated)
Boca Apertada	Corinto (?)	Plan with heights/no scale	1835 Brandt (undated)
Labirinto	Corinto (?)	Plan with heights/no scale	1835 Brandt (undated)
Olho D'Água	Monjolos (?)	Plan with heights/no scale	1835 Brandt (undated)
Santo Hipólito	Santo Hipólito	Plan/profile/no scale	1835 Brandt (undated)
Vargem D'Anta	Monjolos (?)	Plan with heights/no scale	1835 Brandt (undated)
Paroba	Baldim (?)	Plan with heights/no scale	1835 Brandt (undated)
Cortume	Baldim (?)	Plan with heights/no scale	1835 Brandt (undated)
Forquilha	Baldim	Plan with heights/no scale	1835 Brandt (undated)
Nova de Maquiné	Cordisburgo	Plan/profile	1835 Lund (1836)
Cerca Grande	Matozinhos	Plan with heights	1836 Brandt (undated)
Lapa Vermelha	Lagoa Santa	Plan/no scale	? Reinhardt (1888a)

chambers because his team was the first to traverse the narrow passage separating chambers IV and V (Fig. 3.3). In that cave, Lund formulated various hypotheses that he would later modify or elaborate in greater detail as information from new caves being excavated became available.

Interestingly, when he published his first scientific memoir dedicated exclusively to the Maquiné cave (Lund 1836), Lund was perceptibly concerned to reconcile the observed phenomena with Cuvier's catastrophist theory. The fossils that were found there were mixed in with a red soil that also contained saltpeter and fragments of limestone. Lund called that layer of material "diluvium" because he believed it to be the result of a great cataclysm that had led to the disappearance of life on the planet.

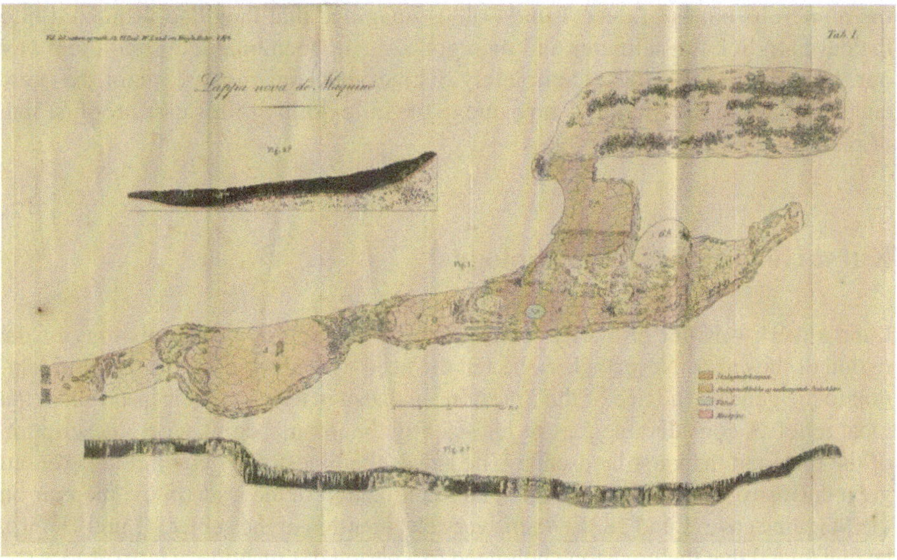

Fig. 3.3 Map of the Lapa Nova de Maquiné (plan and profile) (Produced by Peter Andreas Brandt in 1835. Published in Lund (1836))

Given that the entrance to the cave was quite narrow, it was improbable that even such a violent inundation could have managed to introduce the bones into the cave. That led Lund to hypothesize that the animals of the extinct fauna already found themselves in the vicinity of the entrance, fleeing from the floods, when they were violently introduced into the cavity and tumbled around in the clayey sediments (Lund 1836). The frequent fractures suffered by the bones were also attributed to the cataclysmic event, having supposedly been caused by the dragging of great blocks of limestone over the bones. Lund even inferred the direction of the great flood as having been from north to south because the blocks of limestone are only to be found in those caves with openings to the north such as the Maquiné cave (Lund 1838).

It was in Maquiné that Lund recognized, for the first time, the existence of a stalagmitic crust that covered the fossil material, separating it from the bones of more recent animals. That layer of calcite was to be adopted as a guide layer, because Lund was able to identify it in almost all of the caves he worked in. Lund attributed to that layer the action of preserving the bones and avoiding their degradation through contact with the atmosphere. However, some of the bone remains were found on top of the stalagmitic layer. Given the impossibility of evoking the diluvium to explain those bones of recent animals, Lund (1836) found himself forced to come up with ingenious modern explanations that he would later improve upon. The presence of the bones of small rodents was attributed to the activities of cave owls. Some of the larger rodents, like the pacas, have underground habits so that it is not unusual to find them in caves. However, to explain the numerous bones of deer, Lund (1836) imagined that they had wandered into the caves to lick the saltpeter and then got lost and been unable to find their way out. Lund had to use those completely distinct mechanisms to explain the presence of bones in the Maquiné cave due to the inflexibility of his catastrophist line of reasoning (Auler 2002).

Karst Geology and Morphology

Lund's work was not primarily focused on the geology or geomorphology of the region of the caves. Nevertheless, he felt the need to contextualize the caves in relation to the genesis and age of the limestone and also in terms of their insertion in the local relief. A considerable part of his scientific memoirs begins with a description of the rock and the karst landscape. The age of the limestone, in particular, aroused great curiosity in Lund since his first attempt to address the issue in the memoir on the Maquiné cave (Lund 1836). In his very last mention of the subject (Lund 1845b), he concluded that the limestone belonged to a period of transition, a term used at the time to refer to ancient rocks (of an age somewhere between those of the first rocks ever to form on the planet, the primary rocks, and those with fossils present in them – the secondary rocks), considering that no form of life was ever detected in the limestone.

Lund's ideas about the soil, however, were more detailed because it was essential for him to explain the origin of the fossil-bearing sediments. Lund was a pioneer in his identification of two kinds of soil to be found in the karst region of Lagoa Santa: the inferior layer of yellow soil and the upper layer of red soil, and he also proposed that the soil covering the karst relief of Lagoa Santa originated from alterations to the layers of phyllite rocks that no longer exist in most of the region (Lund 1838). That hypothesis has been corroborated by modern geochemical studies (Piló 1998).

Lund described many of the features that typify karst formations such as swallets, limestone pavements, dolines, limestone cliffs, and caves. He paid special attention to dolines (sinkholes), considering one of the main features of that karst region the presence of depressions that would fill with water seasonally, forming temporary lakes with complex hydrological regimes (Lund 1837).

Right from the outset of his studies at the Maquiné cave, Lund showed how well he understood the dissolution processes that were involved in the formation of caves, insofar as he proposed that the cave had been formed by the infiltration of the waters that flowed into fissures in the rocks right through to the end of the cave, dissolving the limestone as it went through the layers and fractures in the rocks (Lund 1836). Later, however, Lund (1838) proposed that the caves had been formed at a time when the entire region was covered by lakes or was on the seabed. That opinion was put forward in isolation and was not corroborated by his later studies; indeed, in his last work, Lund proposed a very complex model for karst evolution and the genesis of the caves. According to Lund (1845b), the region was originally a plateau with layers of shale (phyllite) overlying the limestone formation. Weathering acted on the superficial phyllite layers, eroding them and eventually reaching the limestone. Water then began to penetrate the cracks in the rock forming the caves and filling them with soil (Lund 1845b). In time cliffs were already formed, the horizontal penetration of water was favored, evacuating the older sediments and depositing new sediment layers (Lund 1845b).

Lund was also a pioneer in the recognition of the complex interactions involved in the processes that deposited sediments and removed them from the caves and which resulted in highly heterogeneous sediments with widely differing ages, conditioned by paleoclimatic variations (Auler et al. 2009).

The Stratigraphy of the Sediments

Establishing the correlation between the sediments in the caves and the fossils required a considerable effort on Lund's part. From the beginning of his work, he had been careful to describe the stratigraphic context of the bone finds, seeking for information to determine their antiquity and the processes that had led them to being found inside the caves. The attention and concern he dedicated to the fossil-bearing sediments were so great that he sent a collection of samples of them to Denmark accompanied by a catalogue (Lund 1845a; Neves and Piló 2005), so that they might serve as references for comparative purposes in the work of future researchers.

The frequent layers of calcite which, in Lund's (1838: 41) words, constituted "a carpet that marks the limit separating the past and the present" were an essential component of the cave stratigraphy. Their genesis intrigued Lund, and he put forward a highly ingenious hypothesis for it. His idea was that their growth was related to the rate at which water dripped. If the water dripped very slowly, there was time enough for limestone to be deposited at the point of dripping, thereby forming a conical stalagmite. If however the water dripped much faster, then it would run off the sides of the stalagmite and extend sideways to form calcite crusts (Lund 1836). According to Lund (1836), there would be a point of equilibrium, because if the dripping were even faster, it would lead to erosion and not precipitation (Lund 1836). Clearly, Lund understood the relations between chemically saturated waters (capable of precipitating calcite) and more aggressive waters (capable of dissolving rather than depositing calcite). The frequent occurrence of stalagmitic layers in caves that were essentially dry was what led him to suggest that there had been an earlier, moister period immediately after the diluvium (Lund 1836) and that was, undoubtedly, a pioneering paleoenvironmental interpretation in Brazil.

Another aspect that aroused the naturalist's curiosity was the presence of the saltpeter deposits, and he was sorry to see that many bone remains had been lost through the removal of those deposits to serve as raw material for gunpowder manufacture. After conducting some experiments of his own, Lund concluded that the saltpeter was not formed inside the caves nor was its origin linked to that of the limestone. Instead, it had been formed by rainwater leaching the soils and carrying mineral salts in solution into the caves where the contact with the limestone provoked the precipitation of the saltpeter sediment in the cave (Lund 1838). At the time, his theory was contrary to the prevailing idea that the saltpeter had been formed by the decomposition of bat guano and other materials inside the caves. It was only at the end of the twentieth century that independent experiments were able to arrive at conclusions similar to Lund's (Shaw 1997). As mentioned above, Lund understood the complexity of what the sedimentary deposits in the caves registered. Sediments with different origins (stemming from the soils above through vertical fractures in the rock or through the actions of rivers or lakes) could mix and give rise to deposits with different ages and characteristics. Lund perceived that in certain situations, stratigraphic inversion could have taken place, resulting in the deposition of older sediments on top of younger sediments. According to Lund (1845b), insofar as the land began to sink, the upper entrances to the cave would have been abandoned and that would have meant that the more recent sediments and fossils would have been located in the lower levels (Fig. 3.4). That phenomenon can take place in a single cavern and has now been confirmed by ^{230}Th dating of stalagmite crusts in Lagoa Santa (Auler et al. 2009).

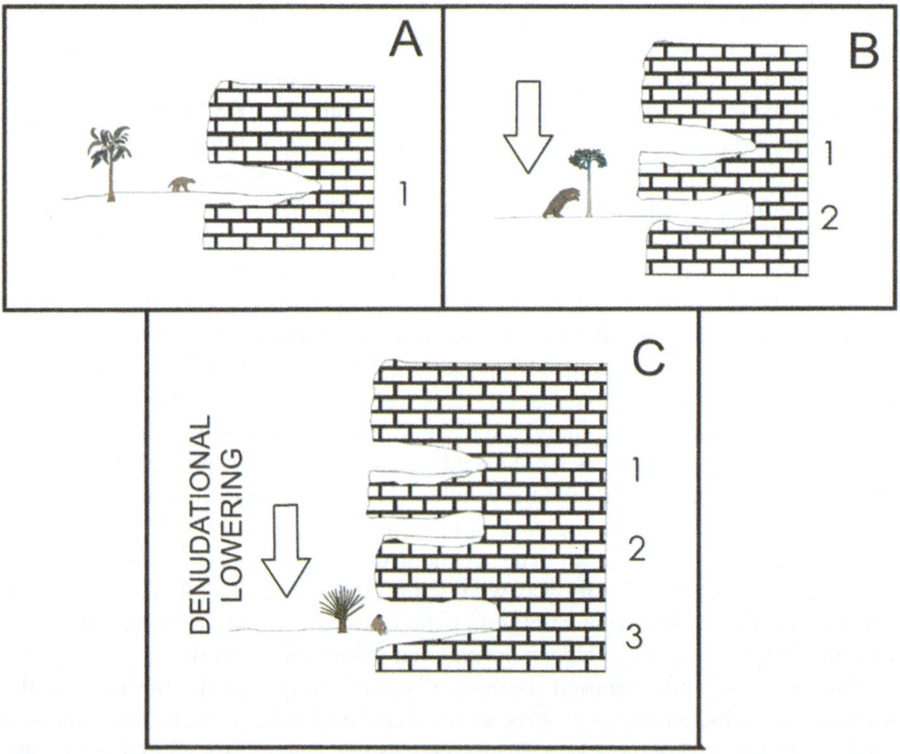

Fig. 3.4 P. W. Lund's inverted stratigraphic model. Caves would have been formed in the limestone rock providing an opportunity for the entry of sediments and fossils (A). As the surface of the Earth gradually lowered due to chemical and physical denudation (B), the higher caves (*1*) would have been abandoned, and their sedimentary deposits would have become preserved. Other caves would be excavated (C) at lower levels (*2* and *3*). Thus the higher caves would come to contain the older deposits, and the more recent deposits would be in the lower ones (Reproduced from Auler 2002)

Lund believed that humans had arrived on the scene late at Lagoa Santa, because he found few human remains in the fossil-bearing caves. Up until March 1844, of the 800 caves he investigated, only six had human bones in them (Lund 1845b). Among the latter, none was more challenging than the Sumidouro cave. In it, human bones were found side by side with the remains of extinct fauna species, thereby enabling Lund to address two of the great enigmas of the time: the great antiquity of the human remains and the possible coexistence of human beings with the extinct fauna species. Lund (1845b) considered that the sediments at Sumidouro, generally speaking, belonged to one of the three great groups (Table 3.2). The darker sediments of organic humus were obviously more recent and their presence was due to floods. The other two facies which contain most of the fossils were, in his view, variants of the same sediment, and their different appearance merely depended on how long they had been immersed in water during flooding. They varied from red to yellow with the yellow ones being those most exposed to the action of water.

The human bones were mixed up with the bones of the extinct fauna, and, because they all had a similar appearance, that is, a metallic, copper-like sheen, Lund (1845b) considered them to be of a similar age. Thus, Lund not only attributed great antiquity to the human bones, but he also defended the idea of their contemporaneity with the extinct fauna.

A large number of researchers investigated that possibility (see the review in Piló et al. 2005), but none of them actually reexamined the sedimentary contexts of the caves. Studies undertaken by Piló et al. (2004, 2005), involving detailed geochemical and stratigraphic analyses using ^{14}C and ^{230}Th dating as controls, showed that the fossil-bearing sedimentary layers are far more complex and ancient than previously believed and that they are present in two distinct sets. The oldest set was deposited prior to 240,000 years ago, and the more recent set around 8,000 years ago (Piló et al. 2005). The more recent set is the one associated to human bones and proves their great antiquity. The data obtained by Piló et al. (2004, 2005) does not actually prove

Table 3.2 Stratigraphy and fossil content, according to Lund (1845b), for the Sumidouro cave

Sediment	Characteristics		Fossils
More recent organic "humus"	Dark when moist and light gray when dry	Deposited periodically by flooding. Contains an abundance of aquatic shells	Man, bats, extinct capybara, wild pig, cave dog, jaguar, extinct pig, deer, armadillo, giant sloth, bones of rodents, monkeys, birds, reptiles, and fish
Original sediment filling the cave modified by floodwater	Grayish-yellow sediment	Partly calcified. Abundance of aquatic shells. Contains most of the human remains	Man, extinct horse, llama, giant sloth, capybara, deer, and wild pig
	Yellowish sediment with darker patches	Contains few aquatic shells	Man, monkey (marmoset), cave dog, llama, giant sloth, extinct horse, and capybara
Original sediment filling the cave	Reddish sediment	No aquatic shells. Hard in parts	Deer

Adapted from Piló et al. (2005)

Lund's assertion about the coexistence of humans and the megafauna of the Sumidouro cave. However, ^{14}C dating of material from other caves in the Lagoa Santa region supports his proposal because there is overlapping of the dating of the oldest human remains and the most recent megafauna remains (Neves and Piló 2003).

The impossibility of determining the age of the collected material must have been highly frustrating for Lund. In his last work, Lund describes his formidable attempt to evaluate the age of a deposit of small bones of rodents in the Escrivânia cave. Lund ordered his assistants to remove the entire contents of the cave. The work took 3.5 months and removed a layer of sediments 12 m thick (Lund 1846). More than 6,000 barrels of material were extracted and yielded an extraordinary number of small bones that corresponded to seven million individuals (Lund 1846). They were presumed to have been regurgitated by owls inhabiting the cave. Presuming an average consumption of four small rodents a day, and that the cave had been inhabited uninterruptedly by owls, Lund calculated that it would take at least 5,000 years to accumulate such a deposit.

How the Fossils Got into the Caves

Lund's earliest interpretation of how the fossils got to be inside the Maquiné cave was strongly influenced by the catastrophist theory, although he already showed signs of a more modern understanding of that process in the case of the entrance of the more recent bones. As time went by, Lund's ideas evolved considerably, so that in the end, with his work at the Sumidouro cave, he came up with a very broad explanation that made no mention whatever of the diluvium. Lund (1845b) identified five processes to explain how the fossils got into the caves: (i) introduction by predators such as the saber-toothed tiger. That idea, defended by Buckland (1823), was supported by the observations of innumerable small bones of rodents regurgitated by the cave owls; (ii) animals accidentally falling into vertical cracks in the rocks. Lund (1845b) used the example of the frequent loss of cattle in deep holes in the rock and again on his observation of a vertical conduit at the Cerca Grande cave, which was at first barren but when reexamined 2 years later contained 24 cadavers; (iii) getting lost in the labyrinthine caves. Some animals with subterranean habits enter the caves naturally to shelter or to lick the saltpeter and then do not manage to find the way out. According to Lund that was an infrequent process; (iv) animals that live in caves and eventually die there too. This group of animals consists of very few species, mostly bats and rodents; and (v) animals swept in by flash floods and torrents. The force of the waters is capable of carrying complete cadavers of animals into the caves. According to Lund, that process was most common in caves besides lakes or dolines, and it was far less important than processes (i) and (ii). Indeed modern papers on taphonomy (Auler et al. 2006; Simms 1994) consider processes (i) and (ii) to be the most significant.

Final Remarks

When analyzing the work of Peter Wilhelm Lund in Lagoa Santa, it is important to bear in mind the short space of time (a mere 10 years) and the scarce resources available to Lund. From being a naturalist focusing on botany and zoology, Lund quickly became an expert on a theme that was completely incipient at that time and almost everything was new. Working in the darkness of the caves, exhuming an extinct fauna unknown to the science of the day and constantly needing to contextualize the finds in terms of stratigraphy and taphonomy were a tremendous challenge. Although he lacked resources in terms of literature on the subject and was isolated from the scientific world at large, Lund masterfully carried out pioneering studies in areas as diverse as botany, zoology, geology, speleology, paleontology, and archaeology. He always based himself on his own observations and applied strict scientific rigor. His early dependence on catastrophist thinking gradually waned and was practically absent from his later works.

The greatest proof of the value of Lund's work is the fact that it is constantly cited in modern scientific papers and that renewed interest in it has resulted in academic theses and biographies. Almost 200 years after he first arrived at the caves of Lagoa Santa, his legacy and those of his successors, like Eugenius Warming, are still very much alive and represent a scientific landmark of international importance.

In spite of the distance between them, few people became as intimate with Lund's work as Herluf Winge (1857–1923) who spent decades working as the reviser of Lund's collection and coordinated the monumental treatise *E Museo Lundii*, published in the period from 1888 to 1915. In the last volume of the work, Winge pays just homage to the value of the countryman he never actually met:

> When Lund began his investigations on the animal remains in the bone-bearing caves of Brazil, he had no prior experience of that type of activity. He had to work in isolated regions in the remote interior of Brazil without the aid of sufficient comparative material or books. The material was only extracted from the caves a little at a time, so that it took a long time before it was possible for him to review it as a whole. Most of the remains that were found were just incoherent fragments, and some of the cave animals were strange creatures, very different from the species of today. It would have been a miracle if Lund had not frequently made mistakes in interpreting the bones. What is nearly a miracle, however, is that it did not happen more often than it actually did, and especially that, generally speaking, Lund himself frequently corrected his own mistakes. From his manuscripts it becomes very clear that in the end he acquired great skill in determining and interpreting the bones. Those who limit themselves to reading his printed scientific memoirs will not gain the right idea of the extent of his knowledge. (Winge 1915: 262)

Acknowledgments Colleagues of the Grupo Bambuí de Pesquisas Espeleológicas helped with fieldwork. Mike Robson has kindly provided the English translation from the Danish original of Winge's note on Lund.

References

Auler, A. S. (2002). Peter Wilhelm Lund e sua visão das cavernas. *O Carste, 14*, 24–30.

Auler, A. S., Piló, L. B., Smart, P. L., Wang, X., Hoffmann, D., Richards, D. A., Edwards, R. L., Neves, W. A., & Cheng, H. (2006). U-series dating and taphonomy of Quaternary vertebrates from Brazilian caves. *Palaeogeography, Palaeoclimatology, Palaeoecology, 240*, 508–522.

Auler, A. S., Smart, P. L., Wang, X., Piló, L. B., Edwards, R. L., & Cheng, H. (2009). Cyclic sedimentation in Brazilian caves: Mechanisms and palaeoenvironmental significance. *Geomorphology, 106*, 142–153.

Brandt, P.A. (n.d.). s/data *Caderno de esboços*. Original nos arquivos do Zoologiske Museum, Copenhague.

Buckland, W. (1823). *Reliquiae Diluvianae or observations on the organic remains contained in caves, fissures and diluvial gravel, and on other geological phenomena attesting the action of an universal deluge*. London: John Murray.

Gomes, M. C. A., & Piló, L. B. (1992). As minas de salitre: a exploração econômica das cavernas em Minas Gerais nos fins do período colonial. *Espeleo-Tema, 16*, 83–93.

Lund, P. W. (1836). Om Huler I Kalksteen I det indre af Brasilien, der tildeels inderholde fossile knoker – Førstte Afhandling. *Det Kongelige Danske Videnskabernes Selskabs Naturvidenskabelige of Mathematiske Afhandlinger, 6*, 207–249.

Lund, P. W. (1837). Om Huler I Kalksteen I det indre af Brasilien, der tildeels inderholde fossile knoker – Anden Afhandling. Lapa de Cerca Grande. *Det Kongelige Danske Videnskabernes Selskabs Naturvidenskabelige of Mathematiske Afhandlinger, 6*, 307–332.

Lund, P. W. (1838). Blik paa Brasiliens dyreverden för sidste jordomvaeltning. Første Afhandling. *Det Kongelige Danske Videnskabernes Selskabs Naturvidenskabelige of Mathematiske Afhandlinger, 8*, 29–60.

Lund, P. W. (1840). Blik paa Brasiliens dyreverden før sidste jordomvaeltning. Tredie Afhandling. *Det Kongelige Danske Videnskabernes Selskabs Naturvidenskabelige of Mathematiske Afhandlinger, 8*, 217–272.

Lund, P. W. (1845a). *Katalog over Professor Dr. P.W. Lund palaeontologiske samling. 2. Afdeling breccier*. Kopenhaguen: Zoologiske Museum.

Lund, P.W. (1845b). Notice sur des ossements humains fossiles, trouvés dans une caverne du Brésil. *Mémoires de la Societé Royale des Antiquaires du Nord*, 49–77.

Lund, P. W. (1846). Meddelelse af Det Udbytte de i 1844 undersögte knoglehuler Have afgivet til hundskaben om Brasiliens Dyreverden før Sidste Jordomvaeltning. *Det Kongelige Danske Videnskabernes Selskabs Naturvidenskabelige of Mathematiske Afhandlinger, 12*, 1–94.

Neves, W. A., & Piló, L. B. (2003). Solving Lund's dilemma: New AMS dates confirm that humans and megafauna coexisted at Lagoa Santa. *Current Research in the Pleistocene, 20*, 57–60.

Neves, W. A., & Piló, L. B. (2005). E Museo Lundii. *O Carste, 17*, 2–5.

Piló, L.B. (1998). Morfologia cárstica e materiais constituintes: Dinâmica e evolução da depressão poligonal Macacos-Baú – Carste de Lagoa Santa, MG. Ph.D. Dissertation, Universidade de São Paulo.

Piló, L. B. (2002). P.W. Lund e a geomorfologia cárstica de Lagoa Santa. *O Carste, 14*, 12–17.

Piló, L. B., Auler, A. S., Neves, W. A., Wang, X., Cheng, H., & Edwards, R. L. (2004). Revisitando a Lapa do Sumidouro: marco paleoantropológico do Quaternário americano. *Revista Brasileira de Paleontologia, 7*, 337–348.

Piló, L. B., Auler, A. S., Neves, W. A., Wang, X., Cheng, H., & Edwards, R. L. (2005). Geochronology, sediment provenance, and fossil emplacement at Sumidouro Cave, a classic late Pleistocene/early Holocene paleoanthropological site in eastern Brazil. *Geoarchaeology, 20*, 751–764.

Reinhardt, J. (1888a). De Brasilianke Knoglehuler og de i dem forekommende Dyrelevninger. *E Museo Lundii*, 1–49.

Reinhardt, J. (1888b). Oversigt over Lunds Hulerejser og de vigtigste paa dem besøgte Huler, for største Delen uddraget af hans Dagbogs – Optegnelser. *E Museo Lundii, 1*, 50–56.

Rubbioli, E. L., & Auler, A. S. (2002). Peter Andreas Brandt e os levantamentos topográficos das grutas pesquisadas por Lund. *O Carste, 14*, 18–23.
Shaw, T. R. (1992). *History of cave science*. Sidney: Sidney Speleological Society.
Shaw, T. R. (1997). Historical introduction. In C. A. Hill & P. Forti (Eds.), *Cave minerals of the world* (pp. 27–43). Huntsville: National Speleological Society.
Simms, M. J. (1994). Emplacement and preservation of vertebrates in caves and fissures. *Zoological Journal of the Linnean Society, 112*, 261–283.
Spix, J. B., & Martius, C. F. P. (1828). *Reise in Brasilien*. Lentner: Gedruckt bei I.J.
Winge, H. (1915). Jordfundne og nulevende Gumlere (Edentata) fra Lagoa Santa, Minas Gerais, Brasilien. *E Museo Lundii, 3*, 1–321.

Chapter 4
The Anthropological Studies of Lagoa Santa in the National Museum of Rio de Janeiro: Insertion, Debates, and Scientific Controversies at the Turn of the Nineteenth to the Twentieth Century

Adriana T.A. Martins Keuller

Abstract This chapter sets out to reconstruct the anthropological studies of Lagoa Santa at the National Museum of Rio de Janeiro in the context of the emergence of the discipline in the sphere of the institution in the nineteenth century through the 1930s. I use that reconstruction to gain an understanding of the practices adopted by the Museum's scientists which included undertaking scientific expeditions, increasing its collections, and establishing relations with other comparable museums in Brazil and the world at large. In that light, I present some of the debates and controversies associated to the studies of Lagoa Santa that the National Museum and the scientific community became embroiled in.

Introduction

Lagoa Santa, Peter Lund, and his followers are the object of this book. In this chapter, I have endeavored to delineate a historical slant on the theme in the context of the process of the institutionalization of anthropology at the National Museum of Rio de Janeiro. The National Museum played an important role in the scientific world in the nineteenth century, dedicating itself to studies and research in the Natural Sciences, and in the twentieth century, it became outstanding for its anthropological studies. Anthropologist L. Castro Faria stated that the National Museum "is the only [Brazilian] research institute where anthropology is cultivated in all the sectors of its broad domains" (Faria 1999).

A.T.A. Martins Keuller (✉)
Astronomy and Related Sciences Museum (MAST),
Rua General Bruce, São Cristóvão, Rio de Janeiro, Brazil
e-mail: adriana.keuller@gmail.com

P. Da-Gloria et al. (eds.), *Archaeological and Paleontological Research in Lagoa Santa*, DOI 10.1007/978-3-319-57466-0_4

To address the subject, in addition to harking back to the Museum's anthropological studies in the abovementioned period, I present the dynamics established among scientists in the construction of anthropological knowledge, observing their performances both within and outside of the institution.[1] Regarding the social and scientific interests that involve the actors and the process of constituting a scientific field, it is important to be aware of the networks of exchanges established locally and overseas. I have adopted P. Bourdieu's idea of "field" which he defined as a system of objective relations in a setting in which agents and their social positions are located (Bourdieu 1983).

In that light, the proposal here consists of getting to know the National Museum as it was in the nineteenth and twentieth centuries by analyzing the emergence of the anthropological studies of Lagoa Santa based on the constitution of the respective collections and the scientific production. At a second moment, the chapter will address the debate that sprang up at the turn of the nineteenth to the twentieth century with a brief description of the scientific production developed on the theme presented in a way that reveals the scientific relations constituted between the National Museum and the scientific world at large. Lastly, the chapter explains the controversies that sprang up in the face of new discoveries in the 1920s and 1930s, showing the role the institution played in the broader context of the Brazilian scientific community.

The National Museum in the Nineteenth Century

As soon as the Portuguese Royal Family arrived in Rio de Janeiro in 1808, a series of measures was taken to boost the city's growth and urbanization. Originally known as the Royal Museum of Natural History, the National Museum of Rio de Janeiro was founded in 1818, but its origins date back to the very beginning of the nineteenth century (Lopes 1997). With the increasing entry of new products, ideas, and men in the Court of Rio de Janeiro, the National Museum gradually became a Metropolitan Museum like its counterpart institutions in Europe and America. That environment attracted practitioners of Natural History, especially the traveling naturalists that visited Brazil (Jancsó 1997). It was in the light of that new scientific spirit that the scientists of the National Museum constituted the institution's collection and constructed scientific exchanges between Brazil's territorial regions and the nations of the Portuguese Empire.

Instead of being just another of the chambers of curiosities that typified eighteenth-century museums, the National Museum became a veritable "Cathedral of Science",[2] collecting, cataloguing, and exhibiting its collections and producing

[1] I refer to the A. Pickering's idea whereby scientific knowledge is a construction that is socially established (see Pickering 1992).

[2] According to the ideas of S. Sheets-Pyenson, museums are considered to be temples of Science in the terms of the prevailing scientific concepts of the day; each with its own organization and classification (see Lopes 1997).

and disseminating knowledge with programs of investigation that were typical of the Science of its day. In that context, its directors created regulations instituting administrative councils and norms regulating the functioning of the institution such as contracting naturalists and facilitating the work of traveling naturalists that conducted scientific expeditions and visited Brazil.

At the time, it was expected that in return for the imperial protection and assistance, those foreign expeditions would send part of the material they collected to Rio de Janeiro, where it would form part of the collection of the National Museum (Lopes 1997).

During the mandates of successive administrative boards, the National Museum engaged in the task of diffusing the lights of civilization, progress, and scientific development on behalf of the new Empire of Brazil. A complex interplay of forces that conjugated the personal, institutional, and scientific interests of the action of its "men of Science" (Schwarcz 1993) made the process of constructing the Imperial State feasible (Lopes 1997; Mattos 1986).

The National Museum, Ethnology, Anthropology, and Scientific Relations in the Mid-nineteenth Century

In 1842, the director of the institution, Custódio Alves Serrão, created a new regulation entitled "Numismatics, Liberal Arts, Archaeology, uses and customs of the Indigenous nations." That statute provided for an administrative council for the Museum, defined regulations for its internal functioning, and stimulated contact with other provinces of the country and with other museums in Europe (Keuller 2012a).

Some years later, the practice of ethnography was introduced in another imperial institution: the Brazilian Historical and Geographic Institute (*Instituto Histórico e Geográfico Brasileiro* – IHGB). Created in 1838, the Institute was intended to colligate, publish, and file documents on the History and Geography of Brazil. The Institution's Journal was created in 1839 and in it were published various studies on indigenous peoples and on the possibilities of exploring the Brazilian hinterland. Many of those studies were personally encouraged by the Emperor Pedro II who always underscored the importance of getting to know and collect information on the Brazilian people and Brazilian territory itself (Domingues 1996).

Insofar as their researchers shared the same interests, both institutions were part of the local scientific community. Such men belonged to what was known as the "Good Society" of the Brazilian Empire, and as scientists, they fostered, expanded, and affirmed the construction of the Imperial State and elevated the figure of Pedro II (Mattos 1986). The actions of the directors involved personal and scientific initiatives that congregated institutional interests with those of the Imperial State they represented (Lopes 1997; Mattos 1986). As an example, the director of the National Museum's Ethnographic section, Manoel de Araújo Porto Alegre (1806–1879), was also a member of the IGHB and of the Academy of Fine Arts (*Academia de Belas Artes*).

In that context, both institutions vigorously engaged in sponsoring a scientific expedition to the north and northeast of Brazil with the participation of Brazilian naturalists. The Project was proposed as early as the 1840s, but it only became feasible in 1856 when the Emperor Pedro II decided to finance it. It was called "The Ceará Scientific Commission" (*Comissão Científica do Ceará*) and nicknamed "the Butterfly Commission" (*Comissão das Borboletas*). Its aim was to collect geographic, astronomical, botanical, zoological, and ethnographic information from Brazil's northern and northeastern regions. From 1859 to 1861, the Commission traveled in the province of Ceará and part of Maranhão and the Amazon region (Lopes and Corrêa 1995; Pinheiro 2002). A member of the IHGB, A. Gonçalves Dias (1823–1864), was responsible for the ethnographic studies and was instructed by Manoel de Araújo Porto Alegre to register everything he could about the Indians: their physical characteristics, their moral and intellectual characters, the languages they spoke, and their historical traditions while at the same time collecting fossils in the regions visited (Keuller 2012a).

In keeping with the leading interests of the time, they were anxious to find evidence of the common origin of the continents and of humanity itself and to identify elements that differentiated human races from one another. The reference framework for Dias's work was the theory of racial hierarchies which was in vogue at the time, particularly the idea of racial decadence disseminated by Brazilian scholars and by A. Quatrefages de Breu (1810–1892), a friend of the Emperor.[3] In the debates on the origin of humanity, Quatrefages was a defender of monogenesis. He believed that humans had a single origin, but the races and peoples became hierarchized according to their different mental and moral levels.[4] Concern that the Indians would be exterminated after their contact with civilization had intensified interest in primitive tribes (Keuller 2012a).

Together with that extant idea, there was also interest generated by the Romantic movement of the nineteenth century which exalted the nation's riches, its grandiose proportions and its exoticism, and consequently the natural treasures of the Earth and the habits and way of life of the Indians as well. Thus the exploratory excursions expanded frontiers and carried the lights of civilization to the country's wildest and most backward regions. Explorers and their sponsors dreamed of the progress of Brazil. The museological collections represented the dimension of the nation's assets with the natural riches and the Indians of Brazil forming an image that mixed grandiosity, exoticism, and the dream of progress. That was the singular aspect of the Brazilian Empire in the face of civilized nations like France and England. With its movement of political expansion and affirmation, the Imperial Government sought to gain knowledge of the peoples in the nation's hinterland, explore the wealth there, discover new products to trade, and implement a policy of

[3] Quatrefages based himself on the ideas of Count A. de Gobineau (1816–1882) in his *Essai sur l 'inegalite des races humaines* of 1853.

[4] The discussion between monogenesists and polygenesists also took place in Brazil as the National Museum's studies show (see Gould 1999; Keuller 2012a).

populating the interior associated to the questions of immigration and colonization and replacing slave labor with non-slave labor (Costa 2007; Mattos 1986).

In the research work and the instructions that orientated the studies of the day, the use of adequate instruments and the forms of measurement used were evidence of awareness of the anthropometric methods and techniques developed in Europe around 1850 (Blanckaert 2001). In his travel log and in the correspondence A. Gonçalves Dias sent to the Emperor Pedro II and to other members of the Commission like Guilherme Süch Capanema (1824–1906) and Manuel Ferreira Lagos (1816–1871), it can be seen that the practice the Commission adopted was designed to gain knowledge of the natural history of Man. Even though Dias was engaged in the field of ethnography, was charged with describing the indigenous peoples, and was highly interested in the indigenous languages, he also presented references to the anatomical studies of P. Camper (1722–1789) and the phrenology of F. J. Gall (1758–1828). In the opinion of Lagos, who was in charge of the Commission's zoology section, Dias's exclusive responsibility was Anthropology, "in spite of Man being at the top of the chain of succession of human creation" (Dias 1971; Keuller 2012a). That was the first mention ever made of the term Anthropology in Brazil, and its practice was not the same as that adopted by P. Broca (1824–1880) and the Anthropological Society of Paris, founded in the very year that the Commission began its work (Faria 1998).[5]

The Ceará Scientific Commission faced various hardships and setbacks among its members during the years of the expedition. Among them was the supposed shipwreck with the loss of the material collected and the travel diaries The National Museum was the institution that benefitted the most from the Commission, and its material expanded Brazil's scientific collections. The changes brought about by the Museum's new directors at the end of the nineteenth century show how the institution branched out into new paths and became a showcase of knowledge. They also show the pathways trodden by the various branches of human natural history within the institution (Keuller 2012a).

Anthropology and the Reform of the National Museum at the End of the Nineteenth Century

In 1876, the museum's new director, botanist Ladislau Netto (1838–1894), carried out an institutional reform establishing the beginning of anthropological studies besides those of the natural sciences that had prevailed up until then. The botanist's interest in the practice of anthropology and archaeology dated back to the 1860s when he finished his studies in Europe. Accompanying the novelties that were all the rage on the European continent such as the repercussions of the publication of

[5] The anthropological studies carried out by the French doctor Paul Broca were based on anatomical and physiological knowledge from the point of view of the natural history of Man (see Faria 1998).

Charles Darwin's (1809–1882) *The Origin of Species*, in 1859, and J. Boucher de Perthes (1788–1868) discovery of fossil remains in 1863, Netto became interested in the origin of Man and, more particularly, that of the primitive inhabitants of Brazil, and so he came to intervene in an area that was not his field of expertise (Netto 1885).

During his mandate, Netto decided to create a "special Museum" that would be furnished with an anthropology section. Together with the Imperial Government, the botanist restructured the institution's old statutes and brought about a series of changes all designed to favor the progress of the sciences, qualify individuals (active citizens of the Empire), and promote Brazil in the eyes of the civilized nations.

Under the new regulations, the National Museum set its purpose as being "the study of natural history, particularly in Brazil, and the teaching of the natural and physical sciences especially in the aspect of their application to agriculture, industry and the arts." In that context, Anthropology was placed alongside Zoology, Anatomy, and Paleontology in alignment with the concept adopted by the Anthropological Society of Paris, created by P. Broca (1824–1880) in 1859. Archaeology, Ethnography, and Numismatic studies were accommodated in an annex to the Museum (Keuller 2012a).

Among Netto's other decisions was the institution of weekly courses for the public. The Anthropology course (Lacerda 1877), for example, given by J. B. Lacerda (1846–1915), subdirector of the first section, was the first ever in Brazil. Netto also created a specialized journal entitled Archives of the National Museum, publishing original works in the field of Natural Sciences and regular studies of Anthropology. The journal circulated free of charge to libraries and similar institutions in the Empire and overseas, thereby facilitating scientific dialogue and publicizing the scientific work being undertaken.

It is important to underscore the honorable mention that the journal of the Anthropological Society of Paris made in 1877 of J. B. Lacerda and J. Rodrigues Peixoto's paper entitled Contribution to the anthropological study of Brazil's indigenous races (*Contribuição ao estudo anthropológico das raças indígenas do Brasil*) (Lacerda and Peixoto 1876).

In compliance with the new parameters established for the Natural Sciences, a Physiology laboratory was set up in 1880 under the direction of French physiologist Luis Couty (1854–1884) and with the cooperation of J. B. Lacerda who would later become its director. The laboratory had a broad research program and introduced the so-called experimental studies.[6] The practice of anthropology as developed by the Anthropological Association of Paris at the time stressed the importance of creating a laboratory given the fact that anthropology was based on scientific deductions and inductions established by observation and experimentation (Keuller 2012a). According

[6] In that laboratory, the very first physiological experiments ever undertaken in South America took place. Among the subjects studied were animal venoms, toxic and nutritional plants, and the physiology of climate, yerba mate (*erva-mate*), and coffee as well as human and animal diseases. As the experimental studies developed, further attention focused on microscopic beings so that the laboratory also adopted a bacteriological focus (see Lopes 1997).

to the foundations proposed for it by P. Broca, Anthropology was understood to be the Natural History of Man, and its studies were based in Anatomy and Physiology.

Concerned with the question of qualifying its professional staff in a country where only doctors, engineers, and lawyers prevailed, the National Museum attracted to its staff various Brazilian naturalists that had studied overseas. That was the case of Ladislau Netto, a well-trained botanist and naturalist. Many of the traveling naturalists that the Museum hired for its external work were foreigners as, for example, the German naturalist Herman von Ihering (1850–1930) or the Swiss naturalist, Emilio A. Goeldi (1859–1917).[7] As regards the practice of Anthropology, given its focus on anatomical and physiological knowledge of the human being, several doctors became interested in the field, among them J. B. Lacerda, J. Rodrigues Peixoto, and E. Roquette-Pinto (Corrêa 1999; Faria 1999; Schwarcz 1993; Seyferth 1985).

In Brazil and in other countries, there was a constant interest in the scientific milieu in obtaining fossil material with which to carry out new experiments. Emperor Pedro II was particularly interested in the American Man; he took an active part in the debates on the question and sponsored expeditions. Together with L. Netto, he even took part in international events like the International Congress of Americanists. His correspondence with the French naturalist A. Quatrefages de Breu portrays that situation very clearly.

Given the growing importance attributed to anthropological studies in Europe and the United States, the complexity of that knowledge domain also increased, and it came to require more and more specialization to address it. Anthropologists J. B. Lacerda and J. Rodrigues Peixoto requested that the Museum's director Ladislau Netto should create a section that was separate from the Museum's other Science areas, declaring that the domains of anthropology "are vast and complex." In 1888, a new statute reorganized the National Museum giving expression to the new specialties that were constituting themselves and gaining space within the institution. Thus, the Museum's fourth section was created embracing the fields of Anthropology, Ethnology, and Archaeology (Keuller 2012a).

Anthropology and Scientific Exhibitions and Expeditions at the Turn of the Nineteenth to the Twentieth Century

The advent of the Republic brought with it some important changes to the functional structure of the National Museum, reallocating it to a building that today is the Quinta da Boa Vista. The courses for the public were replaced by conferences on topics that concerned each one of the Museum's sections, and the staff was reorganized.

[7] New regulations established in 1890 prohibited staff from accumulating more than one post in the institution and implanted the signing of an attendance registry. That led to several naturalists being dismissed. Ihreng went to what was to become the Paulista Museum (*Museu Paulista*), and Goeldi became the director of the Pará Museum (*Museu Paraense*) (see Keuller 2012a).

This last measure was the one that affected the Museum most. The decree of 1890 meant that many naturalists left the Museum and the staff was reduced considerably. In 1876, anthropological practice could count on five posts for professional staff. In 1892, Decree 1.179 reduced that to four, and by 1910, there were only two staff members (Keuller 2012a). Again in 1892, all the posts of directors of sections became the object of competitive public selection exams. Other positions that existed were that of subdirector, naturalist, and preparer.

By then it had become normal for local scientific institutions to share their staff and materials. Many professionals opted to ingress in the Faculty of Medicine of Rio de Janeiro, among them, João Joaquim Pizarro (1842–1906), former director of the Anthropology, Zoology, and Paleontology section in the period 1876–1885.

The difficulty found to fill the new posts was due to the lack of suitably qualified professionals and to the Republican government's requirement that candidates must possess a scientific degree only served to accentuate the National Museum's problems. As a result, the Institute began to boost its educational role of qualifying and diffusing information. The public conferences of the beginning of the twentieth century were reactivated, and the collections exhibited to the public at large were enhanced with a more instructive presentation.

At the turn of the nineteenth to the twentieth century, the functions of preparer and naturalist[8] began to be shared by the different sections of the museum. In regard to field trips, they were undertaken to explore and collect fossil plant and animal material and collect ethnographic and anthropological objects. For internal work, the Museum's practice was directed toward qualifying young people in the practice of anthropology, making them fit to conduct the procedures and meet the requirements of the Natural Sciences, such as description, classification, and conservation of the objects, and the cataloguing and exhibition of material under the guidance of one of the teachers in charge.

The Republican government had a policy of encouraging and facilitating the constitution of networks of laboratories, offices, and materials of counterpart institutions, especially with the regional museums of São Paulo and Pará, insofar as they all had the same basic purpose. The overriding idea was to foster the country's development, especially in the field of agriculture.

Throughout the nineteenth century, the role the National Museum performed in Natural History research was well known in foreign scientific centers. One outstanding example worth mentioning is the Brazilian Anthropological Exhibition at the National Museum organized by Ladislau Netto in 1882. The importance of anthropological practice within the institution can be measured by the remarkable preparations for that grandiose event which enjoyed the interest and support of the Imperial Government and of the provinces of the Brazilian Empire.

[8] The work of the traveling naturalist was to make excursions for the purpose of acquiring indigenous artifacts or other items the institution needed. It was also the naturalist's responsibility to conduct studies, classify the collections under the supervision of the directors, and assist in the scientific exchanges. The function of the preparer was to prepare and conserve the section's objects and assist in various other activities. He was also supposed to attend the Museum's conferences to learn more about natural sciences (see Keuller 2012a).

The quest to identify the origin of American Man was a recurrent concern in scientific circles and also the purpose of many of the scientific expeditions that the National Museum promoted in the northern and southern regions of Brazil at the end of the nineteenth century. The research work sought to establish the antiquity of the indigenous Brazilian. Bones, funeral urns, and pottery artifacts collected by naturalists were among the objects that appeared on the lists sent out by the government to all the Brazilian provinces by means of circulars and instructions addressed to the provincial presidents.

Ladislau Netto stressed the importance of the study of American Man in the aspect of anthropological knowledge, both physical and moral, by means of his remains in the form of bones, pottery, and the languages. In an official correspondence addressed to the Ministry of Agriculture at the time, he argued that "only through the study of the skeletons of our aborigines or the examination of their artifacts and languages is it possible to obtain such definitive knowledge of their nature and of their physical development (...)" (Keuller 2012a).

A considerable part of the material collected by the research efforts was taken from piles of shells situated in coastal regions or near lakes, rivers, bays, or land inundated by river waters. In them, there were the remains of fish and bones, skeletons, funeral urns, as well as ornaments, weapons, and utensils of various kinds that had once belonged to ancient indigenous tribes. The material collected at that time was studied intensely in an endeavor to obtain information on past and present tribes. Similar deposits had been found in Europe and North America, where they were known as shell middens or shell mounds.[9] The instructions that were sent out made the procedures and norms of anthropological practice of the day very clear. They informed the collectors and the presidents of the provinces as to which objects were of greatest interest to the institution's scientists. The anthropological objects were skeletons or single bones but only those belonging to aboriginal races and especially craniums, mummies, body ornaments, collars made of teeth, and human bones. The instructions also referred to the need for care and attention when excavating the terrain. According to the instructions, indigenous bones and skulls could be obtained in the cemeteries of indigenous tribes, hidden in natural caves and hollows in the rock or in the so-called middens. In addition to detailing how the objects should be described and classified, the document outlined the precautions to be taken in packing them and transporting the boxes to the National Museum.

Very few of the expeditions in those days could actually count on the presence of the researcher responsible for them in the field. With the aim of enlarging the Museum's collections for the holding of the Brazilian Anthropological Exhibition, L. Netto himself traveled to the northern region accompanied by Domingos Soares Ferreira Penna (1818–1888), Francisco da Silva Castro, and Vicente Chermont de Miranda e Assis. They carried out their field work on the island of Marajó in the region of Pacoval and the Capim River in 1881 and 1882, exploring burial sites, funeral urns, and pottery objects among other materials (Keuller 2012a).

[9] See shell middens studies in the *Archivos do Museu Nacional*, vol. I, 1876.

Donations arrived at the National Museum from various parts of the country, and there were exchanges of objects on the part of governments, institutions, and even pieces from private individuals all to go on display at the Anthropological Exhibition.

The event itself took place in the Museum's old building at the Campo da Aclamação, and it was organized in eight rooms, each one paying homage to a given scientist. The Anthropology room was dedicated to Peter W. Lund (1801–1889) and acclaimed the Danish naturalist for his studies of fossils in Lagoa Santa in Minas Gerais in the 1830s and 1840s. On display in the room were indigenous skeletons and skulls of the Tembé and Turuyara tribes exhumed by L. Netto on the banks of the Capim River, three skeletons obtained by Duarte Paranhos Schutel (1837–1901), many craniums of Botocudos, bones from the shell mounds of the province of Santa Catarina, and photographs of Botocudo Indians taken by the Imperial Geological Expedition (1875–1878), led by C. Hartt (1840–1878). There were also commemorative diplomas awarded by the Paris Exhibition of 1878 to B. Lacerda and J. Rodrigues Peixoto for their research into indigenous races.[10]

In addition to the various indigenous objects brought together in the ethnographic and archaeological displays, an actual family of Botocudo Indians and three Xerentes were brought to the event where they caused quite a stir (Faria 1993). The exhibition was a great success with many visitors and a certain degree of national and international repercussion. In view of that, L. Netto wished to organize a new event in 1884, the American Anthropological Exhibition, but it never came to fruition (Faria 1998).

The institution organized various other exhibitions in Brazil's northern and southern regions toward the end of the nineteenth century. In the region of Lagoa Santa, in Minas Gerais, which was where Peter Lund had lived and done his research work, fossils were found of the greatest importance for the studies that were the National Museum's focus of interest.[11] Of the vast quantities of material collected, including fossilized remains of mammals and human craniums, only one specimen remained in Brazil, and that was the cranium donated to the IHGB. The rest Lund sent to the then recently inaugurated Copenhagen Museum and one cranium was sold to the British museum by Peter Claussen, a local explorer (Marchesotti 2011; Martinez 2012).

In 1873, French scientist A. Quatrefages, in a letter to the Emperor, inquired about the work of the National Museum and included in his letter a comment on his own studies of the races. Highly interested in the scientific work being carried out in Brazil, Quatrefages was a well-known author in Brazilian intellectual circles.

[10] Divided into four sections: anthropology, P. Lund Room; archaeology, Jean de Levy and C. Hartt Rooms; ethnology, Pero Vaz de Caminha, José de Anchieta, and Alexandre Rodrigues Ferreira Rooms; and ethnography and archaeology, Gabriel Soares and Von Martius Rooms (see Filho 1882).

[11] The exchanges among the local scientific institutions could also give rise to disputes. That was the case with one cranium of Lund's collection that was donated to the IHGB. A plaster model of it was lent to the National Museum for research purposes, as confirmed by the request of A. Quatrefages in 1879 to the director of the Museum, who intended to take it to Moscow. On the death of the director in office, L. Netto, the IHGB finally managed to get the cranium back (see Keuller 2012a).

Making use of the prestige he enjoyed, in 1874, Quatrefages appealed to the Emperor to organize new expeditions to the caves of Lagoa Santa (Domingues and Sá 2003).

Considering all the efforts being made to study the origins of mankind in the Americas and the human races, at the request of Pedro II, the National Museum sent a collection of Botocudo craniums and two complete skeletons to A. Quatrefages in Paris and to the German anthropologist R. Virchow (1821–1902) in Berlin (Lacerda and Peixoto 1876).

The subdirector of the section Domingos Sergio de Carvalho (1866–1924) made a new attempt to organize an expedition to Lagoa Santa, but his request to the board of the National Museum was unsuccessful. The paucity of financial resources and the reduced staff numbers at the beginning of the Republican period were probably the reason for that impediment (Keuller 2012a).[12]

The Debate on American Man: Dialogues Between the National Museum and the Scientific World

The question of the origin of American Man had been present in the National Museum ever since the anthropological studies were implemented alongside the natural sciences in 1876. That interest was corroborated by the director of the institution at the time L. Netto, by J. B. Lacerda himself, and by J. Rodrigues Peixoto in the relations established with anthropologists A. Quatrefages and R. Virchow. Other Brazilian institutions were also interested in the matter, such as the IHGB and the Rio de Janeiro Faculty of Medicine as well as the museum's counterpart institutions, the Paulista Museum and the E. Goeldi Museum.

P.W. Lund's fossil discoveries of the American Man – the Lagoa Santa race – were documented in correspondence he sent to the IHGB during the 1840s in the Memoirs of the Royal Society of Antiquarians of the North, in a letter to the director of that Society, C. C. Rafn (1795–1864), and in the Institute itself (Lund 1842; Lund 1844; Vignaud 1913). Indeed, Lund's discovery had a strong effect on the European and North American scientific world.[13]

In the midst of the various studies on shell middens that were outstanding in the 1870s,[14] attention was strongly drawn to those ancient inhabitants. Besides the

[12] D. S. Carvalho suggested to the National Museum Congregation that naturalist C. Moreira could be sent to Lagoa Santa to explore the caves in the region. The Museum's director J. B. Lacerda (1888–1915), however, rejected the request (see Keuller 2012a).

[13] To exemplify, North American explorer Isaac G. Strain (1821–1857), who was on an expedition in Brazil at the time as a corresponding member of the Philadelphia Academy of Natural Sciences and of the IHGB, sent a letter to North American doctor and anthropologist Samuel G. Morton (1799–1851), then vice-president of the Academy, telling him about Lund's finds. The author remarks that it may have been Morton who communicated the discovery (see Starin 1846).

[14] See *Archivos do Museu Nacional*, 1876.

human bones, the descriptions indicate the use and purpose of material objects associated to them such as funeral urns, pottery objects, and stone tools.

In 1876, National Museum scientists J. B. Lacerda and J. Rodrigues Peixoto made a detailed study of Brazil's indigenous races based on the work of contemporary French and German anthropologists like P. Broca, F. Pruner-Bay (1808–1882), A. Quatrefages, P. Topinard (1830–1911), and R. Virchow. The subject was a leading concern in North American circles, as exemplified in the work of North American anthropologist Samuel G. Morton (1799–1851) and his vast collection of skulls. To the south of the continent, Francisco P. Moreno was conducting (1852–1919) new studies in Argentina, researching skulls in Patagonia (Lacerda and Peixoto 1876).

Based on the craniometric studies of the French school of P. Broca, Lacerda and Peixoto analyzed a collection of Botocudo skulls at the National Museum and the skull that P. Lund had donated to the IHGB as well as analyzing teeth observations and measurements made in loco by traveling naturalists. In their conclusions, Lacerda and Peixoto show that the primitive race in Brazil was dolichocephalic; that contemporary indigenous races represented a mixture of two different types; that the Botocudos were closest to the primitive race; that in more remote times, there had been a race with an extreme depression of the forehead; and that most of the Brazilian indigenous races were not in the habit of creating artificial cranial deformations. Thus, the authors demonstrated that the skull found at Lagoa Santa was similar to that of the Botocudos.

The two authors defended the idea that "the American Indian is the product of American soil," and they believed in P. W. Lund's proposition that the formation of the new continent had preceded that of the Old World. They also believed in the basic similarities among the peoples dispersed on the Americas, even though their beliefs, rites, customs, and languages presented slight differences (Lacerda and Peixoto 1876). Lacerda and Peixoto's considerations aroused great interest in the scientific world. The research that followed considered all the measurements they had made and the impressions that had been taken, especially from the specimen preserved at the IHGB.

In a study from 1881, Quatrefages commented on the similarities that the two Brazilian authors had described between the Botocudos and the Lagoa Santa race. In the light of their work, the French anthropologist compared the Lagoa Santa Man with European fossil humans. Following that, he made a comparative analysis of the collection in the Natural History Museum of Paris, comparing craniums from Brazil with others from Peru. In his conclusions, he remarks that the fossil Man of Lagoa Santa differed from the fossils of European Man, insofar as they presented dolichocephaly and hypsistenocephaly. He also affirmed that in Brazil as in Europe, the fossil Man had left descendants who had contributed toward the formation of the current population, thereby confirming Lacerda and Peixoto's results, when they considered the Botocudo race to be the result of the crossing of the Lagoa Santa type with other ethnological elements. What was yet to be determined was the number and nature of those "elements," but one of them, at least, was identified, namely, brachycephaly. It was also confirmed that the type of fossil Man at Lagoa Santa was part of the composition of the Peruvian population and was more or less detectable on the coast of the Pacific. In Peru and Bolivia, the ethnic element of Lagoa Santa marked its presence in a way that that was more or less clear in relation to its presence in Brazil, although this element seems to have had a lesser influence in Peru than in Brazil. The same ethnic element was found in other places and not just in Brazil and Peru (Quatrefages 1881).

Other scholars dedicated their efforts to the same question, basing themselves on the analyses of Lund and Lacerda and Peixoto. That was the case with Dutch anthropologists H. Ten Kate (1858–1931), who in his 1885 study of the Californian race used the methods outlined by the National Museum scientists and reproduced by Quatrefages. Regarding the rest of the collection, Ten Kate's studies made use of the Danish naturalist Ch. Fr. Lütken's (1827–1901) observations of the Lund collection in Copenhagen (Lütken 1883),[15] as well as his own assessment of the Danish collection and the evaluation of J. Kollmann, even though the latter only analyzed four skulls. The conclusions indicated a similarity between the race at Lagoa Santa and the California race and were in alignment with Quatrefages's studies regarding the analogy of the Lagoa Santa and the Papua race (Ten Kate 1885).

In a later work in 1890, Ten Kate discussed the question of the plurality of the American race without however getting involved in the theory of mono- or polygenesis. His belief was that the study of races should be based on sets of physical characteristics such as eye color, hair, nose, etc., and not on any single aspect or feature. He acknowledged the similarity of aspects of the American population (Virchow 1888) with the Asian Mongols, but in his view, there was no confirmation as to the ancestry of the Americans, and he therefore underscored the need for further studies to get to know their origin (Ten Kate 1890).[16]

Along the same lines, another important work was that of the curator of the Copenhagen Museum, Danish anthropologist Sören Hansen. In 1888, he analyzed the entire collection of Lagoa Santa skulls (16 altogether). In addition to his own museum's collection, he included two skulls belonging to the British Museum. He based himself on the analysis and measurement of the IHBG cranium carried out by Lacerda and Rodrigues Peixoto. Up until then, the collection that the scholar of Copenhagen analyzed had only been undertaken by the Danish zoologist J. T. Reinhardt (1816–1882), the first curator of Lund's collection at the Copenhagen Museum, then by Gervais, by Kollmann, and later by Ten Kate. In his conclusions, Hansen stated that the Lagoa Santa skulls (with one brachycephalic exception) presented uniformity. In the terminology of the day, the specimens are dolichocephalic,

[15] Ch. Lütken gives a description of Lund's discoveries in the cave known as *Lapa da Lagoa do Sumidouro* in the Lagoa Santa region of Minas Gerais. According to that author, it was the only cave in which numerous skulls and fossilized human bones were found in a partially or totally petrified condition together with the fossil remains of animals, especially of mammal species of the post-Pliocene or even Pliocene epoch (see Lütken 1883).

[16] Ten Kate corroborates German anthropologist R. Virchow's thesis on the plurality of the American race. In his study of American craniology, Virchow identifies the existence of three different types of crania: the brachycephalic, the dolichocephalic, and the mesocephalic. As such, he states that "the construction of a common type of American Indian should be definitively renounced" (see Ten Kate 1890).

hipsistenocephalic, prognathate, medofacial, and so on. The similarity with the Papua type identified by Quatrefages was confirmed, and that led this last scientist to admit the existence of a primitive race situated in a considerable part of the South American region and mixed with other (brachycephalic) elements (Hansen 1888).

Interest in and attention to the question of the origin of Man in the Americas redoubled among North American anthropologists at the turn of the nineteenth to the twentieth century. The Czech-born anthropologist domiciled in the United States, Ales Hrdlička (1869–1943), engaged in various studies addressing the question of Man's antiquity in North America, but the fossil discoveries in Argentina led him to switch his interest to South America (Hrdlička 1907). Hrdlička and two other scientists, W. H. Holmes (1846–1933) of the Smithsonian Institute and Bailes Willis of the United States Geological Survey, made an expedition to Argentina and Brazil to study fossil material related to American Man. Their excellent analysis of the studies addressing the theme became an essential reference source for research, and their report on their visit to the IHGB in Rio de Janeiro to see the cranium deposited there was especially outstanding (Hrdlička et al. 1912).

The special interest in the South American continent led it to become the object of other anthropological research projects of North American societies, which made the issue the theme of a research plan submitted to the Carnegie Institution of Washington in 1907, proposing the creation of a Department of Anthropology to investigate the ethnic relations of the South American continent with other continents. The plan was signed by F. W. Putnam (1839–1915), Roland B. Dixon (1875–1935), W. H. Holmes, A. L. Kroeber (1876–1960), and F. Boas (1858–1942), and it proposed that there should be four distinct lines of research dedicated to the investigation of the antiquity of the presence of Man in South America. The first would consist of analyses of research in the area of the Pampas, then there would be a search for archaeological evidence of the influence of Andean culture in Central America, and that would lead to observations of the influence of the Arawaks of Brazil on regions of the United States, embracing the Caribbean and the Gulf of Mexico, and lastly an examination of the (then) recent discoveries on the South America Pacific coast of evidence of contact between the Polynesian islands and the South American continent (Putnam et al. 1908).

That expansion of research into the South American continent led French ethnologist Paul Rivet (1876–1958) to address the question as well. As a doctor attached to the French Geodesic Mission to Ecuador, Rivet researched into the pre-Columbian races of that country and demonstrated that they were similar to those of Lagoa Santa. In his conclusions, he admitted that the Lagoa Santa race had been present not only in the historical region of Ecuador but also in other parts of South America, in areas extending from Ecuador and Brazil to as far south as Tierra del Fuego in Argentina (Rivet 1908).

As has been shown above, the great interest of the scientific world in Man in the Americas led to the development of much research in various parts of the continent. All of them cited the work of Lacerda and Peixoto of the National Museum as a reference.

The Scientific Controversies of the 1930s

In the 1920s and 1930s, the National Museum developed a research plan for the region of Lagoa Santa in Minas Gerais. That was the period when the Anthropology and Ethnology section and later the entire Museum came under the direction of, first, Edgard Roquette-Pinto (1883–1954) and then Heloísa Alberto Torres (1895–1977). It was the moment when the prestige of the institution and its scientists in anthropological matters affirmed itself before all other similar museums. The work carried out under the leadership of anthropologist and anatomist Álvaro Fróes da Fonseca (1890–1988) at the Anthropology Laboratory contributed strongly toward that success.

Fróes managed to gather round him what were mostly young doctors from the Faculty of Medicine of Rio de Janeiro, like José Bastos de Ávila (Keuller 2012b), who would eventually succeed him as head of the section.[17] With the suggestions and guidance of E. Roquette-Pinto and A. Fróes da Fonseca for the research being undertaken there, the Anthropometry Service started by E. Roquette-Pinto became reference for the anthropological study and characterization of the Brazilian population (Fróes da Fonseca 1926; Roquette-Pinto 1933).

Among the new admissions to the staff that took place via public competitive entrance examinations in 1925 were Heloísa Alberto Torres and Jorge H. A. Padberg-Drenkpol. The latter graduated from the universities of Freiburg (1912–1922) and Munich (1922–1924) in Germany, and he became an assistant and later a preparer in the Physical Anthropology Division in 1932, when it was under the command of A. Fróes da Fonseca.

With adequate funding and professional staff trained for fieldwork by the Museum, Padberg-Drenkpol was sent to carry on Lund's research in the region of Lagoa Santa. The research was conducted in 1926. During the journey of exploration, the scientist visited the entire east bank of the Velhas River from Lagoa Santa to Sete Lagoas. He made geographic sketches and registered geological observations, identified the main caves in the southern part of the region, visited the Maquiné cave that Lund had described, and, among other finds, discovered new sites such as Lapa do Caetano, Lapa da Limeira, Lapa D'Água, Lapa da Amoreira, Lapa do Tombo, and Lapa da Vargem de Baixo. This last site he referred to as Lapa Mortuária (Mortuary Cave), and in it the scientist found the remains of approximately 80 human individuals and many fossils of animals. In a second visit to the region, made in the same year, he discovered a skull in the Lapa da Limeira and other fragments in the Lapa D'Água and Lapa da Amoreira (Keuller 2012a).

Padberg-Drenkpohl headed other expeditions to the Lagoa Santa region at the behest of the National Museum, but none of them was as successful as the first one

[17] To qualify and prepare professional for its staff, the National Museum organized the second Anthropology course given by E. Roquette-Pinto in 1915. The third course in Anthropometry took place in 1932 and was given by J. Bastos de Ávila. Among the students attending was the young Maria Júlia Pourchet (1906–1993) (see Keuller 2012a).

(Padberg-Drenkpol, J.A; Ribeiro, P. M. 1929; Padberg-Drenkpol 1926/1931). Generally speaking, the material that he collected consisted of fossil material, photographs, and rough drafts of maps. According to the section's report for 1936, not all of it was catalogued or studied (Keuller 2012a). However, in another report, he refers to a craniometric study of the Lagoa Santa race along with the other previous studies. He elaborated a practical method based on sexual differences in the way indicated by German anthropologist R. Martin. He also made a comparative analysis of cranial capacity. Lastly, he established a synopsis of the racial characteristics associated to the skull of the Lagoa Santa race (see Chap. 5 of this volume).

Padberg-Drenkpol was always a controversial figure, and during the period of his activity, he had many clashes with the institution. The height of such situations was the polemical competitive public selection examinations of 1933, when the scientist requested to be exempted from the need to take them in order to occupy the chair of Stratigraphy and Paleontology in the Museum's first section,[18] a position that he aspired to. He had graduated and qualified himself at German universities and had studied under eminent European professors.[19] Having been interim chair professor since 1932, Padberg-Drenkpol forwarded a summary of his work and studies to justify his performance in the respective field. In addition to various studies focused on the European world, he emphasized the paleontological research conducted in the limestone region of Lagoa Santa and highlighted his conferences and a University Extension course on Stratigraphy and Paleontology (Padberg-Drenkpol 1932) given through an agreement with the University of Rio de Janeiro in 1932. He had also given a series of conferences for the Museum on topics such as the geology of the Lagoa Santa region, the paleontological finds in the Lagoa Santa caves (at a solemn session to mark the Museum's anniversary), and the geomorphology of the Lagoa Santa caves, given at the Pan-American History and Geography Congress in 1932 (Keuller 2012a).

The fierce discussion with the Museum Congregation was an example of the disputes among the branches of science, eager to establish their specialized fields.[20] The Museum Congregation alleged that the scientist's work was in Prehistory and not Paleontology and argued that his placement should be in the Anthropology and Ethnology section and, furthermore, that any candidate for the Paleontology chair should be a specialist in invertebrates. Padberg-Drenkpol retorted that Prehistory

[18] In 1931, the National Museum had five sections: first section, Minerology and Petrography and Stratigraphy and Paleontology; second section, Botany; third section, Zoology; fourth section, Anthropology and Ethnography (and Archaeological collection); fifth section, Natural History (Educational Assistance Service) (Keuller 2012a).

[19] Padberg-Drenkpol studied with the following professors: Wilhem Deecke in Geology; Ludwig Neumann in Geographic Morphology; E. Wepfer, Karl Deninger, and Max Schlosser in Paleontology; Franz Doflein in Zoology; Friedr. Oltmanns in Botany; Franz Keibel in Comparative Anatomy; Eugene Fischer in Osteology and Anthropology; and Koch-Grunberg in Ethnology (Keuller 2012a).

[20] The "field" I refer to here is in the sense of P. Bourdieu's definition of "scientific field" (see Bourdieu 1983).

was a branch of Paleontology. The scientist's request was denied, and he then moved a legal action against the government to have the public selection process annulled (Keuller 2012a).

From then on, he was constantly complaining about persecution and the problems he was facing at the Museum. In an article published in the December 2, 1934, issue of the newspaper *Folha de Minas*, Padberg-Drenkpol spoke about his speleological research, his discoveries, and the course he had administered. He also made references to his brilliant performance and the public praise that had been bestowed on him. The article also referred to his antagonists at the Museum and remarked that he had been the one who had suggested to the director of the Museum at the time, Arthur Neiva, that he, Padberg-Drenkpol, should take up and continue Lund's research (Padberg-Drenkpol 1934).

In a letter to the director of the newspaper *Folha de Min*as, the National Museum professors declared that Padberg-Drenkpol had been hired in 1925 on the indication of E. Roquette-Pinto and that the same Roquette-Pinto had appointed him, without due public selection process, to the post of preparer for the Anthropology section in 1930. Again, his appointment to an interim post in the first section was a proposal of the head of the section, A. Fróes de Fonseca, at the instigation of Roquette-Pinto. They also declared that on that occasion, the same director issued a favorable opinion in support of Padberg-Drenkpol's request to the Provisional Government to receive a permanent appointment. The Museum's Congregation, however, denied the request and preferred to hold the public selection process (Keuller 2012a).

The newspaper report underscored Padberg-Drenkpol's many merits, among them the discovery of the Mortuary Cave, and the distinction the scientist made among the caves in tropical regions according to their period in contact with water, discussing that there are juvenile caves (with water in them at present day levels), mature caves (from 10 to 100 m above the water table), and senile caves, high up and in an advanced state of degradation. According to Padberg-Drenkpol, P. Lund studied juvenile caves, and he himself studied the mature caves (more intact and undisturbed) like the Mortuary Cave, whose base was 14 m higher than the water table, with the presence of a lot of human material and also rich in extinct fauna and microfauna remains. The Sumidouro Cave, on the other hand, was totally flooded. In his studies, the scientist defended the idea that Paleoamerican Man was a relatively recent arrival in relation to the post-Diluvium or post-Pleistocene periods and therefore could not have been the contemporary of the extinct species which were in an advanced state of fossilization and decomposition (Padberg-Drenkpol 1934).

Other expeditions to the region took place in the 1930s undertaken by a Minas Gerais Academy of Science team consisting of Harold Walter (1897–1976), Arnaldo Cathoud, and Aníbal Mattos (1889–1969), but their results indicated a different conclusion (see Chap. 6 of this volume). The discrepancy between the two points of view turned the issue into a veritable scientific controversy.[21]

[21] Here we take the definition of controversy to be that defined by Kostas Gavroglou: in a controversy, the parties involved publicly seek to formulate their arguments and contra-arguments in such a way that the scientific community will accept them and seek to impose new elements or trans-

In 1935, the Minas group got in touch with the National Museum regarding the Lagoa Santa research. According to physical anthropologist Maria Julia Pourchet's (1906–1993) account in her study conducted in 1958, Walter, Cathoud, and Mattos found in 1935 a human skeleton in the Confins Cave, one which had already been explored by Padberg-Drenkpol, who had called it the Mortuary Cave. They called the skeleton Confins Man. Pourchet explains that in 1937, those scientists, with the support of the Minas Gerais Academy of Sciences, presented a communication about their discovery at the International Symposium on Early Man in Philadelphia (Pourchet 1958; Cathoud et al. 1937).

Argentinian scientist Antonio Serrano, director of the Paraná Museum, also took part in the debate. During the research he conducted in Brazil in 1936, Serrano visited the National Museum and later declared, in one of his publications, that the question of the contemporaneity of Lagoa Santa Man with the extinct fauna species of the region affirmed by the Minas Gerais Academy of Sciences group had been denied by the scholars at the National Museum (Avila 1941a; Serrano 1938).

Furthermore, in 1936, J. Bastos de Ávila suggested in a report that they should go to Belo Horizonte, Minas Gerais, to discuss the question (in the terms established by Padberg-Drenkpol) and a proposal for future research into Lagoa Santa Man.

The craniometric studies and the studies of the mandibles collected at the Mortuary Cave continued at the National Museum right through to the end of the 1930s (Keuller 2012a). Meanwhile, the Minas group carried on with their studies of the Confins Man. The national and international scientific communities' interest in the new discoveries led the Minas Gerais Academy of Sciences to request technical opinions from the National Museum regarding Confins Man to be issued by J. Bastos de Ávila and J. Padberg-Drenkpol. The replies were published within another study by those authors entitled "In regard to Confins Man" (*A propósito do Homem de Confins*) in which they presented and discussed their technical opinions (Cathoud et al. 1939).

According to the craniometric evaluation made by Bastos de Ávila and Padberg-Drenkpol, "from the anthropological point of view, Confins Man is identical in every way to Lagoa Santa Man – *Homo sapiens lagoanus* (E. F. Eickstedt 1937)" (Avila 1941b; Mattos 1941).

In a scientific excursion to the Velhas River basin (Lapa de Carrancas) in 1936, Bastos de Ávila concluded from his studies there that *Homo sapiens lagoanus* was not a contemporary of the extinct megafauna. He therefore clashed with the opinions of Walter, Cathoud, and Mattos of the Minas Gerais Academy of Sciences.

To settle the question of Lagoa Santa Man, Bastos de Ávila suggested that new Stratigraphic and Paleontological studies should be conducted in the region where Lund had carried out his research, and he added that they should employ scientific methods and rigor in them. The information forwarded by the National Museum in the name of the two scientists show the institution's important role in Anthropology and in the scientific community.

---

form their practice (see Gavroglu 2004).

Conclusion

The studies of Lagoa Santa in the National Museum are closely associated to the development of the practice of Anthropology itself within that institution. Given its nature as a space dedicated to the study of the natural sciences, the National Museum incorporated the new specialty to its structure in 1876 when it included zoological and anatomical studies among its activities. The development of that practice, despite its specialization, was very much associated to the administration of the Museum directors, obtaining greater emphasis as the area was becoming increasingly specialized and was gaining more space within the institution.

Research into Lagoa Santa had been an activity present in the institution ever since the 1860s and gained in volume with the work of J. B. Lacerda and J. Rodrigues Peixoto with their detailed analyses of the Lagoa Santa cranium. Their analysis was to become a reference work for later studies on the same theme, and it facilitated increased exchanges with foreign scientists like A. Quatrefages de Breu, while at the same time, it highlighted the research work that was being done in Brazil.

The use of laboratories as research instruments made it possible for the National Museum to conduct microbiological and genetic studies, sharing interests, objects, and even the scientists themselves with other institutions.

Changes in the political and social contexts at the turn of the nineteenth to the twentieth century affected the institution, impairing the functioning of its scientific activities and reducing the number of scientific expeditions. The new rules and regulations implanted at the time maintained interest in the discussion on the origin of American Man, on the question of race and mixtures of races, and the insertion of Brazil among the civilized nations, like England and France.

The expansion of the field of Anthropology in Brazil starting in the years 1910 and the institutionalization of the practice of Anthropology in other regional museums, in government circles, in the academic worlds, and in other institutes made the period a promising one. The institution's scientific collection was enriched by the exchanges it engaged in, and the anthropological practice it developed made the Museum a national reference.

The Museum's directors always emphasized its educational aspect and organized public courses and conferences to compensate for the nonexistence of Higher Education courses in Natural Sciences. In that way the Museum qualified its own scientists. The first three Anthropology courses in the nineteenth century were given there by B. Lacerda, and in the twentieth century, such courses were given by E. Roquette-Pinto and later by J. Bastos de Ávila.

The renewed research activities at Lagoa Santa became more intense in the 1920s and 1930s, when the Museum was under the direction of E. Roquette-Pinto and, later, under Heloísa Alberto Torres. During that period, a network of scientific relations in Brazil and abroad was established that included contact with leading French, German, North American, and Latin American scientists among others.

The scientific prestige achieved by National Museum scientists mirrors the trajectory of the anthropological studies developed at the institution, whose specificities involved both social and political demands. Scientists like J. B. Lacerda and J. R. Peixoto in the nineteenth century and afterward and J. Padberg-Drenkpol and J. Bastos de Ávila in the twentieth century were evidence of leadership and scientific authority in regard to the anthropological studies of Lagoa Santa of their day.

References

Ávila, J. B. (1941a). A propósito do Homen de Confins. In *A Raça de Lagoa Santa* (pp. 411–421). São Paulo: Editora Nacional.

Ávila, J. B. (1941b). Os Achados de restos fósseis da Raça de Lagoa Santa na Lapa de Carrancas. In *A Raça de Lagoa Santa* (pp. 387–391). São Paulo: Editora Nacional.

Blanckaert, C. (2001). Lógicas da Antropotecnia: mensuração do homem e bio-sociologia (1860–1920). *Revista Brasileira de História, 21*, 145–156.

Bourdieu, P. (1983). Campo Científico. In P. Ortiz (Ed.), *Pierre Bourdieu* (pp. 122–155). São Paulo: Ática.

Cathoud, A., Walter, H. V., & Mattos, A. (1937). *The Confins Man. A contribution to the study of early man in South America*. International Symposium, on Early Man, Academy of Natural Sciences, Philadelphia.

Cathoud, A., Walter, H. V., & Mattos, A. (1939). *A propósito do Homem de Confins*. Ed. Apollo, Belo Horizonte.

Corrêa, M. (1999). *As Ilusões da Liberdade. A escola de Nina Rodrigues e a antropologia no Brasil*. São Paulo: CDAPH.

Costa, E. V. (2007). *Da Monarquia a República*. São Paulo: UNESP.

Dias, A. G. (1971). Correspondência ativa de A. G. Dias. Anais da Biblioteca Nacional 84.

Domingues, H. M. B. (1996). As ciências naturais e a construção da nação brasileira. *Revista Brasileira de História, 135*, 41–59.

Domingues, H. M. B., & Sá, M. R. (2003). Controvérsias evolucionistas no Brasil do século XIX. In H. M. B. Domingues (Ed.), *A Recepção do Darwinismo no Brasil* (pp. 97–123). Rio de Janeiro: Ed. Fiocruz.

Faria, L. C. (1993). *Antropologia – espetáculo e excelência*. Rio de Janeiro: EdUFRJ.

Faria, L. C. (1998). *Antropologia – escritos exumados 1*. Niterói: EdUFF.

Faria, L. C. (1999). *Antropologia – escritos exumados 2*. Niterói: EdUFF.

Filho, M. M. (1882). *Guia da Exposição de 1882*. Rio de Janeiro: Typ. G. Leuzinger & Filhos.

Fróes da Fonseca, A. (1926). As novas fichas antropológicas do Museu Nacional. *Boletim do Museu Nacional III, 3*, 13–30.

Eickstedt, E. F. Rassenkund und Rassengeshiche der Menchkeite. 1937.

Gavroglu, K. (2004). *O Passado da História como Ciência*. Porto: Porto Editora.

Gobineau, A. (1853). *Essai sur l'inégalité des races humaines*. Paris: Firmin-Didot.

Gould, S. J. (1999). *A Falsa Medida do Homem*. São Paulo: Martins Fontes.

Hansen S. (1888). Lagoa Santa racen. En Anthropologisk Undersogelse af Jordfundne Menneskelevninger fra Brasilianske Huler. Med et Tillaeg om det Jordfundne Menneske fra Pontimelo, Rio de arrecifes, La Plata. En Samling af Afhandlinger, e Museo Lundii, I, 5, Kjobenhavn (pp. 1–37).

Hrdlička, A. (1907). *Skeletal remains suggesting or attributed to early man in North America*. Washington, DC: U. S. Govt. Printed Official.

Hrdlička, A., et al. (1912). *Early man in South America*. Washington, DC: Washington: Govt. Printed Official.

Jancsó, I. (1997). A sedução da liberdade. In F. Novais & L. M. Souza (Eds.), *História da Vida Privada no Brasil* (pp. 387–437). São Paulo, I: Cia. das Letras.

Keuller, A. T. A. M. (2012a). *Os estudos físicos de Antropologia no Museu Nacional do Rio de Janeiro: cientistas, idéias, objetos e instrumentos (1876–1939)*. São Paulo: Humanitas.

Keuller, A. T. A. M. (2012b). Entre Antropologia e Medicina: uma análise dos estudos antropológicos de Álvaro Fróes da Fonseca nas décadas de 1920 e 1930. Boletim do Museu Paraense Emílio Goeldi. *Ciências Humanas, 7*, 687–704.

Lacerda, J. B. (1877). Curso de Antropologia. *Archivos do Museu Nacional, II*, 165–169.

Lacerda, J. B., & Peixoto, J. R. (1876). Contribuição ao estudo anthropológico das raças indígenas do Brasil. *Archivos do Museu Nacional, I*, 47–75.

Lopes, M. M. (1997). *O Brasil descobre a pesquisa científica: os museus e as ciências naturais no século XIX*. São Paulo: Hucitec.

Lopes, M. M., Corrêa, M. (1995). As Aves que aqui gorjeam. Mimeograph.

Lund, P. W. (1842). Carta escrita de Lagoa Santa por P. Lund. *Revista do IHGB, 4*, 80–87.

Lund, P. W. (1844). Correspondência do mesmo P. Lund. *Revista do IHGB, 6*, 326–334.

Lütken, C. F. (1883). L'Exposition de quelques-uns des crânes et des autres ossements humains de Minas-Gerais dans le Brésil central découverts et déterrés par le feu Pofessor P. W. Lund". Congreès International des Américanistes. Compte-rendu de la cinquième session (pp. 20–48). Copenhagen.

Marchesotti, A. P. A. (2011). P. W. Lund: o naturalista que revelou ao mundo a pre -história brasileira. E-papers, Rio de Janeiro.

Martinez, P. H. (2012). A nação como pedra: a coleção paleontológica no Brasil. 1836–1844. *História, Ciência, Manguinhos, 19*, 1155–1170.

Mattos, A. (1941). *A Raça de Lagoa Santa: velhos e novos estudos sobre o homem fóssil americano*. São Paulo: Editora Nacional.

Mattos, I. R. (1986). *O Tempo Saquarema*. São Paulo: Hucitec.

Netto, L. (1885). Investigações sobre a archeologia brasileira. *Archivos do Museu Nacional, VI*, 257–544.

Padberg-Drenkpol, J. (1926/1931). Relatório de excursão científica à Lagoa Santa. Museu Nacional (SemearMN).

Padberg-Drenkpol, J. (1932). Curso de Estratigrafia e Paleontologia, com especial aplicação à Geologia do Brasil e à evolução dos organismos. *Revista da Universidade do Rio de Janeiro. se rie II, 2*, 249–260.

Padberg-Drenkpol, J. (1934). Lund e as recentes pesquisas espeleológicas do Museu Nacional. Jornal Folha de Minas. December 2.

Padberg-Drenkpol, J. A., Ribeiro, P. M. (1929). Catálogo do Material coligido em Lagoa Santa – Minas. Manuscrito, Museu Nacional – RJ.

Pickering, A. (1992). *Science as practice and culture*. Chicago: University of Chicago Press.

Pinheiro, R. (2002). As histórias da Comissão Científica de Exploração (1856) na correspondência de Guilherme Schüch Capanema. Dissertação de mestrado, UNIICAMP.

Pourchet, M. J. (1958). Crânios fósseis de Lagoa Santa (Brasil). Contribuição bibliográfica ao estudo do "Homem Sapiens Lagoanus". Miscelllanea Paul Rivet Octogenario Decata. Universidad Nacional Autónoma de Mexico, Cidade do Mexico, T II:463–469.

Putnam, F. W., Dixon, R. B., Holmes, W. H., Kroeber, A. L., & Boas, F. (1908). Anthropological research – petition from representatives of the Archaeological Institute of America, the American Folk-lore Society, the Anthropological Society of Washington, the American Anthropological Association, and the American Association for the Advancement of Science; relative to the proposed establishment of a Department of Anthropological Science. Carnegie Institution of Washington, 3–7. Printed Leaflet.

Quatrefages, A. (1881). L'Homme fossile de Lagoa Santa (Brésil) et ses descendants actuels. Note de M. De Quatrefages. (Extrait). *Comptes Rendus de l'Académie de Sciences de Paris, 93*, 882–884. Available at http://gallica.bnf.fr/ark:/12148/bpt6k3049g

Rivet, P. (1908). A Raça de Lagoa Santa entre os pré-colombianos do Equador. Bulletin de la Société d'Anthropologie de Paris 9(9): 209–274. Available at: http://www.persee.fr/web/revues/home/prescript/article/bmsap_0037-8984_1908_num_9_1_7057 consulted on 15 Dec 2012.

Roquette-Pinto, E. (1933). *Ensaios da Antropologia Brasileira*. São Paulo: Cia. Ed. Nacional.

Schwarcz, L. M. (1993). *O Espetáculo das Raças: cientistas, instituições e questão racial no Brasil (1870–1930)*. São Paulo: Cia. das Letras.

Serrano, A. (1938). *Los Sambaquis o concheros brasileños*. Tucumán: Ed. Univ. Nac.

Seyferth, G. (1985). A antropologia e a teoria do branqueamento da raça no Brasil: a tese de João Batista de Lacerda. *Revista do Museu Paulista, 30*, 81–98.

Starin, I. G. (1846). Correspondência de I. G. Starin para S. G. Morton (07/12/1843). Proceedings of the Academy of Natural Science of Philadelphia 2:11–14. Available at: http://biodiversitylibrary.org/page/6605665 consulted on 12 Nov 2013.

Ten Kate, H. (1885). Sur les crânes de Lagoa Santa. Bulletins de la Société d'Anthropologie de Paris. III Se rie, tome 8: 240–244. Available at: http://www.persee.fr/web/revues/home/prescript/article/bmsap_0301-8644_1885_num_8_1_6375 consulted on 15 Dec 2012.

Ten Kate, H. (1890). Sur la question de la pluralite et de la parente des races en Ame rique. *Congre s International des Ame ricanistes. Compte- rendu de la huitième sesión* (pp. 288–294). Paris.

Vignaud, H. (1913). La question de l'antiquité de l'homme americain. *Journal de la Société des Americanistes Tome, 10* (1), 15–23. Available at: http://www.persee.fr/web/revues/home/prescript/article/jsa_0037- 9174_1913_num_10_1_2845 consulted on 15 Dec 2012.

Virchow, R. (1888). *La Crâniologie Americaine. Congres International des Americanistes* (pp. 251–260). Berlin: Compte-rendu de la septème session.

Chapter 5
The Physical Anthropology Archives of the National Museum of Rio de Janeiro: Lagoa Santa in the First Half of the Twentieth Century

Verlan Valle Gaspar Neto and Claudia Rodrigues-Carvalho

Abstract Ever since the nineteenth century the region of Lagoa Santa in the state of Minas Gerais has been an important center of archaeological, paleoanthropological, and related field studies of a theme dear to the international anthropological community, namely, the arrival of *Homo sapiens* in the Americas. This chapter explores the incursions in that region undertaken by two anthropologists from the National Museum of Rio de Janeiro, Padberg-Drenkpohl and Bastos de Ávila, in the first half of the last century. The (unpublished) written material left by the two researchers, consisting of reports, notes, and letters, is of historical importance not only for the specific theme of the arrival of *Homo sapiens* in the Americas but equally for the history of anthropology in Brazil in the widest sense. Accordingly, the coming pages present those records and registrations so that they may become accessible to a wider audience, in keeping with the proposal of this book. First comes a presentation of documents associated to the campaigns undertaken by Padberg-Drenkpohl in the second half of the 1920s, followed by those referring to the incursion of Bastos de Ávila together with Nei Vidal, of the National Museum's Stratigraphy Division, during the second half of the 1930s. Lastly, the chapter closes with a section dedicated to the debates between the National Museum and the Minas Gerais Academy of Sciences concerning "Confins Man" and the coexistence of humans with the extinct megafauna species of the Americas.

V.V. Gaspar Neto (✉)
Department of Law, Humanities and Letters, Rural Federal University of Rio de Janeiro, Três Rios Institute, Av. Prefeito Alberto da Silva Lavinas, 1847, Centro, Três Rios, RJ, Brazil
e-mail: verlan02@yahoo.com.br

C. Rodrigues-Carvalho
Biological Anthropology Sector of the Anthropology, Department of the National Museum, Federal University of Rio de Janeiro, Quinta da Boa Vista, s/n°, São Cristovão, Rio de Janeiro, RJ, Brazil
e-mail: claudia@mn.ufrj.br

P. Da-Gloria et al. (eds.), *Archaeological and Paleontological Research in Lagoa Santa*, DOI 10.1007/978-3-319-57466-0_5

Introduction

Just as the chapters brought together in this volume show, the region of Lagoa Santa in the state of Minas Gerais, Brazil, has always had an outstanding position in the discussions about the arrival of *Homo sapiens* on the American continent, which has been going on for more than a century and that has received the contributions of researchers from various specialized scientific fields, ranging from paleoanthropology to human population genetics and including archaeology and linguistics, among others (González-José et al. 2008; Neves 2013; Neves and Atui 2004; Neves et al. 1999a, b; Neves and Piló 2008; Powell and Neves 1999; Salzano 2011; Silva and Rodrigues-Carvalho 2006). The importance of Lagoa Santa can be measured along at least two intersecting vectors. One of them is the history of the anthropological and archaeological interventions in Lagoa Santa, which have their starting point in the studies carried out by the Danish naturalist Peter Lund in the first half of the nineteenth century (Neves and Piló 2008; Prous 1992; Silva and Rodrigues-Carvalho 2006). In the course of the years that followed, those studies endowed the region with a portentous international visibility, insofar as they brought to light various examples of the remains of humans and ancient animals; to those finds, others were added by successive generations of researchers. Such notable importance was deservedly acknowledged by the Brazilian government in 2010, when it issued a commemorative postage stamp paying homage to that pioneer in the area of paleoanthropology.

The second vector along which the importance of Lagoa Santa can be observed concerns the biological specificities and the antiquity of the materials found there. Even though that aspect lies outside the scope of this chapter, it is worth noting in passing that those materials, in their condition as the focus of innumerable investigations, from time to time formed the basis of statements that triggered the most heated international discussions about what might be the best explanatory scenario for the settlement of the American continent, as well as for what we could perhaps refer to as an elucidating framework for the biological constitution and ecological context of the so-called Paleo-Americans.[1] The current discussions about the explanatory powers of information obtained in the form of morphological and biomolecular data, for example, not only those that draw us closer to them but also those that place us farther away,[2] cast us back to the reflections elaborated by Latour (2000) and Latour and Woolgar (1997) about the role of controversies in establishing the so-called scientific facts. In short, and by extension, those discussions do not just represent the status awarded to the models and techniques of a morphological versus biomolecular bioanthropology in regard to the settlement of the Americas, they also reveal the play of forces among the various specialties involved in establishing a hierarchic position, in terms of their symbolic capital (scientific authority), within that which we can call a "scientific field" (Bourdieu 1975, 1976, 2004; Gaspar Neto 2012).

[1] For a more detailed approach to this theme see Chaps. 10 and 11 of this volume.

[2] See González-José and Bortolini (2011), González-José et al. (2008), Hubbe et al. (2010), Neves (2006), and others.

Be that as it may, in the light of the reflection introduced in the preceding paragraph, the fact is that Lagoa Santa can be understood as being a symbolic and material nucleus around which orbit not only questions related to the efforts made in the field of bioanthropology and associated fields of knowledge, to elucidate the three-dimensional puzzle that its prehistory represents (Silva and Rodrigues-Carvalho 2006), but also those questions that extrapolate the "purely scientific" altogether. Regarding this last point, one can bear in mind both the political difficulties involved in doing science from the periphery (Neves and Piló 2008) and the various ways in which the paleontological finds can be appropriated outside of the scientific field altogether. That is the case with Luzia (Gaspar Neto and Santos 2010), acknowledged up until the present moment as the Americas' most ancient human remains, with an estimated age of 11,000 years before the present (Neves and Piló 2008).[3]

Much of what has been discussed up to this point is equally applicable to the subject and time frame this chapter addresses, that is, the incursions of researchers from the National Museum of Rio de Janeiro in Lagoa Santa in the first half of the last century, except, of course, specificities that are inherent to the time frame, such as having to bear in mind that we cannot talk about a confrontation of the "morphology versus molecule" type at that moment. Nevertheless, there were other controversies and discussions then. Without a doubt, the most relevant of them, which persisted practically up until the dawn of the twenty-first century, involved the antiquity of Man's arrival on the Americas and the possibility that he might have coexisted with the extinct Pleistocene megafauna. That debate, in which Lund became involved as far back as the first half of the nineteenth century, did not go unnoticed by the two anthropologists that the National Museum sent to conduct investigations in the region in the early decades of the twentieth century, namely, Jorge Augusto Padberg-Drenkpohl and José Bastos de Ávila.

Unfortunately, the written material of those two researchers, consisting of reports, notes, and letters, offers very little in the narrower sense of biological anthropological research other than general impressions on the question of the settlement of the Americas and the importance of Lagoa Santa Man to that issue. Indeed, the paucity of what we might call corollaries of a more profound stratigraphic nature has already been identified on other occasions (Neves and Piló 2008; Silva and Rodrigues-Carvalho 2006). Nevertheless, their historical importance, whether it be about the discussions around the specific issue of the arrival of *Homo sapiens* in the Americas or about the history of anthropology in Brazil, in a broader sense, should not be underrated. The coming pages present a set of records and registrations endowed with a specific historical and scientific content, with the intention of offering elements which, within the scope of the contribution this book is intended to make, can be accessed by a broader audience. That is because, in view of their antiquity and physical fragility, these documents, currently in the custody of the National Museum's bioanthropology sector, have a very limited circulation

[3]For the earliest information concerning Luzia or Hominin N° 1 of the Lapa Vermelha IV, see Neves et al. (1999a, b). For a skeleton with a similar age or older on the Americas, see Chatters et al. (2014).

(Santos and Mello e Silva 2006), and their identification in the body of the text will be made, as far as possible, in alignment with what has been stipulated for the general biographical sources.[4]

The following pages will first present the documents referring to the campaigns conducted by Padberg-Drenkpohl in the second half of the 1920s. After that, it comes those referring to an incursion conducted by Bastos de Ávila in a partnership with Nei Vidal (from the National Museum's stratigraphic division) in the second half of the 1930s. Lastly, the chapter closes with a small section dedicated to the debates that took place between the National Museum and the Minas Gerais Academy of Sciences regarding "Confins Man" and the coexistence of humans and extinct megafauna in the Americas.

Padberg-Drenkpohl and the Dilemma of the Contemporaneity of the "Lagoa Santa Race" and the Extinct Pleistocene Megafauna

Jorge Henrique Augusto Padberg-Drenkpohl was born in 1877, in Osnabruck, Germany, according to the only biography dedicated to him published in Brazil (Rabuske 1981). After graduating in humanities from the college where he studied, he joined the Society of Jesus in Holland where he graduated in theology and philosophy. He was ordained as a missionary priest and came to Brazil in 1902 to work as a teacher at the *Ginásio e Pensionato* (boarding school) *de Nossa Senhora da Imaculada Conceição* in São Leopoldo, Rio Grande do Sul. After naturalizing as a Brazilian, Padberg-Drenkpohl remained in the state of Rio Grande do Sul until 1907 when he went back to Europe to continue his higher education studies in theology within the organization of the Society of Jesus.

Both Rabuske (1981) and Padberg-Drenkpohl (1925) himself, the latter in a text in which he relates how he proceeded during his submission to a practical test as part of the National Museum's competitive selection examinations for the anthropology sector, refer to the fact that he took his degree at the University of Freiburg in the Breisgau region of Germany in the period from 1912 to 1917. Padberg-Drenkpohl declares that he had become specifically qualified for the areas of prehistoric anthropology (an alternative term for paleoanthropology) and had received training in the fields of ethnology, paleontology, geology, and related subjects from renowned exponents such as Eugen Fischer (general physical anthropology, anatomy of human races, anthropological practice, osteology, etc.), Francisco Keibel (comparative anatomy of the vertebrates), Guilherme Deecke (general and special geology, etc.), and others.

[4]Those documents are listed in a separate section entitled "Documental Sources," organized by authors, appearance in the body of the text, and date. Their respective codes in the Analytic Inventory are also available at the end of each entry between square brackets to facilitate future consultation [AI: description].

According to his biographer (Rabuske 1981), Padberg-Drenkpohl withdrew from the Society of Jesus in 1925 after a failed attempt at a civil marriage in Germany, which only came to fruition, with Vatican approval, years later in Brazil. When he returned to Brazil that same year, Padberg-Drenkpohl established himself in Rio de Janeiro, where he started work as a naturalist attached to the fourth section (anthropology and ethnography) of the National Museum and became part of the permanent staff in 1931. Starting in 1935, parallel to his activities at the Museum, he worked as a Professor of Greek and prehistory at the University of the Federal District (now the Federal University of Rio de Janeiro) (Rabuske 1981).[5]

Apart from the biographical details, for the purposes of this chapter, the fact is that 1 year after his considerations about the selection examination, Padberg-Drenkpohl was sent to the Lagoa Santa region with the mission of continuing the studies begun by Peter Lund in the nineteenth century (Padberg-Drenkpohl 1926a, b, c; Padberg-Drenkpol 1937), and 3 years later he undertook yet another excursion to the region (Padberg-Drenkpohl 1929, 1937).[6] One of the reasons for the mission was that the Danish naturalist had shipped all his paleoanthropological finds to Denmark and had only left one cranium of the so-called Lagoa Santa race in Brazil, at the Brazilian Historical and Geographic Institute (Padberg-Drenkpohl 1926c; Rabuske 1981).

According to the report for 1926 (Padberg-Drenkpohl 1926c), at that moment the paleoanthropological investigations in Lagoa Santa would be based on two complementary needs. The Museum's researcher referred to the first as "conservative," and it consisted of concentrating on all the things Peter Lund had done in terms of mapping the caves that he had visited and disseminating information on his finds and his studies to the Brazilian scientific community, given that almost all of Lund's documents were written and published in Danish.[7] The second necessity he called "progressive," and its purpose would be to expand research in the region in order to, if possible, find some intact prehistoric site that could provide proof of the existence of the Lagoa Santa Man in the Quaternary period. It should be observed that, regarding this second "progressive" goal, it is possible to identify the implicit concern that was recurrent among researchers of that time interested in Lagoa Santa's prehistoric past, namely, to discover material proof of human presence on the Americas as far back as the Pleistocene epoch. That concern appears more explicitly expressed in another document from which the quotation that follows was extracted and will be further examined in the final part of this section:

[5] For further information regarding the period when Padberg-Drenkpohl was a Jesuit priest, which was also the period when he published one of his most important works on Brazil's "true" territorial extension, see Rabuske (1981).

[6] At that time Edgard Roquette-Pinto was president of the Museum board as shown by two letters Padberg-Drenkpohl addressed to him in that capacity in 1926 (Padberg-Drenkpohl 1926a, b).

[7] In 1950 a selection of various papers of the Danish naturalist (Lund 1950) organized and translated [into Portuguese] by geologist Carlos de Paula Couto was published with the title *Memórias Sôbre a Paleontologia Brasileira* (Brazilian Paleontology Memoirs). A more recent work about Lund's scientific trajectory, specially related to the Lagoa Santa's investigations, was published by Holten and Sterll (2011).

> The overriding aim of the research undertaken by the National Museum is to clarify, through new finds and original observations, the still unsettled problem of the Lagoa Santa Man, especially his supposed contemporaneity with the extinct quaternary fauna. (Padberg-Drenkpohl 1931: 3)

The 1926 report continues with detailed descriptions of every step of Padberg-Drenkpohl's scientific incursions in the Lagoa Santa region. Regarding the first one, which took place from June 17 to July 29, the researcher narrates that his first move was to visit the house where Peter Lund lived from 1835 to 1880 and also some of the caves nearby, such as the Lapa Vermelha, Lapa do Salitre, Palmital, and Contendas. Very little material was found on those visits: merely a few bones of recent animals and some pottery.

Padberg-Drenkpohl was sure that he would find nothing of importance in Lagoa Santa and so he went to the Mocambo Farm in the neighboring municipality of Pedro Leopoldo and then on to the Sumidouro Cave. According to his report, at the time the cave was inaccessible because of the huge amount of water inside it. He then visited several other caves in the vicinity such as the Cerca Grande, which Lund had excavated and which was apparently worked out.

In another site, the Lapa do Caetano, near to the Mocambo Farm, which had been excavated by the proprietor's son Cassio Lanari, then deceased, a few bones that seemed to be quite recent were all that could be found. From the Lapa do Caetano, Padberg-Drenkpohl was taken to another site on the land of the same property which the scientist named Lapa do Cassio.[8] His investigations there found various human remains (complete crania, maxillae, teeth, and other bones) representing apparently half a dozen individuals and which, in Padberg-Drenkpohl's view, could be identified as belonging to the "Lagoa Santa race" in view of the characteristic dolichocephaly the skulls presented.[9] They were sent off to the National Museum together with other material that was in the possession of the family (Padberg-Drenkpohl 1926c). The part of the report referring to that first incursion in 1926 closes without any discussion of the specificities of the remains found in the Lapa do Caetano and Lapa do Cassio (such as, e.g., anthropological aspects). They are equally absent from the part dedicated to the second excursion and glaringly so from the 1929 report (Padberg-Drenkpohl 1929). The anthropometrics only appeared in a document of 1937 (Padberg-Drenkpol 1937), and even then they were preliminary and prob-

[8] According to Padberg-Drenkpohl (1931), the first person to take up Peter Lund's studies in Brazil was Cassio Umberto Lanari, the oldest son and bearer of the name of his father, the proprietor of the Mocambo Farm, on which there were various rock-shelters and caves. Lanari investigated several of them, especially the Lapa do Caetano, where he had found the human bones of three individuals. In 1909, a year before his death, Cassio Lanari had published an article in the *Annaes da Escola de Minas de Ouro Preto* (v. 11) entitled *Ossadas humanas fósseis encontradas numa caverna calcárea das vizinhanças do Mocambo* (Human bone fossils found in a limestone cave in the vicinity of Mocambo) (Lanari 1909). There are also references to Lanari's investigations, and his interpretation that the bones found in the Lapa do Caetano had been deposited there in Pleistocene times (see Neves and Piló 2008).

[9] Dolichocephalic craniums are long and narrow.

ably in response to an official request made by the director of the National Museum at the time, Alberto Betim Paes Leme (Leme and Betim 1936a, b).[10]

The second excursion to the Lagoa Santa region took place from September 18 to December 8, 1926 (Padberg-Drenkpohl 1926c). First, Padberg-Drenkpohl explored the caves associated to the drainage basin of the river Ribeirão da Mata, a tributary of the das Velhas River, which today lies within the metropolitan area of greater Belo Horizonte. There, however, he obtained no results, so he moved on to the municipality of Confins where he explored the site known as Lapa da Vargem de Baixo and which subsequently came to be known as the Lapa Mortuária.[11] In a letter addressed to Roquette-Pinto (Padberg-Drenkpohl 1926a) dated to October 12 of that same year, the National Museum's man in the field reported that he was in the municipality of Santa Luzia when he came across a human jawbone in the hands of a collector. He tried to acquire it for the Museum but with no success. Nevertheless, his enthusiasm was aroused by the find, and so after asking around, he managed to get to the site in question, where he found several teeth on the surface. Padberg-Drenkpohl wrote that on that very trip he had come across the remains of at least 30 individuals in the form of maxillae and various other kinds of bone, as well as the bones of animals and some artifacts such as knives, scrapers, pottery fragments, and axes made from pebbles.

Although, initially, no complete skeletons were found, that situation was to change later as announced by a letter (Padberg-Drenkpohl 1926b), again addressed to Roquette-Pinto, dated November 23, and the fact was later ratified in the 1926 report (Padberg-Drenkpohl 1926c). In that letter, Padberg-Drenkpohl insists that the Lapa da Vargem de Baixo should be renamed to Lapa Mortuária (mortuary site) because of the great quantity of burials and human and animal bones deposited there, which, he declared, had been found at a depth of two to three meters. He estimated that he had found a quantity of bones belonging to approximately 80 individuals. Among them were 50 maxillae (Padberg-Drenkpohl 1926b, c): complete craniums, only half a dozen of which were in good conditions (meaning susceptible to good craniometric analyses) and, a priori, identifiable as belonging to the "Lagoa Santa race," as well as all kinds of artifacts, albeit they did not make it possible to gain a better understanding of the nature of the location ("As yet, I do not know how to explain such an accumulation of dead people, whether it was a cemetery, a place of sacrifice or where enemies were slaughtered," as he wrote in his

[10] In both documents, Alberto Betim Paes Leme requests information on the Padberg-Drenkpohl's studies related to the Lagoa Santa region insofar as the Museum was interested in renewing research in that region. In one of the documents (Paes Leme 1936b), the Museum director asks Padberg-Drenkpohl whether he would be interested in taking charge of those new incursions himself.

[11] New excavations were carried out more recently at the site under the scope of the Origins project (see Chap. 9 of this volume). The results have not yet been published (Walter A. Neves, personal communication).

second letter), which could readily be interpreted as indications of a very poorly developed culture (Padberg-Drenkpohl 1926b, c).[12] Again in the second letter of 1926 (Padberg-Drenkpohl 1926b), he briefly mentions an excavation carried out at the Lapa da Limeira, five kilometers from the Lapa Mortuária, where yet another measurable cranium was found and a preliminary analysis indicated that it, too, belonged to the "Lagoa Santa race."

In concluding his report (Padberg-Drenkpohl 1926c), the National Museum's researcher states that the two incursions in Minas Gerais had only obtained results on one of the investigative fronts, namely, the progressive front, because it had led to the discovery of human bones in five caves, namely, the Lapa do Caetano, Lapa Mortuária, Lapa da Limeira, Lapa D'Água, and Lapa da Moreira.[13] Thus, in his view, it was imperative to invest in investigations on the conservative front, which could begin with the creation of maps that would provide an accurate idea of Peter Lund's trajectories in the nineteenth century, especially in the territories between Lagoa Santa and Sete Lagoas. That was to take place in the form of a new attempt made in a campaign 3 years later.

It is better to consider the report on the 1926 excursions and the correspondences they refer to as being "ethnographic narratives" of Padberg-Drenkpohl's successive actions rather than analytical texts. The same can be said of the report on the 1929 excursion (Padberg-Drenkpohl 1929) with the obvious exception of the presence of the mathematical calculations that he made in order to be able to create a map that would be the most faithful portrayal possible of the geography of Lagoa Santa at that time. Actually the underlying theme of that third excursion was precisely the confection of cartographic documents. More analytical statements about the paleoanthropological and geological findings appeared 2 years later in an unpublished text dedicated to the history of the scientific interventions in Lagoa Santa since the times of Peter Lund, and the short excerpt presented at the beginning of this section was taken from that text.

That unpublished document not only took up once more Lund's interventions but also those Padberg-Drenkpohl himself had conducted in Lagoa Santa, and it did so basically in narrative terms. In any event, it is worth highlighting at least two small blocks of analytical discussion conducted by the researcher, the first of which was directed at the geological scenario of Lagoa Santa. When he made a comparison between the reality he had visited and some of the international studies of the time, Padberg-Drenkpohl described Lagoa Santa as belonging to a set of karstic phenomena in a similar way to what had been observed in the karstic regions of Italy, Cuba, and the Yucatan peninsula (which embraces parts of Mexico and Guatemala) (Padberg-Drenkpohl 1931).

The second block specifically addresses the bone finds and then registers the quantity of human material (bone remains of 80 distinct individuals as set out in the 1926 report) and nonhuman remains (bones of recent and of Pleistocene fauna

[12] Regarding the animal bones, Padberg-Drenkpohl reports that the majority were of the recent fauna and that specimens of megafauna remains had only been found in the stratum immediately below the one which contained the human bones, which, to him, meant that the humans and the Pleistocene animals had not been contemporaries.

[13] In none of the written material, reports, or correspondence is there any detailed information about the interventions in the three last-mentioned sites.

species such as the mastodons and the giant sloth) collected during the three incursions, with special mention to the Lapa Mortuária. Padberg-Drenkpohl closes his text with the following considerations: (1) the human bone remains were indicative of the existence of a specific race, namely, "the Lagoa Santa race," typified by its pronounced cranial features marked by dolichocephaly and a certain hypsistenocephaly,[14] and (2) there was no way to prove that the Lagoa Santa race had been a contemporary of the extinct Pleistocene megafauna, especially not based on stratigraphic analyses. That stance, contradicting the hypotheses that Lund put forward, can be detected in the following excerpts (Padberg-Drenkpohl 1931: 6):[15]

> What is important is that we found the remains of those extinct animals, probably belonging to the Quaternary period, in the red clay inside the site, that is to say, at a lower and more ancient level, which would seem to indicate that they are older than the human remains!
>
> Altogether, we only found human remains, such as those that Lund found almost exclusively in the Sumidouro Cave, still invaded by the waters and therefore incapable of guaranteeing undisturbed deposits, in five caves, namely (1) Lapa Mortuária, (2) Lapa da Limeira, (3) Lapa do Caetano, (4) Lapa D'Água, (5) Lapa da Moreira [...] in none of them did we find proof of the contemporaneity of the Lagoa Santa race (to which all the manifestations of human bones belong, given their notable dolichocephaly and hypsistenocephaly, and so on) with the Quaternary fauna, so that, up until now it has not been proved" (underlined in the original).

José Bastos de Ávila and the Carrancas Caves

Seven years had passed since Padberg-Drenkpohl's last incursion in the Lagoa Santa region, and the National Museum was fostering a new research campaign in the region, this time with José Bastos de Ávila as the main protagonist, in a partnership with Nei Vidal of the stratigraphic division of the same institution. There only exists a very descriptive report (de Ávila and Vidal 1937a) referring to this incursion and one which is somewhat superficial in terms of anthropological analyses as such, compared to those legated by Padberg-Drenkpohl. Thus, as was the case with Padberg-Drenkpohl, the results were never actually published (Neves and Piló 2008).

In common with other historical personalities of the physical and biological anthropology fields in Brazil (Gaspar Neto 2012; Santos 2012), a more elucidative biography of José Bastos de Ávila is still pending (Gonçalves 2011). In any event, what is known is that Bastos de Ávila was born in the city of Petrópolis in 1888 in the state of Rio de Janeiro and that he entered the National Museum in 1932, only to withdraw from the institution 6 years later in order to dedicate himself

[14] Hypsistenocephalic skulls are high and narrow.

[15] In that text, Padberg-Drenkpohl tells how he had also found two complete skeletons, something that had not been registered in the 1926 report (Padberg-Drenkpohl 1926c) or in the correspondence of that year (Padberg-Drenkpohl 1926a, b).

exclusively to the investigations underway in the sphere of the Educational Research Institute of the Federal District (Rio de Janeiro at the time), where he had been active since 1918 (Gonçalves 2011; Gonçalves et al. 2012).

Bastos de Ávila had a degree in medicine, and accordingly he ended up dedicating his anthropological studies to the question of the school-age children's development and to analyses of the collection of bones in the National Museum, as demonstrated in a few of the available documents, some publications, and the papers of Gonçalves (2011) and Gonçalves et al. (2012).[16]

Regarding the incursion in the region of Lagoa Santa, the 1937 report (de Ávila and Vidal 1937a) gives an account of the directives determined by the National Museum board, namely, the execution of research in the caves to be found on the Nova Granja Farm in the municipality of Santa Luzia in the region of the das Velhas River, which encompasses Lagoa Santa and its vicinities. That Farm was indicated because the son of its owner, a certain Dr. José Machado, had informed the Museum that human bones had been found there, and later Bastos de Ávila and Nei Vidal were able to confirm the finds.

The report is divided into two parts. The first consists of a set of general remarks about all the material that was needed for the expedition, such as photographic material, lighting equipment, digging tools, surveying instruments, and assistants. The second part narrates all the steps taken by the researchers from the moment they left Rio de Janeiro on February 16, 1937 up to the time they visited the Lapa Mortuária in Confins (no date given), where Padberg-Drenkpohl had been before them.

The second part of the report begins with some considerations regarding topographic aspects of the region, its geological constitution, and its vegetation. In the words of Ávila and Vidal (1937a: 7), "not only the region's topographic relief but various other aspects that we refer to led us to believe that formerly there were huge lakes or seas in the region which today is practically dry."

In any event, they were to investigate two caves at the same time. They called them Carrancas N°1 and Carrancas N°2.[17] It seemed that Carrancas N°1 had suffered some kind of recent intervention. The researchers first give a detailed account of the geological aspects of the cave (nature of the soil and the rock formations) and then go on to their archaeological and anthropological endeavors. They describe how they only found human bones in the main entrance to the cave and those had apparently belonged to at least four individuals. The bones were practically at the surface, totally disarticulated, and in a very poor state of conservation. It seemed clear to the researchers that in spite of the evidence of the passage of water inside the cave at some time in the past, the bodies had been deliberately buried there,

[16] In the work of Gonçalves (2011) and Gonçalves et al. (2012), there are also analyses of Bastos de Ávila's publications, especially those dedicated to the anthropometric analyses of children as well as to the revisitation of a novel *No Pacoval de Carimbé* published by him in 1993, in which, among other topics, he discusses the scientific, social, political economic, and cultural situation of Brazil in a frank dialogue with Edgard Roquette-Pinto and Euclides da Cunha. Lastly, it must be noted that Maria Júlia Pourchet was Bastos de Ávila's successor in the studies of the physical growth of schoolchildren, although the specificities of that relationship deserve further study.

[17] Both sites have been recently destroyed by the extraction of limestone (Walter A. Neves, personal communication).

because there were stone artifacts, such as arrowheads and bone breakers, together with the bones. Also, the way they had been buried was indicative of that fact, as the following excerpt from the report shows:

> (...) it is indeed an authentic burial of each group of bones because in each one, more or less fragmented, there are all the pieces of the support apparatus. The impression one gets is that the bones had been previously fractured so that by reducing the volume they could be more easily buried in a squatting position with knees up to the jaw and the lower limbs positioned next to the tip of the cranium, and that when the cranium appeared to stick out, it had been unhesitatingly beaten in and broken up, so it is easy to see why it is so hard to find a cranium with its anatomy intact. (de Ávila and Vidal 1937a: 11)

Just 100 m away from Carrancas N°1, Carrancas N°2 showed signs that it too had been the target of recent excavations. Indeed, in their report Bastos de Ávila and Nei Vidal stated that one of the workmen they hired had previously worked in a similar campaign conducted by the Minas Gerais Academy of Sciences and that on that occasion various bones and stone artifacts had been removed and shipped to Belo Horizonte. In the end, the National Museum team's efforts only brought in a few fragments of bone and some stone artifacts, which they considered to be indicative that the cave had been used as a cemetery by indigenous peoples.

When the investigations in the two caves, Carrancas N°1 and N°2, had finished, Bastos de Ávila and Nei Vidal went on through another great gap in the solid rock in which there were two caves, and there they came across vestiges of Lund's studies. They were informed that a huge tooth of a carnivorous animal had been extracted from there and that it was in the hands of the Minas Gerais Academy of Sciences.

After dealing with the successive steps of the research endeavor, the report presents a list of the material that was collected: one fragmented cranium, which was being reconstructed at the National Museum's premises, several fragments of bone, a polished stone axe, fragments of arrowheads, a diabase millstone probably used to fracture the bones, geological samples for analysis, and a second cranium almost intact offered by the Farm owner and which had been exhumed from one of the caves. The only conclusion they arrived at, and even then it was provisional, was that the craniums were probably those of Botocudo Indians:

> This cranium and the one that we are reconstructing, after being submitted to preliminary examination lead us to believe that they belonged to "Botocudos". However, more detailed studies, which we are undertaking, will enable us to issue a more accurate opinion. (de Ávila and Vidal 1937a: 16)

The report narrates how, when they departed from the Nova Granja Farm, they travelled to Confins in order to investigate three other caves, one of which was the famous Lapa Mortuária, which Padberg-Drenkpohl had discovered and excavated in the second half of the 1920s. Their investigations did not bring in any material results, and they noticed that of the three caves they visited, the Lapa Mortuária had been the one most explored, and the Minas Gerais Academy of Sciences was among those that had explored it. From that point on, Bastos de Ávila and Nei Vidal register a series of criticisms regarding the Academy's exploration methods (such as their use of explosives), and they finalize the document with the conclusion that the best thing the National Museum could do would be to redirect its anthropological studies to the regions of Montes Claros and Serra do Cabral. First, according to various reports there were caves there with great potential for such investigations, and some of them even had rock paintings. Second, considering that the region of Lagoa Santa had been devastated by incursions of researchers and adventurers, especially foreigners, none of whom had any serious commitment to scientific quality and all of them less dedicated to science than to getting away with human and natural materials to supply foreign institutions and the international black market. Indeed, those very same observations appear in two other documents: a letter (de Ávila and Vidal 1937b) and a kind of memorandum (de Ávila and Vidal 1937c) addressed to the director of the National Museum at the time, Alberto Betim Paes Leme. The excerpts from that report that follow below reveal how critical Bastos de Ávila and Nei Vidal were when expressing their discontent with the state of conservation of the Nova Granja and Confins caves and with the behavior of the foreign researchers in Lagoa Santa.[18]

> Another reason, no less appreciable, that reinforces our suggestion [to move the research activities to Montes Claros and the Serra do Cabral regions] is the fact we have observed that there are many depredations resulting from the use of inadequate, ruinous research methods on the part of lay individuals – most of them foreigners – untrained in any kind of scientific subject, whose activities merely address the commercial aspect or the egotistic desire to possess a rarity. (de Ávila and Vidal 1937a: 18)

> Numerous foreigner naturalists are travelling everywhere in the interior of Brazil collecting material not only to supply scientific institutions abroad but also to sell in the markets for such items. To achieve their ends, they stop at nothing, disrespecting the laws that protect our natural riches, and they perform such devastation that they make our own research activities impossible or deficient. Those are the facts that have led us to propose that our research be shifted from Lagoa Santa to the Montes Claros and Serra do Cabral zones. (de Ávila and Vidal 1937a: 18–19)

The Controversy Between the National Museum and the Minas Gerais Academy of Sciences

As announced in the introduction to this chapter, this third and last section consists of some paragraphs dedicated to the discordance between the National Museum researchers and those of the Minas Gerais Academy of Sciences regarding the contemporaneity of humans and the extinct megafauna in the Lagoa Santa region. That discussion is alluded to in the work of Neves and Piló (2008) with special attention

[18] In the Physical Anthropology Archives at the National Museum there is a cutting from a newspaper with the title "The National Museum and the Caves in Lagoa Santa" possibly dated to May 28, 1937. The cutting does not identify the name of the newspaper. What is interesting is the text, probably written by Bastos de Ávila and Nei Vidal, insofar as it is a mixture of texts that can be found in the report and in the memorandum addressed to the director of the National Museum (Ávila and Vidal 1937a, c, respectively). In that case, could that cutting be a draft of the text to be published in the newspaper, worked on previously? Would it have been submitted for the prior approval of the director of the National Museum? The cutting is registered in the *Inventário Analítico do Arquivo de Antropologia Física do Museu Nacional* (Analytical Inventory of the Physical Anthropology Archives of the National Museum) (Santos e Mello e Silva 2006) and the entry code is AF.T.1.1.013.

paid to the discoveries of Harold V. Walter of the Minas Gerais Academy of Sciences. In 1935, when excavating the Lapa Mortuária in Confins, investigated by Padberg-Drenkpohl in 1926, Walter had found (underneath a stalagmitic crust) the incomplete remains of a human skeleton (Confins Man) together with fragments of a horse cranium as well as the teeth and part of the femur of a small mastodon" (Neves and Piló 2008: 126; see Chap. 6 of this volume). That had led him and other researchers from the same institution to conclude that the large extinct mammals and the human being had been contemporaries. Decades later there would be proof of that contemporaneity (Neves and Piló 2003), but that discussion lies outside the scope established for this chapter.

As Neves and Piló have underscored, the National Museum researchers were radically opposed to that proposition. Proof of that opposition can be found in three documents in the Physical Anthropology Archives of the National Museum, two of them signed by Padberg-Drenkpohl (1937, 1939) and another by Bastos de Ávila (1939).

As mentioned above, in his notes of 1937, possibly in response to a request from the then director of the National Museum (Paes Leme 1936a, b), Padberg-Drenkpohl (1937) alludes to the Minas Gerais Academy of Sciences' discovery of the so-called Confins Man in the Lapa Mortuária, which, according to that institution's representatives, was typified by very low prognathism (66.30° in accordance with the method proposed by Paul Rivet). Thus, according to Padberg-Drenkpohl, for its discoverers Confins Man would have been a representative of a pre-mongoloid race, the most primitive of the Americas. Highly discomfited by that suggestion, the National Museum anthropologist set himself to conduct a critical review not only of the idea of the supposed prognathism as such but also the way it was measured according to Paul Rivet.

Mathematical, anthropometric, and methodological details apart, one of the first things Padberg-Drenkpohl does in that document is to question Paul Rivet's methods, stating firmly that, not only in the case of the Lagoa Santa skulls studied but also in those of other populations such as the Papuas (originally typified as showing low prognathism), what could be seen was actually a high degree of prognathism. Padberg-Drenkpohl had proceeded to measure all the craniums in the National Museum (that he had found) using the method proposed by Rivet, and in all of them he found angles greater than 70°.[19]

With those measurements done and a brochure listing all the measurements related to Confins Man, he detected serious flaws in the final results, especially considering that the cranium in question had been fractured into ten pieces and been imperfectly reconstituted with the upper portion of the maxilla displaced forward, resulting in

[19] The terminology adopted by Padberg-Drenkpohl seems a little confusing in the text, as does some of the information (such as when he states that he measured ten skulls found by Lund, whereas in other documents he states that the reason for the National Museum's taking up the research at Lagoa Santa once more was partly because, with the exception of one skull donated to the IHGB, all the material that the Danish naturalist had found had been shipped to his country of origin). In any case, he makes it clear that the "Lagoa Santa race" would be one that presents a clear mesognathism or even a weak orthognathism, that is, the face is not projected anteriorly to the neurocranium.

what he called a "fantastic prognathism." Making all the measurements over again, Padberg-Drenkpohl concluded that Confins Man was actually nothing more than the simple cranium of an individual of the "Lagoa Santa race," and, accordingly, there was nothing about it to situate the skull as a contemporary of the extinct megafauna.[20]

As mentioned elsewhere, that contraposition to the proposals of the Minas Gerais Academy of Sciences was to reappear 2 years later, in 1939, in the form of a letter signed by Arnoldo Cathoud (1939), in his capacity as secretary general of the institution, addressed to Heloísa Alberto Torres and dated to March 15 of that same year. Cathoud asks Torres about a paper presented by an Argentinean researcher, Antonio Serrano, which states that, contrary to what the researchers of the Minas Gerais Academy of Sciences propose, the researchers at the National Museum argued that "Confins Man" was not a contemporary of the extinct megafauna. To address Cathoud's request, on May 11 the director of the National Museum sent an official letter to Padberg-Drenkpohl (Torres 1939) requesting more detailed information in response to the questioning of the Secretary General of the Minas Gerais Academy of Sciences, and that was the aforementioned official request that Padberg-Drenkpohl (1939) and Bastos de Ávila (1939) replied to.

Padberg-Drenkpohl's (1939) reply was actually just a textual re-elaboration of a document written in 1937 (Padberg-Drenkpol 1937) in which he ratified his position of not considering "Confins Man" to be anything other than a specimen of Lagoa Santa Man:

> today, I can add, succinctly, the following: this find, abusively denominated "Confins Man" is, with all certainty and evidence one more example of the Lagoa Santa race, a "lagoan" if you will (or more correctly a Lacid, as the University suggested in 1935), or in a more rigorous scientific denomination a representative of *Homo sapiens lagoanus* Eickstedt 1937, such as almost half a dozen of which I myself removed from the same cave after it had been discovered and scientifically explored. Apart from the 10 I brought in, we have knowledge of 16 of Lund's (one of them is here in the Historical Institute) and others from Ecuador studied by P. Rivet, etc. (Padberg-Drenkpohl 1939: 3; underlined in the original)

It follows from that first collocation of Padberg-Drenkpohl's that in his view, the "Confins Man" or any other bone specimen coming from the Lagoa Santa region did not coexist with specimens belonging to the extinct Pleistocene megafauna. That is because, according to him, remains of the former are situated in layers that antecede those in which the bones of large mammals were found, and in addition there is no evidence of any anthropic treatment of those animal remains. Nevertheless, in his reply, Padberg-Drenkpohl does not dismiss the possibility that in the future evidence indicative of human presence on the Americas in the Pleistocene epoch may be found and maybe even in Lagoa Santa. To him, what was happening was that paleoanthropology in Brazil had not managed, so far, to gather sufficient indi-

[20] In the same document, there is a reference to the identification and classification of more than 300 teeth from Lagoa Santa caves, and his verdict is that "these teeth of the Lagoa Santa race, rigorously classified and marked, constitute an odonatological standard/reference, displaying the racial particularities and enabling them to be reliably compared to others" (Padberg-Drenkpohl 1937: 2).

cations of that far greater antiquity, and that fact called for a calmer attitude in addressing the issue. In his own words, tinged with a certain acidity directed at the Minas Gerais Academy of Sciences:

> Obviously that does not imply the impossibility of there having been an American Pleistocene Man; that is something which, as in the other continents, can only be stated after it has been proved. Gratuitous assertions, however, not founded on the impartial dictates of Science, cannot guarantee it. (Padberg-Drenkpohl 1939: 5, underlined in the original)

Bastos de Ávila's reply (1939), even though it contains a whole set of craniometric analyses comparing two "Lagoa Santa race" craniums with the "Confins Man" skull, is more succinct. In it, he corroborates Padberg-Drenkpohl's collocations, namely, that from the anthropometric standpoint there was absolutely no justification for considering the "Confins Man" to be a specimen apart from the "Lagoa Santa race" (*Homo sapiens lagoanus*) and, again, that the question of its antiquity, especially in regard to coexistence with the Pleistocene megafauna, although it had been left open by Lund, had now been settled. That was because, in this last case, a satisfactory conclusion would only come to light when the entire Lagoa Santa region had been the object of more intense and continuous investigations.

Final Remarks

As the historical retrieval achieved by Neves and Piló (2008), the articles brought together in Silva and Rodrigues-Carvalho (2006), and the dozens of articles published in specialized journals in the last few decades have allowed us to perceive, the question of the human settlement of the Americas not only necessarily involves Lagoa Santa but also continues to provoke heated discussions. They are discussions which, despite the nuances they may have acquired in different periods, have contributed one way or another to the efforts being made on various fronts of scientific investigation to reconstitute the prehistoric past of our continent and, by extension (why not), to achieve an increasingly accurate map of the dispersion of *Homo sapiens* ever since we first appeared on the African continent around 200,000 years ago. The importance of the documents that have been addressed here is that, based on the contributions of a specific institution, the National Museum, they do at least serve for us to reconstitute a given historical period. Despite that remark, however, their importance, as stated at the beginning of this chapter, goes well beyond the limits of what we might call an internal dimension of science (the anthropological research into the arrival of Man in the Americas), given that they also show how, in common with what we can see in other facets of social life, clashes and controversies are the driving force of science, or rather, of its progress. Today, we have "morphology" versus "molecule"; in the first half of the last century it was "Confins Man" versus "Lagoa Santa Man." Although this aspect has not been fully explored, it could well be so on other occasions, and in that context the documental registrations of Padberg-Drenkpohl and Bastos de Ávila continue to be interesting sources for addressing it.

References

Bourdieu, P. (1975). La spécificité du champ scientifique et les conditions sociales du progrès de la raison. *Sociologie et Societes, VII*, 91–118.
Bourdieu, P. (1976). Le champ scientifique. *Actes de la Recherche en Sciences Sociales, 2*, 88–104.
Bourdieu, P. (2004). *Os usos sociais da ciência: por uma sociologia clínica do campo cientifico*. São Paulo: Editora UNESP.
Chatters, J. C., Douglas, J. K., Asmeron, Y., Kemp, B. M., Polyak, V., Blank, A. N., Beddows, P. A., Reinhardt, E., Arroyo-Cabrales, J., Bolnick, D. A., Malhi, R. S., Culleton, B. J., Erreguerena, P. L., Rissolo, D., Morell-Hart, S., & Stafford Jr., T. W. (2014). Late Pleistocene human skeleton and mtDNA link Paleoamericans and modern native Americans. *Science, 344*, 750–754.
Gaspar Neto, V. V. (2012) A outra face do crânio: antropologia biológica no Brasil hoje. PhD. Dissertation. Universidade Federal Fluminense.
Gaspar Neto, V. V., & Santos, R. V. (2010). *The colour of the bones: scientific narratives and cultural appropriations of 'Luzia', a prehistoric skull from Brazil*. Mana 5, Selected Edition.
Gonçalves, A. S. (2011). *José Bastos de Ávila e as pesquisas em antropologia física no Museu Nacional (1928–1938)*. Master Thesis. COC/FIOCRUZ.
Gonçalves, A. S., Maio, M. C., & Santos, R. V. (2012). Entre o laboratório de antropometria e a escola: a antropologia fisica de José Bastos de Ávila nas decadas de 1920 e 1930. *Boletim do Museu Paraense Emilio Goeldi – Ciências Humanas, 37*, 671–686.
González-José, R., & Bortolini, M. C. (2011). Integrating different biological evidence around some microevolutionary processes: Bottlenecks and Asian-American artic gene flow in the new world settlement. *Evolution: Education and Outreach, 4*, 232–243.
González-José, R., Bortolini, M. C., Santos, F. R., & Bonatto, S. L. (2008). The peopling of America: Craniofacial shape variation on a continental scale and its interpretation from an interdisciplinary view. *American Journal of Physical Anthropology, 137*, 175–187.
Holten, B., & Sterll, M. (2011). *P. W. Lund e as grutas com ossos em Lagoa Santa*. Belo Horizonte: Editora UFMG.
Hubbe, M., Neves, W. A., & Harvati, K. (2010). Testing evolutionary scenarios for the settlement of the new world. *PloS One, 5*, E11105.
Lanari, C. U. (1909). Ossadas humanas fósseis encontradas numa caverna calcárea das vizinhanças do Mocambo. *Annaes da Escola de Minas de Ouro Preto, 11*, 15–35.
Latour, B. (2000). *Ciência em ação: como seguir cientistas e engenheiros sociedade afora*. São Paulo: Editora UNESP.
Latour, B., & Woolgar, S. (1997). *Vida de laboratório: a produção dos fatos cientificos*. Rio de Janeiro: Relume-Dumará.
Lund, P. W. (1950). *Memórias sôbre a paleontologia brasileira (revistas, organizadas e traduzidas por Carlos de Paula Couto)*. Rio de Janeiro: Instituto Nacional do Livro.
Neves, W. A. (2006). Origem do homem nas Americas: fósseis *versus* moleculas. In H. P. Silva & C. Rodrigues-Carvalho (Eds.), *Nossa origem: o povoamento das Américas – visões multidisciplinares* (pp. 45–76). Rio de Janeiro: Vieira & Lent.
Neves, W. A. (2013). *Um esqueleto incomoda muita gente*. Campinas: Editora da Unicamp.
Neves, W. A., & Atui, J. P. V. (2004). O mito da homogeneidade biológica na população paleoindia de Lagoa Santa: implicações antropológicas. *Revista de Antropologia, 47*, 159–206.
Neves, W. A., & Piló, L. B. (2003). Solving Lund's dilemma: New AMS dates confirm that humans and megafauna coexisted at Lagoa Santa. *Current Research in the Pleistocene, 20*, 57–60.
Neves, W. A., & Piló, L. B. (2008). *O povo de Luzia: em busca dos primeiros americanos*. São Paulo: Editora Globo.
Neves, W. A., Powell, J. F., & Ozolins, E. G. (1999a). Modern human origins as seen from the peripheries. *Journal of Human Evolution, 37*, 129–133.
Neves, W. A., Powell, J. F., Prous, A., Ozolins, E. G., & Blum, M. (1999b). Lapa Vermelha IV Hominid 1: Morphological affinities of the earliest known American. *Genetics and Molecular Biology, 22*, 461–469.

Powell, J. F., & Neves, W. A. (1999). Craniofacial morphology of the first Americans: Pattern and process in the peopling of the new world. *Yearbook of Physical Anthropology, 42*, 153.
Prous, A. (1992). *Arqueologia brasileira*. Brasilia: Editora da Universidade de Brasilia.
Rabuske, P. A. (1981). J. H. Augusto Padberg-Drenkpol: o homem que mediu o Brasil. *Pesquisas, 20*, 1–127.
Salzano, F. M. (2011). The prehistoric colonization of the Americas: Evidence and models. *Evolution: Education and Outreach, 4*, 199–204.
Santos, G. (2012). The birth of physical anthropology in late imperial Portugal. *Current Anthropology, 53*, 33–45.
Santos, R. V., & Mello e Silva, M. C. S. (2006). *Inventário Analitico do Arquivo de Antropologia Física do National Museum*. Rio de Janeiro: National Museum.
Silva, H. P., & Rodrigues-Carvalho, C. (Eds.). (2006). *Nossa origem: o povoamento das Americas – visões multidisciplinares* (Vol. 119). Rio de Janeiro: Vieira & Lent.

Documental Sources

de Ávila, J. B. (1939). *Letter from José Bastos de Ávila addressed to Heloisa Alberto Torres, dated June 15, 1939* (typed) (IA: AF.T.1.1.017].
de Ávila, J. B., & Vidal, N. (1937a). *Report on the exporatory work in the caves of Carrancas in Nova Granja, Minas Gerais* (1937) (typed) (IA: AF.T.1.1.012].
de Ávila, J. B., & Vidal, N. (1937b). *Letter addressed to the director of the National Museum, Alberto Betim Paes Leme, dated February 26, 1937* (hand written) [IA: AF.T.1.1.013].
de Ávila, J. B., & Vidal, N. (1937c). *Memorandum addressed to the director of the National Museum, Alberto Betim Paes Leme, dated May 19, 1937* (typed) [IA: AF.T.1.1.013].
Cathoud, A. (1939). Letter addressed to the director of the National Museum, Heloisa Alberto Torres, dated March 1, 1939 (typed) [IA: AF.T.1.1.017].
Paes Leme, A. B. (1936a). *Official letter addressed to José Bastos de Ávila inquiring about Padberg-Drenkpohl's research in Lagoa Santa* (typed) [IA: AF.T.1.1.011].
Paes Leme, A. B. (1936b). *Request addressed to Padberg-Drenkpol inquiring about his research in Lagoa Santa* (typed) [IA: AF.T.1.1.011].
Padberg-Drenkpohl, J. H. A. (1925). *Declaration* (typed) [IA: AF.T.6.1.002].
Padberg-Drenkpohl, J. H. A. (1926a). *Letter from J.A. Padberg-Drenkpohl to Edgard Roquette-Pinto, dated December 12, 1926* (hand written) [IA: AF.T.1.1.003].
Padberg-Drenkpohl, J. H. A. (1926b). *Letter from J.A. Padberg-Drenkpohl to Edgard Roquette-Pinto, dated November 23, 1926* (hand written) [IA: AF.T.1.1.003].
Padberg-Drenkpohl, J. H. A. (1926c). *Report on two excursions to the limestone region of Lagoa Santa in 1926* (typed) [IA: AF.T.1.1.003]
Padberg-Drenkpohl, J. H. A. (1929). *Report on the 1929 excursion to the limestone region of Lagoa Santa* (typed) [IA: AF.T.1.1.003]
Padberg-Drenkpohl, J. H. A. (1931). *Dr. Lund and the National Museum* (typed) [AF.T.1.1.007].
Padberg-Drenkpohl, J. H. A. (1939). *Letter from J.A. Padberg-Drenkpol to Heloisa Alberto Torres, dated May 15, 1939* (typed) [IA: AF.T.1.1.017]
Padberg-Drenkpol, J. H. A. (1937). *Notes for a working report of Dr. Padberg-Drenkpohl in 1937* (typed) [IA: AF.T.1.1.016].
Torres, H. A. (1939). *Official letter addressed to Jorge Henrique Augusto Padberg-Drenkpol, dated May 11*, 1939 (typed) [IA: AF.T.1.1.017].

Chapter 6
The Minas Gerais Academy of Sciences: Lund's Inheritors

Fernando Walter da Silva Costa

Abstract In the decades from the 1930s to the 1960s, the Minas Gerais Academy of Sciences was the protagonist of archaeological and paleontological studies in the state of Minas Gerais. The Academy was a private institution made up of amateurs; it never enjoyed any form of public support, and its central goal was to ensure the continuation of the studies started by Peter Lund in the Lagoa Santa region. Despite its notable contribution to Brazilian archaeology, the Academy's research efforts were never taken seriously by the professional archaeologists who worked in Minas Gerais in the 1970s, and consequently it has never occupied a respected position in the national scientific memory.

Introduction

From the 1930s to the 1960s, a private institution made up of amateurs and with no form of government support was the protagonist of archaeological and paleontological research in the state of Minas Gerais. The overriding objective of the Minas Gerais Academy of Sciences was to continue the research conducted by Peter Lund in the region of Lagoa Santa.

The Academy never enjoyed any kind of public sponsorship, although there was always an expectation that it eventually receives this type of support. All its activities were financed and conducted by a small group of members (its founders), and, despite all the obstacles they had to face, they managed to carry out important activities in the region and regularly published their results, often in articles and books in both Portuguese and English.

F.W. da Silva Costa (✉)
Archaeological Consultant, Vitória, ES, Brazil
e-mail: fwscosta65@gmail.com

P. Da-Gloria et al. (eds.), *Archaeological and Paleontological Research in Lagoa Santa*, DOI 10.1007/978-3-319-57466-0_6

Historical Context

At the beginning of the 1930s in Belo Horizonte, preparations were underway to celebrate the hundredth anniversary of Lund's work in the Lagoa Santa region. The main person in charge of organizing them was Professor Anibal Mattos[1] of the Belo Horizonte Model School (*Escola Modelo de Belo Horizonte*), who had taken on the chairmanship of the committee responsible for the events and ceremonies and for making a complete inventory of all the research carried out by that "wise Dane" (Scientia 1947).

One of Mattos's most distinguished collaborators was the university professor and doctor Arnaldo Cathoud, who took on the task of writing about the naturalist's studies and later publishing *A Raça de Lagoa Santa no Pleistoceno Americano* (The Lagoa Santa Race in the American Pleistocene) (Cathoud 1940).

The main result of those events, apart from the celebrations themselves, was the creation of the Minas Gerais Academy of Sciences. According to Lúcio dos Santos, official orator of the ceremony on August 10, 1935, in the auditorium of the Belo Horizonte Normal School,[2] to mark the founding of the association,[3] the Academy would be:

> [The] congregation of people in an institution dedicated to the disinterested study of the sciences and [more specifically in taking up once more] Brazilian paleontological studies finalized by Dr. Lund in the limestone zone of the Velhas River watershed as well as Brazilian archaeological studies. Its mission would also be to train elites in the scientific terrain [with] the support of the Government of Minas Gerais, which, as the repository of the trust of the people, defender of the legitimate interests of our land, will not fail to give the new institution the support it needs. (Scientia 1947:3)

Another fact that greatly encouraged the organization of the Academy and definitively reinforced the links of the institution with the work of the Danish naturalist was the conclusion arrived at by Padberg-Drenkpohl, who carried out excavations in the region of Lagoa Santa in 1928 and 1929. His results contested directly Lund's hypothesis and provoked many negative reactions in Belo Horizonte's intellectual and academic circles and even in the federal capital[4] (Padberg-Drenkpohl 1926, 1929). The words of Roquette-Pinto summed up the general feeling very well:

> It was the technical insufficiency of the person in charge of that work that led to the expedition achieving so little [...] not one step was taken towards clarifying the problem: were Lagoa Santa Man and the [extinct] mammals found there contemporaries? (Roquette-Pinto 1934)

In short, the Academy was born with a well-defined objective: to continue the work of Lund in Lagoa Santa, defending his heritage by proving his ideas right.

---

[1] In the 1930s, Mattos published various articles and books about Lund (Mattos 1935a, b, 1939).

[2] Normal School in Brazil was a place aimed to train new teachers.

[3] Almost 1 year after the actual founding, which took place on October 24, 1934.

[4] Rio de Janeiro, in those days.

Another reason why Aníbal Mattos, Arnaldo Cathoud, and Harold Walter organized themselves into a nonprofit institution was their awareness that their individual efforts, as was the fashion among Brazilian amateurs of the day, would be insufficient to guarantee researches' credibility, attract the interest and support of the state, or even denounce the degradation of the deposits at Lagoa Santa (Mattos 1961).

In that sense, the institution sought to legitimize itself in the eyes of the Belo Horizonte elite by enrolling illustrious members belonging to the most varied sectors (intellectuals, professors, and politicians) because it was structured in the same way as the Historical Institutes and Academies of Letters of the day, that is to say, as an association sponsored by the state, and made up of prominent residents of the capital city of Minas Gerais capable of interceding with the authorities in favor of the association. The members, in common with other institutions of the same kind, occupied chairs that were named after a patron.[5]

As an example, the list of founder members includes engineers Lúcio José dos Santos, Mário Werneck, and Lourenço Baeta Neves, in addition to the doctors Otávio Magalhães and Octaviano Neves, among others. In 1947, the governor of Minas Gerais, Milton Campos,[6] occupied the office of president of honor (Scientia 1947). None of them ever had any outstanding participation in the Academy, and apparently, they did not help much in obtaining any kind of official support either. Not even the Academy's executive president, Mattos, who had ready access to official circles, achieved any success in the incessant quest for any kind of public funding.

It is hard to explain the total omission of the authorities in regard to the Academy, especially that of a state government that was accustomed to sponsoring cultural, literary, and educational institutions and activities. Probably that attitude was due to the Minas Gerais political elite's total lack of interest in, and ignorance of, the nature of the Academy's work.

The lack of any regular funding greatly weakened the Academy and meant that its very existence was always at risk. The lack of funds for research was a harsh reality for the institution and little by little undermined its capacity to carry on the work, as some of Walter and Mattos's statements made clear. The former declared that "The explorations and excavations became increasingly costly because, despite our collection's having become highly valuable, there was no prospect of any financial help for the Academy's work" (Walter 1976:43). The latter remembers that "[…] since the year 1934, the Academy has never received any assistance from the public authorities despite its notable scientific efforts" (Mattos 1950b:290).

The Academy depended exclusively on the contributions of its main members who met all the expenses from their own incomes, as Walter remembers when he states that "the work lacked continuity because very often the money ran out; at other times the rains made excavation difficult, and furthermore the constant rise in

[5] A curious fact is that the patron of chair number one, occupied by Mattos, was Baron Homem de Melo and not Lund as would have been expected. The latter was patron of chair number two.

[6] Milton Soares Campos was governor of Minas Gerais from March 1947 to January 1951.

Fig. 6.1 H. V. Walter and W. Tansley, Confins Cave, 1947 (Photo: Josaphat Penna)

the cost of living made the excursions to the interior increasingly more expensive to the point where they practically stopped altogether" (Walter 1976:37).

That was another highly important aspect of the institution and perhaps what most distinguishes it from the Historical Institutes and the Academies of Letters, namely, the willingness of the main members to finance its activities to avoid the total paralysis of the research. The chronic lack of funds, apart from the fact that it ended up modeling an unparalleled institution, definitively limited the Academy's sphere of action to its original region, that is, Lagoa Santa.[7]

Considering the Academy's almost three decades of activity, the huge sums involved in making the field work feasible, the expense with publications, and even the costs of financing the visit of geologist W. Tansley[8] (University of Chicago/ McGill University; Fig. 6.1) in the 1940s and 1950s give some idea of the degree of the main members' strong commitment to the research (Mattos 1950a, b), especially bearing in mind that they were not wealthy men.

[7] Harold Walter carried out investigations in a small cave in the municipality of Lagoa da Prata, where he found paleontological remains (Walter 1976:40). Josaphat Penna undertook an expedition to register the rock paintings in the Cabral Mountains (*Serra do Cabral*) (Penna 1964:421).

[8] He died in the early 1970s and was buried at the Cemitério da Paz (cemetery), in Belo Horizonte.

The Academy's Performance

The richest and most productive phase of the Academy's trajectory, during which the main paleontological and archaeological excavations were carried out and the respective information was published, lasted up until 1956. Over that period, the three main members complemented one another in their activities; Mattos, because of his respectable position in the artistic and intellectual circles of the Minas Gerais capital, performed the role of the Academy president and chief interlocutor with society and official circles; Cathoud looked after the laboratory analyses of the bone remains and was responsible for the reports; and Walter, thanks to his intimate knowledge of the Lagoa Santa region, was responsible for the archaeological and paleontological fieldwork. Later, the dentist Josaphat Penna joined the Academy and took on the responsibility for all the work related to rock paintings as well as being a great collaborator of Walter's in the fieldwork (Fig. 6.2). They formed the core group that guaranteed the Academy's existence with their campaigns, exhibitions, and publications right through to when activities ceased altogether (Mattos 1961).

Fig. 6.2 Lapa do Baú, 1947, standing: W. Tansley, J. Penna, and D. Walter; sitting: H. V. Walter and two unidentified individuals (Photography by an unidentified author)

The work that Walter coordinated in Confins in the years from 1933 to 1935, at the same site that the National Museum team led by Drenkpohl had excavated in 1926, was what really marked the beginning of the Academy's existence. The results of those excavations and the discovery of a human cranium together with the bones of a mastodon (*Mastodon*) and an extinct species of horse (*Perissodactyla*) resulted in publications that had considerable repercussions, especially in the scientific community of the day (Stewart and Walter 1955; Walter et al. 1937).

Furthermore, their findings raised various questions that made them return to the site "at the beginning of 1956 when tests were made in both the conglomerate in the posterior end and in the hard alluvial ground next to the left wall. However, nothing of note was found" (Walter 1958:89). Actually, a collection of items that testified to human occupation was found in that second campaign, but none of them were reliably related to the cranium found there years before. The only thing that could be stated for certain was that the discovery of human bones associated to Pleistocene animal remains in the Lapa Mortuária of Confins was the Academy's high point, and it ensured a position of reasonable prominence for the Minas Gerais institution in the national sphere.

The research activities did not stop when the cranium was discovered, and they went on to discover various species of the Pleistocene megafauna. In 1939/1940, in the Lapa da Lagoa Funda, the fossil cranium of a short-faced bear (*Arctotherium brasiliensis*) was found and in 1944, in the Lapa dos Borges, an almost complete skeleton of a *Glyptodon* (*Glyptodon clavipes*), in addition to other finds at the Lapa do Galinheiro, in the Abrigo da Lagoa do Sumidouro, and in the Lapa Mortuária of Confins itself (Walter 1940). However, archaeological research soon became the institution's core activity, and excavations were carried out at the Lapa Vermelha (from 1937 to 1940 and from 1943 to 1944; Fig. 6.3), Sumidouro (1939, 1942, 1947), Samambaia (1946), Marciano (1946/1947 and 1949), Mãe Rosa (1944, 1949/1950), Eucalipto (1948/1949 and 1954; Fig. 6.4), Limeira (1952/1953), and Cerca Grande, which had been the object of rock art research in the 1940s and had been excavated in 1955/1956 (Walter 1948, 1958, 1976).

The excavation reports and the lists of the items found together with observations on the stratigraphic levels, or sometimes more specific aspects of the excavated soil, have all been lost. Walter used to keep them in his home as Clifford Evans from the Smithsonian Institute noted. The future coordinator of the Brazilian National Program of Archaeological Research (*Programa Nacional de Pesquisas Arqueológicas*), who visited Walter in 1949, remarked that:

> Mr. Walter has taken upon himself the added task of their excavations. The specimens, catalogued with data on the circumstances of discovery, he has preserved along with his extensive fossil collections in the basement of his home in Belo Horizonte due to a complete lack of local, state or federal interest in the scientific or educational importance of his material. (Evans 1950a:341)

After Walter's death, almost nothing was preserved. The only information referring to his fieldwork refers to the more important items like human remains and polished

Fig. 6.3 Lapa Vermelha I, 1948 (Photo: H. V. Walter)

stone axes, arrowheads, and others. That information can be found in the various publications in the period from 1934 to 1956.[9]

That was one of the Academy's great merits, publishing the results of all its members' work with a certain regularity and very shortly after the excavations themselves. Again, the Academy's publications were not restricted to the scientific community because, in addition to articles, chapters of books, and books in the academic sphere, its members always made every effort to publicize their research projects as much as possible in the press. It is important to bear in mind that the good relations the leading members of the Academy had with the local press and the fact that Mattos himself had a column in the newspaper *Diário de Minas*, in addition to having once run the publishing companies *Apollo* and *Biblioteca Mineira de Cultura*, both located in Belo Horizonte, greatly contributed toward the dissemination

[9]The work and the publications were interrupted by the World War II.

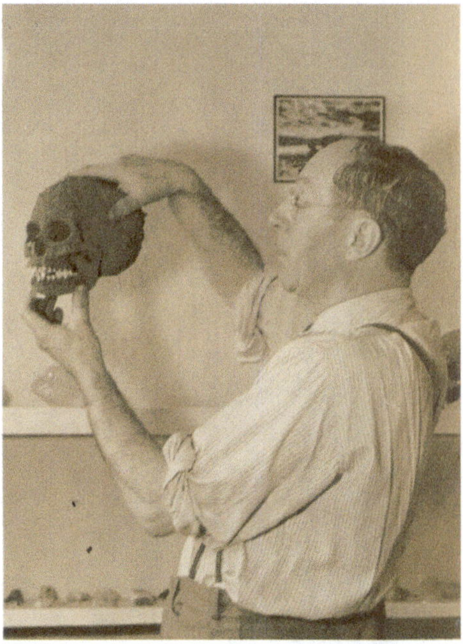

Fig. 6.4 H. V. Walter with a human cranium from the Lapa do Eucalipto. Pedro Leopoldo, 1948–1949 (Photo: Derek Walter)

and publicity enjoyed by the Academy's works. In that aspect, those contacts may have compensated to some extent for the chronic shortage of funds by giving them much cheaper access to the editorial medium.

The production of books and articles in Portuguese and English, written or translated by Walter, also contributed toward endowing the publications with a certain repercussion, which quite often went beyond the frontiers of Latin America. Good examples of such publications are: The Confins Man: A Contribution to the Study of Early Man in South America, by Walter, Cathoud, and Mattos, published in Philadelphia, in 1937 (Walter et al. 1937); Lagoa Santa Man, by Mattos, published in the Handbook of South American Indians in 1944 (Mattos 1944); and the Portuguese-English book *A Pré-História da Região de Lagoa Santa – The Prehistory of the Lagoa Santa Region*, which was actually on sale in the USA in the 1950s at the Cadmus Book Shop Inc. in New York (Walter 1958). In Evans's opinion:

> [T]he monography contains a variety of interesting data written in readable style, and has 71 good illustrations of artifacts, caves, fossils and human skeletal material. The information gathered indicates that the region will be a fruitful one for extensive archaeological and paleontological researches in the history of early man in South America. (Evans 1950b:101)

Despite the considerable volume of publications, the Academy was unable to publish a periodical that would offer space for the presentation and diffusion of its members' research work. There was an attempt to do so 13 years after its foundation, when the journal *Scientia* was launched with the proposal that it should be instrumental in making the works of the Academy more widely known. The journal did not last beyond the first issue, and it is unknown how many copies were printed.

The volume itself is a collection of texts referring to the Academy's activities during the early years of its existence, and 8 of its 11 articles addressed the archaeological and paleontological studies carried out in Lagoa Santa, two are in the field of history, and one in the field of physics (Scientia 1947).

From 1957 on, the excavations gave way to exhibitions, ceremonies, and participation in meetings. Consequently, the number of publications gradually dropped. From 1959 to 1961, the Academy organized three exhibitions in Belo Horizonte with considerable repercussions.

In 1959, the "first Exhibition of rock paintings ever held in Brazil took place to mark the 25th anniversary of the Minas Gerais Academy of Sciences" (Penna 1964:420). During that ceremonial event, various new members were inducted, the most outstanding being Aloísio Pimenta, Ângelo Machado, and José Israel Vargas who, albeit they did not actually take any part in the work of the Institution, were to perform important roles on their own account in the Brazilian academic and scientific scenarios in the decades that followed.

In 1960, the most important pieces of the archaeological and paleontological collections of the Academy were exhibited for some months in the auditorium of the *Cultura Inglesa*, which was run by Walter in the Guimarães building in the downtown of Belo Horizonte (D. Walter, personal communication in 1996). Another exhibition of the rock painting collection took place during the fifth Brazilian Anthropological Meeting in June 1961, in Belo Horizonte (Penna 1964).

The Academy's trajectory was always dependent on the total dedication of a small group of members. The figures of Mattos and Walter (Fig. 6.3) are of overriding importance, given that the very existence of the institution was the fruit of their actions. Thus, it is easy to understand that the rhythm of the research, meetings, and publications depended exclusively on the availability, physical disposition, and, above all, the financial capacity of those two men.

While it is possible to identify precisely the date and venue of the Minas Gerais Academy of Sciences's founding, the same is not true for its demise. At the beginning of the 1960s, its activities began to wind down because, in addition to the fact that Cathoud had died, Mattos and Walter were 74 and 65 years old, respectively. Even with the presence of Josaphat Penna, who had been a member since 1940 and was much younger than the others, the work never achieved the same scope as in former days.

The Academy did not manage to mobilize collaborators who could ensure the continuity of the work at Lagoa Santa after the death or retirement of its leading members, and that left the rest of them apprehensive as Penna narrates:

> We would be glad to hand over this material (the archaeological and paleontological collections) to an institution that could organize it more efficiently and more educationally and avoid the loss of such valuable work on the death of the author. (Penna 1964:420)

Another factor that contributed to the extinction of the activities was the arrival of research teams attached to foreign research institutions that were able to count on large-scale financing with long-term projects and the added intention of qualifying professionals to work with universities and other public institutions (usually museums). Such initiatives altered the profile of Brazilian archaeology, and it began to become dominated by professional archaeologists, leaving no space for amateurs or dilettantes. The main consequence of that process was the vilification of the generation of amateurs in many parts of Brazil and that came at a time when the differences between the work of the amateurs and that of the professionals were almost imperceptible[10] (Gaspar 2000).

In Belo Horizonte, the young members of the French Mission, representatives of the new generation of professionals, not only questioned Walter's competence but even his very honesty. At that time, he was over 70 and was the last survivor of the Academy's founders. Some of them even suggested that the aging member of the Academy had been trading in archaeological specimens.[11] There is a shocking gap between the attitudes of the professional archaeologists of the Smithsonian Institute and those of the members of the French Mission regarding their attitudes to Walter and the Academy.

In more than 30 years of research activities, the Academy had the merit of having kept interest in the Lagoa Santa region alive even in the face of innumerable difficulties. Furthermore, it played the role which today is performed by the museums and universities, as far as research, publicizing, and protection are concerned. Considering that the institution never had a headquarters building of its own or installations suitable for exhibitions or for storing the respectable collection it had managed to bring together, the collections ended up being exposed to the worst possible conditions of conservation and manipulation. The fact that Belo Horizonte did not have a museum or other suitable space also contributed to the deterioration of many of the collected pieces.

After Walter's death in 1976, his son donated the remaining items of the collection, including the specimens of the Pleistocene fauna, to the Federal University of Minas Gerais. There was no money involved in the transaction, quite different from the example of the collection of another amateur, Guilherme Tiburtius, which was sold to the Municipal Authority of Joinville in 1963 to compose the collection that originated the *Museu Arqueológico de Sambaqui* (Archaeological Museum of Sambaqui) (Tiburtius 1996:28).

Final Remarks

The Academies of Letter and the Historical Institutes in general would never survive without government subsidies (Schwarcz 1989).[12] It could hardly be any different for the Minas Gerais Academy of Sciences, especially when the particularly high

[10] Other countries settled the conflict between professional and amateurs in a more creative way by attracting the interest of the public at large, which then helped in the surveillance of archaeological sites, participating in research activities and financing studies (Gaspar 2000:19).

[11] A. Prous, personal communication in 1994.

[12] See also *Revista do Instituto Historico e Geografico de Minas Gerais*, Belo Horizonte, Vol. II,

costs of archaeological and paleontological research are considered. However, the Academy's daily round of great difficulties and lack of funds ended up by molding it into a completely different institution from others, and if it were to exist today, it would be classified as a nongovernmental organization, especially in the light of the modern regulations that govern such institutions: nonprofit voluntary associations of citizens who organize themselves around a specific interest and which are capable of acting nationally or internationally. Such organizations, according to the specific [Brazilian] legislation (Act n° 9.790 23/03/1999) are legally constituted entities governed by private law with objectives associated to fostering culture and the defense and conservation of historical, artistic, and environmental heritage. They may also conduct studies and research and produce and diffuse scientific information and knowledge. The same Act also mentions the question of the donation of physical, human, and financial resources.

Based on that definition, it can be said that the Minas Gerais Academy of Sciences was not only one of the very first NGOs in the state of Minas Gerais but was also undeniably the most "nongovernmental" of them all, because never, at any stage, did it receive public money to finance its activities.

In addition to failing to support the Academy's initiatives, the government authorities were completely remiss in regard to the protection of the archaeological heritage. At that time, the protection of archaeological assets was not entirely regulated, and there was no formally structured legislation covering such heritage assets.

Penna revealed his concern for that situation when, at the UNESCO-sponsored the second Intellectual Meetings of São Paulo (*II Encontros Intelectuais de Sao Paulo)*, he denounced the precarious situation of "the precious artistic heritage, lamentably disdained by the public authorities who continue to be deaf and indifferent to the Academy's appeals to protect the most ancient of Brazilian artistic manifestations". He also mentioned the situation of the rock paintings, which he said were being "systematically damaged by unscrupulous visitors" (Penna 1964:419).

The polemic as to whether the members of the Academy were scientists or dilettantes is innocuous. Walter's own words remove any doubts on that score: "for more than thirty years, my hobby has been to conduct field research and to study the prehistory of the Lagoa Santa region" (Walter 1976:13). With that, he showed that he and his colleagues had never considered themselves to be professional but, above all, enthusiasts of the prehistory of Lagoa Santa and the surrounding region.

The fact that the group was made up of amateurs and dilettantes does not diminish the importance of its contributions to Brazilian paleontology and archaeology, especially in regard to the question of the antiquity of human presence on Brazilian soil and its possible contemporaneity with the extinct Pleistocene fauna.

Despite the lack of adequate infrastructure, of well-trained technical staff, and, worst of all, of permanent financial support, its members managed to publish many of the results of their research efforts with great frequency for the standards of their day. Furthermore, much of the data in the records of their work are extremely valu-

1945, pp. 258–260, and Vol. IV 1957, pp. 291–293.

able for present-day researchers because they are actually the only records of research conducted at sites with great potential, but which today have been destroyed.

The Academy also managed to achieve one of its primary objectives and that may have been its biggest contribution in the 30 years of its activities. It was to strictly comply with the determinations of its founding document, namely, to see "the paleontological studies finalized by Dr. Lund in the limestone region of the Rio das Velhas river taken up once more and Brazilian archaeological studies reactivated," thereby continuing the Danish naturalist's work and publicizing his discoveries (Scientia 1947).

While the academics, on the one hand, contributed toward keeping Lund's name in evidence in the national and international scientific scenario, on the other hand, ironically enough, they did not manage to do the same for themselves. Despite the long time during which they were active and their considerable contributions to archaeology and paleontology, the works of Aníbal Mattos, Harold Walter, Arnaldo Cathoud, and Josaphat Penna have remained practically unknown to the general public and are hardly respected at all by the specialized community.

Even though they performed the work with great dedication, had no interest in gaining either fame or fortune, and managed to publish many results in Brazil and abroad, the Academy was very unfairly treated.

The fact that it was composed of amateurs in no way diminishes the importance of the Minas Gerais Academy of Sciences, but it was always regarded depreciatively by the Brazilian archaeological community as if it were a mere brotherhood of dilettantes. That was largely due to the attitudes of the professional archaeologists who arrived in Minas in the 1970s and who made a considerable contribution to constructing a negative view of the Institution and especially of the person of Harold Walter himself. During all his activities, he never received a penny of public money, and after he ceased, he has never been allotted a place of respect in the national scientific memory.

References

Cathoud, A. (1940). *A Raça de Lagoa Santa no Pleistoceno Americano*. Belo Horizonte: ED. Apollo/Biblioteca Mineira de Cultura.

Evans, C. (1950a). A report in recent archaeological investigations in the Lagoa Santa Region Minas Gerais. *American Anthropologist, 15*, 341–343.

Evans, C. (1950b). Book note on H. V. Walter: The prehistory of the Lagoa Santa Region Minas Gerais. *American Anthropologist, 15*, 101.

Gaspar, M. (2000). *Sambaqui: arqueologia do litoral Brasileiro*. Rio de Janeiro: Jorge Zahar.

Mattos, A. (1935a). *O sábio Dr. Lund e estudos sobre a pré-história brasileira*. Belo Horizonte: Biblioteca Mineira de Cultura 5, Editora Apollo.

Mattos, A. (1935b). *Coletânea Peter W. Lund*. Belo Horizonte: Biblioteca Mineira de Cultura 5, Editora Apollo.

Mattos, A. (1939). *Peter Wilhelm Lund no Brasil. Problemas de Paleontologia Brasileira, Serie Brasiliana*. Sao Paulo: Nacional Biblioteca Pedagogica Brasileira 5.

Mattos, A. (1944). Lagoa Santa Man. In J.H. Steward (Ed.), *Handbook of South American Indians* (pp. 399–400). Bureau of American Ethnology Bulletin 143, Washington, DC.

Mattos, A. (1950a). *A raça de Lagoa Santa – velhos e novos estudos sobre o homem fóssil americano*. Sao Paulo: Editora Nacional.

Mattos, A. (1950b). Provas da contemporaneidade da raça de Lagoa Santa com as especies extintas do Pleistoceno da regiao de Lagoa Santa. *Kriterion, 8*, 275–291.

Mattos, A. (1961). *O Homem das Cavernas de Minas Gerais*. Belo Horizonte: Editora Itatiaia Ltda.

Padberg-Drenkpohl, J. H. (1926). *Relatório de duas excursões à região calcária de Lagoa Santa em 1926*. Rio de Janeiro: Museu Nacional.

Padberg-Drenkpohl, J. H. (1929). *Relatório da excursão de 1929 à região calcária de Lagoa Santa*. Rio de Janeiro: Museu Nacional.

Penna, J. (1964). Pinturas e gravuras rupestres de Minas Gerais. In *Origens do Homem Americano* (pp. 419–421). São Paulo: Instituto de Pré-História da Universidade de São Paulo.

Roquette-Pinto, E. Edição de 28 de outubro de 1934. Folha de Minas, Belo Horizonte.

Schwarcz, L. M. (1989). *Os Guardiões da Nossa História Oficial: Os Institutos Históricos e Geograficos Brasileiros*. São Paulo: Idesp.

Scientia. (1947). *Estudos da Academia de Ciências de Minas Gerais. 1° volume. Comemorativo do Cinquentenario de Belo Horizonte*. Belo Horizonte: Editora Apollo.

Stewart, T. D., & Walter, H. V. (1955). Fluorine analysis of putatively ancient human and animal bones from Confins Cave, Minas Gerais, Brazil. *Proceedings of the 31st International Congress of Americanist, 2*, 925–927.

Tiburtius, G. A. E. (1996). *Arquivos de Guilherme Tiburtius/Guilherme Augusto Emílio Tiburtius*. Joinville: Museu Arqueológico de Sambaqui de Joinville.

Walter, H. V. (1940). *Urso Extinto "Arctotherium brasiliensis" da caverna de Lagoa Funda, Minas Gerais. Contribuição à Paleontologia Brasileira*. Rio de Janeiro: Gráfica Guarany.

Walter, H. V. (1948). *The prehistory of the Lagoa Santa Region (Minas Gerais)*. Belo Horizonte: Veloso & Cia. Ltda.

Walter, H. V. (1958). *Arqueologia da região de Lagoa Santa*. Rio de Janeiro: Sociedade Editora e Gráfica Ltda.

Walter, H. V. (1976). *50 anos de excursões e explorações em Minas Gerais: uma autobiografia de H. V. Walter*. Belo Horizonte: Vigília.

Walter, H. V., Cathoud, A., & Mattos, A. (1937). The Confins man. A contribution to the study of early man in South America. In G. G. MacCurdy (Ed.), *1937 early man* (pp. 341–348). Philadelphia: J. B. Lippincott.

Chapter 7
Archaeological Missions to the Lagoa Santa Region in the Second Half of the Twentieth Century

André Prous

Abstract In the 1840s, Danish naturalist P. W. Lund found human remains mixed with the bones of extinct fauna species in the Sumidouro Cave near to the city of Lagoa Santa. After analyzing the conditions in which the bones had been deposited, he formulated the hypothesis of a possible coexistence of the "antediluvian" fauna and Man. At the time, that hypothesis was not taken seriously by European or North American researchers. In the twentieth century, however, various expeditions were sent to the Lagoa Santa karst to verify that possibility. None of those organized by the National Museum in the 1920s achieved any success regarding that point. However, amateurs from the nearby city of Belo Horizonte continued to discover human remains there (the Confins Man) and kept alive the international community's interest in the region. In 1954 and 1955, W. Hurt conducted excavations at Cerca Grande and dated the charcoal material he found to about 10,000 years ago. From 1971 to 1976, the Franco-Brazilian Mission headed by A. Laming-Emperaire excavated the Lapa Vermelha IV rockshelter, where it uncovered human remains (of a woman later named Luzia) that were dated to as 11,000 years old, under the remains of megafauna species that were a little more recent. In the following years, the excavations in the surroundings of Lagoa Santa were abandoned, although a team from the Federal University of Minas Gerais (UFMG), that had begun to research new regions, continued to make sporadic surveys of the rock paintings in the karst.

Introduction

Ever since P. W. Lund's early research, the possibility of the great antiquity of human presence in the Americas has been the focus of intense debate. Until radioactive isotope dating became available in the mid-twentieth century, that antiquity

A. Prous (✉)
Federal University of Minas Gerais (MHNJB/UFMG),
Rua Gustavo da Silveira, 1.035, Santa Inês, Belo Horizonte, MG, Brazil
e-mail: aprous80@gmail.com

P. Da-Gloria et al. (eds.), *Archaeological and Paleontological Research in Lagoa Santa*, DOI 10.1007/978-3-319-57466-0_7

could only be estimated by associating anthropic remains with the fossilized vestiges of extinct large fauna species that had putatively disappeared towards the end of the Pleistocene. Successive expeditions undertaken in the first half of the twentieth century (headed by Bastos de Ávila and Padberg-Drenkpohl)[1] failed to provide any indications of such contemporaneity; indeed, their reports were never published (see Chap. 5 of this volume for more details). In turn, the finds of the members of the Minas Gerais Academy of Sciences, although publicized in Brazil (Walter 1958) and abroad (a chapter in the Handbook of South American Indians – Mattos 1946), enjoyed little credibility, due to their lack of stratigraphic observations.

That being so, American prehistorian H. Wormington, an expert in the study of the settlement of the Americas, suggested to archaeologist and museologist W. Hurt, then recently nominated to the position of Museum Director at the University of South Dakota, that he should test P. Lund's hypotheses by carrying out systematic excavations in the rockshelters of the Lagoa Santa region. The results of the research led by the North-American in 1955 and 1956, and only published 14 years later (Hurt and Blasi 1969), led to the organization of another mission in the 1970s, headed by a French archaeologist, A. Laming-Emperaire.

The American-Brazilian Mission (1955/1956)

W. Hurt associated himself with the National Museum of the Federal University of Rio de Janeiro, whose representative on the team was the paleontologist Carlos de Paula Couto (the Museum did not have an archaeologist at the time), later to become a leading expert on South American quaternary mammal fauna. Other members of the team were Oldemar Blasi from the state of Paraná and Altenfelder Silva from São Paulo, who both would continue to be active in archaeology. In 1955, a prospecting campaign was carried out to identify the sites that would be excavated in the following year.

As usually happens, the contacts of the North-American "professional" with most of the "amateurs" from the Minas Gerais Academy of Sciences was not at all cordial. Only Josaphat Penna, whose interest was essentially in the rock paintings, actually collaborated with the mission and helped to guide the researchers in the region.

In 1956, excavations began in some rockshelters in the Cerca Grande rocky mass (Fig. 7.1), at Lapa do Ballet (included in the group of sites that the researchers called Lapa do Chapéu) and at the Lapa das Boleiras rockshelter. As the researchers failed to find any bones of Pleistocene megafauna associated to any kind of human remains or material of anthropic origin, they considered that the human occupation of the site had occurred after the disappearance of the big mammals. So, they only published a brief note on their research at the time (Hurt 1960). In those days, radiocarbon dating was still a rare and costly technique, and they delayed the submission of any of the carbon samples they had extracted from the lower levels for testing.

[1] His name is correctly spelt with that "h", although in Brazil it has often been incorrectly spelt without it.

Fig. 7.1 W. Hurt's excavation in Cerca Grande (1956)

When they did so, they were highly surprised by the results: 9028 ± 120 and 9720 ± 128 BP (Hurt 1964) – the oldest dates known for Brazil up until then. In view of that, they published a synthesis of their field and laboratory research activities (Hurt and Blasi 1969).

In addition to presenting a critical history of the research that had been carried out in the region up till then and presenting information on the radiocarbon dating, the report described the 24 burials removed from four rockshelters of the Cerca Grande rocky mass and from the Lapa das Boleiras site. It also presented a typology of the lithic industries, of the bones, and of the rare fragments of pottery found in the rockshelters. Even though the remains are not described according to their stratigraphic levels, most of them came from levels two to seven of rockshelter IV at Cerca Grande (those dated as being from 9,000 to 9,700 BP). Typical of what the authors of the report refer to as the "Cerca Grande Complex" are the burials in a flexed position in a space marked out by small stones and covered by a larger one. Outstanding among the assemblages of the lithic industry are the totally or partially polished axe blades made from sillimanite, compact hematite, dolomitic limestone, quartzite, amphibolite, and basic rocks from the level dated to 9,028 ± 120 BP. They are ellipsoid, some short, some long, have no grooves, reentrances, or other modifications suggesting some form of handle mounting. There were also 37 hammerstones and anvilstones made of quartz or quartzite. Almost all of the latter have a depression on one or two of their faces. The report also mentions 11 choppers or chopping tools – pebbles with one end flaked. The chipped lithic industry includes 7,379 flakes, most of them (71.25%) of quartz, but some (17.75%) of quartzite and others (10%) of jasper or chalcedony. We believe that the majority of the rare flakes of diabase (0.9% of the total) are the result of the manufacture or reforming of axe blades. Hurt and Blasi counted four dozen reworked pieces of quartz and flint, more than half of them in the form of two-sided projectile points of quartz. The rest consisted of end scrapers, side scrapers with a reworked edge, bolas, or awls. Another notable aspect was the presence of some pieces of quartzite, hematite, and sillimanite that must have been imported from regions dozens of kilometers from there, whereas small quartz crystals could readily be found around 10 km away from the excavated rockshelters (Prous 1978). Given my knowledge of the industries of the region acquired in studies of the material from Santana do Riacho in 1986, I decided to review a considerable part of the material collected by the American-Brazilian mission conserved at the National Museum. I observed that what they classified as "choppers" were actually hammerstones that had broken when they were being used. On the other hand, the quartz "bolas" and various pieces that were supposedly scrapers, were actually bipolar cores, and not artifacts that had been either shaped or retouched. In fact, a considerable number of the quartz items had been struck on an anvil. Some of the pieces described as "scrapers with one side elongated" are actually flakes with signs of having possibly been used, but they do not present any signs of deliberate modification. There do exist, however, few true scrapers of quartz and flint, including tiny, nail-shaped pieces of carefully retouched quartz, similar to specimens from the early Holocene that we had found in the Serra do Cipó. They are the only objects that have been deliberately shaped, but they are extremely rare.

The excavations also uncovered points or awls made of bird and mammal bones. Another notable find in the intermediate level of rockshelter 2 at Cerca Grande was the perforated shell of an *Olivella*, which is an exclusively marine genus, so its presence suggests the existence of exchange networks attaining great distances. Unfortunately, I did not find this piece in the Museum's collection.

The American-Brazilian mission's overriding aim was to verify Lund's hypotheses regarding the antiquity of human presence. That being so, the little team did not pay much attention to the rock art, which at the time would have been impossible to date, and which consequently could not be correlated with the buried archaeological levels. Accordingly, they merely mentioned the graphisms at the Cerca Grande and Lapa do Ballet sites, referring to the latter in their report as the Lapa do Chapéu. In it, they identified two distinct sets of paintings: the older one of which "has a style similar to that at Cerca Grande" (Hurt and Blasi 1969:39), that is to say, with zoomorphic representations; and the other, more recent set, with "a series of human figures" (ibidem).

Even though it failed to demonstrate archaeologically the coexistence of humans and the megafauna, the American-Brazilian Mission proved that the lower levels of the Lagoa Santa rockshelters preserved some of the oldest vestiges in South America.

The Franco-Brazilian Mission (1971–1977)

Origins

In October 1962, at the invitation of Josaphat Penna, whom she had met at a meeting on The Origins of American Man (*Origens do Homem Americano*), organized in 1961 by Paulo Duarte in São Paulo, A. Laming-Emperaire visited some of the

Lagoa Santa sites. However, she then went to Chile with her husband, who died during an excavation. In 1970, the French archaeologist was finalizing a research program in southern Chile, where she had been studying the first human occupations. The publishing of W. Hurt and O. Blasi's report encouraged her to redirect her attention to Brazil. Having observed the limitations of the results presented in the publication of the *Arquivos do Museu Paranaense* (Hurt and Blasi 1969), A. Emperaire intended to carry out research at sites whose stratigraphy would make it possible to obtain a more detailed chronological subdivision than the one observed in the rockshelters at Cerca Grande. Above all, she wished to situate the archaeological remains in a paleoenvironmental panorama, addressing regional climate, geomorphological evolution, fauna, and flora changes with the help of experts. Accordingly, she set up a research project through the *Commission des Fouilles* do *Ministère des Affaires Etrangères* (Excavations Committee of the French Ministry of Foreign Affairs) in a partnership arrangement with the National Museum of Rio de Janeiro. Museum archaeologist M. Beltrão was in charge of the project's administrative coordination in Brazil. In France, the study of the documentation would be carried out by researchers of the *Unité de Recherche Archéologique* (Archaeological Research Unit) N° 5 of the *Centre National de la Recherche Scientifique* (CNRS; National Center of Scientific Research) to which I was attached.

The Team and the History of Operations

Annette Emperaire brought together a large number of collaborators to cover all the areas of research she intended to address. To ensure the feasibility of the regional geomorphological studies and the studies on the formation of the sites, she got in touch with Aziz Ab'Saber of the University of São Paulo (USP) who even visited the site with us during the first campaigns. In 1975, she articulated a partnership with pedologist J. P. Queiroz Neto (also attached to the USP) and the student he was tutoring, H. Kohler. In France, she approached the CNRS's *Centre de Géomorphologie* (Geomorphology Center) in Caen, specialized in the study of surface formations, and secured the collaboration of its director André Journeaux. The director and his collaborators took part in the fieldwork and a Ph.D. student (Dominique Alduc) was designated to accompany the research work at the Lapa Vermelha rockshelter, which would be the theme of his doctoral thesis. Fausto Cunha and Martha Locks Guimarães (National Museum of Rio de Janeiro) were called in for the studies of the Holocene fauna, and the latter eventually collected specimens of the local fauna to facilitate the identification of the specimens obtained from the excavations. The French paleontologist R. Hoffstetter was put in charge of identifying the Pleistocene fauna remains. José Luis Leme from the USP's Zoology Museum studied the malacofauna (mollusks). In 1975, we handed over to Maria Aparecida Vulcano (Biosciences Institute at USP) insects' nests that had been found in the excavations and some observations on the lairs of prehistoric mammals, while Ulisses Confalonieri, of the Osvaldo Cruz Institute (Rio de Janeiro) received the coprolites. Miya Pereira travelled to Denmark to study the skeletons that Lund had collected, and she was supposed to study any human remains that might be found during the excavations. With access to an excellent xylotheque, Calvino Mainieri of the USP's Technological Research Institute agreed to carry out the analyses of the charcoal removed from the hearths – until this moment anthracology had not yet even been introduced in Brazil. In 1976, Pascale Prous was charged with the responsibility of setting up an herbarium for the region and studying the microscopic plant remains in the Palynology Laboratory of the then recently created Archaeology Sector of the Federal University of Minas Gerais's Natural History Museum. We also made an unsuccessful attempt to contact Salgado Labouriau in the hopes of engaging him in the study of the phytoliths.

From 1974 on, the Mission's photographer was Sydney Anthonioz – nowadays Picasso. In 1973, the head of the Mission had allotted me the task of making a survey/inventory of all the rock paintings in the region (Prous 1977). As I did not feel that I was sufficiently qualified for that, I requested her to bring in another more experienced person from France. So, it was that in the following year Pierre Colombel joined the team. He was a CNRS technical staff member who had worked for many years with the Tassili n'Ajjer Cave paintings. To analyze the rhythmic alignments of the designs painted at the Sumidouro Cave, the project sought the collaboration of astronomer Dominique Ballereau (*Observatoire de Paris*).

The first prospections were carried out in 1971 under the coordination of A. Emperaire (Fig. 7.2) with the participation of French team members Danielle Lavallée and André Prous, and Brazilian researchers from the Paulista Museum (Luciana Pallestrini, Filomena Chiara, Margarida Andreatta, Agueda Vilhena de Moraes – now Vilhena Vialou) and V. Pentado Coelho from USP. At the last moment, N. Guidon, also attached to the USP, decided to travel to Piauí state to begin her research in the Serra de Capivara. Right away, we could see that most of the rock-shelters with a favorable morphology that we visited within the planned radius of activities had been exploited already, either by members of the Minas Gerais Academy of Sciences or by local rural inhabitants that they used to hire to excavate the sediments richly laden with remains. Other sites had been destroyed by economic exploitation, namely, the extraction of saltpeter or limestone. As for the terraces of Velhas River valley, which we also prospected intensely, they had all been exploited in the quest for gold ever since the eighteenth century. So, what appeared to be the most interesting location was the rockshelter N° IV of the Lapa Vermelha (it was in Pedro Leopoldo county, but it is now in the recently created municipality of Confins; Fig. 7.3). It had an extraordinary stratigraphy: the surveys made in 1971 registered a depth of more than 2 m without even reaching the base, and it was possible to accompany the annual deposition of stratigraphic layers and that would obviously allow for an exceptional control of the slightest chronological differences (Fig. 7.4). On the other hand, the low density of remains per cubic meter, because of the rapidity of the sedimentation, seemed to have protected the site from the greed of the amateurs, who had at one stage opened a small excavation pit there (Walter 1947).

The permanent excavation team consisted of Annette Laming-Emperaire, her daughter Laure, Joaquina Pavia, Francisco (Paco) Pavia, in charge of topography and with whom I learned the trick of how to get by in prospecting when resources

Fig. 7.2 Anette Laming-Emperaire in 1976

Fig. 7.3 Excavation in the Lapa Vermelha IV rockshelter

had to be improvised or were absolutely minimal, and Edna Luísa de Melo Tavares. A. Prous and the priest João Alfredo Rohr, who had participated intensely in the research activities in previous years did not excavate at the Lapa Vermelha in 1976. On the other hand, Pascale Prous, Paulo Junqueira, Osvaldo Heredia, and Solange Caldarelli did participate in this last campaign. There was also important participation on the part of the Brazilian Archaeological Institute (*Instituto Arqueológico*

Fig. 7.4 Holocene sedimentary deposits

Brasileiro – IAB) of Rio de Janeiro, not only in the prospections to the northeast of the Lagoa Santa region (coordinated by Ondemar Dias), but also in the form of supplying excavators (Paco and Quina Pavia, Lílian Cheuiche, Eliane Carvalho, Alícia Coirolo, etc.).

The National Historical and Artistic Heritage department (*Secretaria do Patrimônio Histórico e Artístico Nacional*), nowadays known as IPHAN, had demanded that the excavations should involve the training of a certain number of Brazilian archaeologists. So, in 1973 and 1974, several people officially undertook archaeological field activities, even though some of them were not yet qualified to do so. Among them were the brothers Akos and Lehel de Silimon, attached to the Mato Grosso Development Company (*Companhia de Desenvolvimento do Mato Grosso* – CODEMAT), as well as several students, researchers, and other people selected by the coordinator. At one point the team had 25 members, comprising individuals who were mere beginners and others already accustomed to performing independently. That situation made coordination very difficult, especially bearing in mind that A. Emperaire, having spent so many years in Chile, found it hard to understand the various Brazilian accents and also did not like the role of boss. Among other archaeologists who took part in at least one of the campaigns, we can highlight A. Kern, F. La Salvia, C. Perota, F. Araujo, S. Leite, and E. Malherbi. Among the foreign researchers, Zulema Seughel (Universitad de Concepción, Chile), Miguel Mendes (Director of the Museum in Popayán, Colombia), Sidney Anthonioz (USA), Suzana Monzon, and *O. Heredia* (Argentina; who sometime later would become Director of the Archaeology and Museology course of the Estacio de Sá United Faculties in Rio de Janeiro), N. Orloff (Belgium), A. Wesolovska (Poland), and Mirtis, a researcher from Ecuador. Some of those personalities were to leave their mark on the memoirs of the Mission. Father Rohr is obviously one; he used to conduct his morning gymnastics cutting down a tree before going up to the Lapa

Vermelha rockshelter. His enthusiasm for excavating down to the depths never achieved before made him always keep his sector some meters lower than the level the others had managed to achieve in other parts of the site. In that way, we actually had a preview of what we could expect when we completed the general excavation. His enthusiasm and rapidity were to the detriment of his observation of the natural stratigraphy – that was a concern that was beyond the researcher from Santa Catarina state – it was also something that did not fail to create difficulties for me, charged with working around them. In compensation, however, experienced as he was in all the domains of natural sciences, he always had something he could teach to everyone. I remember having spent two days sculpting a strange combustion structure with all possible care and attention; it was a circle of burned earth from which what appeared to be tubes of burned earth irradiated. I thought that I was discovering some new form of prehistoric hearth. After he had passed by me several times in silence and with an obvious air of commiseration for the naive excavator, Father Rohr took pity on me and said "André, don't you see that you are just excavating the foot of a tree that was struck by lightning?".

Another outstanding figure was that of the Hungarian Akos de Silimon. He was unusually strong, and he carried Paco Pavia on his shoulders, supported on tiptoe on the narrow edges of the Poções cornice, so that Paco could make a trace on a high spot. His brother, the petulant Lehel, always with a revolver in its holster, was always carrying a bottle of alcohol, which he used to throw flames from his mouth to burn the nests of the African bees that made our lives a misery. When he was informed of the crash of a Varig flight near Paris in which the powerful Filinto Müller (Getulio Vargas's former head of the political police) had died, he became pale and prophetically declared "I am a dead man". One week later, when he had just got back from Lagoa Santa, he was assassinated in Cuiabá. We took advantage of the presence of those two Hungarians in the team to try and coopt their countryman Mihaly Banyai, a photographer of Lagoa Santa who actually engaged in illegal excavations in order to create a collection, which he was installing in his *Castelinho* (little castle), which he had built near to the Lapinha Cave. What we hoped was that he would become interested in modern practices and would transform himself from being a depredator into a collaborator (in France there was a long tradition of supporting amateurs in that field). The photographer actually spent a week with us at the Lapa Vermelha, participating in the excavations alongside his two countrymen, but he never became convinced that our scrapings and brushings with fine paintbrushes and trowels could ever be useful. In a book published in 1997, he ridiculed the way the French made such great efforts for such a poor crop of items (Banyai 1997).

Results of the Prospections (Laming-Emperaire et al. 1974)

The surveys made during the initial prospections confirmed the presence of tupiguarani remains at the base of the painted rock at the Sumidouro site (as previously observed by H. Walter), and also the existence of rare, isolated, lithic remains around the lake – nothing that would justify an investment within the scope of the Mission's established objective. The small caves tested around the Confins pond and the Lapinha Cave also showed the presence of painted pottery and human remains, but we did not find the space or the deep stratigraphy suitable for an ample excavation. In the upper site of Caieiras, the lithic remains and a mandible were dated to 9,600 ± 200 BP, but the small space available not only in the small cave but also in the rockshelter did not suggest that there was any chance of obtaining interesting results. At the Lapa São José of Confins, also known as the Lapa do Galinheiro, we found a notable collection of worked gastropod shells, some of which were even painted, but the small size of the gallery, still with its sediment intact, did not classify it as a possible habitat (even so it contained two burials which were later removed by M. Banyai when the owner of the place decided to turn the site into a garage). The sediments in rockshelter N° 6 at the Cerca Grande and those of the Lapa Mortuária had been completely removed, and at the beautiful Lapa Vermelha I, only a few remains were left in situ. At the Lapa do Ballet rockshelter, the surveys only identified recent levels of occupation, and the external pottery site failed to yield detailed information about its occupants in ancient times. Other sites that were of little interest to us were the open-air ones attributed to the Sapucaí tradition, which were excavated by part of the team in the lacustrian terrace of the nested sinkholes (*uvala*) of Lapa Vermelha or prospected on the other side of the Velhas River. Another site located in the area of the geomorphological surveys that were carried out to understand changes in the landscape was the Tiãozinho Fernandes site in the channel of the Ribeirão da Mata river, to the southeast of the Doutor Lund village. A deposit of tens of meters of sediment had carried with it a huge quantity of plant material, including whole tree trunks with ages ranging from 3,000 to 5,000 BP (Kohler and Malta 1991). At the top of the sequence, prehistoric pottery remains were found in disturbed position. The discussions about the events that produced the deposits there led A. Emperaire and A. Ab'Saber to bet a case of champagne on the result. Actually, the geologist eventually won the bet, but by then his French colleague had died, so he never received its prize. The study of the Velhas River Valley raised the hypothesis that at one time it had flowed in the channel of what is now the Ribeirão da Mata river. After 1974, it became clear that to increase the possibility of finding undisturbed sites with promising features and capable of complementing the Lapa Vermelha IV, in whose lower levels, then being excavated, anthropic remains were very scarce, we needed to increase the radius of the area being prospected. We therefore expanded the radius of the research area northward, towards the foothills of the Serra do Cipó mountains. With the location of the rockshelter known as Lapinha de Jaboticatubas, a small team were sent there to carry out some survey pits. Unfortunately, when we arrived, we realized that Rosalino (H. Walter's informant and collector of specimens), hearing of our project, had arrived 1 week before and, like a real Attila, had left only emptiness behind him. All that was possible was to collect the thousands of quartz flakes and the small fragments of human bones that the Confins village resident had despised and to check if there was any portion of the original sediment left undisturbed and that could be taken advantage of. The huge Sucupira site and the small Gentio cave had also been the targets of depredation. Finally, A. Emperaire and I managed to visit a magnificent undisturbed site, the great Santana do Riacho rockshelter. It was then decided that I should be in charge of the studies there, in 1976, making use of the team I was putting together at the Natural History Museum of Federal University of Minas Gerais with the support of specialists in rock paintings attached to the French Mission. Parallel to that, it was planned to extend the research activities to embrace the extreme north of the state and the municipality of Montalvânia, where the mayor had invited us to explore the rich undisturbed archaeological deposits for which he had cultivated a passionate interest. In 1976, A. Emperaire put me in charge of carrying out the first prospections in that municipality and they led to the elaboration of an excavation project in that region for the year 1977.

Meanwhile, the excavation at the Lapa Vermelha IV would carry on working with a team with fewer members led by A. Emperaire herself.

The Excavations

The Holocene Levels of Lapa Vermelha IV

The most important excavations carried out by the French mission were those at Lapa Vermelha IV. They embraced a surface area of 300 m^2 and in certain sectors went down to a depth of 10 m below the floor of the previous excavations in 1971 (14.5 m below *datum*) (Laming-Emperaire 1979).

An occupation by pottery users had left its remains, especially on the external slopes, where we found signs of postholes in the form of circles of soft sediment and blocks of shoring up material as well as large pottery fragments. However, the main pottery occupation (of the Sapucaí type) was found a few hundred meters away from the rockshelter, in the open, on a small rise overlooking the lake. In rockshelter IV, the Holocene pre-pottery levels contained more than a hundred small combustion structures (less than 50 cm in diameter) and some larger ones (up to 2 m across) dated to within the last 6,000 years. Many of them were grouped behind two large stalagmitic blocks that occupied the limit between the sheltered zone and the external one. Apparently, their authors had used it as a shelter (perhaps from the wind) for their fires. Various small fireplaces had their bases paved with limestone plaques – something difficult to explain because the sediment was not rich in organic material, and so there was no risk of it catching fire. As there were often snail shells of Strophocheilideae family associated to the hearths, I thought at first that those stones had been heated and used to make the snails expel their mucus before eating them (something that we did ourselves during our experimentations at the UFMG). If that were the case, however, why were the stone plaques inside the hearth areas and not beside them? It must be questioned as to whether the snails had not gotten there on their own to bury themselves in the ashes, but no signs of snail holes were found in the sediments and, furthermore, many of the shells showed signs of having been heated. As they were immediately above and not below the levels of charcoal

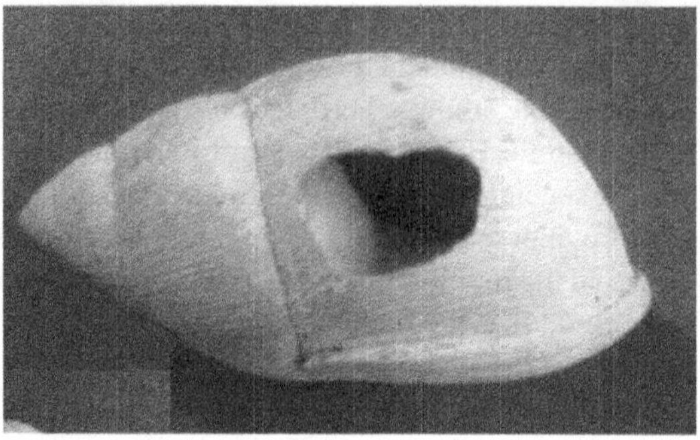

Fig. 7.5 Perforated shell

and ashes, the shells could not have been burned accidentally *postmortem*. Indeed, the "cemetery" of big snails was the retraction gap that forms between the sediment that fills the shelter and the rock walls. Therefore, the ancient inhabitants of Lagoa Santa themselves must have brought in the mollusks found in the combustion structures; it seems that they ate them and then used the shells. I inventoried more than 50 of them that had been perforated by humans and transformed into planes, and some were also colored red (Fig. 7.5). The humans that frequented the shelter also brought in shells of local freshwater bivalves of the *Diplodon* genus and used them as instruments; one of them had been worked to form a kind of gouge. At the Lapa do Urubu (near to Vespasiano) a similar shell had been used as a palette paint holder and was full of a rusty red iron pigment.

Next to the hearths there were some quartz crystal flakes – material that can be found outside the karst in the vicinity of Pedro Leopoldo, just a few kilometers away. Although many of them had been identified as "scrapers" in the first screening conducted by National Museum of Rio de Janeiro trainees, our review of most of that material in 1982 showed that they were almost all unipolar or bipolar flakes, fragments and debris (*cassons* and shard), and, more rarely, cores and bipolar core pieces. These remains hardly ever formed coherent assemblages, as those that are found in diversified and perhaps more durable occupation, where there would be indications of stone knapping carried out at the location. That can be found, for example, in a modest concentration in the northern sector at a depth of around 5.9 m, where four bipolar cores, a dozen flakes (many of them showing evidence of transversal knapping of crystals and of small pebbles), and a lot of waste from chipping activities. Or again, more to the south, the structure "F4" of the 1976 excavation (on the surface of the "clay 6.08" level) contained a millstone of basic rock found near to a sandstone grindstone and to various good-sized crystals and some pebbles of hyaline quartz. These artefacts were scattered around the same soil of occupation along with dozens of quartz flakes and blocks of lode quartz. At the same place, a limestone blade seems to have been extracted deliberately, while many rhombohedral calcite crystals may represent a mere collection of curiosities. Nearby, in the soil of occupation "F5" of 1976 (installed above "clay 7.5 m"), dozens of pieces of flaked quartz made from pebbles or crystals were associated to various blocks of green rock including a hammerstone and a probable pre-form of an axe blade. The absence of any anvil at the site, despite the clear signs of bipolar knapping, could be explained by the fact that the blocks of fallen limestone must have been used as anvils for that purpose. There were no marks of impact on them, but the marks could have been blurred by the rapid alteration of that kind of rock under the sediments.

In various Holocene levels the knapped quartz was accompanied by limestone plaques taken from the walls. Some fragments seem to have been trimmed to form fine rectangular pieces with sharp edges, sometimes formed by a calcite vein or a quartz lode, whereas others appeared to be unworked. They were not concentrated near to hearths, so sometimes there was a doubt as to their anthropic origin. Our experience when breaking up blocks of limestone that were in the way of the excavations showed that sharp edges of limestone cut very well when they are newly formed, but they become blunt very quickly.

In short, most of the Holocene remains could correspond to an extremely intermittent visitation by small groups of two or three individuals, who would spend a night in the rockshelter, perhaps on their journeys between the valleys of the Ribeirão da Mata and Velhas River. No heavy objects such as those that are normally found in other rockshelters in the region (palm nut crackers, anvils, etc.) were brought in to the Lapa Vermelha or abandoned there. Indeed, even medium-sized objects (hammerstones, big cores, and axe blades) are extremely rare at this site.

A few traces of subsistence remains can be found near the combustion structures, they include fish bones, and they are concentrated in the recent pre-pottery levels (around 3,500–4,000 BP) – which seems to be consistent with the presence of a bone fish hook dated to the same period (Fig. 7.6). Some of the fish (*Doradidae*)

Fig. 7.6 Bone hook

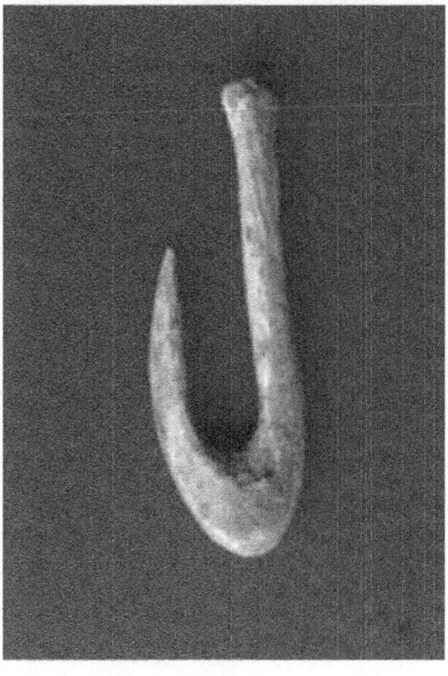

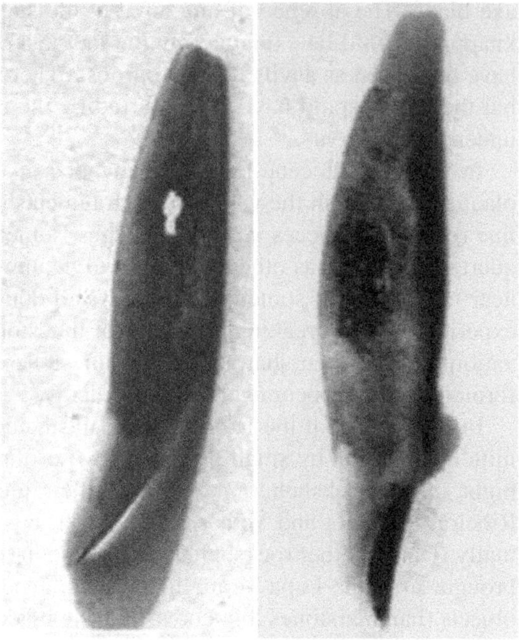

Fig. 7.7 Perforated deer tooth

were of a good size and may have come into the sinkholes through the network of underground water courses or been brought in by a fisherman. Anyway, the most part of the approximately 60,000 bones that were collected at all levels were those of marsupials, bats, and small rodents. They were animals that lived in the caves or that were eaten by owls and found in regurgitation pellets (Cunha and Guimarães 1978). The bones of some slightly larger mammals (armadillos) were found associated to a few of the hearths. In addition to food remains, an ornament was found (a perforated deer tooth, Fig. 7.7). The fauna inventory of the Lapa Vermelha site made it possible to identify a new species of the *Cavia* genus and also to register climate change aspects. For example, in the Early Holocene *Megalobulimus oblongus* gradually replaced *M. yporangus*, the latter being associated to a formerly colder climate in the region.

The Pleistocene Levels of Lapa Vermelha IV

The levels corresponding to the transition from the Pleistocene to the Holocene began to occur at a depth of 8 m below the *datum* to the north, and from 10 to 11 m to the south. Even before those deep levels were buried down, the space available for occupation was very small, just a narrow corridor running from the internal wall of the rockshelter and an external outcrop to which a pronounced ramp gave access (Fig. 7.8). The place would hardly attract human habitation, although it could serve

Fig. 7.8 J. Pavia and A. Prous in the "corridor", excavating Late Pleistocene levels

for more furtive activities or as a den for animals. Indeed, remains of anthropic origin become increasingly scarce and generally doubtful. However, this is not the place to explain the complicated local stratigraphy which we described in other publications (Prous 1986). We would merely underscore that the quartz remains disappear almost completely, although some limestone items that are probably instruments have been found in the "yellow" levels (dated to more than 15.000 years BP) at a depth of 12–13.5 m. Only a single one of those pieces is unquestionably of human fabrication, but its precise stratigraphic location was not clearly established. In the more recent "red" level (dated to between 9,000 and over 14,000 years – non-calibrated BP) there were giant sloth remains (*Scelidotherium* claws and its characteristic spherical coprolites). In slightly older levels, A. Emperaire found a small structure with charcoal and ashes in an irregularity in the rock wall with some animal bones, which she interpreted as being a small fireplace of anthropic origin. Right underneath, at a level dated to 11,000 BP, we found various human bones (long bones, hip bones, and a jawbone) of the skeleton of a young woman scattered on a single depositional floor and following its natural slope. It may have been that the woman suffered a fatal accident and her body remained unburied for some time. At a place where the skull (Fig. 7.9) and one of the long bones had rolled to, there was a depression which naturally took them to a greater depth along with various blocks that had fallen in there naturally. Neves and Piló (2008:277) published a

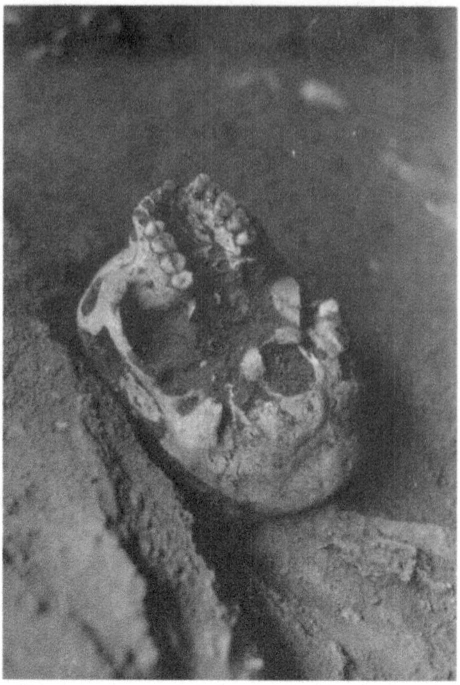

Fig. 7.9 Skull of skeleton n°1 in situ in the sunken pocket of *red sediment*

sketch that makes use of my original drawing, made for the Brazil 500 exhibition, explaining the process (Fig. 7.10). The field plans show the divisions between the sectors with the more recent "red" sediments, the older "yellow" ones, and also the sunken pocket (Fig. 7.11).

In the pocket of sunken material (which had possibly been drawn down towards a swallet that the excavation did not reach), there were charcoal dated to 12,000 BP, but they should not be used to date the rest of the bones because they were all in disturbed positions. The age of the bone deposits must be determined by the age of the sedimentary strata immediately above and immediately below the one which contains the post-cranium skeleton, therefore not going beyond 11,000 BP. A coprolite that was probably human (Fig. 7.12) was found close to a leg bone. Below the level of the skeleton, later nicknamed Luzia, the only things found were limestone platelets possibly utilized and some rounded charcoals.

Survey of the Rock Paintings

Despite the importance of her earlier contribution to the study of European Paleolithic art, at first, A. Emperaire showed little interest in the graphic registrations in Lagoa Santa, which she and most of the other researchers imagined to be

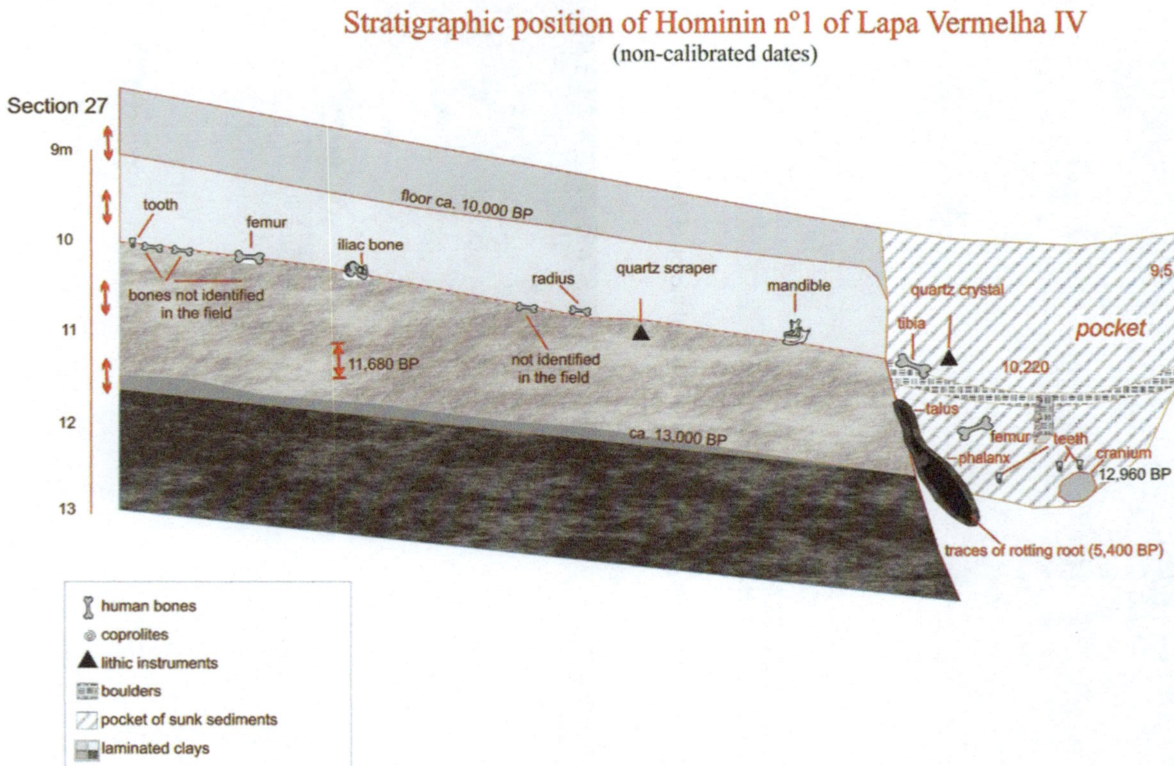

Fig. 7.10 Scheme showing the layout of the sedimentary contents of Lapa Vermelha IV

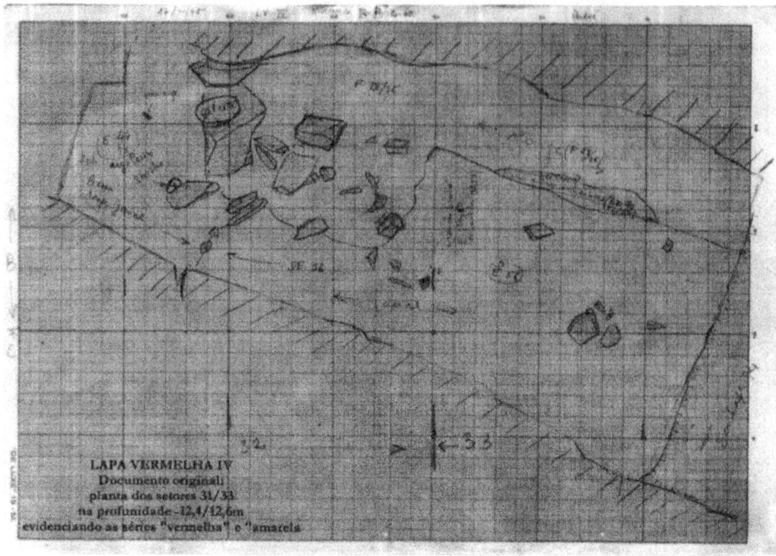

Fig. 7.11 Sketch of the *yellow* and *red* sediment series, and the top of the *red pocket*

Fig. 7.12 Coprolite, probably human

quite recent. That changed in 1974, when we discovered paintings buried down in the Lapa Vermelha IV: a small schematized quadruped in profile, some geometric designs, and a yellow stain. The lowest figure at the time was at the same sedimentary level as a fireplace dating back to 3,720 ± 120 BP, so that the figure must have been at least that old. That happened to be the very first chronological indication for a rock painting in Brazil. In 1976, a network of cross-hatched incisions was uncovered in another sector, and years later I noticed buried vestiges of rock paintings at the entrance to a small cave buried at a depth of more than 5 m, which implied that they were at least between 5,000 and 7,000 years old. That estimate is reinforced by the presence of blocks of hematite and prepared red pigments that were collected on the floor of the cave entrance. More to the south, at the level of fireplace F5, excavated in 1976, various objects stained red also indicate that there had been painting activities there during the Middle Holocene.

With that information in mind and further stimulated by her contact with G. Reichel-Dolmatoff, A. Emperaire decide to invest in a survey of the region's painted and engraved assemblages. Initially, a dozen sites were copied entirely creating a collection of thousands of figures. Unfortunately, the quality of the registrations was very uneven due to the inexperience of many of the tracers and the constant attacks of the African bees that had recently invaded the region and had not yet hybridized with the European-type bees.

We were obliged to work using gloves, high collared shirts and hats covered with gauze. At Sumidouro, we traced perched on trestles up until 11:30 am, when the sun used to reach the entrance to the bees' nest in its inaccessible position in the rocks, and then we had to flee and work at the Samambaia site in the afternoon. Such attacks were also part of the daily routine at the Lapa Vermelha rockshelter, although they were usually less violent as the nest was farther away and we made smoke by burning cattle dung to keep them away.

Conclusion

The Franco-Brazilian research activities in Lagoa Santa can be viewed in the context of a much broader effort to understand the most ancient of the American colonizers. In 1976, A. Emperaire went to the Soviet Union at the invitation of A. P. Okladnikov to analyze the industries of Siberia. She also went to Canada to see the industries of the extreme north of the New World. With her accidental death in Curitiba in 1977, the death of C. Mainieiri, and the withdrawal of D. Alduc, this first French Mission lost its spirit. I no longer wished to carry on the research at the Lapa Vermelha, which to me became something of a memorial to the person who had been my tutor. Furthermore, I lacked the enthusiasm to continue pursuing her dream of getting down to the sedimentary base, above all because I considered that material more ancient than 15.000 BP did not appear to be unquestionably of anthropic origin. At the time, I had just created the Archaeology Sector at the UFMG, and I had to look after all aspects of the prehistory of a state (Minas Gerais) that was larger than my country of origin. I could not remain indefinitely hypnotized by a single location that only offered doubtful occurrences. I had just begun the excavation of a very rich site that offered an impressive variety and volume of data, and I was getting ready to work in parallel, and in a comparative manner, in the north of Minas Gerais. Even so, I did make some attempts, together with the remaining members of the French Mission, to recover and organize the documents stemming from the research. I made a review of all the handwritten material, made a rapid analysis of the lithic and shell industries, and did my best to see that some of the laboratory studies continued to make progress. Those who have been through similar situations know how difficult it is for a young inexperienced person to replace a famous coordinator. So, I ended up abandoning the Lapa Vermelha, although, even today, that weighs on my conscience. However, A. Emperaire's performance was by no means useless. Theme 5 of the Franco-Brazilian Colloquium concerning the application of the study and cartography of superficial formations (at that time the term geoarchaeology had not yet come into use) was maintained, brought to Belo Horizonte (Coutard et al. 1978; Prous 1978) and published in our journal *Arquivos do Museu de História Natural* (Archives of the Natural History Museum). The creation of the Archaeology Sector of the UFMG stemmed directly from the repercussions of the French Mission among the authorities of Minas Gerais. A. Emperaires's insistence that I should work with rock paintings gave me sufficient experience to develop that field of study in the state of Minas Gerais. The experience lived through at the Lapa Vermelha site was probably an important influence on the decision of the IAB researchers to undertake large scale excavations at the coastland sites of Corondó and Malhada. When A. Kern began his doctoral studies with the Mission coordinator, he took with him to Rio Grande do Sul state a questioning of the lithic traditions proposed for southern Brazil. The French prehistorian directly and indirectly bequeathed to many Brazilian researchers, some of whom did not know her personally, a valuable scientific, methodological, and ethical legacy.

W. Hurt himself later returned to work in Brazil teaching at the Federal University of Paraná and excavating the shell middens of Paraná and Santa Catarina. Towards the end of his life he put forward considerations about the relations between populations in the Brazilian central plateau and those in the Brazilian coastlands. He was also a teacher and a person of the highest worth who contributed to the qualification of southern Brazil's first generation of archaeologists.

References

Banyai, M. (1997). Minhas pesquisas arqueológicas na região de Lagoa Santa. Symbiose, Budapest.

Coutard, J.P., Kohler, H., & Journaux, A. (1978). *Carte du karst, région de Pedro Leopoldo, Lagoa Santa - Minas Gerais, Brésil*, Caen.

Cunha, F. L. S., & Guimarães, M. L. (1978). A fauna sub-recente de vertebrados do Grande Abrigo da Lapa Vermelha (P. L.) de Pedro Leopoldo, Minas Gerais. *Arquivos do Museu de História Natural da Universidade Federal de Minas Gerais, Belo Horizonte, 3*, 201–240.

Hurt, W. (1960). The cultural complexes from the Lagoa Santa region, Brazil. *American Anthropologist, 62*, 569–585.

Hurt, W. (1964). Recent radiocarbon dates for Central and Southern Brazil. *American Antiquity, 113*, 3–10.

Hurt, W., & Blasi, O. (1969). O projeto arqueológico Lagoa Santa, Minas Gerais, Brasil. *Arquivos do Museu Paranaense, NS, Arqueologia, 4*, 1–63.

Kohler, H., & Malta, I.M. (1991) O cenário cárstico: guardião da fauna extinta aos legados bandeirantes. Paper presented at the "3° Congresso da ABEQUA", Belo Horizonte.

Laming-Emperaire, A. (1979). Missions archéologiques franco-brésiliennes de Lagoa Santa, Minas Gerais, Brésil: le grand abri de Lapa Vermelha. *Revista de Pré-História, 1*, 54–89.

Laming-Emperaire, A., Prous, A., Moraes, A.V., & Beltrão, M.C. (1974). Grottes et abris de la région de Lagoa Santa, Minas Gerais, Brésil. Cahiers d'Archéologie d'Amérique du Sud 1. EPHE, Paris.

Mattos, A. (1946). Lagoa Santa Man. In J. Steward (Ed.), *Handbook of South American Indians 1* (pp. 399–400). Washington, DC: Smithsonian Institution.

Neves, W. A., & Piló, L. B. (2008). *O povo de Luzia: em busca dos primeiros americanos*. São Paulo: Editora Globo.

Prous, A. (1977). Missão de estudo de arte rupestre de Lagoa Santa. Arquivos do Museu de História Natural da UFMG 2:51–65.

Prous, A. (1978). L'Homme et la nature dans la région de Lagoa Santa (Brésil). *Arquivos do Museu de História Natural da UFMG, 3*, 65–92.

Prous, A. (1986). L'archéologie au Brésil – 300 siecles d'occupation humaine. *L'Anthropologie, 90*, 257–306.

Walter, H. V. (1947). Arqueologia da Lapa Vermelha de Pedro Leopoldo. *Scientia, 1*, 34–37.

Walter, H. (1958). *Arqueologia da região de Lagoa Santa, Minas Gerais; índios pré-colombianos dos abrigos-rochedos*. Rio de Janeiro: Sedegra.

Chapter 8
The National Museum's Contributions to Lagoa Santa Research in the Second Half of the Twentieth Century

Sheila Mendonça de Souza and Andersen Liryo

Abstract This chapter addresses the importance of the National Museum of Rio de Janeiro for Lagoa Santa archaeological research during the second half of the twentieth century in two perspectives: as a repository for the collections that stemmed from the work of internal and external researchers and in its role as a research institute, participating in projects and contributing toward the qualification of new researchers. Regarding the topic of Lagoa Santa, during the period in question, the National Museum experienced great changes and took part in fieldwork as well as played an outstanding role in the academic discussions and the respective production of knowledge, even though it was not the major protagonist of the various archaeological research projects unfolded in the region. Based on the history of the National Museum's participations and contributions, the chapter describes the changes in the focus of research determined by the institution and the country's varying situations and whose impacts are mirrored in the Museum's outstanding role in the national scenario. The institution's contributions in the form of the activities and the production of its leading professionals are discussed, focusing on the form of their participation in field research in Lagoa Santa and on the scholars that dedicated long periods of laboratory research and office work.

S. Mendonça de Souza (✉)
Department of Endemic Diseases Samuel Pessoa at the National School of Public Health Sergio Arouca of the Oswaldo Cruz Foundation, Rua Leopoldo Bulhões, 1480 – sala 606, Manguinhos, Rio de Janeiro 21041-210, RJ, Brazil
e-mail: sferraz@ensp.fiocruz.br

A. Liryo
Department of Anthropology of the National Museum of the Federal University of Rio de Janeiro, Quinta da Boa Vista, s/n°, São Cristovão, Rio de Janeiro, RJ, Brazil
e-mail: liryo@hotmail.com

P. Da-Gloria et al. (eds.), *Archaeological and Paleontological Research in Lagoa Santa*, DOI 10.1007/978-3-319-57466-0_8

Introduction

The National Museum can be considered Brazil's oldest scientific institution and museum (Safra 2007). As such, it has taken part in the discussions related to the archaeology of the primitive groups of Lagoa Santa ever since the nineteenth century. Despite not having been the main protagonist of most archaeological research endeavors undertaken in the region, it has nevertheless played an important role in the discussions of them. In addition, it is the repository of a significant portion of the collection that resulted from the research in Lagoa Santa. The Museum has also contributed toward an important historical and scientific production on the subject, and for that reason, it has been a reference for various different projects and research efforts dedicated to Lagoa Santa, a position it managed to maintain throughout the second half of the twentieth century.

Looking back over its institutional history, we can see how, when it was still the Royal Museum, it received collections stemming from the earliest excavations at those sites. Its specialists took part in the discussions of the findings and the evolutionary interpretations involving the presence of human beings in those paleontological and archaeological sites, as was the case with a fragment of a lower jawbone registered as the number 00114 in the Registry Book of the collection of the Biological Anthropology Sector, received from Peter Lund in the nineteenth century. When Royal Museum, which was founded in 1818, became the National Museum in 1892, it received its entire legacy and became the repository of many kinds of material and the destination of many donations that included archaeological material.

In the first half of the twentieth century, the discussion about the contemporaneity of Man with the extinct fauna was one of the outstanding themes associated to Lagoa Santa. The Museum, at that time in the process of consolidating itself as a vanguard scientific institution in Brazil, organized field missions involving some of its researchers (see Chaps. 4 and 5 of this volume). Such efforts contributed toward expanding the collection of objects coming from the archaeological area of Lagoa Santa, bringing in specimens of human remains that motivated anthropologists like Jose Bastos de Ávila to propose certain interpretations about the human group to which the famous bones, that had aroused so much interest ever since Lund's time, belonged.

However, the main focus of institutional interest at the time was on the constitution of the Brazilian population, its racial composition and admixture, especially present in Edgard Roquette Pinto's work in collaboration with other researchers connected to the National Museum, like Heloisa Alberto Torres and Álvaro Fróes da Fonseca (Schwarcz 1993; Souza 2012). The study of prehistoric collections like those of Lagoa Santa took a secondary position in the anthropological discussions, and important paleontological studies at the Museum were only carried out much later. As Castro Faria (2000) records, the study of Negros, mulattoes, and Brazilians in a somatological perspective dominated the work, and as a consequence, the studies of the archaeological skeletal remains were relegated to a secondary plane. Even so, Heloísa Alberto Torres's administration, in the period of transition from the first to the second half of the twentieth century, managed to achieve a strong influx of important Brazilian and international names in the sphere of anthropology, such as Eduardo Galvão, Alfred Métraux, Paul Rivet, and Claude Lévi-Strauss, and that fostered the maintenance of the archaeological research topics in the sphere of important scientific interests (Miglievich-Ribeiro 2000).

After the end of the 1940s, all eyes focused once again on Brazil's indigenous and prehistoric groups, including Lagoa Santa, stimulated in this case by the results obtained by the Minas Gerais Academy of Sciences (see Chap. 6 of this volume). Effective changes in Brazil, accompanying the new post-war atmosphere, were destined to boost Brazilian science, while at the same time the institution's incorporation to the University of Brazil on June 16, 1946 (Decree 8,689) brought about a significant change in its trajectory. During the 1940s and 1950s, new concepts in university outreach, associated with a complete reform of the exhibitions and other investments, boosted the growth and vainglory of a Museum that still occupied a central and hegemonic position in the country, leading the field in areas of scientific research and representing scientific interests nationally and internationally. In addition to the undeniable competence of its professional staff, its leadership was most certainly favored by its historical condition and its location in what had been the administrative capital of the nation for almost two centuries.

The 1950s saw the topic of Lagoa Santa back on the scientific agenda once more. The decade began with an opportune contribution made by National Museum paleontologist Carlos de Paula Couto, with his edition of *Peter Wilhelm Lund, Memórias sobre a Paleontologia Brasileira* (Peter Wilhelm Lund, Memoirs of Brazilian Paleontology), published by the *Instituto Nacional do Livro/Ministério da Educação e Saúde* (National Book Institute/Ministry of Health and Education) (Lund 1950). That very year, the American Antiquity review (Volume XV [4]) published an article by Clifford Evans entitled "A Report on Recent Archaeological Investigations in the Lagoa Santa Region, Minas Gerais, Brazil," thereby reawakening the interest and the scientific community's mobilization around the topic (Evans 1950). Wesley Hurt's interest in the question of human occupation and the discussion on its contemporaneity with the extinct Pleistocene fauna motivated a proposal that a scientific expedition should be undertaken to that archaeological area within the same decade.

New paradigms and techniques, innovative methods, and the possibility of absolute dating by means of radiocarbon isotopes were among the conditions that reinvigorated archaeological fieldwork and laboratory. The possibility of an American-Brazilian research mission to the region aroused new interest and stimulated researchers of the National Museum's Anthropology and Ethnography Division and its Geology and Mineralogy Division. While, on the one hand, the archaeologists and anthropologists were discussing the Lagoa Santa "race" and culture and searching for evidence of its relation with animal species that had disappeared, on the other hand, paleontologists were searching for the answer to that same enigma in a study of the Pleistocene fauna. At the beginning of the 1950s, the Museum was represented by figures like Luis de Castro Faria, in anthropology and archaeology, and Carlos de Paula Couto, in geology and paleontology.

Heloísa Alberto Torres was the head of the Museum in the 1950s, and in the course of the five decades, that is the timeframe of this chapter, there were 13 directors. José Candido de Carvalho succeeded Heloísa, still in the 1950s, and then came José L. de Araújo Feio and Dalcy de Oliveira Albuquerque, whose mandates covered the 1970s. They accompanied the Museum's partners in two missions to Lagoa Santa with foreign participants: the American mission and the French mission. While in the 1950s the Museum had occupied an outstanding position in the national sphere that was further boosted by its new university status, in the 1970s, it was suffering the consequences of changes that came one after the other over the years: the transfer of the Brazil capital to Brasília, investment redistribution, teaching and research decentralization, and so on.

On the other hand, the institutional role of the Museum seems to have been boosted by the enactment of Law n° 3,924, dated July 26, 1961, which made provisions regarding archaeological and prehistoric monuments, and which had repercussions in the field of archaeology, that included the research in Lagoa Santa. In the view of Mendonça de Souza (1991), that Law's emphasis on professional competences, distinguishing "amateurs" from "professionals," contributed toward driving away various traditional groups, small institutions, and individual researchers who were passed over in favor of the academic institution's researchers. The same author states that the regulations established by groups anxious to accredit representatives of the higher State institutions to the detriment of private institutions and autonomous professionals, who were relegated to the category of "amateurs" that put the national heritage at risk, received the important support of Luiz de Castro Faria of the National Museum. In that context, traditional archaeological groups like the Minas Gerais Academy of Sciences and the Minas Gerais Historical Institute (*Instituto Histórico e Geográfico de Minas Gerais*), despite their traditions, former works, and notoriety, were relegated to the role of mere amateurs (Walter 1958) 1958). Accordingly, the National Museum had its prerogative of receiving the material originating from archaeological sites guaranteed, at least for some time, and also that of representing archaeology institutionally in the national sphere.

The beginning of the second half of the twentieth century provided the National Museum with an opportunity to project itself as an institution on the national scene and legitimized it to take part in research activities all over the country and receive collections from them (Mendonça de Souza 1991).

Thus, regarding Lagoa Santa, the second half of the twentieth century was a time when the National Museum experienced considerable changes. Reinvigorated in the first decades of the period, it was able to make various contributions to the question of Lagoa Santa's prehistory. It is precisely the archaeology of Lagoa Santa over that period that is the focus of this chapter.

One Institution: Various Roles

Against the background succinctly described above, the National Museum performed, and indeed still performs, a series of different academic-scientific roles in regard to Lagoa Santa research.

As a repository of collections, whether of its own researchers and staff or of other professionals, it has the primary role of guarding and preserving the archaeological, paleontological, biological, and cultural material stemming from that archaeological area. In that aspect, it is responsible for its adequate handling and for satisfying the increasing demand from the Museum's own researchers, as well as from other researchers from outside the institution, for an opportunity to research its collections. As a research institution, it produces knowledge through the participation of its teams and students in research projects in the field and in the laboratory, in addition to contributing as one of the leading institutions in the training of biological anthropologists in Brazil.

The National Museum as Repository of Collections: Contributions to the Custody of the Lagoa Santa Heritage

There are currently around 100,000 items in the collection storage spaces of the Archaeology Sector of the National Museum, distributed in 220 collections of general archaeology, pre-Columbian archaeology, and Brazilian archaeology. Among the items, 1500 are from archaeological sites in the Lagoa Santa region. The overall collection includes pieces from a variety of sites and occupations and consists of lithic, pottery, crockery, paleofauna (bone and malacological), archaeobotany (seeds, wood, and charcoal), and geological (rocks and sediments) materials. Many of the sites identified as belonging to the complex of ancient occupations of Lagoa Santa, such as Cerca Grande, Lapa das Boleiras, Lapa Mortuária, Lapa do Sumidouro, Lapa dos Confins, Caieiras, and Lapa Vermelha IV, are represented in the collection. A significant part of it was officially received by the institution in the 1950s. However, like so many other old collections, information on the origins of the pieces is not very detailed, and some of them originated from individual donations. During the five decades in question, the outstanding additions were the materials excavated or recovered in prospections at Cerca Grande, carried out by Wesley Hurt of the American-Brazilian mission (Hurt and Blasi 1969) and those brought in from different sites investigated by Annette Laming-Emperaire during the French-Brazilian mission (Laming-Emperaire et al. 1975).

In the collection storage spaces of the Biological Anthropology sector, there are more than 2,300 deposited pieces, items that may be as simple as a single bone or may represent a complete skeleton. Of that total, 236 items are associated to the archaeological region of Lagoa Santa or nominally to one or another of its archaeological sites. Most of them are human bones coming from the excavations at the

Lapa Mortuária of Confins (61 entries) and from the Lapa do Caetano de Matozinhos cave (54 entries). Of the former, 42.8% correspond to material that came during the second half of the twentieth century and mainly consist of material that was excavated at Cerca Grande, which accounts for 36.4% of all the items originating from Lagoa Santa.

The entries for that period begin in 1957 with the Lagoa Santa cranium number 01055 donated by Josaphat de Paula Penna (a dentist and an amateur archaeologist) to Paula Couto, who, in turn, passed it on to Castro Faria and, accordingly, to the Biological Anthropology Sector's collection. After that, there are 86 registrations of the ingress of material brought in by the American-Brazilian mission, still in the 1950s. In 1973, another ten additions refer to Lagoa Santa, and seven of them originated from collections made by Lilia Leite.[1] The others are registered as having come from the collections made by Annette Laming-Emperaire and the Franco-Brazilian mission.

Finally, after all those incorporations of material to the collection, there appears the registration of an individual, later to be nicknamed "Luzia," represented by the cranium and some other bone fragments, discovered in 1975 in the Lapa Vermelha IV, also by the Franco-Brazilian mission, and which was undoubtedly the most outstanding find of that period. Accordingly, even though the Franco-Brazilian mission does not represent the greatest accrual of material to the Biological Anthropology Sector in terms of numbers, nevertheless it does represent by far the most important entries from the scientific point of view. In addition to the polemic associated to the contemporaneity of the finds, in other words, regarding the chronology of the various different bones of a single individual, and also the dynamics of the formation of the Lapa Vermelha site (Cunha and Guimarães 1978b), the study of that cranium had a considerable repercussion in the following century, and it was one of the stays that supported the revision of the extant model for the peopling of the Americas (Neves et al. 1996, 1998, 1999a, b).

In the Geology and Paleontology departments, where the eight spaces dedicated to such collections are found (viz., Minerology, Petrography, Stratigraphy, Meteoritics, Economic Geology, Paleobotany, Invertebrate Paleontology, and Vertebrate Paleontology), there are more than 7,000 entries representing items ranging from isolated fragments to sets of pieces of different origins. More specifically, it is in the Paleovertebrate Sector where the bones of the extinct fauna are deposited and where there are 41 entries specifically related to Lagoa Santa and more than half of them (24 entries) are materials that were collected by Paula Couto during the period in focus.

A considerable part of that collection, even though only partially published, presents a huge potential for future studies. Unfortunately, with the years, the institutional

---

[1]The only reference we have to Lilia Leite is the register of official incorporation of the Biological Anthropology Sector, which merely describes the fact that she had collected the respective bone and fragments, which were hidden in a hole in the Lapa de Confins, on August 3, 1973, just a few days after Madame Emperaire had collected fragments of a femur, a humerus, and a cranium in that very same cave on July 31, 1973.

conditions have not favored the development of a program of custody that would accompany the growth of the deposits and of the new conditions stemming from the integration of the Museum with the University. It was only at the end of the period encompassed by the present research, specifically the 1990s, that new ways of capturing resources, the replacement of those in professional posts, the convergence of supporting programs and lines of finance specifically targeting the costs of curation, and other opportunities promoted a new enthusiasm for the collections, among them, those of Lagoa Santa. Different departments of the Museum launched their new programs of documentation, digitalization, reaccommodation, improved security, and reform of the collections' spaces and as a consequence many of the collections became more secure and offer now better conditions for research.

That long overdue contribution brought with it better access to the materials and enhanced the possibility of meeting the research demands. During the process, which went on into the first decade of the twenty-first century, it also became possible for the Museum to retrieve material that had been out on loan or whose licensed exit for study purposes was still in force. That was the case of materials from the Lagoa Santa collection which returned to the Biological Anthropology Sector after having been described in special studies conducted by Ferigolo (1987). Others that had not been handled for a long time became the object of study again or were sent back to the sector where they belonged. That was the case of the so-called *calota de hominídeo* (hominid calotte) published by Alan Bryan (1978) and Beattie and Bryan (1984) as being a probable find from the archaeological region of Lagoa Santa. Bryan came across it in 1975, mixed in with Harold Walter's paleontological collection in Minas Gerais. Much later it was proven to be just a piece mounted by sticking fragments of bone together (Anjos et al. 2005).

A Research Institution: Physical Anthropology, Archaeology, and Paleontology of Lagoa Santa

Research Missions and Projects

The Museum took part in two big archaeological research projects in the Lagoa Santa region in the period in question. The first was a research mission that brought together Americans and Brazilians with the participation of the *Museu Paranaense* (Paranaense Museum) and the *Instituto de Sociologia de São Paulo* (São Paulo Institute of Sociology), and its leader was Wesley Hurt from the University of South Dakota (Hurt 1960, 1964; Hurt and Blasi 1969). The second was a mission that united French and Brazilian researchers with the participation of the *Museu Paulista* (Paulista Museum) and the *Museu de História Natural da Universidade Federal de Minas Gerais* (Museum of Natural History of the Federal University of Minas Gerais). It was coordinated by Annette Laming-Emperaire, of the *Centre National de la Recherche Scientifique* (CNRS; National Center of Scientific Research) (Laming-Emperaire et al. 1975). Their importance for the study of that archaeological region was so great that they are the subject of an entire chapter of this book (see Chap. 7). Accordingly, they will only be mentioned in passing here for the purpose of articulating facts and names associated to the National Museum.

Both missions were the fruit of international collaboration efforts and they prospected and excavated Lagoa Santa in the period being studied. Naturally, the collections produced by their fieldwork were placed in the custody of the National Museum, and the formal act of their inclusion in the Museum's collection is duly registered as described earlier in this chapter.

Museum researchers or students helped make up the teams of both missions, and some of them produced material on the subject. Among those that took part in the American-Brazilian mission in the 1950s were Luiz de Castro Faria and Carlos de Paula Couto. Among those that took part in the Franco-Brazilian mission in the 1970s were Maria da Conceição Beltrão, Osvaldo Heredia, Fausto Luiz de Souza Cunha, and Martha Locks Guimarães. Other Museum researchers did not participate in the fieldwork at Lagoa Santa, but dedicated themselves to laboratory research during those periods, and they too produced material on the topic. Outstanding among them were Marília Carvalho de Mello e Alvim, Tarcísio Torres Messias, and João Carlos de Oliveira Gomes.

Marília Carvalho de Mello e Alvim was without any doubt the reference name in Physical Anthropology at the time. She got her degree in Geography and History from the Lafayette Faculty in 1952 and her teaching degree in the following year from the Faculty of Philosophy, Science, Language and Literature of the University of the Federal District. She was passionate about anthropology and after being admitted to the National Museum in 1957 went on to spend the 1960s dedicating herself to the study of human bones (Powell et al. 2006). Right from the beginning of her anthropological research, she directed her attention to the topic of Lagoa Santa. Her first paper published together with Tarcísio Messias (Messias and Mello e Alvim 1962) compared the Lagoa Santa skeletons with those of the Botocudo, an indigenous group represented in a small nineteenth century collection that belonged to the Museum. That was followed by an individual publication on the same topic (Mello e Alvim 1963a).

Her next work was produced to support her candidature for the qualification of "*Livre Docência*"[2] in the Chair of Anthropology and Ethnology at the University of Guanabara (Mello e Alvim 1963b). Under the tutorship of the anatomists Benjamin Vinelli Baptista and Álvaro Fróes da Fonseca, she described the morphology of the talus bones in the Museum's Lagoa Santa collection, discussing their possible morphological-functional and evolutionary implications. Again, in that same line of osteometric studies comparing Lagoa Santa groups and other indigenous and prehistoric groups in Brazil, she developed her doctoral project as part of her attempt to obtain a permanent position at the National Museum, and she published the first comparisons between Lagoa Santa and the builders of the shell mounds (*sambaquis*) (Mello e Alvim and Mello Filho 1965).

---

[2] Thesis defense required for those intending to become chair professors.

In the 1970s, she continued her comparative studies still within the sphere of classical osteometric classification. With the financial support of the National Scientific and Technological Development Council (*Conselho Nacional de Desenvolvimento Científico e Tecnológico* – CNPq) and the collaboration of her team, made up of Lilia Cheuiche Machado, José Flávio Pessoa de Barros (CNPq scholarship holder), and Marcus Infante Vieira, she also analyzed the collection at the Natural History Museum of the Federal University of Minas Gerais and private collections belonging to Hélio Diniz, Mihály Banyai, and Harold Walter. Based on the study of around 200 individuals from 13 different archaeological sites, she produced what she considered to be her great synthesis of Lagoa Santa (Mello e Alvim 1977).

Pursuing similar goals to those of her predecessor, she indicated a classification for those individuals and agreed that they were inserted in what was then referred to as the "Lacida race" in the older anthropo-physical classifications (Ávila 1950). Still bound to a classificatory intention, she described the Lagoa Santa groups as having craniums that were "very elongated, ovoid, with prominent zygomatic arches, tectiform in posterior view, medium sized, and with moderately thick bones [...]." Despite the obvious differences in relation to the morphologically Mongoloid groups that constructed the shell mounds (Mello e Alvim and Mello Filho 1965), Marília clung faithfully to the paradigm of her day which defended the proposal that the Lagoa Santa groups were a morphological variation of American indigenous populations (Mello e Alvim 1977).

Marília defended the much debated "homogeneity" for the Lagoa Santa groups, and with that, she justified the proposition of an anthropo-physical unity mainly based on cranial morphology. In the reflections on the morphology of the group, she took into account the apparent isolation of the Lagoa Santa population in national territory and its long temporal persistence in an attempt to explain the findings. In her text (Mello e Alvim 1977:161), she presents the following in her conclusions:

> Once the anthropo-physical unity of "Lagoa Santa Man" is accepted, the permanence of a single morphological pattern for such a long time is only admissible if the following factors are considered: (a) Low demographic density and relatively homogeneous genetic potential of the primeval group; (b) absence, in the area, of morphologically distinct indigenous groups; (c) geographic confinement; (d) relative environmental stability.

The model this author proposed for the Lagoa Santa, consolidated by two decades of her publications, classes, and lectures, became hegemonic, and, for years on end, Brazilian historians of prehistory considered it to be the last word on the subject.

Nevertheless, the 1980s were to bring in changes and the overdue exhaustion of osteometry as it was still being practiced at the National Museum. During that decade, Marília Mello e Alvim began a series of diversified studies in co-authorships and, focusing once more on Lagoa Santa, ran a series of genetic distance studies based on the epigenetic characteristics of the cranium (Mello e Alvim et al. 1983/1984; Mello e Alvim and Souza 1990). She also conducted radiological studies of the facial sinuses of the Lagoa Santa skulls (Bertolazzo and Mello e Alvim 1985). The last time she addressed the topic of Lagoa Santa was in the 1990s, when she had the opportunity to study the skeletons from the Serra do Cipó (Mello e Alvim 1992/1993) excavated by André Prous and his team and in the custody of the Natural History Museum of the Federal University of Minas Gerais. For decades, she was considered the leading authority on the physical anthropology of the Lagoa Santa primeval groups.

Regarding the human bones recovered by the Franco-Brazilian Mission in the 1970s, she confirmed that they belonged to a single individual who was later to become the famous "Luzia." That interpretation, which contradicted the archaeologists, stemmed from her extremely experienced morphological "eye," which enabled her to bring together apparently disperse bones and unite them as a single specimen. It meant that there had to be a whole geomorphological and archaeological rereading of the great shelter of Lapa Vermelha IV (Cunha and Guimarães 1978b).

When asked to issue a technical opinion, she examined some bones and two crania that Castor Cartelle, of the Catholic University of Minas Gerais (*Pontifícia Universidade Católica de Minas Gerais*), had found in a cave called Gruta das Onças, in Jacobina in the state of Bahia (Beltrão et al. 1991). According to Mello e Alvim's description, the crania were of a child around 9 years old and a youth, and they were associated to other fragments of the bone. She described them as having a delicate constitution, being dolicocranial with a curved forehead and prominent occiput, oval vertical shape, short wide face, not very deep glenoid cavity, and shovel-shaped lateral incisors as well as other features she observed in the long bones, such as the presence of a third trochanter in the femur. Although Marília never published those analyses, the material she referred to, found more than 1,500 km from the nuclear area of Lagoa Santa, reinforced the possible existence of the Lagoa Santa morphology in different parts of Brazil as well.

Acknowledging that individual variations would be of little importance in that particular series, Marília defended the concept of the homogeneity of the findings from different sites, justifying that possibility based on the groups' relative isolation, a supposed low-population density, relative stability of the environment, and the absence of other known groups in the archaeological region in the same period as cited above (Mello e Alvim 1977). Her work was always based on conventional osteometric techniques and methods, and that fact together with a reactive resistance to any change in the paradigm probably contributed to her maintaining the defense of the morphological homogeneity of the Lagoa Santa series throughout her career. That interpretation supported what Alfredo Castellanos (1932) and José Bastos de Ávila (1950) had proposed in the first half of the century, when they defended the existence of an anthropo-physical unit or "race" known as "Lagoa Santa Man." The theory seemed to be confirmed when Marília carried out studies of the traces of epigenetics in those series. A more recent contestation by Neves and Atui (2004) is based on a more extensive series, more accurate dating, and especially on a more modern osteometric analysis.

Despite having always called attention to the peculiar aspect of the Lagoa Santa crania, whose absence of mongoloid features even Peter Lund had remarked on (Lund 1842, 1844; Mendonça de Souza et al. 2006), Marília never considered the possibility of a non-Asiatic origin for the Lagoa Santa groups. Even though she defended their great antiquity, she never thought they could be associated to what she considered to be the first entrance of Man in the Americas, that is, that they might be earlier than the Clovis culture. To her, the possibility of cases of extreme variation among the Amerindians of Asian origin always seemed to be good enough to explain the non-Mongoloid morphology encountered in Lagoa Santa.

On the other hand, regarding the supposed similarity of the Lagoa Santa craniums with those of Botocudo Indians, a thesis defended by João Baptista de Lacerda, Marília presented results of a comparative craniometrics study that led her to disagree with her predecessor at the National Museum (Mello e Alvim 1963a).

Maria da Conceição de Moraes Coutinho Beltrão, or, more simply, Maria Beltrão, was the head of the archaeology team at the National Museum who most involved herself in the Lagoa Santa research activities. Her group carried out fieldwork and studied archaeological and paleontological collections. After the abrupt end of the Franco-Brazilian mission, she set up another field research project investigating caves and rock-shelters in the area of Central in the state of Bahia, where she studied sites with signs of very ancient prehistoric occupation; she believed she had found occupations equivalent to, or even older than those of Lagoa Santa. The project lasted for many years and uncovered a new, rich archaeological area, eventually producing a great amount of information including aspects of rock paintings.

Her scientific production associated to Lagoa Santa began in the 1970s with her participation in the Franco-Brazilian mission, which resulted in scientific communications and publications. Among them were analyses of lithic material and pottery (Beltrão 1974) brought in by the field missions to the Lagoa Santa region that had been occurring ever since 1956, such as the lithic material recovered from controlled archaeological levels (Laming-Emperaire et al. 1975).

In the 1980s, she also presented a communication as co-author regarding the find at the Gruta das Onças, in Jacobina, Bahia, discussing the morphological similarities of the specimens to those of Lagoa Santa (Cartelle and Beltrão 1985). That find was within the limits of the area of the Central project. It was cited in an international publication as being the first time that a "Lagoa Santa type" had been found outside Minas Gerais (Weber 1986). This theme was taken up again more recently in a book on the settlement of Americas in which the author published a photograph of one of the craniums and reviewed the data (Beltrão 2008). In the same communication, Beltrão refers to a dating done in California by R. Taylor of a fragment of human skullcap (4830 ± 70 BP, UCR-3187/CAMS-10909) and defends the idea of the continuity of the Lagoa Santa occupations with primeval groups in the Bahia region, where the finds in question date back to more recent periods. Reviewing her initial ideas, shared with paleontologist Castor Cartelle, she confirmed the absence of any contemporaneity of similar human bones found in Bahia with those of the extinct fauna discovered in the same cave, because the former were in a situation of redeposition.

In the 1990s, Beltrão published two more papers and referred to Lagoa Santa once more (Beltrão 1996; Dilehay et al. 1992), and more recently she took it up again to defend the idea of the association between Bahia and Lagoa Santa. In the results of her various other projects, including those in Central, Bahia, she continues to defend the existence of those associations. In doing so, she associates the dating and the archaeological context in that region (Beltrão et al. 2005) to the settlement of the Americas (Beltrão 2008), and departing from there, she refers back to the Lagoa Santa findings, proposing new interpretations and possible ethno-archaeological relations. Beltrão (2008) also relates Lagoa Santa and the Tukano creation myth through an association of the Tukano people's iconography and pottery with the rock paintings of the Central archaeological region, which, in turn, is associated to Lagoa Santa through the craniums from the Gruta das Onças and also through the rock paintings. According to the researcher, in the period of the *optimum climaticum* (ca. 5,000 BP), the population of Lagoa Santa moved to Bahia as confirmed by the dating of the Gruta das Onças craniums (4,830 ± 70 AP) and by the evident similarity of the Minas Gerais paintings with those of Bahia. Then they moved to the region of Uaupes in the Amazon, a region between Colombia and Brazil, in the last retreating movement of the Tukano people.

Luiz de Castro Faria was an outstanding figure in the 1950s because of his activities as a field archaeologist and his constant efforts to obtain proper legal protection for the archaeological sites. His presence as representative of the National Museum in the excavations of the American-Brazilian mission in Lagoa Santa led by Wesley Hurt was extremely important in the aspect of making the mission feasible. His intense concern for the preservation of archaeological sites, especially focused on the shell mounds of the Brazilian coastlands and on Lagoa Santa, gave him a certain international notoriety, and at one point, he elaborated a technical report for UNESCO on the regulation of archaeological excavations, stressing the need for specialists and opening the way for international cooperation (Simão 2009).

Despite his important presence in international cooperation, Castro Faria published very little about the results of the excavation work in Lagoa Santa. His only publication as co-author with Wesley Hurt and Oldemar Blasi was a manuscript summarizing the excavation work (Castro Faria et al. 1956).

Carlos de Paula Couto is widely acknowledged to have been one of the greatest Brazilian paleontologists and played important roles in the polemic regarding the contemporaneity of the extinct fauna and the human groups in Lagoa Santa. He addressed the question from a paleontological standpoint from the 1950s on. At the National Museum, he dedicated himself to systematizing and classifying the Pleistocene fauna, with a special emphasis on mammals. His commented review of Lund's memoirs (Lund 1950) provided readily accessible material for the debates on Lund's finds in Lagoa Santa.

During that same decade, he undertook field research in the Lagoa Santa region (Paula Couto 1958b), putting together geomorphological, geological, and paleontological data to construct a model that would reinforce the interpretation of the prehistoric settlement of the area and its relation to the extinct fauna. Among the aspects he discussed was the formation of lakes in the Cerca Grande area, their relation to the archaeological sites, and their estimated chronologies. During the 1950s and the decades that followed, he published papers and took part in the debate on the topic on various occasions (Paula Couto 1953, 1958a, b, 1964a, b, 1968, 1970, 1974).

Together with Luiz de Castro Faria and other colleagues, he took part in the American-Brazilian mission headed by Wesley Hurt. That field project led to the conclusion that there was no evidence for the contemporaneity between the extinct fauna and the humans in the area, an opinion that Paula Couto subscribed to. He was also of the opinion that the skeletons found in Lagoa Santa were relatively recent due to the type of burials, very similar to those of contemporary indigenous groups, and due to the rock paintings which only represented modern animals (Paula Couto 1958a).

Fausto Luiz de Souza Cunha was a geologist and paleontologist dedicated to the study of mammals, and in the 1950s, he was already attached to the National Museum. As a paleontologist, he had studied the *Hippidion* (extinct horses) based on the material from the Lapa Mortuária in the Museum collection, and it was on the basis of those studies that he presented his candidature for the higher qualification of "*Livre Docência*" to the Federal University of Rio de Janeiro (Cunha 1960). He took part in the Franco-Brazilian mission in Lagoa Santa, dedicating himself to excavation and coordinating very thorough, meticulous zooarchaeological work, as can be seen from the contributions he made as a result of that participation (Cunha and Guimarães 1978a).

His outstanding role became clear when he and Martha Locks Guimarães published a new interpretation of the formation of the deposits excavated at the Lapa Vermelha by the Franco-Brazilian mission. That stratigraphic and contextual reinterpretation made sense of the human skeletal remains found in Lapa Vermelha IV, which, despite having been found in widely separated stratigraphic levels, actually belonged to a single individual (Cunha and Guimarães 1978b). Fausto did not publish anything further regarding the human skeletons of Lagoa Santa or the antiquity of the population, but in the 1980s, he did publish two papers on the question that had placed him in the Franco-Brazilian mission, namely, the extinct fauna of the Lapa Vermelha and Cerca Grande sites (Cunha and Guimarães 1981; Cunha and Magalhães 1986).

For *Martha Locks Guimarães*, better known simply as Martha Locks, the Franco-Brazilian mission was her beginner's school. She later went on to undertake research with Fausto Cunha and published with him. After joining the Anthropology Department of the National Museum, she became a member of Maria Beltrão's team, taking responsibility for projects linked to Lagoa Santa and later for a considerable part of the activities of the Central Project, for decades.

Both Martha Locks and Maria Beltrão continued to work with the question of Lagoa Santa beyond the period being studied here. Besides the two, other researchers who had entered the National Museum more recently, like Claudia Rodrigues Ferreira de Carvalho, Hilton Pereira da Silva, and Ricardo Ventura Santos, joined the discussion regarding the antiquity of the Lagoa Santa population, especially because of the new debate that had arisen around "Luzia" (Silva 2000; Silva et al. 2001).

Considering that the National Museum continues to be an important repository of Lagoa Santa collections, the researchers of the Museum's anthropology area have always had close contact with research on the topic and have addressed related issues. Even those that are not directly involved in a research project in that archaeological region are permanently active in receiving colleagues from other institutions and taking part in the scientific discussions, either as representatives of the Museum or by contributing to the work of their peers (Silva 2009).

In the period that followed, for example, Claudia Rodrigues, the Museum's director since 2010, and Hilton Silva, who was a researcher in the Biological Anthropology Sector for some years, organized a series of lectures to update the discussions in course on the topic of the physical characteristics of the ancient human remains of Lagoa Santa, their genetic relation to other American groups, and to the settlement of the Americas. The result of those encounters was consolidated in a book dedicated to discussing what progress had been made in regard to determining the antiquity of the peopling of Lagoa Santa and the pathways that research should take in the new century (Mendonça de Souza et al. 2006; Rodrigues-Carvalho and Silva 2006; Silva and Rodrigues-Carvalho 2006a, b). Both devoted part of their time to participating as much in recent research projects as outreach projects regarding Lagoa Santa and its people. Ricardo Ventura, taking a more historical-social line, published some papers as author and as co-author addressing the question of society's appropriation of the image of "Luzia," highlighting the extant discussion around racism and affirmative policies (Gaspar-Neto and Santos 2009; Maio and Santos 2010).

Thus, at the National Museum, against that background of the national and institutional transformations of the second half of the twentieth century, new projects were developed, new knowledge was produced, and the collections stemming from the archaeological and paleontological sites of Lagoa Santa were expanded. New motivation for research on the topic or the review of old themes and questions not satisfactorily answered, international missions, and the ingress of new names that were to emerge as leaders in their fields of knowledge all left their mark on the institution's most important anthropological and paleontological production in the decades from 1950 to 2000.

References

Anjos, M. J., Lopes, R. T., Mendonça De Souza, S. M. F., & De Jesus Efo, D. J. (2005). Investigation of a fossilized calotte from Lagoa Santa, Brazil, by EDXRF. *X-Ray Spectrometry, 34*, 189–193.

Ávila, J. B. (1950). Anthropometry of the Indians of Brazil. Handbook of South American Indians, Smithsonian Institution, Bureau of American Ethnology. *Bulletin, 143*, 71–84. Smithsonian Institute, Washington.

Beattie, O., & Bryan, A. (1984). A fossilized calotte with prominent browridges from Lagoa Santa, Brazil. *Current Anthropology, 25*, 345–346.

Beltrão, M.C.M.C. (1974). Industrie litiques trouvées au de la prospection 1971. In: *Grottes et Abris de la Région de Lagoa Santa – Minas Gerais, Brésil – Cahiers d'Archéologie d'Amérique du Sud 1* (pp. 123–133). Paris: Imprimerie Nationale.

Beltrão, M. C. M. C. (1996). O Homem de Lagoa Santa: Uma Nova Abordagem. *Revista do Instituto Histórico e Geográfico Brasileiro, 156*, 53–78.

Beltrão, M. C. M. C. (2008). *Lê peuplement de l'Amérique du Sud. Essai d'archéogéologie. Une approche interdisciplinaire*. Paris: Riveneuve éditions.

Beltrão, M. C. M. C., Neme, S. M. N., Andrade, C. O. L. C., & Joria, F. A. (1991). Projeto Central: primeiros resultados. *Clio, Série Arqueologia, 4*, 39–47.

Beltrão, M., Taylor, E. R., Kinder, D. L., & Southon, J. R. (2005). Historical context and radiocarbon age of Lagoa Santa human populations in Brazil. In B. D. Dillon & M. A. Boxt (Eds.), *Archaeology without limits: Papers in honor of Clement W. Meighan* (pp. 257–264). Lancaster: Labyrinthos.

Bertolazzo, W., & Mello e Alvim, M. C. (1985). Os Seios Frontais em Grupos Indígenas Brasileiros – Homem de Lagoa Santa, Construtores de Sambaquis e Bocotudos. *Clio – Série Arqueológica, 2*, 113–130.

Bryan, A. (1978). An overview of Paleoamerican prehistory from a circum-pacific perspective. In A. L. Bryan (Ed.), *Early man in America from a circum-pacific perspective* (Vol. 1, pp. 306–327). Edmonton: University of Alberta, Occasional Papers.

Cartelle, C., & Beltrão, M.C.M.C. (1985). Notícia Prévia sobre o Achado do Homem de Lagoa Santa na Bahia. Abstract in the Comunicações do IX Congresso Brasileiro de Paleontologia, Fortaleza-CE, p. 148.

Castellanos, A. (1932). La raza sudamericana de Lagoa Santa y las migraciones malayo-polinesias. Quid Novi, Revista de las Asociaciones de Ex-Alumnas y Padres de la Escuela Normal N° 2, Año 1, Nros. 3–4. Rosario, Argentina.

Castro Faria, L., Hurt, W.R., & Blasi, O. (1956). Escavações Arqueológicas da Região de Lagoa Santa, Minas Gerais, Brasil. Manuscript at the Museu Nacional.

Cunha, F.L.S. (1960). Sôbre o Hippidion da Lapa Mortuaria de Confins, Lagoa Santa, Minas Gerais: estudos geo-paleontológicos baseados na lapa mortuaria e na coleção "Padberg-Drenkpol" do Museu Nacional. Livre Docência Thesis, Universidade do Rio de Janeiro.

Cunha, F. L. S., & Guimarães, M. L. (1978a). A fauna sub-recente de vertebrados do abrigo da Lapa Vermelha (P.L.) de Pedro Leopoldo, Minas Gerais. *Arquivos do Museu Nacional de História Natural da Universidade Federal de Minas Gerais, 3*, 201–240.

Cunha, F.L.S., & Guimarães, M.L. (1978b). Posição geológica do homem de Lagoa Santa no grande abrigo da Lapa Vermelha Emperaire (PL) Pedro Leopoldo – Estado de Minas Gerais. In *Coletânea de Estudos em Homenagem a Annette Laming-Emperaire/Serie Ensaios* (pp. 275–305). Museu Paulista da Universidade de São Paulo.

Cunha, F. L. S., & Guimarães, M. L. (1981). A fauna sub-recente de vertebrados do grande abrigo da Lapa Vermelha; Emperaire (P.L.), Pedro Leopoldo, Estado de Minas Gerais. *Revista da Universidade de São Paulo, 28*, 235–272.

Cunha, F. L. S., & Magalhães, R. M. M. (1986). *A fauna de vertebrados sub-recentes de Cerca Grande* (pp. 119–131). Matozinhos, Minas Gerais: Publicações Avulsas do Museu Nacional.

Dilehay, T., Calderon, G. A., Politis, G., & Beltrão, M. C. M. C. (1992). Earliest hunters and gatherers of South America. *Journal of World Prehistory, 6*, 145–204.

Evans, C. (1950). A report on recent archaeological investigations in the Lagoa Santa Region of Minas Gerais, Brazil. *American Antiquity, 15*, 341–343.

Faria, L. C. (2000). Antropologia. Escritos Exumados: dimensões do conhecimento antropológico. Niterói: Eduff, v.

Ferigolo, J. (1987). Paleopatologia Comparada de Vertebrados: 'Homem de Lagoa Santa', 'Homem do Sambaqui de Cabeçuda' e mamíferos pleistocênicos. Ph.D. Dissertation, Universidade Federal do Rio Grande do Sul.

Gaspar-Neto, V. V., & Santos, R. V. (2009). A cor dos ossos: narrativas científicas e apropriações culturais sobre Luzia, um crânio pré-historico do Brasil. *MANA, 15*, 449–480.

Hurt, W. R. (1960). The cultural complexes from the Lagoa Santa Region, Brazil. *American Antiquity, 62*, 569–585.

Hurt, W. R. (1964). Recent radiocarbon dates for Central and Southern Brazil. *American Antiquity, 30*, 25–33.

Hurt, W. R., & Blasi, O. (1969). O Projeto Arqueológico "Lagoa Santa", Minas Gerais, Brasil (nota final). *Arquivos do Museu Paranaense – Nova Série Arqueologia, 4*, 1–63.

Laming-Emperaire, A., Prous, A., Vilhena de Moraes, A., Beltrão, M. C. M. C., & Lema, J. L. (1975). *Grottes et abris de la Région de Lagoa Santa – Minas Gerais, Brésil. Série Cahiers d'Archéologie d'Amérique du Sud 1* (pp. 123–133). Paris: Imprimerie Nationale.

Lund, P. W. (1842). Carta escripta da Lagoa Santa ao senhor secretario do Instituto. *Revista do Instituto Histórico e Geográfico Brasileiro, 4*, 30–87.
Lund, P. W. (1844). Carta escripta de Lagoa Santa a 21 de abril de 1844. *Revista do Instituto Histórico e Geográfico Brasileiro, 6*, 334–342.
Lund, P. W. (1950). *Memórias sobre a paleontologia brasileira. Revistas e comentadas por Carlos de Paula-Couto*. Rio de Janeiro: Instituto Nacional do Livro, 591p.
Maio, M. C., & Santos, R. V. (2010). *Raça como Questão. História, Ciência e Identidades no Brasil*. Rio de Janeiro: Editora Fiocruz.
Mello e Alvim, M. C. (1963a). Diversidade morfologica entre os índios "Botocudos" do leste brasileiro (sec. XIX) e o Homem de Lagoa Santa. *Boletim do Museu Nacional – Antropologia-Nova Série, 23*, 1–70.
Mello e Alvim, M. C. (1963b). Estudo do Homem de Lagoa Santa, Minas Gerais, Brasil – morfologia do astrágalo. *Boletim do Museu Nacional – Antropologia Nova Série, 21*, 1–42.
Mello e Alvim, M. C. (1977). Os antigos habitantes da area arqueológica de Lagoa Santa, MG-Brasil – Estudo morfológico. *Arquivos do Museu de Historia Natural da Universidade Federal de Minas Gerais, 2*, 119–174.
Mello e Alvim, M.C. (1992/1993). Os antigos habitantes da Serra do Cipó (MG – Brasil): estudo morfológico preliminar. *Arquivos do Museu Nacional de História Natural da Universidade Federal de Minas Gerais* 13/14:107–128.
Mello e Alvim, M.C., & Mello Filho, D.P. (1965). Morfologia craniana da população do sambaqui da Cabeçuda (Laguna-Santa Catarina) e sua relação com outras populações de paleoameríndios do Brasil. In: *Homenage a Juan Comas en su 65 Aniversario (v.II), Mexico*. Anais 2:37–42.
Mello e Alvim, M. C., & Mendonça de Souza, S. M. F. (1990). Relações biologicas entre populações indígenas atuais e pré-históricas do Brasil. *Clio – Série Arqueológica, 1*, 69–79.
Mello e Alvim, M.C., Soares, M.C., & Cunha, P.S.P. (1983/1984). Traços não-métricos cranianos e distância biológica em grupos indígenas interioranos e do litoral do Brasil. 'Homem de Lagoa Santa', índios Botocudos e construtores de sambaquis. Arquivos do Museu de Historia Natural da Universidade Federal de Minas Gerais 8/9:323–338.
Mendonça de Souza, A. A. C. (1991). História da Arqueologia Brasileira. *Pesquisas, Antropologia, 46*, 1.
Mendonça de Souza, S., Rodrigues-Carvalho, C., Silva, H. P., & Guimarães, M. L. (2006). Revisitando a discussão sobre o Quaternário de Lagoa Santa e o povoamento das Américas: 160 anos de debate científico. In H. P. Silva & C. Rodrigues Carvalho (Eds.), *Nossa Origem – O Povoamento das Américas: Visões Multidisciplinares* (pp. 19–43). Rio de Janeiro: Vieira & Lent.
Messias, T. T., & Mello e Alvim, M. C. (1962). Contribuições ao estudo do Homem de Lagoa Santa; material das escavações arqueológicas na região de Lagoa Santa, Minas Gerais, Brasil (1956). *Boletim do Museu Nacional (Nova Série – Antropologia), 20*, 1–55.
Miglievich-Ribeiro, A.M. (2000). Heloisa Alberto Torres e Marina São Paulo de Vasconcellos: entrelaçamento de círculos e formação das Ciências Sociais na cidade do Rio de Janeiro. Ph.D. Dissertation, Universidade Federal do Rio de Janeiro.
Neves, W. A., & Atui, J. P. V. (2004). O mito da homogeneidade biológica na população paleoíndia de Lagoa Santa: implicações antropológicas. *Revista de Antropologia, 47*, 159–205.
Neves, W. A., Munford, D., & Zanini, M. C. (1996). Cranial morphological variation and the colonization of the new world: Towards a four migration model. *American Journal of Physical Anthropology, 22*(Suppl), 176.
Neves, W. A., Powell, J. G., Prous, A., & Ozolins, E. G. (1998). Lapa Vermelha IV hominid 1: Morphological affinities for the earliest known American. *American Journal of Physical Anthropology, 26*(Suppl), 169.
Neves, W. A., Munford, D., Zanini, M. C., & Pucciarelli, H. M. (1999a). Cranial morphological variation and the colonization of the new world: Towards a four migration model? *Ciência e Cultura, 51*, 151–165.
Neves, W. A., Powell, J. F., & Ozolins, E. G. (1999b). Extracontinental morphological affinities of Lapa Vermelha IV, Hominid 1: A multivariate analysis with progressive numbers of variables. *Homo, 50*, 263–282.

Paula Couto, C. (1953). Os primeiros homens americanos. O homem de Lagoa Santa. In *Paleontologia Brasileira – mamíferos* (pp. 414–428). Rio de Janeiro: Instituto Nacional do Livro.

Paula Couto, C. (1958a). Homem de Lagoa Santa Não É Nenhum Ancestral Americano: trata-se de remanescente de uma raça que teria começado a povoar a região – nao há conexão dos restos fósseis de animais extintos com os de esqueletos humanos – as pinturas encontradas representam apenas animais da fauna recente [02 de abril de 1958]. Jornal Correio da Manha, Rio de Janeiro.

Paula Couto, C. (1958b). Notas à margem de uma expedição científica a Minas Gerais. *Kriterion, 11*, 401–423.

Paula Couto, C. (1964a). O Homem de Lagoa Santa e o Pleistoceno Sul-Americano. *Revista Brasileira de Arqueologia, 1*, 27–29.

Paula Couto, C. (1964b). *O Pleistoceno e a antropologia do homem americano. Origens do Homem Americano* (pp. 36–54). São Paulo: Instituto de Pré-história/USP.

Paula Couto, C. (1968). O Pleistoceno Sul-Americano e as migrações humanas pré-históricas. In *Pré-História Brasileira* (pp. 3–42). São Paulo: Instituto de Pré-história/USP.

Paula Couto, C. (1970). Paleontologia da região de Lagoa Santa, Minas Gerais, Brasil. *Boletim do Museu de Historia Natural da Universidade Federal de Minas Gerais, Serie Geologia, 1*, 1–21.

Paula Couto, C. (1974). O homem e o meio Pleistoceno. *Revista do Instituto de Filosofia e Ciências Humanas da Universidade Federal do Rio Grande do Sul, 2*, 66–76.

Powell, M. L., Cook, D. C., Bogdan, G., Buikstra, J. E., Castro, M. M., Horne, P. D., Hunt, D. R., Koritzer, R. T., Mendonça De Souza, S. F., Sandford, M. K., Saunders, L., Sene, G. A. M., Sullivan, L. P., & Swetnam, J. J. (2006). Invisible hands: Women in bioarchaeology. In J. E. Buikstra & L. A. Beck (Eds.), *Bioarchaeology: The contextual analysis of human remains* (pp. 131–194). Oxford: Elselvier.

Rodrigues-Carvalho, C., & Silva, H. P. (2006). Questões não respondidas, novas indagações e perspectivas futuras. In H. P. Silva & C. Rodrigues-Carvalho (Eds.), *Nossa Origem – O Povoamento das Américas: Visões Multidisciplinares* (pp. 187–195). Rio de Janeiro: Vieira & Lent.

Safra. (2007). *O Museu Nacional*. São Paulo: Banco Safra.

Schwarcz, L. M. (1993). *O Espetáculo das Raças – cientistas, instituições e questao racial no Brasil 1870–1930*. São Paulo: Companhia das Letras.

Silva, H.P. (2000). Luzia: Como o Brasil pode mudar a História da Ocupação Humana no Continente Americano. A Notícia, Campos, 21 nov., p. 4.

Silva, H. P. (2009). Resenha do livro O Povo de Luzia, em busca dos primeiros americanos, de Walter Alves Neves & Luis Beethoven Pilo. *Amazônica, 1*, 568–574.

Silva, H. P., & Rodrigues-Carvalho, C. (2006a). A Busca pelos primeiros americanos. In H. P. Silva & C. Rodrigues-Carvalho (Eds.), *Nossa Origem – O Povoamento das Américas: Visões Multidisciplinares* (pp. 11–17). Rio de Janeiro: Vieira & Lent.

Silva, H. P., & Rodrigues-Carvalho, C. (Eds.). (2006b). *Nossa Origem – O Povoamento das Américas: Visões Multidisciplinares*. Rio de Janeiro: Vieira & Lent.

Silva, H. P., Rodrigues-Carvalho, C., Souza, S. M., & Santos, R. V. (2001). E, então apareceu Luzia. Com 11 mil anos de idade, a primeira brasileira questiona o roteiro do povoamento das Américas. *Rio Artes, 30*, 6–8.

Simão, L. M. (2009). Elos do patrimônio: Luiz de Castro Faria e a preservação dos monumentos arqueológicos no Brasil. *Boletim do Museu Paraense Emílio Goeldi, Ciências Humanas, 4*, 421–435.

Souza, V. S. (2012). Retratos da nação: os 'tipos antropológicos' do Brasil nos estudos de Edgard Roquette-Pinto, 1910-1920. *Boletim do Museu Paraense Emílio Goeldi, Ciências Humanas, 7*, 645–669.

Walter HV (1958). Arqueologia da Região de Lagoa Santa, Minas Gerais (Índios Pré-Colombianos dos Abrigos-Rochedos). Rio de Janeiro, Sedegra. 227pp.; 61 il.

Weber, R. L. (1986). Current research: The Amazon, Eastern Brazil, and the Orinoco. *American Antiquity, 51*(1), 169–171.

Chapter 9
The Origins Project and the First Americans' Controversy

Danilo Vicensotto Bernardo, Walter A. Neves, and Renato Kipnis

Abstract The arrival of Man in the New World has long been occupied a prominent position in studies of the fields of archaeology and related sciences, such as biology and biological anthropology. The debates on the subject were intensified, however, since the discovery of the oldest cultural manifestations of the Americas, the Clovis points, in the late 1920s. In South America, the study of the human occupation of the Lagoa Santa region has generated controversy since the early works of Lund in the nineteenth century. Recently, the project "Origins" deepened the archaeological research in Lagoa Santa, focusing its actions in thematic axes that resulted in extensive scientific production about the origins of the first Native Americans.

Introduction: The Origins of the Controversy

For a long time now, in the world's major scientific communities, the question of Man's appearance in the New World has occupied an outstanding position in anthropological studies and in those of related sciences like paleontology and biological anthropology (Silva and Rodrigues-Carvalho 2006).

Generally speaking, however, up until the middle of the seventeenth century, all the thinking about the origins of Man in the Americas was of a strictly speculative nature and almost always associated to religious, mythical, or folkloric

D.V. Bernardo (✉)
Laboratory for Studies of Biological Anthropology, Bioarchaeology and Human Evolution, Human Sciences and Information Institute, Federal University of Rio Grande, Av. Itália, Km 8, Carreiros, Rio Grande, 96201-900 RS, Brazil
e-mail: danilobernardo@furg.br

W.A. Neves
Laboratory for Human Evolutionary and Ecological Studies, Department of Evolutionary Biology and Genetics in the Biosciences Institute, University of São Paulo, R. do Matão 277, sala 218, São Paulo, SP, Brazil
e-mail: waneves@ib.usp.br

R. Kipnis
Scientia Scientific Consulting, São Paulo, SP, Brazil
e-mail: rkipnis@scientiaconsultoria.com.br

P. Da-Gloria et al. (eds.), *Archaeological and Paleontological Research in Lagoa Santa*, DOI 10.1007/978-3-319-57466-0_9

explanations interpreted on the basis of sixteenth- and seventeenth-century chroniclers' entirely ethnocentric visions. The first discussions of the settlement of the American continent from an archaeological point of view, at the beginning of the twentieth century, were considerably biased toward the anthropological vision of the day which, in the light of the entirely modern features of the first Americans, considered that antiquity of Man's arrival in the New World could not surpass the mark of 4000 BC (Arensburg and Nès El 1995; Bernardo 2007; Lima 2006; Prous 2006).

In the mid-1920s, the discovery of projectile points associated with the remains of extinct Pleistocene animal species at an archaeological site near Folsom in New Mexico (USA) led to the first questioning of that scenario. Indeed, the antiquity of the human presence on the American continent was established in 1927 with discovery of the Folsom fluted points associated with the skeletal remains of an extinct bison species of the late Pleistocene era. Those archaeological items delineated two different but similar technological (or cultural) traditions, the Folsom tradition and the Clovis tradition, both representative of the Paleoindian culture (Fagan 1987, 1999; Figgins 1927; Kilby 2011; Meltzer 2011; Meltzer et al. 1997; Pedler and Adovasio 2011; Sellet 2011).

The advent of absolute dating techniques (i.e., radiocarbon dating) in the 1950s revealed that the oldest archaeological stratum of the Clovis culture horizon dated back to 11,400 BP. Examples of its lithic industry began to be found at sites from coast to coast in all of North America and in latitudes ranging from Canada to Central America (Bradley and Stanford 2004), confirming the Clovis-First model, that is, that Clovis Paleoindian culture was the cultural framework of the American continent's first colonizers (Gruhn and Bryan 2011; Waters and Stafford 2007).

Nevertheless, despite the expressive domination of that paradigm in the debates on the archaeology and the origins of Man on the Americas, some alternative hypotheses suggesting differences in chronologies and routes of entry and dispersion, and differences in the archaeological and cultural contexts, gradually became part of the discussions (for a synthesis, see Lima 2006).

Some of those new hypotheses concerning the settlement of the Americas have only recently taken their place in the debate. Tom Dillehay's emblematic work at the Monte Verde site in southern Chile, indicating dates of 11,500–12,700 BP, the research undertaken at Pedra Pintada site in the Amazon, and the work carried out over four decades at the Boqueirão da Pedra Furada site in the Serra da Capivara National Park, Piauí, Brazil, are all examples of recent research that are indicative of a pre-Clovis occupation in the New World (Boëda et al. 2014; Dillehay 1997, 1999, 2000, 2003; Dillehay et al. 1992; Dixon 2001a, b; Guidon and Delibrias 1986; Roosevelt et al. 2002; Prous 2006). Other hypotheses, however, have had their ups and downs at the center of the debate on the origins of American Man for over a century. An example of this last situation is the archaeology of the Lagoa Santa region in the state of Minas Gerais, Brazil (Mendonça de Souza et al. 2006; Prous 1991).

Lagoa Santa in Perspective

The Lagoa Santa region, in the state of Minas Gerais, Brazil, is situated in a tropical, karstic area found in the upper pre-Cambrian geological formation of the Bambuí group, represented by limestone of the Sete Lagoas formation, and in it there are many rockshelters and caves (known locally as "Lapas") containing extensive evidence of paleontological and archaeological material (Mendonça de Souza et al. 2006).

Danish naturalist Peter Wilhelm Lund (1801–1880) settled in the region and was the first to carry out systematic investigations there (see Chaps. 2 and 3 of this volume). He dedicated himself to exploring fossils and visited many caves, discovering and describing a large number of Pleistocene mammal taxa (Kelner and Campos 1999). His finds were made widely known in the European community because of his reports and his correspondences with exponents of paleontology like Charles Darwin (1809–1892) and because of the large shipments of collections he sent to Europe (Faria 2008; Luna Filho 2007), describing Lagoa Santa as a rich source of fossils of the Pleistocene/Holocene transition period.

Apart from the fossils of paleontological interest, however, there were other outstanding vestiges among Lund's finds, including 30 human skeletons discovered during his excavations of the Sumidouro cave (Neves and Piló 2008) in 1842 and 1843.

Given the great antiquity attributed to them by Lund because of the association of the human bones with the fossils of extinct megafauna species, the collection of specimens from the Sumidouro cave attracted great attention from those interested in investigating the main features of the first humans in South America, leading to many different interpretations (Hansen 1888; Hrdlička 1912; Kollmann 1884; Lacerda and Peixoto 1876; Quatrefages 1887; Ten Kate 1885; Virchow 1892; see Chap. 10 of this volume). From then on, Lagoa Santa became Brazil's most outstanding archaeological area (Morais 1999), and without doubt the origin and antiquity of the human specimens from Lagoa Santa were among the aspects that were later considered to be the most polemical and commented questions associated with Lund's work.

Such great visibility stimulated the organization of many expeditions and research activities in the area and generated a large collection of human skeletons and archaeological artifacts recovered from the region. The archaeological data generated by those regional interventions, allied to the abundance of fossils recovered by Lund and later by Harold Walter (1897–1976), were a strong justification for launching an extensive, systematic, scientific effort, combining archaeology and paleontology (Evans 1950).

However, many of those interventions were carried out by amateur archaeologists during the prescientific stage of Brazilian archaeology (Prous 1991, 1997), resulting in excavations conducted in the 1940s and 1950s without any strict stratigraphic control. Even so, they produced an enormous quantity of material, including dozens of human skeletons (Walter 1958), and that material served as the basis for the archaeological research in the region up until the 1970s (Mendonça de Souza et al. 2006; Prous 1991).

Against that background, two extensive scientific expeditions undertaken in the Lagoa Santa region were outstanding (Meggers 1985; see Chap. 7 of this volume). The first, in the mid-1950s, was conducted by Wesley Hurt (1917–1997) and Oldemar Blasi (1920–2013), and it produced an abundance of human skeletal material (Hurt and Blasi 1969). The second was the famous Franco-Brazilian mission which took place in the 1970s, involving French and Brazilian researchers. It was organized and headed by Annette Laming-Emperaire (1917–1977). Perhaps the most notable product of the latter mission was a human skeleton recovered from the Lapa Vermelha IV cave (Laming-Emperaire 1979), later nicknamed Luzia by one of us (WAN) and dated from 11,200 to 11,680 BP (Feathers et al. 2010; Fontugne 2013; Prous and Fogaça 1999). It constituted the oldest Brazilian human remains, converging with Lund's original expectation that Man was present in the Lagoa Santa region at the time of Pleistocene/Holocene transition (Bryan 1973; Hurt and Blasi 1969).

Nevertheless, the geomorphological complexity of the Karst and the chronological and spatial discontinuity of the occupation of the region seem to constantly renew the discussion on the archaeology of Lagoa Santa (Araujo et al. 2005a, 2012a; Auler et al. 2006; Piló et al. 2005; Prous and Fogaça 1999). While the richness of the remains recovered in Lagoa Santa has been unquestionable, ever since Lund's time, the scarcity of reliable and systematic dates for them and the lack of an accurate and spatially wide stratigraphic contextualization together with the gaps in the spatial distribution of those finds have meant that, for over a century, no overarching hypotheses about the region's prehistory could be formulated and correctly tested (Mendonça de Souza et al. 2006).

Among the many discussions about the prehistory of Lagoa Santa, those addressing the biological origins of the first Americans are certainly among the most polemical (Mendonça de Souza et al. 2006). Against that background, classical physical anthropology research (see Chap. 11 of this volume) connected with molecular evidence of the genetic origin of the first settlers of Americas, such as Hrdlička's proposal of the existence of an American holotype (Hrdlička 1912, 1932) has reinforced the ideas of homogeneity, a common geographic origin, and the emergence of diversity among native Americans only after the first migrations (Armelagos et al. 1982; Cann 1994).

Outstanding in those discussions is the model proposed by Neves and collaborators (see Chaps. 10 and 11 of this volume) first disseminated in the late 1980s (Neves and Pucciarelli 1989). Based on the biological associations of the craniums of the first Americans and those of populations in other parts of the world, those authors proposed that the New World was settled by two distinct population stocks. That was in flagrant disagreement with the three-migration model predominant at the time (Greenberg et al. 1986) (for other models see, e.g., González-José et al. 2008; Powell 2005; Seguchi et al. 2010).

Neves and his collaborators' model, later to be known as the two main biological components model (TMBCM) (Neves et al. 2007c), placed the Lagoa Santa archaeology once more at the center of the debates on the first Americans (Bernardo 2007). Based on the morphological diachrony observed in South American craniums (Hubbe and Strauss 2012a; Neves and Hubbe 2005a), Neves and Pucciarelli (1989, 1990, 1991) suggested that the Americas had been colonized at two separate "moments": first, by a pre-Mongoloid population that migrated earlier than those whom they believed had arrived on the Americas at a second, later "moment", and had originated the present-day indigenous populations.

According to Neves and his collaborators, the first Americans had accentuated morphological affinities with those that arrived in the South Pacific at the end of the Pleistocene period (Neves and Hubbe 2005a). They also showed morphological similarities with the Zhoukoudian fossils found in China (Neves and Pucciarelli 1998) and with more recent Australian-Melanesian and African populations (Neves et al. 1999a, b; Neves and Hubbe 2005a), thereby suggesting that both migratory waves, the first Americans and the ancestors of present-day Americans, used Beringia as their gateway to the New World (Neves and Pucciarelli 1991). Other analyses involving different samples from North America (González-José et al. 2003, 2005, 2006; Powell and Steele 1992; Steele and Powell 1992, 1993) and South America (Munford 1999; Munford et al. 1995; Neves et al. 2003, 2007a) led, independently, to the same results originally obtained by Neves and his collaborators (Bernardo 2007).

Although, at the outset of their proposition, Neves, Pucciarelli, and their collaborators' morphological observations on the skeletal population of Lagoa Santa, their biology, and the implications of those finds for the question of Man's origin on the Americas had been at least partially corroborated by results that other researchers had obtained (see examples in Bernardo 2007; González-José et al. 2005, 2006; Munford et al. 1995; Powell and Steele 1992; Steele and Powell 1992, 1993), a series of criticisms regarding the nature of the specimens, including criticism of their chronologies, their archaeological contextualization (or lack of it), and the methodologies employed in the analyses became increasingly frequent in more conservative circles of researchers dedicated to studying the origins of Man in the Americas (for a few examples, see Brace et al. 2001; Dillehay 2000; Seguchi et al. 2005, 2010; Van Vark et al. 2003). Thus, to test the TMBCM and its associated ideas, it became essential to search for new data on the biology of the first Americans, that is, to say, to increase the number and the quality of the skeletal series from the Lagoa Santa region. Furthermore, they needed to be accompanied by reliable dating, systematic, correct, and broad archaeological contextualization. Those requirements could only be met by deepening the investigations in the archaeological region of Lagoa Santa.

New Events in Lagoa Santa: The Origins Project

With the aim of taking up, once more, archaeological studies in a region with notorious potential for research into the question of the appearance of the first inhabitants of the New World, in the years from 2000 to 2009, under the general coordination of one of us (WAN) and attached to the Laboratory of Ecological and Evolutionary

Human Studies (LEEHS) at the University of São Paulo's Biosciences Institute, the project *Origins and Microevolution of Man in America: a Paleoanthropological approach* was developed and became known more simply as "Origins" project.

It was carried out with the expressive support and stimulus of the São Paulo Research Foundation (FAPESP). Basically, it endeavored to overcome some of the deficiencies historically present in the research undertaken in the Lagoa Santa region, such as in aspects of the archaeological and chronological characterization. In that way, it was intended to make a more profound investigation of the proposed settlement model based on the biological affinities presented by the pristine populations of the region.

The first version of the project, carried out in the years 2000–2004 (Neves et al. 2004b), was delineated and executed on the basis of an exploratory approach with the initial expectations of:

Increasing the sample of skeletal remains of the Lagoa Santa region with a chronology reaching very far back in time
Establishing a systematic program for analyzing the osteological material exhumed from there and already institutionalized
Refining the region's chronology by generating intensive dating of different contexts and regions
Visiting sites and rockshelters already excavated to enrich the archaeological contextualization of the collections produced from them
Surveying and excavating areas of interest for the investigation of the occurrence of occupations in the Pleistocene/Holocene transition period
Enriching the stock of knowledge concerning settlement patterns, resilience, and technology of the archaeological region's most ancient populations

The second version of the "Origins" project was based on a series of embryos generated by the first version. To achieve the depth of the research begun and indicated in the preceding versions and meet the expectations initially generated, the new version of the project adopted a structure in accordance with the activities of the four fields of knowledge that supported the main approaches used to achieve the research objectives, namely, archaeology, biological anthropology, paleontology, and paleobotany. Obviously that division did not imply that there was an innate separation of the areas of research; in fact, they invariably presented interfaces of investigation. Those interfaces led to six well-defined lines along which to build the research problem in the second stage of the project. They were as follows:

Clovis First/Clovis-like
Biological origins of the first Americans
Resilience of Neotropical human foragers: subsistence, material culture, mobility, and social change
Site formation processes and taphonomy in tropical areas
Paleoclimates and paleoenvironments at the end of the Pleistocene period and during the Holocene in the Lagoa Santa region
Man and the megafauna in the Pleistocene/Holocene transition period

Along those lines the Origins project carried out 5 years of investigations of general and specific hypotheses concerning the archaeology of Lagoa Santa.

The aim of the first thematic line, Clovis First/Clovis-like, was to test the hypothesis that the first human settlements of the Americas are not in fact represented by the Clovis culture and that the pioneers actually arrived in the New World well before the appearance of that particular culture. That was to be achieved mainly based on the characterization of the chronological horizons associated with the earliest human occupations in Lagoa Santa by expanding the number of dates for the region using samples of material generated by the project, not only from the rockshelters but also from open-air sites that were surveyed or excavated or both. Although only six sites were excavated (three open-air sites and three caves) during the Origins project, the dates obtained made it very clear that humans began to settle the region somewhere around 10.4 thousand years BP and that by 9.7 thousand BP several rockshelters had already been occupied. Thus, in the sphere of the research undertaken by the Origins project, it was not possible to demonstrate that there had been an occupation either before or even contemporary with the Clovis.

The second thematic axis of research, namely, "The Biological Origins of the First Americans," was intended to achieve the international consolidation of the hypothesis formulated by Neves and his collaborators that the first inhabitants of South America had a similar cranial morphology to that observed in present-day Australian-Melanesian groups and therefore different from that observed in Native Americans and their direct Asian ancestors. That would be accomplished by three main planned actions:

(a) Increasing the earlier skeletal collection of Lagoa Santa by curating the institutionalized remains in addition to those coming from the new archaeological interventions
(b) Refining the chronology of the human bone remains of the Lagoa Santa region by expanding the number of dates of skeletons exhumed under the scope of the project, in addition to those of already institutionalized remains
(c) Analyzing cranial variation by means of morphological comparisons of skeletons from the archaeological region of Lagoa Santa with those of other autochthonous populations in the world

Along this reasoning, the results obtained by each one of the planned actions can be considered expressive. The number of measurable ancient specimens was greatly increased by the various curating activities undertaken with the institutionalized collections and by the exhumation of new specimens in the excavations conducted under the scope of the project. All those activities now enable us to state that there are now around 100 known craniums from different sites in the region, which sets Lagoa Santa in an outstanding position in the world in terms of its skeletal collection.

Given the dire need to construct a solid chronology for the Lagoa Santa human skeletons, 143 samples of human remains (including bones, teeth, and other organic material) were sent overseas to be dated in the course of the "Origins" project. In only 41 of those samples (28.7%), however, was there any collagen present that enabled dating to be done. Attachment 9.1 displays a table with the synthetic data of the 624 samples that were submitted to dating attempts under the scope of the Origins project, and attachments 9.2 and 9.3 present the interpretation of the data in graph form. The dates that were successful leave no doubt that the ages of the specimens exhumed in Lagoa Santa are concentrated in the range of 8.7–7.5 thousand years BP. The date for the oldest specimen coincides with the time when the local population adopted the behavior of burying their dead in rockshelters that were already being occupied at least as far back as 10.2 thousand years BP.

Again, the comparisons made between the cranial morphology of the Lagoa Santa skeletons and the morphologies of other populations in the world absolutely confirmed the results obtained ever since the late 1980s by the project coordinator (WAN) with help of various collaborators, even though they worked with far from perfect samples especially in chronological terms. The first Americans, do indeed, present a cranial morphology far more similar to that of the Australian-Melanesians than to that of the Asian Mongoloids and the present-day Native Americans. Neves and Hubbe (2005a) came to those conclusions based on a comparative analysis of the morphologies of a very large skeletal series, of great age, from the Lagoa Santa region and other series from different parts of the world classically used in craniometric studies (Howells 1973, 1989, 1999).

The same results were obtained when the cranial morphology of the consolidated Lagoa Santa series was compared with another database (Hanihara 1993a, b, 1996, 2000) far more expressive numerically than the one traditionally used (Bernardo 2007). Thus, this aspect is no longer questioned by the experts in this field, even though there are some differences among the models proposed to explain the differences and similarities. Unfortunately, considering the fragility of the Paleoamerican samples from other regions of South America, those models are unlikely to be tested in the immediate future because such testing requires numerically significant temporal series

The third thematic line, "Resilience of Neotropical Human Foragers: Subsistence, Material Culture, Mobility and Social Change," achieved enormous progress during the 9 years of project activity, especially regarding the information stemming from the zooarchaeology and the analyses of the lithic industry. Nevertheless, one goal was not achieved, namely, a modeling that would make it possible to investigate different resilience patterns displayed by the region's occupants. That was because it could only be achieved in the long term when all types of remains from the excavated sites have been completely processed. Even so, preliminary analyses together with the curating and cataloging stages and sporadic investigations have made it feasible to discuss some aspects of the information.

Regarding subsistence, the analysis of animal remains shows quite clearly that only small- and medium-sized animals were part of the diet of Lagoa Santa's ancient hunter-gatherers, with a special emphasis on small rodents, deer, and peccaries (Kipnis et al. 2010a; Kipnis and Da-Gloria 2013; Neves et al. 2004b, 2008; Perez 2009). Notably, there is a complete absence of any bones of the megafauna in the archaeological strata.

Another tendency that became clear was that the numbers of identifiable fauna remains dwindled from the up bottom. Thus, it might have been possible that groups that lived there chose, or were forced, to adopt a predominantly vegetarian diet to the detriment of the higher protein animal diet. However, analyses of part of the data suggest that the phenomenon might be better explained by taphonomic considerations and actually have nothing to do with the diet of the populations that lived there (Kipnis et al. 2010a; Neves et al. 2004b, 2008; Perez 2009). Even so, the phenomenon remains to be systematically investigated. Gaining a proper understanding of it will contribute toward explaining why the human skeletons from Lagoa Santos exhibit such a high level of caries. Unfortunately, it is impossible to make comparisons of caries incidences at very different moments in time because practically all the skeletal material from the region is dated as being from 8.7 to 7.5 thousand years BP.

Observation of the characteristics of the lithic industry at the Lapa das Boleiras, Lapa do Santo, and Lapa Grande de Taquaraçu sites suggests there were important concomitant changes in the lithic industry in the course of the millennia (Neves et al. 2004b, 2008; Pugliese 2008). One of them was in the use of raw materials, insofar as flint, which has excellent physical properties for knapping purposes, only occurs in the lower levels of those sites whose vestiges have already been analyzed (Neves et al. 2004b, 2008).

From the dates that have been obtained, the change must have taken place around 9.0 thousand BP. That strongly suggests that, during the first 1500 years of human occupation in the region, greater emphasis must have been placed on making and using flakes with worked edges and on the search for flint. What can be seen is the maintenance of the same knapping technology and the preference for quartz, especially hyaline quartz throughout the period of occupation, suggesting a strong temporal persistence (Araujo 2004a; Araujo et al. 2005a, 2012a; Neves et al. 2004b, 2008).

It must be stressed that artifacts with formal features are not entirely nonexistent in the region. At the open-air site of Coqueirinho, a fragmented flint point was found that had been carefully reworked. In the Lapa Grande de Taquaraçu rockshelter, a plano-convex artifact was discovered similar to those associated with what is known as the "Itaparica tradition" (Neves et al. 2008; Araujo et al. 2012a).

The fourth line, "Site Formation and Taphonomy in Tropical Areas," was intended to describe the sedimentary deposition process and the formation of the archaeological and paleontological matrixes of the sites that were worked on, to test the hypothesis that the sediments found in the rockshelters are mainly of anthropic origin and that anthropic and non-anthropic disturbances affected the special disposition of artifacts and natural features in the rockshelters in such a way that they are no longer associated with the floor they originally occupied (see Chap. 17 of this volume).

Two main actions were designed for that purpose: (a) spatial analysis of the archaeological remains found in the excavations conducted under the scope of the project and (b) expansion of the archaeological excavations with a focus on the exposure of extensive surfaces to gain a better understanding of the spatial organization (Neves et al. 2004b).

Up to now, it has not been possible to test that second hypothesis systematically because detecting occupation floors inside and outside of the rockshelters involves a systematic program of refitting of lithic artifacts discovered and a detailed analysis of the structures uncovered and documented, as well as establishing the relations among them. In all archaeological research of this kind, this last analysis takes the longest, and sometimes the results are only published years after the excavations have finished (Neves et al. 2008).

As for the first hypothesis that most of the sedimentary matrix found in the Lagoa Santa caves is of anthropic origin, the project achieved far better results than could have been expected. From the outset of the Origins project, we were always surprised to find that the superficial layers of the rockshelters that were excavated or tested tended to be around 7.5 thousand years old (Neves et al. 2008).

Several hypotheses were proposed to explain that phenomenon. The most encouraging one tried to establish a direct relation between the rate of sedimentation and the climate. According to this hypothesis, something happened in the local climate around 7.5 thousand BP which was unfavorable for carrying geogenic sediments into the rockshelters and depositing and accumulating them there. Another hypothesis entertained by the Origins project team was that the phenomenon could be due to the exhaustion of the geogenic sediments that once existed on top of the limestone massifs so that there was no longer any source of the abundant sediments that had formerly ended up in the caves.

The fact that humans apparently left the Karst somewhere between 8.0 and 7.5 thousand BP is a strong argument in favor of the first hypothesis. The climate event that made the humans move away from the region may have been the same one that was unfavorable for the movement of geogenic sediments into the caves.

The second hypothesis, in our view, is more unlikely because it is practically impossible that all the limestone massifs had exactly the same thickness of loose sediments overlying them, and furthermore it was equally unlikely that they had suffered exactly the same rate of erosion for thousands of years.

One stratigraphic aspect that constantly caught our attention was the grayish color of the sediments in the rockshelters which were always very rich in charcoal. The first attempt to explain the color and the abundant presence of charcoal sought an answer in the regional geology and geomorphology marked by the strong presence of limestone, which is gray. According to the model proposed, the thick strata of gray sediments were derived from the times when the sedimentation originating from the rockshelter ceilings and walls predominated. That sediment would have become mixed with the charcoal remains left by occasional anthropic fires. The thinner brown or orange colored layers, on the other hand, would be related to moments when the movement of geogenic sediments derived from the tops and walls of the limestone massif predominated. That kind of scenario emerged most markedly from the excavations at the Lapa das Boleiras rockshelter, the first to be intensively investigated by the project.

When other excavations and test pits began to be carried out in Lagoa Santa, two pieces of information regarding this aspect came to light: first, that in some rock-shelters the accumulation of gray sediment was uninterrupted, even though it was very thick, and second, that several of the sites showed no evidence of any lateral injection cones of sediment (Araujo 2005; Araujo et al. 2008, 2010a). A good example of that is the Lapa do Santo rockshelter which has a sedimentary deposit more than four meters deep, rarely interrupted by any sediments that are undeniably geogenic (Neves et al. 2004b, 2008).

In view of all that information, by the end of the first stage of the excavations, the anthropic hypothesis had emerged as the most likely alternative for explaining most, if not all, sediment accumulated in the rockshelters. Accordingly, the project invested heavily in two strategies, namely, studies of the micromorphology of the sediments accumulated in the rockshelters and determining their detailed chemical composition.

The results of both strategies pointed to the same conclusion, namely, that almost all the gray strata rich in charcoal from the Lagoa Santa caves were produced by the daily activities of those human beings that settled there at the beginning of the Holocene period, noted for making big fires whose ashes became scattered, intentionally or not, all over the floor of the rockshelters (Araujo et al. 2008). At first sight, those results might seem trivial given that everybody knows that the matrix of an archaeological site is always made up of geogenic sediments mixed with organic and inorganic remains introduced by Man. Nevertheless, our results revealed an extreme situation, insofar as human activity on its own, with little or no contribution of geogenic sediments, can account for the accumulation of extremely thick sedimentary matrixes, as is the case with Lapa do Santo (Neves et al. 2004b, 2008). There, the archaeological deposit is as deep as 4 m, corresponding to 2000 years of occupation, even though it is formed mainly of the remains of fires, macrobotanical remains, fragments of bones of small- and medium-sized animals, remains of human burials, and debris left by stone knapping activities, with the first two being by far the most abundant (Neves et al. 2008).

The study entitled "Paleoclimates and Paleoenvironments at the End of the Pleistocene and in the Holocene in the Lagoa Santa Region" was the fifth thematic axis of the research project. It sets out to gather evidence that might eventually explain the two great hiatuses in human occupation of the region. One was at the end of the Pleistocene era and the other in the Middle Holocene. The actions conceived to achieve that objective were: (a) investigation of deposits at the bottom of local lakes; (b) identification, selection, and dating of samples of stalagmitic crusts; and (c) description of regional floristics and phytosociology from 12.0 thousand BP to the present day.

The first activity involved the elaboration of a lacustrine sedimentary column with dating for each 1000-year period. Unfortunately, only two lakes in the region proved to be fertile in paleopalynological terms. Accordingly, a lot of effort was dedicated to the study of the macrobotanical remains and fossil pollen in a column with a high organic content discovered on the banks of one of the water courses that delimit the Lagoa Santa Karst, known as Ribeirão da Mata (Freire 2011; Nakamura 2011).

The project adopted another strategy designed to obtain a reliable paleoenvironmental reconstitution for Central Brazil. It consisted of digging trenches in the bed of the two main lakes in the region, the Cerca Grande and Sumidouro lakes. Two highly organic strata were found: one dated at 9.7 thousand BP and the other at around 2.0 thousand BP (Neves et al. 2008). Both were indicative of periods of a more humid climate that led to the formation of thick vegetation around the lakes and also to the transport of organic material to the beds of the lakes.

In short, it can be said that the information obtained from the lake trenches in the Lagoa Santa region wholly support the proposed model for human settlement in the region, that is, two peaks of high humidity, one around the beginning of the Holocene era and the other in the late Holocene separated by a long dry period from 7.5 thousand BP to 2.0 thousand BP (Araujo et al. 2005a, b; Neves et al. 2004b, 2008). The analyses of the pollen from the bottom of the lakes did not exactly indicate a drier climate for the same period but, instead, a climate marked by great instability (Neves et al. 2008; Rackza et al. 2013).

Another line of paleoenvironmental investigation within the scope of the Origins project was the analysis of oxygen isotopes in the stalagmites. This technique is considered the best indicator of ancient climate conditions, but it presents two considerable problems. The first is to find stalagmites that were formed during the period being studied, and the second is dating at least every 0.5 cm of them by uranium/thorium dating (Cruz et al. 2005). Both activities are extremely costly because neither the dating nor the isotope analyses are done in Brazil. Unfortunately, up until now, only one stalagmite covering the last 7000 years in the Lagoa Santa region has been identified, making it very difficult to discuss the question based on this isolated result. New efforts are expected to clarify the environmental conditions in the region during the period targeted by the studies.

The results of the analysis of the botanical macroremains found on the banks of the Ribeirão da Mata water course suggest that the ancient vegetation in the basin of that small river was not much different from that of today, namely, semi-deciduous forest and savannah (Freire 2011; Nakamura 2011; Neves et al. 2008). Different interpretations, however, favor the idea of a gradual change in the vegetation cover as time went by and, therefore, a different history for the evolution of the paleoenvironment in the macroregion (De Oliveira 1992; De Oliveira et al. 1999). Given the scenario delineated above, the results of the isotope analyses are awaited with great anxiety as they will stimulate the discussion as to whether dry climate or instable climate conditions prevailed in Lagoa Santa during the mid-Holocene period. In any event, either one of them would have been limiting factors for permanent human settlement in the Karst and, perhaps, in most of central Brazil.

It must be underscored that, to support the Project, an exhaustive survey of plant species in the region was carried out to provide information for the identification of pollen and botanical macroremains obtained from the archaeological sites and from the banks of the Ribeirão da Mata.

The sixth thematic axis "Man and the Megafauna in the Pleistocene/Holocene Transition" was intended to provide a chronological contextualization of the megafauna that formerly inhabited the Lagoa Santa region and test the hypothesis that there had been a period when the big mammals and the first human groups in the region were contemporaries. To that end, three actions were proposed: (a) selecting new samples for ^{14}C-AMS dating of the Lagoa Santa megafauna remains to broaden knowledge of the chronological span of their existence, (b) conducting complementary stable isotope analyses of the human and megafauna bone remains in an attempt to improve the reliability of the dates, and (c) carrying out a paleontological excavation at the Gruta Cuvieri using archaeological techniques (Neves et al. 2004b).

The hypothesis of the coexistence of Man and the megafauna in Lagoa Santa goes back to the time of Lund and the scientific missions that later investigated the area. Incredible as it may seem, even when ^{14}C dating became available in the mid-1950s, until the advent of the Origins project, not one single piece of Lagoa Santa megafauna fragment had ever been sent abroad to be dated, or, if one had, it had not been reported in the literature (Neves and Piló 2003; Neves et al. 2004b).

Basically, all the missions that preceded ours had tried to find the answer through stratigraphic associations, in this case between megafauna bone remains and anthropic material, including human skeletons. With the exception of the Lapa Vermelha IV rockshelter, where the remains of a giant sloth had been found associated with a stratum dated at around 9.5 thousand AP (some meters above a stratum that contained the human skeleton that was later to be nicknamed "Luzia"), but which could have been reworked material or even brought in from the surrounding landscape where it had lain for thousands of years, no direct evidence of the hypothetical association was ever found in any of Lagoa Santa's archaeological sites and not even by the project we are reporting on here (Hubbe et al. 2013a).

In view of that statement, the Origins project dedicated its efforts to a systematic program of dating, mainly but not exclusively, of the megafauna bone remains found in the targeted region. The project obtained the Lagoa Santa samples from museums and similar institutions in Brazil and abroad and from our own excavations at the Gruta Cuvieri.

Of the 11 dates obtained for Lagoa Santa material, three (10.8 thousand BP, 9.9 thousand BP, and 9.2 mil thousand AP) give unequivocal support to Lund's theory that Man and the megafauna did in fact coexist there, provided the antiquity of Luzia is accepted as being the mark of Man's presence in the region or near to it, however sporadic that presence may have been. Even if we only accept the oldest dates for human settlement generated by the project, that is, 10.4 thousand BP, two of the three dates set out above are contemporary with human presence. Furthermore, the date of the megafauna in the Ribeira Valley in São Paulo state (10.8 thousand BP) also indicates the contemporaneity of Man and the megamammals to the south of Lagoa Santa (Hubbe et al. 2013a; Piló and Neves 2003).

Given that in the three rockshelters that the project excavated (Lapa das Boleiras, Lapa do Santo, and Lapa Grande de Taquaraçu), thousands of fragments of the fauna that were consumed were found, then the complete absence of any trace of megamammal material among those remains raises a burning question: if the megafauna species were still present in the Lagoa Santa region when the first humans established their presence there and if they coexisted for about 2000 years, why did they not exploit the resource represented by those fauna species now extinct?

Academic Production and Final Remarks

To sum up, the 9 years of intense archaeological, bioanthropological, paleontological, geological, geomorphological, and paleoenvironmental research carried out by the Origins project into Lagoa Santa generated results that had impacts on various areas of knowledge and various aspects of Lagoa Santa's prehistory and, above all, on the knowledge concerning the origins and adaptations of the first Americans. The academic-scientific production that emerged from this multidisciplinary venture (in the years from 2000 to 2013) is set out, separated by year and by area of research in Table 9.1. Figures 9.1, 9.2, and 9.3 represent, respectively, the total of the academic-scientific products generated by years, by areas of knowledge, and by type of production for the same period embraced by Table 9.1.

It must be stressed that, as far as we are aware, no other archaeological or paleontological project headed by Brazilians in this country has given rise to so many publications in important high-impact periodicals or to so many presentations in national and international congresses as the one we have been describing here, and that fact has endowed Brazilian science with visibility, however modest, in the eyes of American and European science. Figures 9.4 and 9.5 represent the proportions of publications in national and international periodicals and the proportions of presentations at national and international conferences, respectively, for the period 2000–2013.

It so happens that the Origins project also produced results with equally strong impacts in other spheres of academic-scientific research such as the training of human resources, constitution of collections, environmental protection, heritage education, and scientific dissemination, most of them in the local and national spheres. Figures 9.6, 9.7, and 9.8 provide a graphic illustration of that information. They display, respectively, the proportions of national to international articles of scientific dissemination, chapters of books, and academic products such as dissertations and theses produced in the period covered by Table 9.1.

Finally, it is worth highlighting that the Origins project achieved another objective specifically related to the training of human resources. Today, projects being undertaken by personnel qualified during the period of execution of the Origins project are continuing to enrich the questions surrounding the study of the origins and adaptability of New World Man. Examples are those coordinated by André Strauss, who took up the excavations at the Lapa do Santo cave to continue researching biological origins and funerary practices of the first Americans, and Pedro da Glória, investigating the health and lifestyles of Lagoa Santa Paleoamericans by adopting an ethnobioarchaeological approach.

Ever since the nineteenth century, the archaeological research undertaken has shown the importance of the Lagoa Santa region, and the results obtained in the last quarter of the twentieth century, which put the supremacy of the Clovis-First model in check, recuperated its importance for the study of New World Man's origins. Thus, the historical context in which Origins project was unfolded seems

Table 9.1 Academic-scientific production generated by the Origins project up to December 2013, classified by year and by research field

Year	Archaeology	Biological anthropology	Paleontology	Paleobotany
2000	(0)	(2) Blum et al. (2000); Neves and Blum (2000)	(0)	(0)
2001	(0)	(4) Blum et al. (2001); Neves and Blum (2001); Neves et al. (2001a); Neves et al. (2001b)	(0)	(0)
2002	(4) Araujo (2002); Araujo et al. (2002); Piló (2002); Piló and Neves (2002)	(3) Blum and Neves (2002); Neves and Blum (2002); Neves et al. (2002)	(0)	(0)
2003	(6) Araujo (2003); Araujo and Marcelino (2003); Araujo et al. (2003a); Araujo et al. (2003b); Neves and Piló (2003); Piló and Neves (2003)	(4) Hubbe et al. (2003a); Hubbe M et al. (2003b); Neves and Hubbe (2003); Neves et al. (2003a)	(2) Auler et al. (2003); Kipnis (2003)	(0)
2004	(5) Araujo (2004a); Araujo (2004b); Araujo et al. (2004); Piló et al. (2004a); Piló et al. (2004b)	(6) Hubbe et al. (2004a); Hubbe et al. (2004b); Neves et al. (2004a); Neves and Kipnis (2003); Neves and Atuí (2004); Neves and Hubbe (2004)	(0)	(2) Araujo (2004c); Silva and Kipnis (2004)
2005	(5) Araujo (2005); Araujo et al. (2005a); Araujo et al. (2005b); Neves and Piló (2005); Piló et al. (2005)	(7) Atuí (2005); González-José et al. (2005); Neves and Hubbe (2005a); Neves and Hubbe (2005b); Neves and Hubbe (2005c); Neves et al. (2005a); Neves et al. (2005b)	(1) Camolez et al. (2005)	(1) Kipnis and Silva (2005)
2006	(1) Neves and Arujo (2006)	(4) González-José et al. (2006); Neves (2006); Neves et al. (2006); Pucciarelli et al. (2006)	(1) Auler et al. (2006)	(0)
2007	(6) Araujo (2007a); Araujo (2007b); Araujo and Neves (2007); Arroyo-Kalin et al. (2007); Jangelme et al. (2007); Neves et al. (2007d)	(4) Bernardo (2007); Neves et al. (2007a); Neves et al. (2007b); Neves et al. (2007c)	(7) Bissaro and Kipnis (2007); Hubbe et al. (2007a); Hubbe et al. (2007b); Luna Filho (2007); Neves et al. (2007e); Perez and Kipnis (2007); Perez et al. (2007)	(0)
2008	(5) Araujo (2008); Araujo and Feathers (2008); Araujo et al. (2008); Kipnis (2008); Neves and Piló (2008)	(7) Almeida and Bernardo (2008); Bernardo et al. (2008); Hubbe et al. (2008a); Hubbe et al. (2008b); Neves and Hubbe (2008); Pucciarelli et al. (2008); Strauss and Hubbe (2008)	(4) Alvarenga et al. (2008); Bissaro (2008); Hubbe (2008); Hubbe et al. (2008)	(0)

(continued)

Table 9.1 (continued)

Year	Archaeology	Biological anthropology	Paleontology	Paleobotany
2009	(3) Araujo (2009); Araujo and Pugliese (2009); Strauss and Araujo (2009)	(7) Bernardo and Neves (2009); Bernardo et al. (2009); Harvati et al. (2009); Hubbe et al. (2009a); Hubbe et al. (2009b); Strauss and Hubbe (2009); Strauss et al. (2009)	(4) Hubbe et al. (2009); Mayer et al. (2009a); Mayer et al. (2009b); Perez (2009)	(0)
2010	(10) Araujo (2010a); Araujo (2010b); Araujo and Neves (2010a); Araujo and Neves (2010b); Araujo and Oliveira (2010); Araujo and Pugliese (2010); Araujo et al. (2010); Bueno; Feathers et al. (2010); Porsani et al. (2010)	(16) Bernardo et al. (2010a); Bernardo et al. (2010b); Da-Gloria and Neves (2010); Hubbe M et al. (2010a); Hubbe et al. (2010a); Hubbe M et al. (2010b); Hubbe M et al. (2010c); Neves et al. (2010); Nunes (2010); Oliveira et al. (2010); Reis et al. (2010); Strauss (2010); Strauss and Hubbe (2010); Strauss et al. (2010a); Strauss et al. (2010b); Strauss et al. (2010c)	(5) Hubbe (2010); Hubbe et al. (2010); Kipnis et al. (2010a); Kipnis et al. (2010b); Mayer and Hubbe (2010)	(0)
2011	(5) Araujo (2011a); Araujo (2011b); Araujo (2011c); Bueno (2011); Okumura and Araujo (2011)	(19) Almeida (2011); Almeida et al. (2011); Bernardo (2011); Bernardo et al. (2011a); Bernardo et al. (2011b); Bernardo et al. (2011c); Da-Glória et al. (2011); Hubbe et al. (2011a); Hubbe et al. (2011a); Hubbe et al. (2011b); Hubbe et al. (2011c); Neves et al. (2011a); Neves et al. (2011b); Oliveira (2011); Strauss (2011); Strauss et al. (2011a); Strauss et al. (2011b); Strauss et al. (2011c); Strauss et al. (2011d)	(3) Hubbe et al. (2011b); Hubbe et al. (2011c); Mayer (2011)	(2) Freire (2011); Nakamura (2011)
2012	(10) Araujo (2012); Araujo and Neves (2012); Araujo et al. (2012a); Araujo et al. (2012b); Bernardo (2012); Bueno (2012); Neves et al. (2012a); Neves et al. (2012c); Strauss and Araujo (2012); Strauss et al. (2012)	(12) Almeida and Bernardo (2012); Bernardo et al. (2012a); Bernardo et al. (2012b); Bernardo et al. (2012c); Da-Glória (2012); Hubbe (2012); Hubbe and Strauss (2012a); Hubbe and Strauss (2012b); Hubbe et al. (2012); Oliveira et al. (2012a); Oliveira et al. (2012b); Strauss (2012)	(4) Hubbe and Auler (2012); Hubbe et al. (2012); Mayer et al. (2012); Neves et al. (2012b)	(0)
2013	(2) Araujo et al. (2013); Dias and Bueno (2013)	(9) Almeida et al. (2013); Bernardo (2013); Hubbe (2013); Hubbe et al. (2013a); Hubbe et al. (2013b); Kipnis and Da-Glória (2013); Neves; Neves et al. (2013); Strauss et al. (2013)	(2) Hubbe et al. (2013a); Hubbe et al. (2013b)	(0)
Total	62	104	33	5
Grand Total	204			

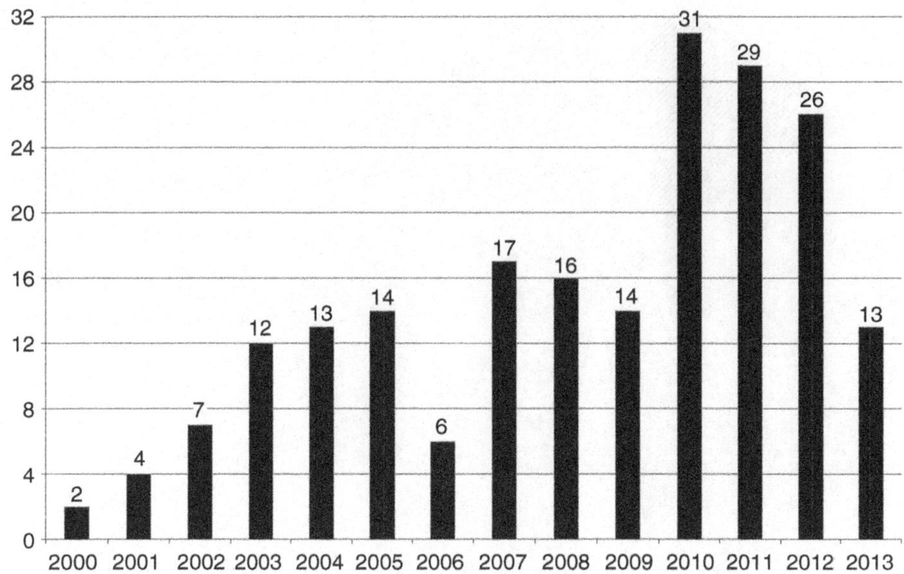

Fig. 9.1 Academic-scientific products generated in the sphere of the Origins project from 2000 to 2013, totals by year

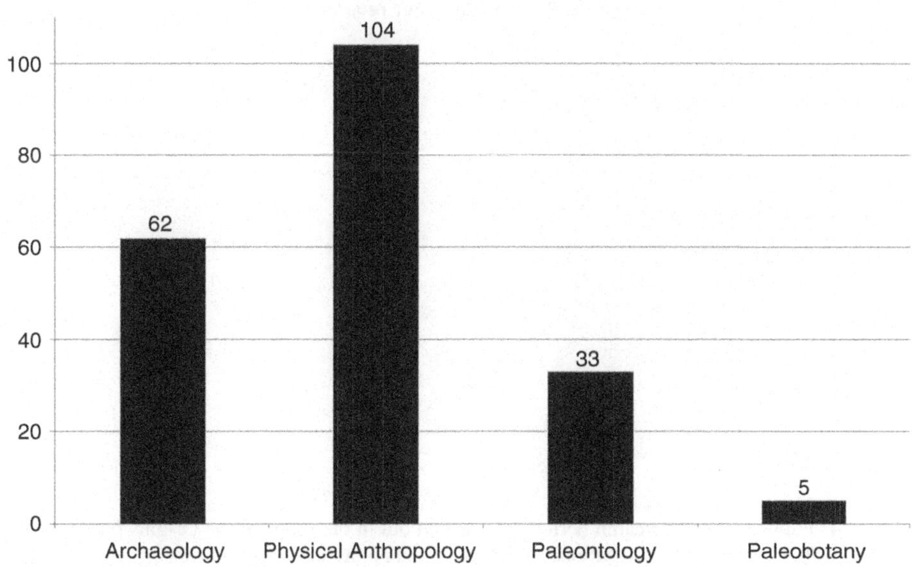

Fig. 9.2 Academic-scientific products generated in the scope of the Origins project from 2000 to 2013, totals by field of research

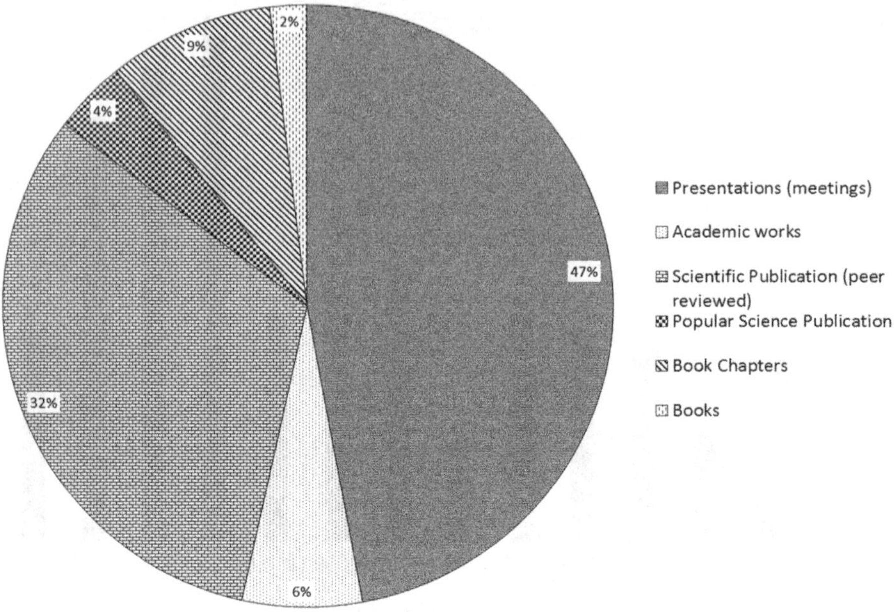

Fig. 9.3 Academic-scientific products generated by the Origins project from 2000 to 2013, classified by product type

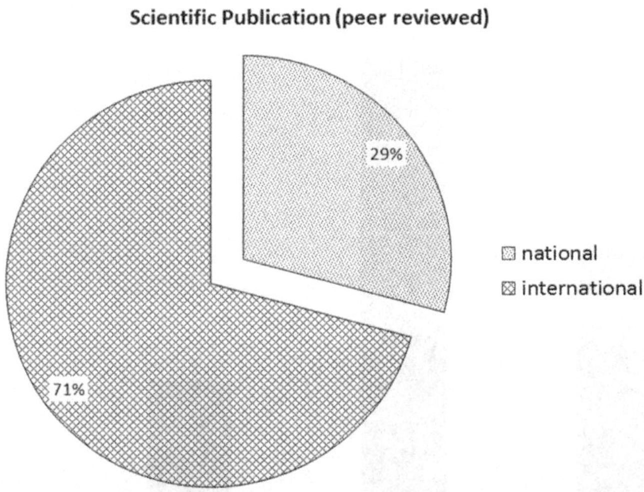

Fig. 9.4 Publications in research periodicals carried out in the scope of the Origins project from 2000 to 2013, classified by the national or international nature of the periodicals

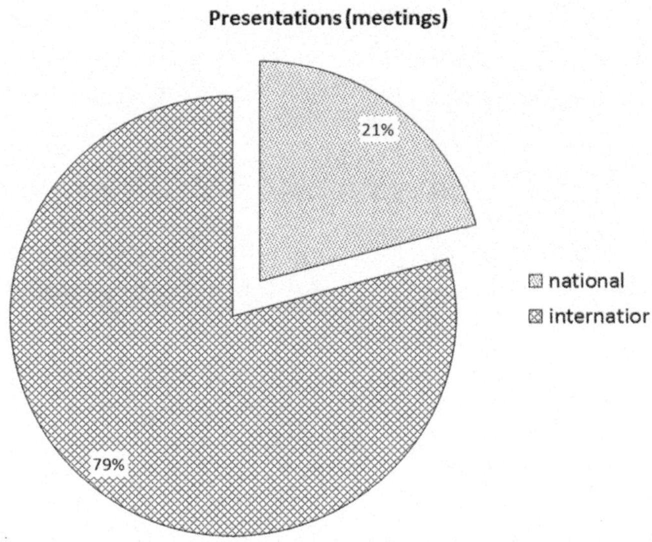

Fig. 9.5 Presentation in conferences of papers stemming from the Origins project in the years from 2000 to 2003, classified by the national or international nature of the event

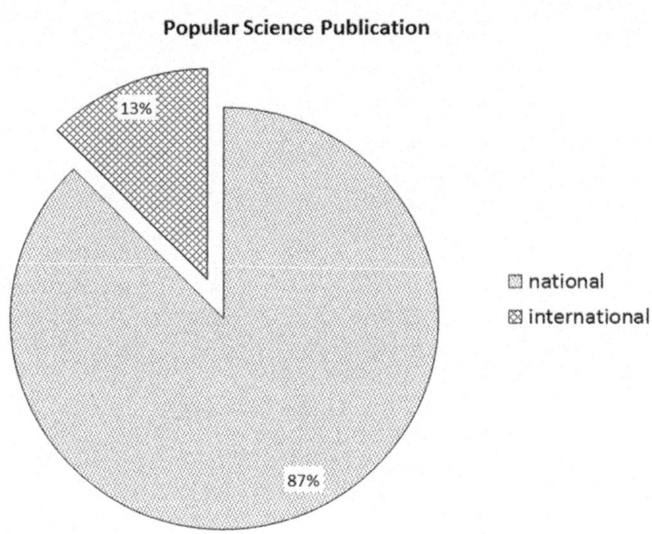

Fig. 9.6 Popular scientific articles stemming from the Origins project in the years from 2000 to 2003 classified by the national or international nature of the publication

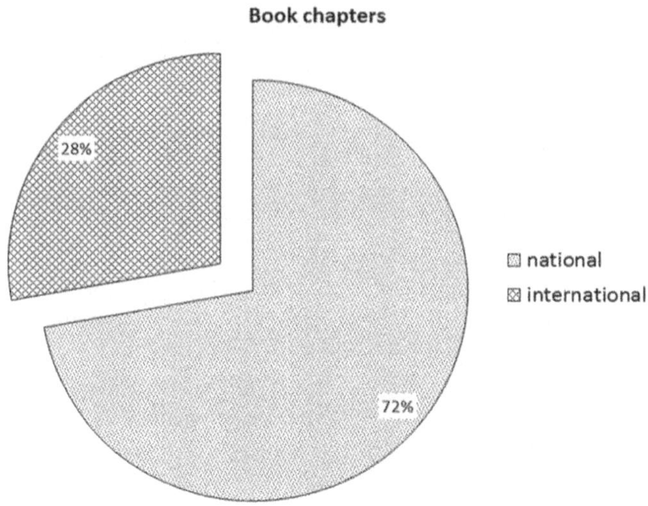

Fig. 9.7 Book chapters stemming from Origins project research in the years from 2000 to 2003, classified by the national or international nature of the publication

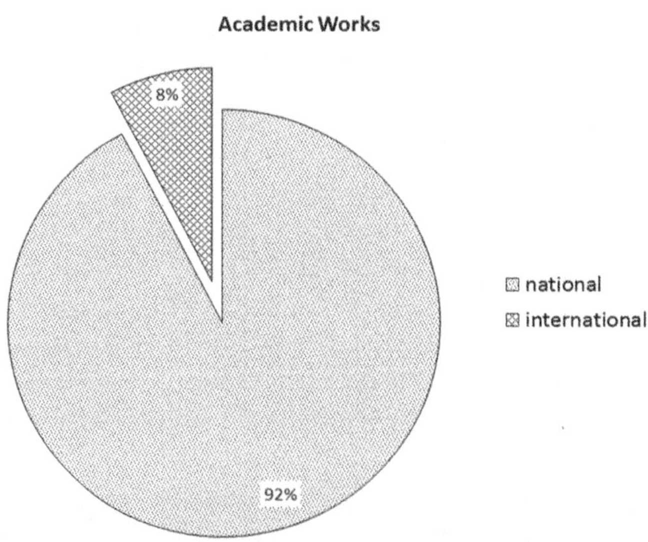

Fig. 9.8 Academic works (dissertations and theses) produced in the period from 2000 to 2013 stemming from Origins project research and classified as national or international

perfectly clear to us, including the field activities finalized in the middle of the year 2009 and the laboratory activities which still have a long way to go. When analyzing all the information, it is important to bear in mind the numbers of students and researchers directly involved in the project development. In absolute numbers, during the 9 years the project lasted, FAPESP awarded six undergraduate, ten masters, and four postdoctorate scholarships, all linked to projects stemming from the Origins project. In relative terms, it would be no exaggeration to state that about 200 students of various academic levels (from high school to postdoctoral students) and various origins (different São Paulo universities, Brazilian states and countries) worked alongside resident and visiting researchers and were trained during the fieldwork campaigns and the laboratory activities. If we are not mistaken, the Origin project's results fully justify the time and the financial and human resources invested in it, and there is still outstanding academic and scientific potential waiting to be exploited in the material generated by the interventions that finalized in 2009.

Acknowledgments It is obvious that for any large-scale project to be feasible, various institutions provide support of different kinds to those executing the project. The project "Microevolution of Man in America: a Paleoanthropological approach" was no different. Thus, we wish to convey our deep gratitude to the following institutions: FAPESP – *Fundação de Amparo à Pesquisa do Estado de São Paulo* (the São Paulo Research Foundation) for the generous financial support it provided (FAPESP Processes 99/00670-7 and 04/01321-6), which enabled us to make competitive research in an area almost entirely monopolized by the North American community. During the period of activity of the Origins project, the FAPESP also granted six undergraduate scholarships (processes 00/02467-3, 04/11485-6, 05/60319-4, 06/61297-7, 07/53185-7, 08/58242-1), ten masters scholarships (processes 02/04581-3, 02/11282-2, 04/11484-0, 05/52273-4, 05/57520-0, 06/51406-3, 07/51947-7, 08/51637-0, 08/51747-0, 09/03753-4), and four postdoctoral scholarships (processes 00/14917-3, 01/06881-1, 06/52188-0, 08/53574-6) all linked to the project; to the Departamento de Genética e Biologia Evolutiva do Instituto de Biociências da Universidade de São Paulo (Department of Genetics and Evolutionary Biology at the Biosciences Institute of the University of São Paulo) for the infrastructure and support it made available to us; to the IPHAN – Instituto do Patrimônio Histórico and Artístico Nacional (National Historic and Artistic Heritage Institute) for the successive authorizations it conceded for conducting archaeological research in the Lagoa Santa region; to the IBAMA – Instituto Brasileiro de Proteção ao Meio Ambiente (Brazilian Institute for Environmental Protection) for granting us permission to work in the Lagoa Santa Karst protected area and for the support of its office in Lagoa Santa; to the IEPHA – Instituto Estadual do Patrimônio Histórico and Artístico de Minas Gerais (the Minas Gerais Historical and Artistic Heritage Institute) for its unrestricted support for the research; to the CPRM – Registro Geológico do Brasil (Mineral Resource Research Company – Geological Registrations of Brazil) for ceding the digitalized versions of the thematic maps of the Lagoa Santa Karst protected area; to the Secretaria de Cultura do Estado de Minas Gerais (Culture Department of the Minas Gerais State government) for the support given when it was most needed; to the Prefeitura Municipal de Matozinhos (Municipal Authority of Matozinhos) for the logistics support and for maintaining the "Peter Lund research base" in that municipality; to the municipal authority of Pedro Leopoldo, for its support in the production of the DVD for publicizing the research undertaken; to the Centro de Arqueologia Annete Laming-Emperaire, Prefeitura de Lagoa Santa (Annette Laming-Emperaire Archaeology Center) maintained by the Lagoa Santa Municipal Authority for its unrestricted support for our research; to the Grupo Bambuí de Pesquisas Espeleológicas (Bambui Speleological Research Group) for its cooperation and the affection of its members toward us; to the Zoological Museum of the University of Copenhagen for granting us access to the Lund Collection and for maintaining it in its custody since 1845; to the Museum of Natural History, London, UK, for the permission granted to study one of the craniums from the Sumidouro cave which has been in its custody since the end of the nineteenth century; to the *Museu de História Natural e Jardim Botânico, Universidade Federal de Minas Gerais* (Natural History Museum and Botanical Garden of the Federal University of Minas Gerais) for giving us access to the institution's bioanthropological collection among which is the Harold Walter collection of human skeletons and fossil animals and for the warm welcome it afforded to all our researchers when working on their collection; to the *Museu Nacional da Universidade Federal do Rio de Janeiro* (National Museum at the Federal University of Rio de Janeiro) for permission to access the institution's bioanthropological collection and the warm welcome our researchers received there when working on that collection; to the *Instituto Histórico e Geográfico do Rio de Janeiro* (Rio de Janeiro Historical and Geographic Institute) for permission granted to study one of the craniums from Sumidouro in its custody since the mid-nineteenth century; to the Royal Library, Copenhagen, for having facilitated our access to various original documents of Peter Lund, including his letters and field notes; and to the *Mineração Lapa Vermelha* company (Lapa Vermelha Mining) for allowing us to carry out activities in lands it owns and at sites in its custody. In addition to all the abovementioned organizations, we cannot fail to mention the individual support given by enthusiasts of our interventions in Lagoa Santa. We thank André Prous (UFMG) and Castor Cartelle (PUCMG) for their encouragement and unrestricted support for the project; Kim Aaris Sorensen (Zoological Museum, University of Copenhagen), Chris Stringer and Robert Krusvinsk (Museum of Natural History, London), Cláudia Cardoso (UFMG), Ricardo Ventura, Claudia Carvalho, and Sheila Mendonça de Souza (National Museum at the UFRJ), and Hilton Pereira da Silva (UFPA) for the support they gave us when we visited their respective institutions; Fernando Walter for having shared with us his knowledge of the Lagoa Santa archaeological sites in addition to his erudition regarding the history of archaeological research in the region; João Bárbara Filho, for the seriousness, efficiency, and care he dedicated to the conception, implantation, and maintenance of various elements of our infrastructure; Eliana Diniz and family for their valuable dedication to accommodating the project team; José Geraldo, Wagner, Diego, Walmir, Renato, Manuel, Roneylson, Clayson, Luciano, and all the workers of Matozinhos and Mocambeiro who spared no efforts to achieve the planned objectives in the fieldwork stages; José Nicolau Hein, for allowing the establishment of various long-term scientific operations on the Cauaia farm which belongs to him and for his empathy with the scientific and biodiversity conservation causes in situ; Augusto Auler of the *Instituto do Carste* (Karst Institute) for his brilliant collaboration with the project at many moments of its execution; independent Danish scientist Mogens Trolle for the local fauna survey conducted using camera trapping in the Lagoa Santa region; Reynaldo Lopes, Cláudio Ângelo, Ricardo Bonalume Neto, Maria Célia Prado, Jader Resende, Gustavo Werneck, Roberta Jansen, Marcos Piveta, Maurício Tuffani, Pablo Nogueira, Dennis Russo, Ricardo Zorzetto, Rodrigo Rudiger, and Laura Schalchli, journalists who helped us considerably with their brilliant work in socializing the discoveries made by the project with the public at large. While writing the present chapter, the authors received financing from the CNPq via processes 461122/2014-6 (DVB) and 300917/2010-4 (WAN).

Supplementary Material

Supplemental 9.1 Samples submitted for dating under the aegis of the origins project

Sample identification	Site/region	Material	Institution	Measured radiocarbon age (in years before present)	Method	Laboratory identification
SR1-SEP.VB	Santana do Riacho, Serra do Cipó, MG	Organic material – *Homo sapiens*	MHN-UFMG	2270±50	AMS	Beta-104291
SR1-SEP.VIJ	Santana do Riacho, Serra do Cipó, MG	Organic material – *Homo sapiens*	MHN-UFMG	7840±60	AMS	Beta-104292
UFMG-HW-12	Coleção Harold Walter, Lagoa Santa, MG	Bone – *Homo sapiens*	MHN-UFMG	7270±60	AMS	Beta-108186
UFMG-HW-15	Coleção Harold Walter, Lagoa Santa, MG	Bone – *Homo sapiens*	MHN-UFMG	6800±50	AMS	Beta-108187
MN 856	Lapa do Caetano, Lagoa Santa, MG	Bone – *Homo sapiens*	MN-UFRJ	2150±50	AMS	Beta-155657
MN1389 (SEP 2)	Lapa das Boleiras, Lagoa Santa, MG	Bone – *Homo sapiens*	MN-UFRJ	8300±100	AMS	Beta-155658
MN 1390 (SEP 1)	Lapa das Boleiras, Lagoa Santa, MG	Bone – *Homo sapiens*	MN-UFRJ	8230±50	AMS	Beta-155659
JULIÃO 1	Sítio Julião, Lagoa Santa, MG	Charcoal	LEEH	157,6±0,7	Conventional radiometrics	Beta-157183
BLL25NO.01	Lapa das Boleiras, Lagoa Santa, MG	Charcoal	LEEH	200±40	AMS	Beta-159231
BLL25NO.02	Lapa das Boleiras, Lagoa Santa, MG	Charcoal	LEEH	8240±50	AMS	Beta-159232
BLH20NO.03	Lapa das Boleiras, Lagoa Santa, MG	Charcoal	LEEH	9210±130	Conventional radiometrics	Beta-159233
BLE41NO.04	Lapa das Boleiras, Lagoa Santa, MG	Charcoal	LEEH	210±60	Conventional radiometrics	Beta-159234
BLE41NO.05	Lapa das Boleiras, Lagoa Santa, MG	Charcoal	LEEH	3830±60	Conventional radiometrics	Beta-159235
BLL17NO.06	Lapa das Boleiras, Lagoa Santa, MG	Charcoal	LEEH	9610±60	AMS	Beta-159236
BLD39NO.07	Lapa das Boleiras, Lagoa Santa, MG	Charcoal	LEEH	160±70	Conventional radiometrics	Beta-159237
CGS9NO.08	Cerca Grande, Lagoa Santa, MG	Charcoal	LEEH	2320±60	Conventional radiometrics	Beta-159238

(continued)

Supplemental 9.1 (continued)

Sample identification	Site/region	Material	Institution	Measured radiocarbon age (in years before present)	Method	Laboratory identification
CGS10NO.09	Cerca Grande, Lagoa Santa, MG	Charcoal	LEEH	8690 ± 140	Conventional radiometrics	Beta-159239
CGS0NO.10-OS	Cerca Grande, Lagoa Santa, MG	Organic sediment	LEEH	2670 ± 40	AMS	Beta-159240
CGS0N.11	Cerca Grande, Lagoa Santa, MG	Organic sediment	LEEH	9680 ± 230	AMS	Beta-159241
BLK10NO.12	Lapa das Boleiras, Lagoa Santa, MG	Charcoal	LEEH	8820 ± 150	AMS	Beta-159242
BLK10NO.13	Lapa das Boleiras, Lagoa Santa, MG	Charcoal	LEEH	7560 ± 110	Conventional radiometrics	Beta-159243
BLK12NO.14	Lapa das Boleiras, Lagoa Santa, MG	Charcoal	LEEH	8370 ± 50	Conventional radiometrics	Beta-159244
BL011NO.15	Lapa das Boleiras, Lagoa Santa, MG	Charcoal	LEEH	8370 ± 110	Conventional radiometrics	Beta-159245
LSF13NO.16	Lapa do Santo, Lagoa Santa, MG	Charcoal	LEEH	7960 ± 50	Conventional radiometrics	Beta-159246
LSF13NO.17	Lapa do Santo, Lagoa Santa, MG	Charcoal	LEEH	8900 ± 50	Conventional radiometrics	Beta-159247
CGSONO. 10SHELL	Cerca Grande, Lagoa Santa, MG	Shell	LEEH	4580 ± 60	Conventional radiometrics	Beta-159911
SR1-XX	Santana do Riacho, Serra do Cipó, MG	Bone – *Homo sapiens*	MHN-UFMG	8280 ± 40	AMS	Beta-162014
BL-H20-C1	Lapa das Boleiras, Lagoa Santa, MG	Plant material	LEEH	190 ± 40	AMS	Beta-162529
BL-M17-G	Lapa das Boleiras, Lagoa Santa, MG	Organic sediment	LEEH	670 ± 40	AMS	Beta-162530
LS-115	Lagoa do Sumidouro, Lagoa Santa, MG	Plant material – moss	LEEH	1570 ± 70	AMS	Beta-162531
LS-275	Lagoa do Sumidouro, Lagoa Santa, MG	Plant material – moss	LEEH	2590 ± 40	AMS	Beta-162532
MPEG-ITA-PRE	Itaituba, Pará	Bone – *Eremotherium*	MPEG	11,370 ± 50	AMS	Beta-163531
ALEC-13	Sambaqui de Alecrim, SP	Bone – *Homo sapiens*	SAKAI	5010 ± 40	AMS	Beta-164991
MHN-S/N	Gruta Cuvieri, Lagoa Santa, MG	Bone – *Catonyx cuvieri*	UFMG	9960 ± 40	AMS	Beta-165398

BLK11N.1	Lapa das Boleiras, Lagoa Santa, MG	Charcoal	LEEH	9840 ± 40	AMS	Beta-168449
BLK12N.1	Lapa das Boleiras, Lagoa Santa, MG	Charcoal	LEEH	10,150 ± 130	Conventional radiometrics	Beta-168451
BLL11N.3	Lapa das Boleiras, Lagoa Santa, MG	Charcoal	LEEH	12,240 ± 50	AMS	Beta-168457
LD-518	Sítio Lund, Lagoa Santa, MG	Charcoal	LEEH	2250 ± 40	Conventional radiometrics	Beta-170718
LD-519	Sítio Lund, Lagoa Santa, MG	Charcoal	LEEH	2200 ± 40	Conventional radiometrics	Beta-170719
ST-44	Lapa do Santo, Lagoas Santa, MG	Charcoal	LEEH	8840 ± 60	Conventional radiometrics	Beta-170723
BR789CHARC	Lapa do Sumidouro, Lagoa Santa, MG	Charcoal	ZMUC	7720 ± 50	AMS	Beta-172186
BR789SHELL	Lapa do Sumidouro, Lagoa Santa, MG	Shell	ZMUC	8700 ± 50	AMS	Beta-172187
FUNDICION-HS	La Fundición, Chile	Bone – *Homo sapiens*	IIARMPGLP	4570 ± 50	Conventional radiometrics	Beta-174679
HW-CONFINS	Lapa dos Confins, Lagoa Santa, MG	Tooth – *Homo sapiens*	MHN-UFMG	12,130 ± 50 (contaminated material)	AMS	Beta-174680
MN892-CAET	Lapa do Sumidouro, Lagoa Santa, MG	Bone – *Homo sapiens*	MHN-UFMG	2210 ± 60	AMS	Beta-174682
ZMUC BR247	Lapa do Sumidouro, Lagoa Santa, MG	Charcoal	ZMUC	7700 ± 40	AMS	Beta-174730
ZMUC BR321	Lapa do Sumidouro, Lagoa Santa, MG	Shell	ZMUC	7340 ± 80	AMS	Beta-174732
ZMUC VE HS	Lapa Vermelha, Lagoa Santa, MG	Bone – *Homo sapiens*	ZMUC	6980 ± 200	AMS	Beta-174733
ZMUC ESC3HS	Escrivânia 3, Lagoa Santa, MG	Bone – *Homo sapiens*	ZMUC	7650 ± 80	AMS	Beta-174734
ZMUC 2358	Gruta do Baú 2, Lagoa Santa, MG	Bone – *Homo sapiens*	ZMUC	8730 ± 50	AMS	Beta-174735
ZMUC 4725	Lapa do Braga, Lagoa Santa, MG	Bone – *Homo sapiens*	ZMUC	9680 ± 70	AMS	Beta-174736
BOLOSSO-S3	Lapa das Boleiras, Lagoa Santa, MG	Bone – *Homo sapiens*	LEEH	8090 ± 40	AMS	Beta-178554
DENTO3-SEPBL	Lapa das Boleiras, Lagoa Santa, MG	Tooth – *Homo sapiens*	LEEH	9620 ± 50	AMS	Beta-178556
LAGSANTOB	Lagoa do Santo, Lagoa Santa, MG	Organic sediment	LEEH	430 ± 70	AMS	Beta-178557
LAPSANTOH	Lapa do Santo, Lagoa Santa, MG	Organic sediment	LEEH	810 ± 50	AMS	Beta-178558
RIBMATA-B1	Ribeirão da Mata	Organic sediment	LEEH	5240 ± 60	AMS	Beta-178559

(continued)

Supplemental 9.1 (continued)

Sample identification	Site/region	Material	Institution	Measured radiocarbon age (in years before present)	Method	Laboratory identification
BOLSEP3DENTE	Lapa das Boleiras, Lagoa Santa, MG	Tooth – *Homo sapiens*	LEEH	8250 ± 50	AMS	Beta-179815
BL-101	Lapa das Boleiras, Lagoa Santa, MG	Charcoal	LEEH	8760 ± 50	AMS	Beta-183563
BL-102/105	Lapa das Boleiras, Lagoa Santa, MG	Charcoal	LEEH	8760 ± 50	AMS	Beta-183564
GP/2E-716	Abismo Iguatemi, Vale do Ribeira, SP	Bone – *Smilodon*	IGUSP	14,470 ± 90	AMS	Beta-183566
Tq-268	Taquaraçu	Charcoal	LEEH	9540 ± 60	Conventional radiometrics	Beta-183575
Tq-402	Taquaraçu	Charcoal	LEEH	8210 ± 50	Conventional radiometrics	Beta-183576
Tq-404	Taquaraçu	Charcoal	LEEH	8750 ± 50	Conventional radiometrics	Beta-183577
SUMIDE-B2-115	Lapa do Sumidouro, Lagoa Santa, MG	Charcoal	LEEH	220 ± 30	Conventional radiometrics	Beta-191131
CVL3-CATONYX	Gruta Cuvieri, Lagoa Santa, MG	Bone – *Catonyx cuvieri*	LEEH	17,430 ± 70	AMS	Beta-196717
LST15	Lapa do Santo, Lagoa Santa, MG	Charcoal	LEEH	8630 ± 50	Conventional radiometrics	Beta-202763
LST45	Lapa do Santo, Lagoa Santa, MG	Charcoal	LEEH	920 ± 50	Conventional radiometrics	Beta-202764
LST58	Lapa do Santo, Lagoa Santa, MG	Charcoal	LEEH	3830 ± 100	Extended counting	Beta-202765
LST59	Lapa do Santo, Lagoa Santa, MG	Charcoal	LEEH	3830 ± 50	Conventional radiometrics	Beta-202766
LST63	Lapa do Santo, Lagoa Santa, MG	Charcoal	LEEH	8550 ± 40	AMS	Beta-202767
LST73	Lapa do Santo, Lagoa Santa, MG	Charcoal	LEEH	4300 ± 90	Extended counting	Beta-202768
Hussan-03	Abismo Gemeo, Iporanga, SP	Bone – megamastofauna	MZUSP	13,210 ± 60	AMS	Beta-202771

UFMG3707	Lapa dos Bichos, MG	Charcoal	UFMG	800 ± 50	Conventional radiometrics	Beta-202772
UFMG3755	Lapa dos Bichos, MG	Charcoal	UFMG	3640 ± 60	Conventional radiometrics	Beta-202773
UFMG3758	Lapa dos Bichos, MG	Charcoal	UFMG	820 ± 40	Conventional radiometrics	Beta-202774
UFMG4385	Lapa dos Bichos, MG	Charcoal	UFMG	9480 ± 130	Extended counting	Beta-202775
UFMG4387	Lapa dos Bichos, MG	Charcoal	UFMG	4460 ± 60	Conventional radiometrics	Beta-202776
UFMG4402	Lapa dos Bichos, MG	Charcoal	UFMG	3590 ± 80	Conventional radiometrics	Beta-202777
CVL395	Gruta Cuvieri, Lagoa Santa, MG	Bone – *Tapirus terrestris*	LEEH	12,390 ± 50	AMS	Beta-202778
CVL2P260	Gruta Cuvieri, Lagoa Santa, MG	Bone – *Cervidae*	LEEH	1940 ± 40	AMS	Beta-202779
CVL24630	Gruta Cuvieri, Lagoa Santa, MG	Bone – unidentified	LEEH	5220 ± 50	AMS	Beta-202780
CVL3CATONYX1	Gruta Cuvieri, Lagoa Santa, MG	Bone – *Catonyx cuvieri*	LEEH	12,430 ± 70	AMS	Beta-202782
ZMUCS/N2	Escrivânia	Bone – *Haplomastodon*	ZMUC	1470 ± 90	AMS	Beta-202806
MCL2969	Toca das Onças, BA	Bone – *Homo sapiens*	PUC-MG	4140 ± 40	AMS	Beta-202810
Lapa do Urubu	Tavares, Confins	Bone – *Homo sapiens*	MHN-UFMG	7950 ± 40	AMS	Beta-202812
CVL2P163/196	Gruta Cuvieri, Lagoa Santa, MG	Bone – *Tapirus terrestris*	LEEH	2020 ± 40	AMS	Beta-205334
CVL2P258/234	Gruta Cuvieri, Lagoa Santa, MG	Bone – *Agouti paca*	LEEH	170 ± 40	AMS	Beta-205335
CVSO1(A)10-15	Abrigo Cuvieri, Lagoa Santa, MG	Charcoal	LEEH	140 ± 40	AMS	Beta-205336
CVSOND2-0-10	Gruta Cuvieri, Lagoa Santa, MG	Charcoal	LEEH	170 ± 50	AMS	Beta-205337
M-722	Sambaqui Maratuá, Santos, SP	Bone – *Homo sapiens*	SAKAI	3350 ± 40	AMS	Beta-205339
MN-821	Lapa da Amoreira, Lagoa Santa, MG	Bone – *Homo sapiens*	UFMG	8040 ± 40	AMS	Beta-205340
SUMIDB1-137	Sítio Sumidouro, Lagoa Santa, MG	Charcoal	LEEH	5030 ± 70	Extended counting	Beta-205350
SUMIDB1-160	Sítio Sumidouro, Lagoa Santa, MG	Charcoal	LEEH	8310 ± 40	AMS	Beta-205351
ARCOTRONCO	Peruaçu, MG	Wood	LEEH	170 ± 60	Conventional radiometrics	Beta-208064

(continued)

Supplemental 9.1 (continued)

Sample identification	Site/region	Material	Institution	Measured radiocarbon age (in years before present)	Method	Laboratory identification
BREJALTRONCO	Peruaçu, MG	Wood	LEEH	560 ± 80	Conventional radiometrics	Beta-208065
CONC-03	Lapa Mortuária de Confins, Lagoa Santa, MG	Shell	LEEH	31,480 ± 310	AMS	Beta-208066
GREG-AST – 1	Ribeirão da Mata, Pedro Leopoldo, MG	Wood	LEEH	5190 ± 40	AMS	Beta-208069
GREG-AST – 2	Ribeirão da Mata, Pedro Leopoldo, MG	Wood	LEEH	4960 ± 40	AMS	Beta-208070
GREG-AST – 3	Ribeirão da Mata, Pedro Leopoldo, MG	Wood	LEEH	4550 ± 40	AMS	Beta-208071
GREG-AST – 4	Ribeirão da Mata, Pedro Leopoldo, MG	Wood	LEEH	2070 ± 40	AMS	Beta-208072
GREG-AST – 7	Ribeirão da Mata, Pedro Leopoldo, MG	Wood	LEEH	8840 ± 40	AMS	Beta-208073
HW-294 (bone)	Lagoa Santa, MG	Bone – *Homo sapiens*	UFMG	8520 ± 40	AMS	Beta-208077
MCL3021	Toca das Onças, Bh	Bone –*Homo sapiens*	PUC-MG	4640 ± 40	AMS	Beta-208078
MCL3306F	Toca das Onças, Bh	Bone – *Homo sapiens*	PUC-MG	4530 ± 40	AMS	Beta-208083
MCL3306J	Toca das Onças, Bh	Bone – *Homo sapiens*	PUC-MG	4620 ± 40	AMS	Beta-208084
UFMG3860	Lapa do Boquete, Peruaçu, MG	Bone – *Homo sapiens*	UFMG	9290 ± 40	AMS	Beta-208093
GREG-RMT-1 DR	Ribeirão da Mata, Pedro Leopoldo, MG	Wood	LEEH	1800 ± 70	AMS	Beta-209470
LSPAINSSEPI	Loca do Suim, Pains, MG	Bone – *Homo sapiens*	LEEH	7350 ± 50	AMS	Beta-210400
LSPAINSSEPII	Loca do Suim, Pains, MG	Bone – *Homo sapiens*	LEEH	7460 ± 50	AMS	Beta-210401
FO 15	Lagoa Formosa, Peruaçu, MG	Lacustrine sediment	LEEH	1090 ± 40	AMS	Beta-211472
FO 35	Lagoa Formosa, Peruaçu, MG	Lacustrine sediment	LEEH	1980 ± 40	AMS	Beta-211473
FO 45	Lagoa Formosa, Peruaçu, MG	Lacustrine sediment	LEEH	3670 ± 40	AMS	Beta-211474
FO 75	Lagoa Formosa, Peruaçu, MG	Lacustrine sediment	LEEH	13,170 ± 60	AMS	Beta-211475
FO 90	Lagoa Formosa, Peruaçu, MG	Lacustrine sediment	LEEH	10,680 ± 40	AMS	Beta-211476
FO 105	Lagoa Formosa, Peruaçu, MG	Lacustrine sediment	LEEH	5680 ± 40	AMS	Beta-211477
OL 138	Lagoa Olhos D'Água, Lagoa Santa, MG	Lacustrine sediment	LEEH	3630 ± 40	AMS	Beta-211478
OL 149	Lagoa Olhos D'Água, Lagoa Santa, MG	Lacustrine sediment	LEEH	6880 ± 40	AMS	Beta-211479

OL 152	Lagoa Olhos D'Água, Lagoa Santa, MG	Lacustrine sediment	LEEH	6650±40	AMS	Beta-211480
OL 156	Lagoa Olhos D'Água, Lagoa Santa, MG	Lacustrine sediment	LEEH	7290±40	AMS	Beta-211481
OL 160	Lagoa Olhos D'Água, Lagoa Santa, MG	Lacustrine sediment	LEEH	11,480±60	AMS	Beta-211482
OL 170	Lagoa Olhos D'Água, Lagoa Santa, MG	Lacustrine sediment	LEEH	15,790±80	AMS	Beta-211483
OL 180	Lagoa Olhos D'Água, Lagoa Santa, MG	Lacustrine sediment	LEEH	18,800±100	AMS	Beta-211484
OL 190	Lagoa Olhos D'Água, Lagoa Santa, MG	Lacustrine sediment	LEEH	25,630±150	AMS	Beta-211485
LST6	Lapa do Santo, Lagoa Santa, MG	Wood	LEEH	3990±40	AMS	Beta-214129
LST77	Lapa do Santo, Lagoa Santa, MG	Wood	LEEH	9880±40	AMS	Beta-214130
LST89	Lapa do Santo, Lagoa Santa, MG	Wood	LEEH	8700±40	AMS	Beta-214131
LST94	Lapa do Santo, Lagoa Santa, MG	Wood	LEEH	8790±40	AMS	Beta-214132
LST100	Lapa do Santo, Lagoa Santa, MG	Wood	LEEH	8720±40	AMS	Beta-214133
LST711	Lapa do Santo, Lagoa Santa, MG	Wood	LEEH	8860±100	AMS	Beta-214134
LST712	Lapa do Santo, Lagoa Santa, MG	Wood	LEEH	8760±40	AMS	Beta-214135
LST726	Lapa do Santo, Lagoa Santa, MG	Wood	LEEH	8720±80	AMS	Beta-214136
LST731	Lapa do Santo, Lagoa Santa, MG	Wood	LEEH	8840±40	AMS	Beta-214137
LST734	Lapa do Santo, Lagoa Santa, MG	Wood	LEEH	10,040±100	AMS	Beta-214138
LST738	Lapa do Santo, Lagoa Santa, MG	Wood	LEEH	8940±40	AMS	Beta-214139
LST744	Lapa do Santo, Lagoa Santa, MG	Wood	LEEH	8930±40	AMS	Beta-214140
LST751	Lapa do Santo, Lagoa Santa, MG	Wood	LEEH	9000±40	AMS	Beta-214141
LST760	Lapa do Santo, Lagoa Santa, MG	Wood	LEEH	7890±40	AMS	Beta-214142
LST762	Lapa do Santo, Lagoa Santa, MG	Wood	LEEH	9150±40	AMS	Beta-214143
GUAFIGUEIRSC	São Pedro do Sul, RS	Bone – *Homo sapiens*	LEPA	170±40	AMS	Beta-214144
GUASMARIARS	Cemitério Figueirinha, SC	Bone – *Homo sapiens*	UNESC	140±40	AMS	Beta-214145
BQT-sep 4	Lapa do Boquete, Peruaçu, MG	Bone –*Homo sapiens*	LEEH	430±40	AMS	Beta-215193
LST Sep 07 b	Lapa do Santo, Lagoa Santa, MG	Bone – *Homo sapiens*	LEEH	7300±40	Conventional radiometrics	Beta-215194
LST Sep 11b	Lapa do Santo, Lagoa Santa, MG	Bone – *Homo sapiens*	LEEH	5920±40	Conventional radiometrics	Beta-215195

(continued)

Supplemental 9.1 (continued)

Sample identification	Site/region	Material	Institution	Measured radiocarbon age (in years before present)	Method	Laboratory identification
LST Sep 14	Lapa do Santo, Lagoa Santa, MG	Bone – *Homo sapiens*	LEEH	8190 ± 40	Conventional radiometrics	Beta-215196
LST Sep 19	Lapa do Santo, Lagoa Santa, MG	Bone – *Homo sapiens*	LEEH	7600 ± 40	Conventional radiometrics	Beta-215200
PF997	Abismo Ponta de Flecha, Iporanga, SP	Tooth – *Toxodontidae*	IGC-USP	11,360 ± 40	AMS	Beta-215330
GB-H01	Gruta dos Brejões, Bahia	Bone – *Homo sapiens*	LEEH	6060 ± 40	AMS	Beta-215486
BL-104	Lapa das Boleiras, Lagoa Santa, MG	Charcoal	LEEH	8600 ± 40	AMS	Beta-216516
LSt04	Lapa do Santo, Lagoa Santa, MG	Wood	LEEH	3930 ± 40	AMS	Beta-216517
LSt19	Lapa do Santo, Lagoa Santa, MG	Wood	LEEH	9090 ± 40	AMS	Beta-216518
LSt21	Lapa do Santo, Lagoa Santa, MG	Wood	LEEH	8710 ± 40	AMS	Beta-216519
LSt48	Lapa do Santo, Lagoa Santa, MG	Wood	LEEH	8830 ± 90	AMS	Beta-216520
LSt50	Lapa do Santo, Lagoa Santa, MG	Wood	LEEH	4070 ± 60	AMS	Beta-216521
LSt71	Lapa do Santo, Lagoa Santa, MG	Wood	LEEH	8820 ± 40	AMS	Beta-216522
LSt736	Lapa do Santo, Lagoa Santa, MG	Wood	LEEH	8640 ± 40	AMS	Beta-216523
LSt759	Lapa do Santo, Lagoa Santa, MG	Wood	LEEH	8710 ± 40	AMS	Beta-216524
TQ-295	Lapa Grande do Taquaraçu, Taquaraçu de Minas, MG	Charcoal	LEEH	9640 ± 40	AMS	Beta-216525
TQ-297	Lapa Grande do Taquaraçu, Taquaraçu de Minas, MG	Charcoal	LEEH	9550 ± 90	AMS	Beta-216526
TQ-417	Lapa Grande do Taquaraçu, Taquaraçu de Minas, MG	Charcoal	LEEH	8080 ± 40	AMS	Beta-216527
TQ-421	Lapa Grande do Taquaraçu, Taquaraçu de Minas, MG	Charcoal	LEEH	1240 ± 60	AMS	Beta-216528
TQ-430	Lapa Grande do Taquaraçu, Taquaraçu de Minas, MG	Charcoal	LEEH	8320 ± 40	AMS	Beta-216529

TQ-441	Lapa Grande do Taquaraçu, Taquaraçu de Minas, MG	Charcoal	LEEH	8710±40	AMS	Beta-216530
TQ-454	Lapa Grande do Taquaraçu, Taquaraçu de Minas, MG	Charcoal	LEEH	8920±40	AMS	Beta-216531
TQ-459	Lapa Grande do Taquaraçu, Taquaraçu de Minas, MG	Charcoal	LEEH	9040±40	AMS	Beta-216532
CVL27108	Gruta Cuvieri, Lagoa Santa, MG	Bone – *Mazama*	LEEH	5180±50	AMS	Beta-218173
CVL27402	Gruta Cuvieri, Lagoa Santa, MG	Bone – cervid	LEEH	9440±40	AMS	Beta-218174
PÇAZUL-12	Caverna poço Azul, Bahia	Tooth – *Mazama*	PUC-MG	1260±40	AMS	Beta-218186
PÇAZUL-17	Caverna poço Azul, Bahia	Bone – *Rodentia*	PUC-MG	5320±40	AMS	Beta-218191
PF997*1	Abismo Ponta de Flecha, Iporanga, SP	Tooth – *Toxodontidae*	IGC-USP	11,100±40	AMS	Beta-218193
LO-2MB-1,39	Lagoa dos Olhos, Lagoa Santa, MG	Lacustrine sediment	LEEH	3840±40	AMS	Beta-220303
LO-2MB-1,41	Lagoa dos Olhos, Lagoa Santa, MG	Lacustrine sediment	LEEH	4510±40	AMS	Beta-220304
LO-2MB-1,43	Lagoa dos Olhos, Lagoa Santa, MG	Lacustrine sediment	LEEH	5350±50	AMS	Beta-220305
LO-2MB-1,45	Lagoa dos Olhos, Lagoa Santa, MG	Lacustrine sediment	LEEH	6800±60	AMS	Beta-220306
LO-2MB-1,47	Lagoa dos Olhos, Lagoa Santa, MG	Lacustrine sediment	LEEH	6590±50	AMS	Beta-220307
LO-2MB-1,53	Lagoa dos Olhos, Lagoa Santa, MG	Lacustrine sediment	LEEH	7650±50	AMS	Beta-220308
LO-2MB-1,55	Lagoa dos Olhos, Lagoa Santa, MG	Lacustrine sediment	LEEH	9330±50	AMS	Beta-220309
LO-2MB-1,58	Lagoa dos Olhos, Lagoa Santa, MG	Lacustrine sediment	LEEH	9660±50	AMS	Beta-220310
CVL2-9648	Gruta Cuvieri, Lagoa Santa, MG	Bone – cervid	LEEH	9680±40	AMS	Beta-220396
MCL3306/G	Toca das Onças, BA	Tooth – *Homo sapiens*	PUC-MG	4570±40	AMS	Beta-220399
MCL3341	Toca das Onças, BA	Tooth – *Homo sapiens*	PUC-MG	4660±40	AMS	Beta-220400
MCLS/NHOMO	Toca das Onças, BA	Bone – *Homo sapiens*	PUC-MG	4270±40	AMS	Beta-220412
PEDRAOCA-2	Sambaqui da Pedra Oca, BA	Bone – *Homo sapiens*	PUC-MG	3190±40	AMS	Beta-220419
PIGMENTO-AMA	Toca das Onças, BA	Organic pigment	PUC-MG	6390±40	AMS	Beta-220421
PIGMENTO-VER	Toca das Onças, BA	Organic pigment	PUC-MG	10,560±40	AMS	Beta-220422
HW-6	Lagoa Santa , MG	Tooth – *Homo sapiens*	UFMG	8510±40	AMS	Beta-220426
HW-13	Lagoa Santa , MG	Tooth – *Homo sapiens*	UFMG	8630±40	AMS	Beta-220429

(continued)

Supplemental 9.1 (continued)

Sample identification	Site/region	Material	Institution	Measured radiocarbon age (in years before present)	Method	Laboratory identification
HW-293	Lagoa Santa , MG	Tooth – *Homo sapiens*	UFMG	7930±40	AMS	Beta-220432
HW-294 (dente)	Lagoa Santa , MG	Tooth – *Homo sapiens*	UFMG	8700±40	AMS	Beta-220433
HW-2520-1	Lagoa Santa , MG	Tooth – *Homo sapiens*	UFMG	8220±40	AMS	Beta-220434
HW-2520-4	Lagoa Santa , MG	Tooth – *Homo sapiens*	UFMG	7640±60	AMS	Beta-220437
HW-2520-7	Lagoa Santa , MG	Tooth – *Homo sapiens*	UFMG	8470±40	AMS	Beta-220440
HW-2520-9	Lagoa Santa , MG	Tooth – *Homo sapiens*	UFMG	8380±40	AMS	Beta-220442
HW-2520-10	Lagoa Santa , MG	Tooth – *Homo sapiens*	UFMG	7770±40	AMS	Beta-220443
TRONCO	Toca dos Ossos, BA	Wood	PUC-MG	2820±40	AMS	Beta-220444
LMC-TEST 3	Lapa Mortuáriá de Confins, Confins, MG	Charcoal	LEEH	9760±70	AMS	Beta-221079
LMC-TEST 4	Lapa Mortuáriá de Confins, Confins, MG	Charcoal	LEEH	8130±60	AMS	Beta-221080
LMC-TEST 5	Lapa Mortuáriá de Confins, Confins, MG	Charcoal	LEEH	4150±40	AMS	Beta-221081
BLO106	Lapa das Boleiras, Lagoa Santa, MG	Charcoal	LEEH	9000±60	AMS	Beta-221451
BLO109	Lapa das Boleiras, Lagoa Santa, MG	Charcoal	LEEH	8710±70	AMS	Beta-221452
BLO113	Lapa das Boleiras, Lagoa Santa, MG	Charcoal	LEEH	8540±60	AMS	Beta-221453
BLO114	Lapa das Boleiras, Lagoa Santa, MG	Charcoal	LEEH	8610±60	AMS	Beta-221454
BLO116	Lapa das Boleiras, Lagoa Santa, MG	Charcoal	LEEH	8540±60	AMS	Beta-221456
BLO117	Lapa das Boleiras, Lagoa Santa, MG	Charcoal	LEEH	9090±60	AMS	Beta-221457
BLO118	Lapa das Boleiras, Lagoa Santa, MG	Charcoal	LEEH	9670±60	AMS	Beta-221458
BLO166	Lapa das Boleiras, Lagoa Santa, MG	Charcoal	LEEH	8810±60	AMS	Beta-221459
BLO167	Lapa das Boleiras, Lagoa Santa, MG	Charcoal	LEEH	9450±60	AMS	Beta-221460
RMT-875	Ribeirão da Mata, Pedro Leopoldo, MG	Wood	LEEH	5170±40	Conventional radiometrics	Beta-222578
RMT-938	Ribeirão da Mata, Pedro Leopoldo, MG	Wood	LEEH	2620±40	Conventional radiometrics	Beta-222579

RMT-1952	Ribeirão da Mata, Pedro Leopoldo, MG	Wood	LEEH	1950±50	Conventional radiometrics	Beta-222582
RMT-2029	Ribeirão da Mata, Pedro Leopoldo, MG	Wood	LEEH	1790±50	Conventional radiometrics	Beta-222583
RMT-2033	Ribeirão da Mata, Pedro Leopoldo, MG	Wood	LEEH	1840±40	Conventional radiometrics	Beta-222584
RMT-2055	Ribeirão da Mata, Pedro Leopoldo, MG	Wood	LEEH	4530±40	Conventional radiometrics	Beta-222585
RMT-2056	Ribeirão da Mata, Pedro Leopoldo, MG	Wood	LEEH	4620±40	Conventional radiometrics	Beta-222586
RMT-2098	Ribeirão da Mata, Pedro Leopoldo, MG	Wood	LEEH	4950±50	Conventional radiometrics	Beta-222593
RMT-2125	Ribeirão da Mata, Pedro Leopoldo, MG	Wood	LEEH	2050±40	Conventional radiometrics	Beta-222594
RMT-2143	Ribeirão da Mata, Pedro Leopoldo, MG	Wood	LEEH	1610±40	AMS	Beta-222595
RMT-2147	Ribeirão da Mata, Pedro Leopoldo, MG	Wood	LEEH	2460±40	Conventional radiometrics	Beta-222596
RMT-2153	Ribeirão da Mata, Pedro Leopoldo, MG	Wood	LEEH	2520±40	Conventional radiometrics	Beta-222597
RMT-2172	Ribeirão da Mata, Pedro Leopoldo, MG	Wood	LEEH	5200±40	Conventional radiometrics	Beta-222598
RMT-2185	Ribeirão da Mata, Pedro Leopoldo, MG	Wood	LEEH	4820±40	AMS	Beta-222599
RMT-2200	Ribeirão da Mata, Pedro Leopoldo, MG	Wood	LEEH	4860±40	Conventional radiometrics	Beta-222600
RMT-2205	Ribeirão da Mata, Pedro Leopoldo, MG	Wood	LEEH	1780±40	AMS	Beta-222601
RMT-2215	Ribeirão da Mata, Pedro Leopoldo, MG	Wood	LEEH	5180±40	AMS	Beta-222602
RMT-2247	Ribeirão da Mata, Pedro Leopoldo, MG	Wood	LEEH	4980±40	Conventional radiometrics	Beta-222603
6330	Buritizeiro, Brasil Central	Charcoal	MHN-UFMG	5970±50	AMS	Beta-224842
6370	Buritizeiro, Brasil Central	Charcoal	MHN-UFMG	1300±40	AMS	Beta-224843

(continued)

Supplemental 9.1 (continued)

Sample identification	Site/region	Material	Institution	Measured radiocarbon age (in years before present)	Method	Laboratory identification
6441	Buritizeiro, Brasil Central	Charcoal	MHN-UFMG	8120 ± 50	AMS	Beta-224844
6449	Buritizeiro, Brasil Central	Charcoal	MHN-UFMG	4580 ± 40	AMS	Beta-224845
CVL2-10365	Gruta Cuvieri, Lagoa Santa, MG	Bone – *Mazama*	LEEH	6880 ± 40	AMS	Beta-230973
GP/2E-706	Caverna-abismo do Iguatemi, Vale do Ribeira, SP	Bone – *Catonyx cuvieri*	IGC-USP	10,760 ± 60	AMS	Beta-230974
MOENDASSHELL	Toca das Moendas, PI	Shell	FUMDHAM	23,960 ± 150	AMS	Beta-230976
2350	Gruta do Marinho, Pains, MG	Charcoal	MHN-UFMG	3100 ± 50	Conventional radiometrics	Beta-230979
3737	Gruta do Marinho, Pains, MG	Charcoal	MHN-UFMG	9610 ± 60	Conventional radiometrics	Beta-230980
RM1-C8C0-2CM	Ribeirão da Mata, Pedro Leopoldo, MG	Lacustrine sediment	LEEH	4990 ± 40	AMS	Beta-232670
1019	Lagoa do Sumidouro, Lagoa Santa, MG	Charcoal	LEEH	490 ± 40	AMS	Beta-234506
1101	Lagoa do Sumidouro, Lagoa Santa, MG	Charcoal	LEEH	5870 ± 50	AMS	Beta-234507
1102	Lagoa do Sumidouro, Lagoa Santa, MG	Charcoal	LEEH	4600 ± 50	AMS	Beta-234508
1103	Lagoa do Sumidouro, Lagoa Santa, MG	Charcoal	LEEH	5820 ± 40	AMS	Beta-234509
1104	Lagoa do Sumidouro, Lagoa Santa, MG	Charcoal	LEEH	2190 ± 40	AMS	Beta-234510
1105	Lagoa do Sumidouro, Lagoa Santa, MG	Charcoal	LEEH	4230 ± 40	AMS	Beta-234511
1136	Lagoa do Sumidouro, Lagoa Santa, MG	Charcoal	LEEH	540 ± 40	AMS	Beta-234512
1173	Lagoa do Sumidouro, Lagoa Santa, MG	Charcoal	LEEH	3460 ± 40	AMS	Beta-234513
1174	Lagoa do Sumidouro, Lagoa Santa, MG	Charcoal	LEEH	4450 ± 40	AMS	Beta-234514
1175	Lagoa do Sumidouro, Lagoa Santa, MG	Charcoal	LEEH	5140 ± 40	AMS	Beta-234515
1176	Lagoa do Sumidouro, Lagoa Santa, MG	Charcoal	LEEH	4660 ± 40	AMS	Beta-234516
1180	Lagoa do Sumidouro, Lagoa Santa, MG	Charcoal	LEEH	1370 ± 40	AMS	Beta-234517
1181	Lagoa do Sumidouro, Lagoa Santa, MG	Charcoal	LEEH	350 ± 40	AMS	Beta-234518
CV-L2-13122	Gruta Cuvieri, Lagoa Santa, MG	Bone – *Smilodon*	LEEH	10,630 ± 60	AMS	Beta-234519

Sample#814	Lagoa do Sumidouro, Lagoa Santa, MG	Charcoal	LEEH	430±40	AMS	Beta-234520
CV-L2-4041	Gruta Cuvieri, Lagoa Santa, MG	Bone – amphibian	LEEH	3480±40	AMS	Beta-235460
1190	Coqueirinho	Charcoal	LEEH	10,460±60	AMS	Beta-237346
MZSP-PV454	Abismo do Fóssil, PETAR, São Paulo	Tooth – *Toxodon platensis*	MZUSP	11,820±70	AMS	Beta-237347
MZSP-PV610	Abismo do Fóssil, PETAR, São Paulo	Tooth – *Eremotherium*	MZUSP	12,490±60	AMS	Beta-237348
MZSP-PV642	Abismo do Fóssil, PETAR, São Paulo	Bone – *Catonyx*	MZUSP	15,720±80	AMS	Beta-237349
MZSP-PV660	Abismo do Fóssil, PETAR, São Paulo	Bone – *Glyptodon clavipes*	MZUSP	17,690±70	AMS	Beta-237350
TMOENDAS-S1	Toca das Moendas, PI	Shell	FUMDHAM	26,730±140	AMS	Beta-238680
UNG2371	Ribeirão da Mata, Pedro Leopoldo, MG	Lacustrine sediment	LEEH	5540±40	Conventional radiometrics	Beta-238681
UNG2374	Ribeirão da Mata, Pedro Leopoldo, MG	Lacustrine sediment	LEEH	5530±40	Conventional radiometrics	Beta-238682
UNG2379	Ribeirão da Mata, Pedro Leopoldo, MG	Lacustrine sediment	LEEH	5490±40	Conventional radiometrics	Beta-238683
UNG2407	Ribeirão da Mata, Pedro Leopoldo, MG	Lacustrine sediment	LEEH	5860±40	Conventional radiometrics	Beta-238684
UNG2415	Ribeirão da Mata, Pedro Leopoldo, MG	Lacustrine sediment	LEEH	5710±40	Conventional radiometrics	Beta-238685
UNG2440	Ribeirão da Mata, Pedro Leopoldo, MG	Lacustrine sediment	LEEH	5520±40	Conventional radiometrics	Beta-238686
FHAM142024	Poço Cavado-1115, PI	Charcoal	FUMDHAM	4940±50	AMS	Beta-242136
FHAM142036-1	Poço Cavado-1115, PI	Charcoal	FUMDHAM	7790±70	AMS	Beta-242137
RMT1/1310	Ribeirão da Mata, Pedro Leopoldo, MG	Wood	LEEH	2240±50	Conventional radiometrics	Beta-242566
536	Lapa Grande do Taquaraçu, Taquaraçu de Minas, MG	Charcoal	LEEH	9960±60	AMS	Beta-242714
544	Lapa Grande do Taquaraçu, Taquaraçu de Minas, MG	Charcoal	LEEH	9930±60	AMS	Beta-242715
335	Lapa do Santo, Lagoa Santa, MG	Charcoal	LEEH	4160±40	AMS	Beta-243392
359	Lapa do Santo, Lagoa Santa, MG	Charcoal	LEEH	3920±40	AMS	Beta-243393

(continued)

Supplemental 9.1 (continued)

Sample identification	Site/region	Material	Institution	Measured radiocarbon age (in years before present)	Method	Laboratory identification
MZSPPV773*	Abismo do Fóssil, PETAR, São Paulo	Bone – *Megatheriidae*	MZUSP	15,190 ± 70	AMS	Beta-246245
364*	Lapa do Santo, Lagoa Santa, MG	Charcoal	LEEH	8900 ± 40	AMS	Beta-246246
CV-L2 14310	Gruta Cuvieri, Lagoa Santa, MG	Bone – *Valgipes bucklandii*	LEEH	10860 ± 40	AMS	Beta-248057
CV-L2 14827	Gruta Cuvieri, Lagoa Santa, MG	Bone – *Cervidae*	LEEH	7630 ± 50	AMS	Beta-248058
CV-L2 15266	Gruta Cuvieri, Lagoa Santa, MG	Bone – *Tayassu*	LEEH	10,390 ± 40	AMS	Beta-248059
COQ-1624	Coqueirinho, MG	Charcoal	LEEH	3800 ± 40	AMS	Beta-248885
COQ-1628	Coqueirinho, MG	Charcoal	LEEH	3620 ± 40	AMS	Beta-248886
COQ-1630	Coqueirinho, MG	Charcoal	LEEH	2750 ± 40	AMS	Beta-248887
COQ-1631	Coqueirinho, MG	Charcoal	LEEH	6810 ± 50	AMS	Beta-248888
COQ-1648	Coqueirinho, MG	Charcoal	LEEH	140 ± 40	AMS	Beta-248889
Santo-1431	Lapa do Santo, Lagoa Santa, MG	Charcoal	LEEH	4460 ± 40	AMS	Beta-248891
Santo-1696	Lapa do Santo, Lagoa Santa, MG	Charcoal	LEEH	8170 ± 50	AMS	Beta-248893
CVL2-2290	Gruta Cuvieri, Lagoa Santa, MG	Bone – *Cervidae*	LEEH	2790 ± 40	AMS	Beta-251075
CVL2-5998A	Gruta Cuvieri, Lagoa Santa, MG	Bone – *Cuniculus paca*	LEEH	5040 ± 40	AMS	Beta-251078
CVL2-7456	Gruta Cuvieri, Lagoa Santa, MG	Bone – *Tayassu*	LEEH	5100 ± 50	AMS	Beta-251079
LMO2139141	Lagoa dos Mares, MG	Lacustrine sediment	LEEH	6190 ± 50	AMS	Beta-252974
LMO2172174	Lagoa dos Mares, MG	Lacustrine sediment	LEEH	12,670 ± 80	AMS	Beta-252975
LMO22022	Lagoa dos Mares, MG	Lacustrine sediment	LEEH	620 ± 40	AMS	Beta-252976
LMO2203205	Lagoa dos Mares, MG	Lacustrine sediment	LEEH	3490 ± 40	AMS	Beta-252977
LMO2230232	Lagoa dos Mares, MG	Lacustrine sediment	LEEH	19,670 ± 100	AMS	Beta-252978
LMO26769	Lagoa dos Mares, MG	Lacustrine sediment	LEEH	13,950 ± 70	AMS	Beta-252979
TRM-30-33	Ribeirão da Mata, Pedro Leopoldo, MG	Lacustrine sediment	LEEH	9570 ± 60	AMS	Beta-252980
TRM-139-144	Ribeirão da Mata, Pedro Leopoldo, MG	Lacustrine sediment	LEEH	9650 ± 60	AMS	Beta-252981
LSSep2	Lapa do Santo, Lagoa Santa, MG	Bone – *Homo sapiens*	LEEH	690 ± 40	AMS	Beta-253497

LSSep14	Lapa do Santo, Lagoa Santa, MG	Bone – *Homo sapiens*	LEEH	8640 ± 50	AMS	Beta-253505
LSSep17	Lapa do Santo, Lagoa Santa, MG	Bone – *Homo sapiens*	LEEH	8560 ± 50	AMS	Beta-253507
LSSep26	Lapa do Santo, Lagoa Santa, MG	Bone – *Homo sapiens*	LEEH	8450 ± 50	AMS	Beta-253511
LS1475	Lapa do Santo	Charcoal	LEEH	9520 ± 60	AMS	Beta-256223
LS1488	Lapa do Santo	Charcoal	LEEH	10,180 ± 60	AMS	Beta-256224
MCLAROSDENT1	Lapa Pintada, Montes Claros, SP	Bone – *Homo sapiens*	LEEH	350 ± 40	AMS	Beta-260110
MCLAROSDENT2	Lapa Pintada, Montes Claros, SP	Bone – *Homo sapiens*	LEEH	650 ± 40	AMS	Beta-260111
CV-L2-8040	Gruta Cuvieri, Lagoa Santa, MG	Bone – *Cervidae*	LEEH	6990 ± 50	AMS	Beta-261274
ST-A1724	Lapa do Santo, Lagoa Santa, MG	Charcoal	LEEH	9660 ± 50	AMS	Beta-263882
ST-A1738	Lapa do Santo, Lagoa Santa, MG	Charcoal	LEEH	9650 ± 50	AMS	Beta-263883
ST-A1737	Lapa do Santo, Lagoa Santa, MG	Charcoal	LEEH	9470 ± 50	AMS	Beta-263884
ST-A1740	Lapa do Santo, Lagoa Santa, MG	Charcoal	LEEH	9370 ± 40	AMS	Beta-263885
madeirasvmg	São Vicente de Minas, MG	Wood	LEEH	270 ± 40	AMS	Beta-264704
ST-Sep1Dente	Lapa do Santo, Lagoa Santa, MG	Bone – *Homo sapiens*	LEEH	8730 ± 60	AMS	Beta-271249
ST-A1734	Lapa do Santo, Lagoa Santa, MG	Charcoal	LEEH	8880 ± 40	AMS	Beta-280487
ST-A1739	Lapa do Santo, Lagoa Santa, MG	Charcoal	LEEH	9670 ± 40	AMS	Beta-280488
ST-A1745	Lapa do Santo, Lagoa Santa, MG	Charcoal	LEEH	10,460 ± 50	AMS	Beta-280489
ST-A1751	Lapa do Santo, Lagoa Santa, MG	Charcoal	LEEH	8810 ± 40	AMS	Beta-280490
ST-A1753	Lapa do Santo, Lagoa Santa, MG	Charcoal	LEEH	1110 ± 40	AMS	Beta-280491
MO-SEP28	Sambaqui Morro do Ouro, Joinville	Bone – *Homo sapiens*	MASJ	3850 ± 40	AMS	Beta-93152
BURI-I	Buritizeiro, MG	Bone – *Homo sapiens*	MHN-UFMG	5590 ± 60	AMS	Beta-96754
SRIACHO-II	Santana do Riacho, Serra do Cipó, MG	Organic material – *Homo sapiens*	MHN-UFMG	5340 ± 60	AMS	Beta-96758
SRIACHO-VA	Santana do Riacho, Serra do Cipó, MG	Bone – *Homo sapiens*	MHN-UFMG	12,890 ± 70	AMS	Beta-96759
SRIACHO-XXII	Santana do Riacho, Serra do Cipó, MG	Organic material – *Homo sapiens*	MHN-UFMG	5740 ± 70	AMS	Beta-96760
GUANABAARA-II	Sambaqui Guanabara II, Joinville	Charcoal	MASJ	2350 ± 120	Conventional radiometrics	Beta-967756
99,1-773	Palli Aika, Chile	Bone – *Homo sapiens*	AMNH-USA	7830 ± 60	AMS	Beta-99066

(continued)

Supplemental 9.1 (continued)

Sample identification	Site/region	Material	Institution	Measured radiocarbon age (in years before present)	Method	Laboratory identification
P11	Lapa do Santo, Lagoa Santo, MG	Charcoal	LEEH	8660 ± 40	AMS	Oxford-28665
N11	Lapa do Santo, Lagoa Santo, MG	Charcoal	LEEH	8520 ± 55	AMS	Oxford-28666
N14	Lapa do Santo, Lagoa Santo, MG	Charcoal	LEEH	8723 ± 37	AMS	Oxford-30488
N11	Lapa do Santo, Lagoa Santo, MG	Charcoal	LEEH	8958 ± 39	AMS	Oxford-30489
P11	Lapa do Santo, Lagoa Santo, MG	Charcoal	LEEH	8765 ± 45	AMS	Oxford-30534
N13	Lapa do Santo, Lagoa Santo, MG	Charcoal	LEEH	8853 ± 36	AMS	Oxford-30685
P11	Lapa do Santo, Lagoa Santo, MG	Charcoal	LEEH	4086 ± 28	AMS	Oxford-30686
P11	Lapa do Santo, Lagoa Santo, MG	Charcoal	LEEH	8802 ± 37	AMS	Oxford-30687
N11	Lapa do Santo, Lagoa Santo, MG	Charcoal	LEEH	8959 ± 36	AMS	Oxford-30688
O12	Lapa do Santo, Lagoa Santo, MG	Charcoal	LEEH	8820 ± 38	AMS	Oxford-30689
N11	Lapa do Santo, Lagoa Santo, MG	Charcoal	LEEH	8590 ± 45	AMS	Oxford-30690
O13	Lapa do Santo, Lagoa Santo, MG	Charcoal	LEEH	8950 ± 45	AMS	Oxford-30691
O13	Lapa do Santo, Lagoa Santo, MG	Charcoal	LEEH	8937 ± 38	AMS	Oxford-30692
N14	Lapa do Santo, Lagoa Santo, MG	Charcoal	LEEH	8830 ± 40	AMS	Oxford-31014
Hussan-01	Gruta do Japonês, MT	Bone – snake	MZUSP	Insufficient collagen	AMS	–
MZSP-PV773	Abismo do Fóssil, PETAR, São Paulo	Bone – *Eremotherium*	MZUSP	Insufficient collagen	AMS	–
MZSP-PV802	Abismo do Juvenal, PETAR, São Paulo	Tooth – *Toxodon platensis*	MZUSP	Insufficient collagen	AMS	–
MZSP-PV805	Abismo do Juvenal, PETAR, São Paulo	Bone – *Eremotherium*	MZUSP	Insufficient collagen	AMS	–
MZSP-S/N	Abismo do Osso, PETAR, São Paulo	Bone – *Catonyx*	MZUSP	Insufficient collagen	AMS	–

Hussan-02	Abismo Gemeo, Iporanga, SP	Bone – *Xenarthra*	MZUSP	Insufficient collagen	AMS	–
HUSSAN-4	Abismo Gemeo, Iporanga, SP	Bone – unidentified	MZUSP	Insufficient collagen	AMS	–
PF-1097	Abismo Ponta de Flecha, Iporanga, SP	Bone – *Nothrotherium*	IGC-USP	Insufficient collagen	AMS	–
AQSA20	Abismo Quaternário, MG	Bone – *Catonyx cuvieri*	LEEH	Insufficient collagen	AMS	–
AQSA77	Abismo Quaternário, MG	Bone – *Catonyx cuvieri*	LEEH	Insufficient collagen	AMS	–
PASSAGEM1	Abrigo da Passagem, Matozinhos, MG	Bone – *Catonyx cuvieri*	LEEH	Insufficient collagen	AMS	–
27,150	Antonião, PI	Tooth – *Eremotherium*	FUMDHAM	Insufficient collagen	AMS	–
3578-r9	Antonião, PI	Bone – *Catonyx*	FUMDHAM	Insufficient collagen	AMS	–
FHAM184-27172	Antonião, PI	Bone – *Eremotherium*	FUMDHAM	Insufficient collagen	AMS	–
FHAM-ANT-S/N#1	Antonião, PI	Tooth – *Homo sapiens*	FUMDHAM	Insufficient collagen	AMS	–
FHAM-ANT-S/N32	Antonião, PI	Tooth – *Homo sapiens*	FUMDHAM	Insufficient collagen	AMS	–
ARAXÁ-MASTO1	Araxá, RJ	Tooth – *Haplomastodon*	MHN-UFMG	Insufficient collagen	AMS	–
ARAXÁ-MASTO2	Araxá, RJ	Tooth – *Haplomastodon*	MHN-UFMG	Insufficient collagen	AMS	–
LGP-P0004	Arroio Chuí, Rio Grande do Sul	Tooth – *Stegomastodon waringi*	UFRGS	Insufficient collagen	AMS	–
MOT0050	Arroio Chuí, Rio Grande do Sul	Tooth – Toxodon platensis	UFRGS	Insufficient collagen	AMS	–

(continued)

Supplemental 9.1 (continued)

Sample identification	Site/region	Material	Institution	Measured radiocarbon age (in years before present)	Method	Laboratory identification
FUMDHAM 25630	Barra do Antonião, PI	Bone – *Homo sapiens*	FUMDHAM	Plaster contamination	–	–
ZMUC472	Baú	Bone – *Chlamidotherium humb*	ZMUC	Insufficient collagen	AMS	–
ZMUC674	Baú	Bone – *Hoplophorus euphractus*	ZMUC	Insufficient collagen	AMS	–
ZMUCS/N1	Baú	Bone – *Hoplophorus euphractus*	ZMUC	Insufficient collagen	AMS	–
ZMUC3316	Camelo	Bone – *Scelidotherium magnum*	ZMUC	Insufficient collagen	AMS	–
1PM-POÇOAZUL	Caverna poço Azul, Bahia	Bone – *Homo sapiens*	PUC-MG	Insufficient collagen	AMS	–
PÇAZUL-1	Caverna poço Azul, Bahia	Bone – *Valgipes*	PUC-MG	Insufficient collagen	AMS	–
PÇAZUL-10	Caverna poço Azul, Bahia	Tooth – *Rodentia*	PUC-MG	Insufficient collagen	AMS	–
PÇAZUL-11	Caverna poço Azul, Bahia	Tooth – *Megalonichidae*	PUC-MG	Insufficient collagen	AMS	–
PÇAZUL-13	Caverna poço Azul, Bahia	Tooth – *Nothrotherium*	PUC-MG	Insufficient collagen	AMS	–
PÇAZUL-14	Caverna poço Azul, Bahia	Tooth – *Stegomastodon*	PUC-MG	Insufficient collagen	AMS	–
PÇAZUL-15	Caverna poço Azul, Bahia	Tooth – *Alouatta*	PUC-MG	Insufficient collagen	AMS	–
PÇAZUL-16	Caverna poço Azul, Bahia	Tooth – *Conepatus*	PUC-MG	Insufficient collagen	AMS	–

PÇAZUL-2	Caverna poço Azul, Bahia	Bone – *Toxodon*	PUC-MG	Insufficient collagen	AMS	–
PÇAZUL-3	Caverna poço Azul, Bahia	Bone – *Eremotherium*	PUC-MG	Insufficient collagen	AMS	–
PÇAZUL-4	Caverna poço Azul, Bahia	Tooth – *Eremotherium*	PUC-MG	Insufficient collagen	AMS	–
PÇAZUL-5	Caverna poço Azul, Bahia	Bone – *Pampatherium*	PUC-MG	Insufficient collagen	AMS	–
PÇAZUL-6	Caverna poço Azul, Bahia	Bone – *Smilodon*	PUC-MG	Insufficient collagen	AMS	–
PÇAZUL-7	Caverna poço Azul, Bahia	Bone – *Catonyx*	PUC-MG	Insufficient collagen	AMS	–
PÇAZUL-8	Caverna poço Azul, Bahia	Bone – *Catonyx*	PUC-MG	Insufficient collagen	AMS	–
PÇAZUL-9	Caverna poço Azul, Bahia	Tooth – *Tayassu*	PUC-MG	Insufficient collagen	AMS	–
PÇAZULHS-1	Caverna poço Azul, Bahia	Bone – *Homo sapiens*	PUC-MG	Insufficient collagen	AMS	–
PM-PÇAZUL*1	Caverna poço Azul, Bahia	Tooth – *Homo sapiens*	PUC-MG	Insufficient collagen	AMS	–
ZMUC1958	Cerca Grande	Bone – *Scelidotherium magnum*	ZMUC	Insufficient collagen	AMS	–
MN-1320	Cerca Grande 2, Lagoa Santa, MG	Bone – *Homo sapiens*	UFMG	Insufficient collagen	AMS	–
MN-1334	Cerca Grande 5, Lagoa Santa, MG	Bone – *Homo sapiens*	MN-UFRJ	Insufficient collagen	Conventional radiometrics	–
MN-1336	Cerca Grande 5, Lagoa Santa, MG	Bone – *Homo sapiens*	UFMG	Insufficient collagen	AMS	–
UFMG-HW-12	Coleção Harold Walter, Lagoa Santa, MG	Bone – *Homo sapiens*	MHN-UFMG	Insufficient collagen	AMS	–

(continued)

Supplemental 9.1 (continued)

Sample identification	Site/region	Material	Institution	Measured radiocarbon age (in years before present)	Method	Laboratory identification
UFMG-HW-15	Coleção Harold Walter, Lagoa Santa, MG	Bone – *Homo sapiens*	MHN-UFMG	Insufficient collagen	AMS	–
ZMUC3035	Come-não-bebe	Bone – *Scelidotherium magnum*	ZMUC	Insufficient collagen	AMS	–
ZMUC2376	Escrivânia 1	Bone – *Scelidotherium magnum*	ZMUC	Insufficient collagen	AMS	–
ZMUC3733	Escrivânia 1	Bone – *Auchenia*	ZMUC	Insufficient collagen	AMS	–
ZMUC11581	Escrivânia 11	Bone – *Chlamidotherium humb*	ZMUC	Insufficient collagen	AMS	–
ZMUC7034	Escrivânia 11	Bone – *Auchenia*	ZMUC	Insufficient collagen	AMS	–
ZMUC-ESC2-HS	Escrivânia 2, Lagoa Santa, MG	Bone – *Homo sapiens*	ZMUC	Insufficient collagen	AMS	–
ZMUC5635	Escrivânia 5	Bone – *Coelodon maquinensis*	ZMUC	Insufficient collagen	AMS	–
ZMUC5715	Escrivânia 5	Tooth – *Ursus brasiliensis*	ZMUC	Insufficient collagen	AMS	–
ZMUC5750	Escrivânia 5	Bone – *Ursus brasiliensis*	ZMUC	Insufficient collagen	AMS	–
ZMUC6861	Escrivânia 5	Bone – *Auchenia*	ZMUC	Insufficient collagen	AMS	–
ZMUC8623	Escrivânia 5	Bone – *Toxodon*	ZMUC	Insufficient collagen	AMS	–
CV-L2 15488	Gruta Cuvieri, Lagoa Santa, MG	Bone – *Scelidotheriinae*	LEEH	Insufficient collagen	AMS	–

CV-L2 15526	Gruta Cuvieri, Lagoa Santa, MG	Bone – *Cervidae*	LEEH	Insufficient collagen	AMS	–
CV-L2-14973	Gruta Cuvieri, Lagoa Santa, MG	Tooth – *Scelidotheriinae*	LEEH	Insufficient collagen	AMS	–
CV-L2-15532	Gruta Cuvieri, Lagoa Santa, MG	Bone – *Cervidae*	LEEH	Insufficient collagen	AMS	–
CVL2-2305	Gruta Cuvieri, Lagoa Santa, MG	Bone – *Cervidae*	LEEH	Insufficient collagen	AMS	–
CVL2-5881	Gruta Cuvieri, Lagoa Santa, MG	Bone – *Cuniculus paca*	LEEH	Insufficient collagen	AMS	–
CVL2-7457	Gruta Cuvieri, Lagoa Santa, MG	Bone – *Tayassu*	LEEH	Insufficient collagen	AMS	–
CVL2-8637	Gruta Cuvieri, Lagoa Santa, MG	Bone – unidentified megafauna	LEEH	Insufficient collagen	AMS	–
CV-L3 P1264	Gruta Cuvieri, Lagoa Santa, MG	Tooth – *Catonyx cuvieri*	LEEH	Insufficient collagen	AMS	–
CVL3-2028	Gruta Cuvieri, Lagoa Santa, MG	Bone – *Tayassu* sp.	LEEH	Insufficient collagen	AMS	–
CVL32079	Gruta Cuvieri, Lagoa Santa, MG	Bone – *Tayassu pecari*	LEEH	Insufficient collagen	AMS	–
CVL32125	Gruta Cuvieri, Lagoa Santa, MG	Bone – *Tayassu* sp.	LEEH	Insufficient collagen	AMS	–
CV-L3-2837	Gruta Cuvieri, Lagoa Santa, MG	Tooth – *V. bucklandii*	LEEH	Insufficient collagen	AMS	–
CV-L3-3245	Gruta Cuvieri, Lagoa Santa, MG	Tooth – rodent	LEEH	Insufficient collagen	AMS	–
CV-L3A-EXP68	Gruta Cuvieri, Lagoa Santa, MG	Tooth – rodent	LEEH	Insufficient collagen	AMS	–
CVL3As/n-arg	Gruta Cuvieri, Lagoa Santa, MG	Tooth – rodent	LEEH	Insufficient collagen	AMS	–

(continued)

Supplemental 9.1 (continued)

Sample identification	Site/region	Material	Institution	Measured radiocarbon age (in years before present)	Method	Laboratory identification
CV-L3B 337	Gruta Cuvieri, Lagoa Santa, MG	Tooth – *Catonyx cuvieri*	LEEH	Insufficient collagen	AMS	–
CVL3-EXP30	Gruta Cuvieri, Lagoa Santa, MG	Bone – *Tayassu* sp.	LEEH	Insufficient collagen	AMS	–
CVL3-EXP69	Gruta Cuvieri, Lagoa Santa, MG	Tooth – *Cuniculus* sp.	LEEH	Insufficient collagen	AMS	–
CV-L3-P1582	Gruta Cuvieri, Lagoa Santa, MG	Bone – *T. pecari*	LEEH	Insufficient collagen	AMS	–
GSt-HB 01	Gruta da Lapa do Santo, MG]	Bone – *Homo sapiens*	LEEH	Insufficient collagen	AMS	–
GSt-HB 02	Gruta da Lapa do Santo, MG]	Bone – *Homo sapiens*	LEEH	Insufficient collagen	AMS	–
M-PUC-14/18	Gruta do Baú, Lagoa Santa, MG	Bone – *Pampatherium*	PUC-MG	Insufficient collagen	AMS	–
MCL3225-10	Gruta do Brejões (salão das Aves), BA	Bone – *Homo sapiens*	PUC-MG	Insufficient collagen	AMS	–
GSABUGO-NOTH	Gruta do Sabugo, Lagoa Santa, MG	Bone – *Nothrotherium*	LEEH	Insufficient collagen	AMS	–
MCLS/N-MYLOD	Gruta dos Brejões, BA	Bone – *Mylodonopsis*	PUC-MG	Insufficient collagen	AMS	–
ZMUCSH17B	Gruta Sumidouro, Lagoa Santa, MG	Tooth – *Homo sapiens*	ZMUC	Insufficient collagen	AMS	–
ZMUC500	Indios 1	Bone – *Scelidotherium magnum*	ZMUC	Insufficient collagen	AMS	–
ZMUC506	Indios 4	Bone – *Scelidotherium magnum*	ZMUC	Insufficient collagen	AMS	–

MPEG-ITA-MAS	Itaituba, Pará	Bone – *Haplomastodon*	MPEG	Insufficient collagen	AMS	–
CHIMBA-HS	La Chimba, Chile	Bone – *Homo sapiens*	IIARMPGLP	Insufficient collagen	AMS	–
FHAM108796	Lagoa do Quari, PI	Bone – unidentified megafauna	FUMDHAM	Insufficient collagen	AMS	–
FHAM79803-27	Lagoa do Quari, PI	Bone – *Toxodon*	FUMDHAM	Insufficient collagen	AMS	–
HW-10	Lagoa Santa , MG	Tooth – *Homo sapiens*	UFMG	Insufficient collagen	AMS	–
HW-12	Lagoa Santa , MG	Tooth – *Homo sapiens*	UFMG	Insufficient collagen	AMS	–
HW-24	Lagoa Santa , MG	Tooth – *Homo sapiens*	UFMG	Insufficient collagen	AMS	–
HW-2520-2	Lagoa Santa , MG	Tooth – *Homo sapiens*	UFMG	Insufficient collagen	AMS	–
HW-2520-3	Lagoa Santa , MG	Tooth – *Homo sapiens*	UFMG	Insufficient collagen	AMS	–
HW-2520-5	Lagoa Santa , MG	Tooth – *Homo sapiens*	UFMG	Insufficient collagen	AMS	–
HW-2520-6	Lagoa Santa , MG	Tooth – *Homo sapiens*	UFMG	Insufficient collagen	AMS	–
HW-2520-8	Lagoa Santa , MG	Tooth – *Homo sapiens*	UFMG	Insufficient collagen	AMS	–
HW-292	Lagoa Santa , MG	Tooth – *Homo sapiens*	UFMG	Insufficient collagen	AMS	–
HW-4	Lagoa Santa , MG	Tooth – *Homo sapiens*	UFMG	Insufficient collagen	AMS	–
HW-NA-3	Lagoa Santa , MG	Tooth – *Homo sapiens*	Centro Arqueológico Lagoa Santa	Insufficient collagen	AMS	–

(continued)

Supplemental 9.1 (continued)

Sample identification	Site/region	Material	Institution	Measured radiocarbon age (in years before present)	Method	Laboratory identification
HW-15	Lagoa Santa, MG	Bone – *Homo sapiens*	UFMG	Insufficient collagen	AMS	–
HW-293	Lagoa Santa, MG	Bone – *Homo sapiens*	UFMG	Insufficient collagen	AMS	–
116394-7	Lagoa São Vitor, PI	Bone – *Toxodon*	FUMDHAM	Insufficient collagen	AMS	–
14690-1	Lagoa São Vitor, PI	Bone – *Eremotherium*	FUMDHAM	Insufficient collagen	AMS	–
MCLS/N5	Lapa da Cerca Grande, Matozinhos, MG	Bone – *Tayassu pecari*	PUC-MG	Insufficient collagen	AMS	–
BOLSEP3OSSO	Lapa das Boleiras, Lagoa Santa, MG	Bone – *Homo sapiens*	LEEH	Insufficient collagen	AMS	–
MPUC5262433	Lapa do Baú, Matozinhos, MG	Bone – *Hoplophorus euphractus*	PUC-MG	Insufficient collagen	AMS	–
3d/05	Lapa do Bauzinho, Matozinhos, MG	Bone – *Tayassu pecari*	LEEH	Insufficient collagen	AMS	–
BQT-H11-N7-3868	Lapa do Boquete, Peruaçu, MG	Bone – *Homo sapiens*	MHN-UFMG	Insufficient collagen	AMS	–
BQT-Sep1	Lapa do Boquete, Peruaçu, MG	Bone – *Homo sapiens*	MHN-UFMG	Insufficient collagen	AMS	–
MN-808	Lapa do Caetano, Lagoa Santa, MG	Bone – *Homo sapiens*	MN-UFRJ	Insufficient collagen	Conventional radiometrics	–
MN836	Lapa do Caetano, Lagoa Santa, MG	Bone – *Homo sapiens*	PUC-MG	Insufficient collagen	AMS	–
364	Lapa do Santo, Lagoa Santa, MG	Charcoal	LEEH	Absence of carbon	AMS	–

LSSep1	Lapa do Santo, Lagoa Santa, MG	Bone – *Homo sapiens*	LEEH	Insufficient collagen	AMS	–
LSSep10	Lapa do Santo, Lagoa Santa, MG	Bone – *Homo sapiens*	LEEH	Insufficient collagen	AMS	–
LSSep10-09	Lapa do Santo, Lagoa Santa, MG	Bone – *Homo sapiens*	LEEH	Insufficient collagen	AMS	–
LSSep1-09	Lapa do Santo, Lagoa Santa, MG	Bone – *Homo sapiens*	LEEH	Insufficient collagen	AMS	–
LSSep15	Lapa do Santo, Lagoa Santa, MG	Bone – *Homo sapiens*	LEEH	Insufficient collagen	AMS	–
LSSep15-09	Lapa do Santo, Lagoa Santa, MG	Bone – *Homo sapiens*	LEEH	Insufficient collagen	AMS	–
LSSep20	Lapa do Santo, Lagoa Santa, MG	Bone – *Homo sapiens*	LEEH	Insufficient collagen	AMS	–
LSSep21	Lapa do Santo, Lagoa Santa, MG	Bone – *Homo sapiens*	LEEH	Insufficient collagen	AMS	–
LSSep21-09	Lapa do Santo, Lagoa Santa, MG	Bone – *Homo sapiens*	LEEH	Insufficient collagen	AMS	–
LSSep22	Lapa do Santo, Lagoa Santa, MG	Bone – *Homo sapiens*	LEEH	Insufficient collagen	AMS	–
LSSep22-09	Lapa do Santo, Lagoa Santa, MG	Bone – *Homo sapiens*	LEEH	Insufficient collagen	AMS	–
LSSep3	Lapa do Santo, Lagoa Santa, MG	Bone – *Homo sapiens*	LEEH	Insufficient collagen	AMS	–
LSSep3-09	Lapa do Santo, Lagoa Santa, MG	Bone – *Homo sapiens*	LEEH	Insufficient collagen	AMS	–
LSSep4	Lapa do Santo, Lagoa Santa, MG	Bone – *Homo sapiens*	LEEH	Insufficient collagen	AMS	–
LSSep4-09	Lapa do Santo, Lagoa Santa, MG	Bone – *Homo sapiens*	LEEH	Insufficient collagen	AMS	–

(continued)

Supplemental 9.1 (continued)

Sample identification	Site/region	Material	Institution	Measured radiocarbon age (in years before present)	Method	Laboratory identification
LSSep5	Lapa do Santo, Lagoa Santa, MG	Bone – *Homo sapiens*	LEEH	Insufficient collagen	AMS	–
LSSep5-09	Lapa do Santo, Lagoa Santa, MG	Bone – *Homo sapiens*	LEEH	Insufficient collagen	AMS	–
LSSep6	Lapa do Santo, Lagoa Santa, MG	Bone – *Homo sapiens*	LEEH	Insufficient collagen	AMS	–
LSSep6-09	Lapa do Santo, Lagoa Santa, MG	Bone – *Homo sapiens*	LEEH	Insufficient collagen	AMS	–
LSSep8	Lapa do Santo, Lagoa Santa, MG	Bone – *Homo sapiens*	LEEH	Insufficient collagen	AMS	–
LSSep8-09	Lapa do Santo, Lagoa Santa, MG	Bone – *Homo sapiens*	LEEH	Insufficient collagen	AMS	–
LSSep9	Lapa do Santo, Lagoa Santa, MG	Bone – *Homo sapiens*	LEEH	Insufficient collagen	AMS	–
LSSep9-09	Lapa do Santo, Lagoa Santa, MG	Bone – *Homo sapiens*	LEEH	Insufficient collagen	AMS	–
LST Sep 15	Lapa do Santo, Lagoa Santa, MG	Bone – *Homo sapiens*	LEEH	Insufficient collagen	AMS	–
LST Sep 16	Lapa do Santo, Lagoa Santa, MG	Bone – *Homo sapiens*	LEEH	Insufficient collagen	AMS	–
LST Sep 17	Lapa do Santo, Lagoa Santa, MG	Bone – *Homo sapiens*	LEEH	Insufficient collagen	AMS	–
LST Sep 22	Lapa do Santo, Lagoa Santa, MG	Bone – *Homo sapiens*	LEEH	Insufficient collagen	Conventional radiometrics	–
LSTSep01Set	Lapa do Santo, Lagoa Santa, MG	Bone – *Homo sapiens*	LEEH	Insufficient collagen	AMS	–

LSTSep02Set	Lapa do Santo, Lagoa Santa, MG	Bone – *Homo sapiens*	LEEH	Insufficient collagen	AMS	–
LSTSep17Set	Lapa do Santo, Lagoa Santa, MG	Bone – *Homo sapiens*	LEEH	Insufficient collagen	AMS	–
LSTSep21Set	Lapa do Santo, Lagoa Santa, MG	Bone – *Homo sapiens*	LEEH	Insufficient collagen	AMS	–
ST-Sep XX	Lapa do Santo, Lagoa Santa, MG	Bone – *Homo sapiens*	LEEH	Insufficient collagen	AMS	–
ST-Sep27-COS	Lapa do Santo, Lagoa Santa, MG	Bone – *Homo sapiens*	LEEH	Insufficient collagen	AMS	–
ST-Sep3-DENT	Lapa do Santo, Lagoa Santa, MG	Bone – *Homo sapiens*	LEEH	Insufficient collagen	AMS	–
ST-SEPIII	Lapa do Santo, Lagoa Santa, MG	Bone – *Homo sapiens*	LEEH	Insufficient collagen	AMS	–
ST-SEPIV	Lapa do Santo, Lagoa Santa, MG	Bone – *Homo sapiens*	LEEH	Insufficient collagen	AMS	–
ST-SEPV	Lapa do Santo, Lagoa Santa, MG	Bone – *Homo sapiens*	LEEH	Insufficient collagen	AMS	–
ST-SEPVI	Lapa do Santo, Lagoa Santa, MG	Bone – *Homo sapiens*	LEEH	Insufficient collagen	AMS	–
ST-SEPVII	Lapa do Santo, Lagoa Santa, MG	Bone – *Homo sapiens*	LEEH	Insufficient collagen	AMS	–
ST-SEPX	Lapa do Santo, Lagoa Santa, MG	Bone – *Homo sapiens*	LEEH	Insufficient collagen	AMS	–
ST-SEPXII	Lapa do Santo, Lagoa Santa, MG	Bone – *Homo sapiens*	LEEH	Insufficient collagen	AMS	–
Lapa do Santo SEP S/NO	Lapa do Santo, Lagoa Santa, MG	Bone – *Homo sapiens*	LEEH	Insufficient collagen	Conventional radiometrics	–
SANTO-1	Lapa do Santo, Lagoa Santa, MG	Bone – *Homo sapiens*	LEEH	Insufficient collagen	Conventional radiometrics	–

(continued)

Supplemental 9.1 (continued)

Sample identification	Site/region	Material	Institution	Measured radiocarbon age (in years before present)	Method	Laboratory identification
ST-SEPII	Lapa do Santo, Lagoa Santa, MG	Bone – *Homo sapiens*	LEEH	Insufficient collagen	AMS	–
ST-SEPII	Lapa do Santo, Lagoa Santa, MG	Bone – *Homo sapiens*	LEEH	Insufficient collagen	AMS	–
ST-SEPXI	Lapa do Santo, Lagoa Santa, MG	Bone – *Homo sapiens*	LEEH	Insufficient collagen	AMS	–
ST-Sep XXb	Lapa do Santo, Lagoa Santo, MG	Bone – *Homo sapiens*	LEEH	Insufficient collagen	AMS	–
ST-SepII-B	Lapa do Santo, Lagoa Santo, MG	Bone – *Homo sapiens*	LEEH	Insufficient collagen	AMS	–
LSUMIMISO-1	Lapa do Sumidouro, Lagoa Santa, MG	Bone – *Homo sapiens*	LEEH	Insufficient collagen	Conventional radiometrics	–
LSUMIMISO-10	Lapa do Sumidouro, Lagoa Santa, MG	Bone – *Homo sapiens*	LEEH	Insufficient collagen	Conventional radiometrics	–
LSUMIMISO-2	Lapa do Sumidouro, Lagoa Santa, MG	Bone – *Homo sapiens*	LEEH	Insufficient collagen	Conventional radiometrics	–
LSUMIMISO-3	Lapa do Sumidouro, Lagoa Santa, MG	Bone – *Homo sapiens*	LEEH	Insufficient collagen	Conventional radiometrics	–
LSUMIMISO-4	Lapa do Sumidouro, Lagoa Santa, MG	Bone – *Homo sapiens*	LEEH	Insufficient collagen	Conventional radiometrics	–
LSUMIMISO-5	Lapa do Sumidouro, Lagoa Santa, MG	Bone – *Homo sapiens*	LEEH	Insufficient collagen	Conventional radiometrics	–
LSUMIMISO-6	Lapa do Sumidouro, Lagoa Santa, MG	Bone – *Homo sapiens*	LEEH	Insufficient collagen	Conventional radiometrics	–
LSUMIMISO-7	Lapa do Sumidouro, Lagoa Santa, MG	Bone – *Homo sapiens*	LEEH	Insufficient collagen	Conventional radiometrics	–

LSUMIMISO-8	Lapa do Sumidouro, Lagoa Santa, MG	Bone – *Homo sapiens*	LEEH	Insufficient collagen	Conventional radiometrics	–
LSUMIMISO-9	Lapa do Sumidouro, Lagoa Santa, MG	Bone – *Homo sapiens*	LEEH	Insufficient collagen	Conventional radiometrics	–
CONF-01	Lapa Mortuária de Confins	Bone – unidentified	LEEH	Insufficient collagen	AMS	–
CONFIN I	Lapa Mortuária de Confins, Confins, MG	Bone – *Homo sapiens*	MHN-UFMG	Plaster contamination	Conventional radiometrics	–
LMC-LC1-NEOC	Lapa Mortuária de Confins, Lagoa Santa, MG	Bone – *Neochoerus sulcidens*	LEEH	Insufficient collagen	AMS	–
LMC-TR2-Ccuv	Lapa Mortuária de Confins, Lagoa Santa, MG	Bone – *Catonyx cuvieri*	LEEH	Insufficient collagen	AMS	–
ZMUC1907	Lapa Vermelha 2	Bone – *Scelidotherium bucklandi*	ZMUC	Insufficient collagen	AMS	–
99,1-773	Palli Aika, Chile	Bone – *Homo sapiens*	AMNH-USA	Insufficient collagen	AMS	–
ZMUC10947	Peri-Peri	Bone – *Ursus brasiliensis*	ZMUC	Insufficient collagen	AMS	–
ZMUC3478	Peri-Peri 1	Bone – *Scelidotherium magnum*	ZMUC	Insufficient collagen	AMS	–
ZMUC-1.7.85.1	Pontimelo	Bone – *Homo sapiens*	ZMUC	Insufficient collagen	AMS	–
ZMUC-Lausen 1	Pontimelo	Bone – *Homo sapiens*	ZMUC	Insufficient collagen	AMS	–
TOXOAPIAI	Região de Apiai, SP	Bone – *Toxodon platensis*	IGC-USP	Insufficient collagen	AMS	–
GREG-RMT-1 D	Ribeirão da Mata, Pedro Leopoldo, MG	Wood	LEEH	Absence of carbon	AMS	–
LGP-I0001a	Rio Grande do Sul	Tooth – *Equus neogeus*	UFRGS	Insufficient collagen	AMS	–

(continued)

Supplemental 9.1 (continued)

Sample identification	Site/region	Material	Institution	Measured radiocarbon age (in years before present)	Method	Laboratory identification
LGP-P0003	Rio Grande do Sul	Tooth – *Stegomastodon waringi*	UFRGS	Insufficient collagen	AMS	–
Itaoca-sep1	Sambaqui Fluvial Itaoca, SP	Bone – *Homo sapiens*	LEEH	Insufficient collagen	AMS	–
PEDRAOCA-1	Sambaqui da Pedra Oca, BA	Bone – *Homo sapiens*	PUC-MG	Insufficient collagen	AMS	–
PEDRAOCA-3	Sambaqui da Pedra Oca, BA	Bone – *Homo sapiens*	PUC-MG	Insufficient collagen	AMS	–
SR1-III	Santana do Riacho, Serra do Cipó, MG	Bone – *Homo sapiens*	MHN-UFMG	Insufficient collagen	Conventional radiometrics	–
SR1-IX	Santana do Riacho, Serra do Cipó, MG	Bone – *Homo sapiens*	MHN-UFMG	Insufficient collagen	AMS	–
SR1-VI	Santana do Riacho, Serra do Cipó, MG	Bone – *Homo sapiens*	MHN-UFMG	Insufficient collagen	Conventional radiometrics	–
SR1-VII	Santana do Riacho, Serra do Cipó, MG	Bone – *Homo sapiens*	MHN-UFMG	Insufficient collagen	Conventional radiometrics	–
SR1-X	Santana do Riacho, Serra do Cipó, MG	Bone – *Homo sapiens*	MHN-UFMG	Insufficient collagen	Conventional radiometrics	–
SR1-XI	Santana do Riacho, Serra do Cipó, MG	Bone – *Homo sapiens*	MHN-UFMG	Insufficient collagen	AMS	–
SR1-XII	Santana do Riacho, Serra do Cipó, MG	Bone – *Homo sapiens*	MHN-UFMG	Insufficient collagen	Conventional radiometrics	–
SR1-XIII	Santana do Riacho, Serra do Cipó, MG	Bone – *Homo sapiens*	MHN-UFMG	Insufficient collagen	Conventional radiometrics	–
SR1-XIXA	Santana do Riacho, Serra do Cipó, MG	Bone – *Homo sapiens*	MHN-UFMG	Insufficient collagen	Conventional radiometrics	–

SR1-XVII	Santana do Riacho, Serra do Cipó, MG	Bone – *Homo sapiens*	MHN-UFMG	Insufficient collagen	Conventional radiometrics	–
SR1-XVII	Santana do Riacho, Serra do Cipó, MG	Bone – *Homo sapiens*	MHN-UFMG	Insufficient collagen	Conventional radiometrics	–
SR1-XVIII	Santana do Riacho, Serra do Cipó, MG	Bone – *Homo sapiens*	MHN-UFMG	Insufficient collagen	Conventional radiometrics	–
SR1-XXI	Santana do Riacho, Serra do Cipó, MG	Bone – *Homo sapiens*	MHN-UFMG	Insufficient collagen	Conventional radiometrics	–
SR1-XXIIIB	Santana do Riacho, Serra do Cipó, MG	Bone – *Homo sapiens*	MHN-UFMG	Insufficient collagen	Conventional radiometrics	–
SR1-XXIV	Santana do Riacho, Serra do Cipó, MG	Bone – *Homo sapiens*	MHN-UFMG	Insufficient collagen	Conventional radiometrics	–
SR1-XXVII	Santana do Riacho, Serra do Cipó, MG	Bone – *Homo sapiens*	MHN-UFMG	Insufficient collagen	Conventional radiometrics	–
SR1-XXVII	Santana do Riacho, Serra do Cipó, MG	Bone – *Homo sapiens*	MHN-UFMG	Insufficient collagen	Conventional radiometrics	–
SR1-XXVIII	Santana do Riacho, Serra do Cipó, MG	Bone – *Homo sapiens*	MHN-UFMG	Insufficient collagen	Conventional radiometrics	–
Quari-1	Sítio 607, PI	Bone – unidentified megafauna	FUMDHAM	Insufficient collagen	AMS	–
LD-505	Sítio Lund, Lagoa Santa, MG	Charcoal	LEEH	Absence of carbon	Conventional radiometrics	–
LD-S1-01	Sítio Lund, Lagoa Santa, MG	Charcoal	LEEH	Absence of carbon	AMS	–
SUMIDB1-180	Sítio Sumidouro, Lagoa Santa, MG	Charcoal	LEEH	Insufficient material	AMS	–
TAMBILLO 2	Sítio Tambillo, deserto de Atacama, Chile	Bone – *Homo sapiens*	IIARMPGLP	Insufficient collagen	AMS	–
ZMUC3041	Sumidouro	Bone – *Ursus brasiliensis*	ZMUC	Insufficient collagen	AMS	–

(continued)

Supplemental 9.1 (continued)

Sample identification	Site/region	Material	Institution	Measured radiocarbon age (in years before present)	Method	Laboratory identification
MHNT-ERE1	Tamanduá de Cima, São Bento do Una, PE	Tooth – *Eremotherium*	MHNT	Insufficient collagen	AMS	–
MHNT-ERE2	Tamanduá de Cima, São Bento do Una, PE	Tooth – *Eremotherium*	MHNT	Insufficient collagen	AMS	–
MHNT-ERE3	Tamanduá de Cima, São Bento do Una, PE	Tooth – *Eremotherium*	MHNT	Insufficient collagen	AMS	–
MHNT-I	Tamanduá de Cima, São Bento do Una, PE	Tooth – *Toxodon*	MHNT	Insufficient collagen	AMS	–
MHNT-II	Tamanduá de Cima, São Bento do Una, PE	Bone – *Toxodon*	MHNT	Insufficient collagen	AMS	–
MHNT-III	Tamanduá de Cima, São Bento do Una, PE	Bone – *Toxodon*	MHNT	Insufficient collagen	AMS	–
MHNT-IV	Tamanduá de Cima, São Bento do Una, PE	Bone – *Toxodon*	MHNT	Insufficient collagen	AMS	–
MHNT-IX	Tamanduá de Cima, São Bento do Una, PE	Bone – terrestrial sloth	MHNT	Insufficient collagen	AMS	–
MHNT-MASTO1	Tamanduá de Cima, São Bento do Una, PE	Tooth – *Haplomastodon*	MHNT	Insufficient collagen	AMS	–
MHNT-MASTO2	Tamanduá de Cima, São Bento do Una, PE	Bone – *Haplomastodon*	MHNT	Insufficient collagen	AMS	–
MHNT-TOXO1	Tamanduá de Cima, São Bento do Una, PE	Tooth – *Toxodon*	MHNT	Insufficient collagen	AMS	–
MHNT-TOXO2	Tamanduá de Cima, São Bento do Una, PE	Bone – *Toxodon*	MHNT	Insufficient collagen	AMS	–
MHNT-TOXO3	Tamanduá de Cima, São Bento do Una, PE	Bone – *Toxodon*	MHNT	Insufficient collagen	AMS	–

MHNT-V	Tamanduá de Cima, São Bento do Una, PE	Bone – *Toxodon*	MHNT	Insufficient collagen	AMS	–
MHNT-VI	Tamanduá de Cima, São Bento do Una, PE	Bone – terrestrial sloth	MHNT	Insufficient collagen	AMS	–
MHNT-VII	Tamanduá de Cima, São Bento do Una, PE	Bone – terrestrial sloth	MHNT	Insufficient collagen	AMS	–
MHNT-VIII	Tamanduá de Cima, São Bento do Una, PE	Bone – terrestrial sloth	MHNT	Insufficient collagen	AMS	–
MHNT-X	Tamanduá de Cima, São Bento do Una, PE	Bone – *Haplomastodon*	MHNT	Insufficient collagen	AMS	–
MHNT-XI	Tamanduá de Cima, São Bento do Una, PE	Bone – *Haplomastodon*	MHNT	Insufficient collagen	AMS	–
MHNT-XII	Tamanduá de Cima, São Bento do Una, PE	Bone – *Haplomastodon*	MHNT	Insufficient collagen	AMS	–
MHNT-XIII	Tamanduá de Cima, São Bento do Una, PE	Bone – *Haplomastodon*	MHNT	Insufficient collagen	AMS	–
MHNT-XIV	Tamanduá de Cima, São Bento do Una, PE	Bone – terrestrial sloth	MHNT	Insufficient collagen	AMS	–
MHNT-XV	Tamanduá de Cima, São Bento do Una, PE	Bone – terrestrial sloth	MHNT	Insufficient collagen	AMS	–
MHNT-XVI	Tamanduá de Cima, São Bento do Una, PE	Bone – terrestrial sloth	MHNT	Insufficient collagen	AMS	–
MHNT-XVII	Tamanduá de Cima, São Bento do Una, PE	Bone – *Toxodon*	MHNT	Insufficient collagen	AMS	–
ZMUC7838	Taquaral 3	Bone – *Scelidotherium*	ZMUC	Insufficient collagen	AMS	–
ZMUC9075	Taquaral 3	Bone – *Scelidotherium magnum*	ZMUC	Insufficient collagen	AMS	–
ZMUC2173	Tatus	Bone – *Scelidotherium magnum*	ZMUC	Insufficient collagen	AMS	–

(continued)

Supplemental 9.1 (continued)

Sample identification	Site/region	Material	Institution	Measured radiocarbon age (in years before present)	Method	Laboratory identification
86182	Toca da Barriguda, PI	Bone – *Scelidodon*	FUMDHAM	Insufficient collagen	AMS	–
80630-2	Toca da Barriguda, PI	Bone – *Scelidodon*	FUMDHAM	Insufficient collagen	AMS	–
146365-1	Toca das Moendas, PI	Tooth – *Equus*	FUMDHAM	Insufficient collagen	AMS	–
FHAM123193a	Toca das Moendas, PI	Bone – *Homo sapiens*	FUMDHAM	Insufficient collagen	AMS	–
FHAM123193c	Toca das Moendas, PI	Bone – *Homo sapiens*	FUMDHAM	Insufficient collagen	AMS	–
FHAM123520	Toca das Moendas, PI	Tooth – *Cervidae*	FUMDHAM	Insufficient collagen	AMS	–
FHAM123754	Toca das Moendas, PI	Tooth – *Cervidae*	FUMDHAM	Insufficient collagen	AMS	–
FHAM123780	Toca das Moendas, PI	Bone – *Homo sapiens*	FUMDHAM	Insufficient collagen	AMS	–
FHAM123791	Toca das Moendas, PI	Bone – *Homo sapiens*	FUMDHAM	Insufficient collagen	AMS	–
TMOENDASESQ3	Toca das Moendas, PI	Bone – *Homo sapiens*	FUMDHAM	Insufficient collagen	AMS	–
TMOENDAS-S1	Toca das Moendas, PI	Bone – *Homo sapiens*	FUMDHAM	Insufficient collagen	AMS	–
MCL2944	Toca das Onças, BA	Bone – *Homo sapiens*	PUC-MG	Insufficient collagen	AMS	–
MCL3051	Toca das Onças, BA	Bone – *Homo sapiens*	PUC-MG	Insufficient collagen	AMS	–

MCLS/N-ERE01	Toca das Onças, BA	Bone – *Eremotherium*	PUC-MG	Insufficient collagen	AMS	–
MCLS/N-ERE02	Toca das Onças, BA	Bone – *Eremotherium*	PUC-MG	Insufficient collagen	AMS	–
MCLS/N-ERE03	Toca das Onças, BA	Bone – *Eremotherium*	PUC-MG	Insufficient collagen	AMS	–
MCLS/N-ERE04	Toca das Onças, BA	Bone – *Eremotherium*	PUC-MG	Insufficient collagen	AMS	–
MCLS/N-ERE05	Toca das Onças, BA	Bone – *Eremotherium*	PUC-MG	Insufficient collagen	AMS	–
MCLS/N-ERE06	Toca das Onças, BA	Bone – *Eremotherium*	PUC-MG	Insufficient collagen	AMS	–
MCLS/N-ERE07	Toca das Onças, BA	Bone – *Eremotherium*	PUC-MG	Insufficient collagen	AMS	–
MCLS/N-ERE08	Toca das Onças, BA	Bone – *Eremotherium*	PUC-MG	Insufficient collagen	AMS	–
MCLS/N-ERE09	Toca das Onças, BA	Bone – *Eremotherium*	PUC-MG	Insufficient collagen	AMS	–
MCLS/N-ERE10	Toca das Onças, BA	Bone – *Eremotherium*	PUC-MG	Insufficient collagen	AMS	–
MCLS/N-GLOSS	Toca das Onças, BA	Tooth – *Glossotherium*	PUC-MG	Insufficient collagen	AMS	–
MCLS/N-NEOCH	Toca das Onças, BA	Tooth – *Noeochoerus*	PUC-MG	Insufficient collagen	AMS	–
MCLS/N-PAMPA	Toca das Onças, BA	Bone – *Pampatherium*	PUC-MG	Insufficient collagen	AMS	–
MCL2955	Toca das Onças, Bh	Bone – *Homo sapiens*	PUC-MG	Insufficient collagen	AMS	–
MCLS/N3	Toca das Onças, Bh	Bone – *Homo sapiens*	PUC-MG	Insufficient collagen	AMS	–

(continued)

Supplemental 9.1 (continued)

Sample identification	Site/region	Material	Institution	Measured radiocarbon age (in years before present)	Method	Laboratory identification
45804	Toca do Artur, PI	Bone – *Palaeolama*	FUMDHAM	Insufficient collagen	AMS	–
37610-1	Toca do Garrincho, PI	Bone – *Hippidion*	FUMDHAM	Insufficient collagen	AMS	–
47739-4	Toca do Garrincho, PI	Tooth – *Eremotherium*	FUMDHAM	Insufficient collagen	AMS	–
MCLS/N4	Toca do Pampatherium, Prudente de Morais	Bone – *Pampatherium*	PUC-MG	Insufficient collagen	AMS	–
FUMDHAM-1162	Toca do Paraguaio, Serra da Capivara, Pi	Bone – *Homo sapiens*	FUMDHAM	Insufficient collagen	AMS	–
FUMDHAM-1336	Toca do Paraguaio, Serra da Capivara, Pi	Bone – *Homo sapiens*	FUMDHAM	Insufficient collagen	AMS	–
FUNDHAMTPAE1	Toca do Paraguaio, Serra da Capivara, Pi	Bone – *Homo sapiens*	FUMDHAM	Insufficient collagen	AMS	–
FUNDHAMTPAE2	Toca do Paraguaio, Serra da Capivara, Pi	Bone – *Homo sapiens*	FUMDHAM	Insufficient collagen	AMS	–
FUNDHAMTPE1*	Toca do Paraguaio, Serra da Capivara, Pi	Tooth – *Homo sapiens*	FUMDHAM	Plaster contamination	–	–
FUNDHAMTPE2*	Toca do Paraguaio, Serra da Capivara, Pi	Tooth – *Homo sapiens*	FUMDHAM	Plaster contamination	–	–
19108	Toca do Pilão, PI	Tooth – *Catonyx*	FUMDHAM	Insufficient collagen	AMS	–
FHAM12378047	Toca do Serrote das Moendas, PI	Bone – *Blastocerus dichotomus*	FUMDHAM	Insufficient collagen	AMS	–
MCL3225-12	Toca dos Brejões, Bh	Bone – *Homo sapiens*	PUC-MG	Insufficient collagen	AMS	–

MCL3225-14	Toca dos Brejões, Bh	Bone – *Homo sapiens*	PUC-MG	Insufficient collagen	AMS	–
MCL3225-16	Toca dos Brejões, Bh	Bone – *Homo sapiens*	PUC-MG	Insufficient collagen	AMS	–
MCLS/N2	Toca dos Brejões, Bh	Bone – *Palaeolama*	PUC-MG	Insufficient collagen	AMS	–
MCLS/N-MASTO	Toca dos Ossos, BA	Tooth – *Haplomastodon*	PUC-MG	Insufficient collagen	AMS	–
MCLS/N-XENOR	Toca dos Ossos, BA	Tooth – *Xenorhinotherium*	PUC-MG	Insufficient collagen	AMS	–
MCLS/N1	Toca dos Ossos, BA	Bone – *Toxodon*	PUC-MG	Insufficient collagen	AMS	–

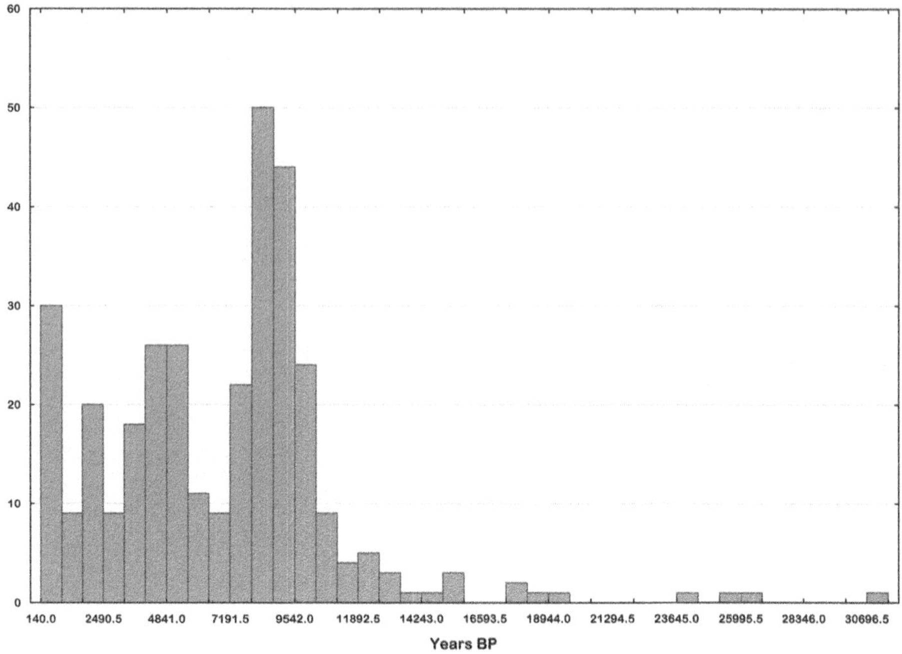

Supplemental 9.2 Distribution of the dates obtained for samples associated to origins project research

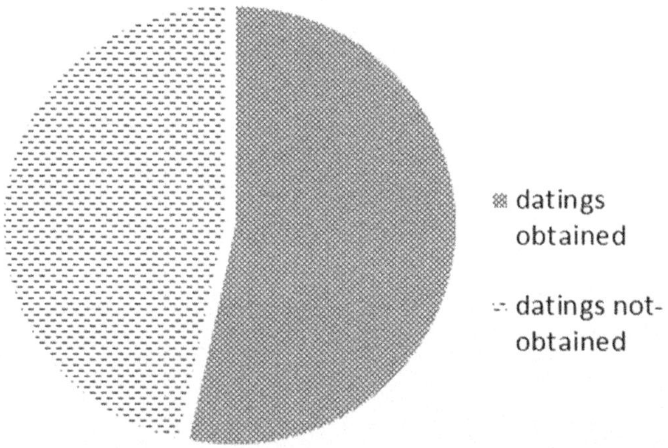

Supplemental 9.3 Proportion of dates obtained (333, or 53.4%) and not obtained (291, or 46.6%) out of the 624 samples submitted for dating under the scope of the origins project

References

Almeida, T. F. (2011). *Análise da dispersão das populações nativas americanas: uma abordagem genético-fisiográfica*. Master Thesis. Departamento de Genética e Biologia Evolutiva, Instituto de Biociências, Universidade de São Paulo, São Paulo.

Almeida, T. F., & Bernardo, D. V. (2008). *Estrutura populacional vista a partir de dados de diversidade craniana e distância geográfica*. Paper presented at the "X Congreso de la Asociación Latinoamericana de Antropología Biológica", La Plata.

Almeida, T. F., & Bernardo, D. V. (2012). *Comparing neutral markers, selective markers, and human cranial morphology: Are human skulls neutral?* Paper presented at the "81st Annual Meeting of the American Association of Physical Anthropologists", Portland.

Almeida, T. F., Neves, W. A., Meyer, D., & Bernardo, D. V. (2011). *The role of biomes in the settlement of the Americas: Testing Beaton's and Dixon's hypotheses*. Paper presented at the "80th Annual Meeting of the American Association of Physical Anthropologists". Minneapolis.

Almeida, T. F., Bernardo, D. V., Neves, W. A., & Inglez, M. (2013). *Sexual dimorphism in human skull: The effect of size correction*. Paper presented at the "82nd Annual Meeting of the American Association of Physical Anthropologists", Knoxville.

Alvarenga, H., Brito, G. R. R., Migotto, R., Hubbe, A., & Hofling, E. (2008). *Pleistovultur nevesi* gen. et sp. nov. (Aves: Vulturidae) and the diversity of condors and vultures in the South American Pleistocene. *Ameghiniana, 45*, 613–618.

Araujo, A. G. M. (2002). *Eventos de seca durante o Holoceno no Brasil: possíveis implicações para o entendimento da variabilidade cultural no período Paleoíndio (11.000–8.000 AP)*. Paper presented at the "XII Congresso da Sociedade de Arqueologia Brasileira", São Paulo.

Araujo, A. G. M. (2003). *Dry Holocene events and human occupation in South America – an archaeological point of view*. Paper presented at the "II Workshop on Extreme Climatic and Paleoclimatic Events in South America: Teleconections Between High and Low Latitudes", São Paulo.

Araujo, A. G. M. (2004a). A variabilidade cultural no Período Paleoíndio (11.000-8.000 AP) no Brasil: algumas hipóteses. *Revista do CEPA, 28*, 111–130.

Araujo, A. G. M. (2004b). *Dryness events during the Holocene and the "Archaic Gap" in Central Brazil*. Paper presented at the "69th Annual Meeting of the Society for American Archaeology", Montreal.

Araujo, A. G. M. (2004c). *Holocene dryness and human occupation in South America: Understanding the 'Archaic Gap'*. Paper presented at the "11th International Palynological Congress", Granada.

Araujo, A. G. M. (2005). *Processos de formação de sítios abrigados na região de Lagoa Santa-MG: o papel da contribuição antrópica e natural no aporte de sedimentos*. Paper presented at the "XIII Congresso da Sociedade de Arqueologia Brasileira", Campo Grande.

Araujo, A. G. M. (2007a). *As primeiras ocupações humanas na América do Sul e seus registros: algumas considerações geomorfológicas e paleoclimáticas*. Paper presented at the "XVI Congresso Mundial da UISPP/XVI Congresso da Sociedade de Arqueologia Brasileira", Florianópolis.

Araujo, A. G. M. (2007b). *Rockshelter geoarchaeology and Paleoindian sites in Central Brazil: A view from the tropics*. Paper presented at the symposium "Developing International Geoarchaeology – DIG – 2007", Cambridge, UK.

Araujo, A. G. M. (2008). *The cultural use of rockshelters by Paleoamericans (12,000–8,000 BP): Three case studies from Central Brazil*. Paper presented at the "6th World Archaeological Congress", Dublin.

Araujo, A. G. M. (2009). Geoarqueologia em sítios abrigados: processos de formação, estratigrafia e potencial informativo. In J. C. R. Rubin & R. T. Silva (Eds.), *Geoarqueologia – Teoria e Prática* (pp. 71–92). Goiânia: Editora da Universidade Católica de Goiás.

Araujo, A. G. M. (2010a). *Paleoenvironments in Brazil and its archaeological implications*. Paper presented at the "V Simposio Internacional-El Hombre Temprano en América: a Cien Años del Debate Ameghino-Hrdlicka (1910–2010)", La Plata.

Araujo, A. G. M. (2010b). Síntese. In A. G. M. Araujo & W. A. Neves (Eds.), *Lapa das Boleiras – Um Sítio Paleoíndio do Carste de Lagoa Santa, MG, Brasil* (pp. 213–219). São Paulo: Editora Annablume/FAPESP.

Araujo, A. G. M. (2011a). *Paleoambientes no Brasil e suas implicações arqueológicas*. Paper presented at the "XVI Congresso Mundial da UISPP/XVI Congresso da Sociedade de Arqueologia Brasileira", Florianópolis.

Araujo, A. G. M. (2011b). Os Primeiros da Fila. *Revista de História da Biblioteca Nacional, 6*, 24–26.

Araujo, A. G. M. (2011c). *Paleoindians in SE Brazil: New data from Lagoa Santa and Rio Claro*. Paper presented at the "76th Annual Meeting of the Society for American Archaeology", Sacramento.

Araujo, A. G. M. (2012). *On vastness and variability: Some thoughts about Paleoindian cultural transmission in Eastern South America*. Paper presented at the "77th Annual Meeting of the Society for American Archaeology", Memphis.

Araujo, A. G. M., & Feathers, J. K. (2008). First notice of open-air Paleoamerican sites at Lagoa Santa: Some geomorphological and paleoenvironmental aspects, and implications for future research. *Current Research in the Pleistocene, 25*, 27–29.

Araujo, A. G. M., & Marcelino, J. C. (2003). The role of armadillos in the movement of archaeological materials: An experimental approach. *Geoarchaeology, 18*, 433–460.

Araujo, A. G. M., & Neves, W. A. (2007). Lapa das Boleiras rockshelter: Geoarchaeological and paleoenvironmental aspects of a Paleoamerican site in the Lagoa Santa region, Brazil. *Current Research in the Pleistocene, 24*, 48–49.

Araujo, A. G. M., & Neves, W. A. (Orgs.). (2010a). Lapa das Boleiras – Um Sítio Paleoíndio do Carste de Lagoa Santa, MG, Brasil. Editora Annablume/FAPESP, São Paulo.

Araujo, A. G. M., & Neves, W. A. (2010b). Métodos de Escavação, Coleta de Amostras e Análises Realizadas. In A. G. M. Araujo & W. A. Neves (Eds.), *Lapa das Boleiras – Um Sítio Paleoíndio do Carste de Lagoa Santa, MG, Brasil* (pp. 23–34). São Paulo: Editora Annablume/FAPESP.

Araujo, A. G. M., & Neves, W. A. (2012). A Ocupação Paleoíndia do Sudeste Brasileiro: Lagoa Santa e Além. In W. F. Morales & F. P. Moi (Eds.), *Tempos Ancestrais* (pp. 47–71). São Paulo: Editora Annablume /NEPAB.

Araujo, A. G. M., & Oliveira, E. C. (2010). A Lapa das Boleiras: inserção ambiental, histórico das pesquisas e caracterização do sítio. In A. G. M. Araujo & W. A. Neves (Eds.), *Lapa das Boleiras – Um Sítio Paleoíndio do Carste de Lagoa Santa, MG, Brasil* (pp. 11–22). São Paulo: Editora Annablume/FAPESP.

Araujo, A. G. M., & Pugliese, F. (2009). The use of non-flint raw materials by Paleoindians in Eastern South America: A Brazilian perspective. In F. Sternke, L. Eigeland, & L. J. Costa (Eds.), *Non-flint raw material use in prehistory – old prejudices and new directions, BAR Series 1939* (pp. 169–175).

Araujo, A. G. M., & Pugliese, F. (2010). A Indústria Lítica. In A. G. M. Araujo & W. A. Neves (Eds.), *Lapa das Boleiras – Um Sítio Paleoíndio do Carste de Lagoa Santa, MG, Brasil* (pp. 75–106). São Paulo: Editora Annablume/FAPESP.

Araujo, A. G. M., Neves, W. A., & Piló, L. B. (2002). Paleoindian subsistence and technology in Central Brazil: Results from new excavations at Boleiras Rockshelter, Lagoa Santa. *Current Research in the Pleistocene, 19*, 4–6.

Araujo, A. G. M., Neves, W. A., & Piló, L. B. (2003a). *Eventos de seca no Holoceno e suas implicações no povoamento pré-histórico do Brasil Central*. Paper presented at the "IX Congresso da Associação Brasileira de Estudos do Quaternário", Recife.

Araujo, A. G. M., Neves, W., & Piló, L. B. (2003b). Eventos de seca no Holoceno e suas implicações no povoamento pré-histórico do Brasil Central. *Anais do "IX Congresso da Associação Brasileira de Estudos do Quaternário"* CD-ROM.

Araujo, A. G. M., Neves, W. A., & Piló, L. B. (2004). Vegetation changes and megafauna extinction in South America: Comments on De-Vivo and Carmignotto. *Journal of Biogeography, 31*, 2039–2040.

Araujo, A. G. M., Neves, W. A., Piló, L. B., & Atui, J. P. V. (2005a). Holocene dryness and human occupation in Brazil during the "Archaic Gap". *Quaternary Research, 64*, 298–307.

Araujo, A. G. M., Piló, L. B., Neves, W. A., & Atui, J. P. (2005b). Human occupation and paleoenvironments in South America: Expanding the notion of an 'Archaic Gap'. *Revista do Museu de Arqueologia e Etnologia, 15*(16), 3–35.

Araujo, A. G. M., Feathers, J. K., Arroyo-Kalin, M., & Tizuka, M. M. (2008). Lapa das Boleiras rockshelter: Stratigraphy and formation processes at a Paleoamerican site in Central Brazil. *Journal of Archaeological Science, 35*, 3186–3202.

Araujo, A. G. M., Piló, L. B., & Neves, W. A. (2010). Estratigrafia e Processos de Formação do Sítio. In A. G. M. Araujo & W. A. Neves (Eds.), *Lapa das Boleiras – Um Sítio Paleoíndio do Carste de Lagoa Santa, MG, Brasil* (pp. 35–74). São Paulo: Editora Annablume/FAPESP.

Araujo, A. G. M., Neves, W. A., & Kipnis, R. (2012a). Lagoa Santa revisited: An overview of the chronology, subsistence, and material culture of Paleoindian sites in Eastern Central Brazil. *Latin American Antiquity, 23*, 533–550.

Araujo, A. G. M., Strauss, A., & Schrage, T. (2012b). *Tropical geomorphology, soils, and archaeology: Towards a tropical geoarchaeology*. Paper presented at the "34th International Geological Congress", Brisbane.

Araujo, A. G. M., Strauss, A., Feathers, J., Paisani, J. C., & Schrage, T. J. (2013). Paleoindian open-air sites in tropical settings: A case study in formation processes, dating methods, and paleoenvironmental models. *Geoarchaeology, 28*, 195–220.

Arensburg, B., & Nès El, M. (1995). On the origin of native Americans: De Acosta versus La Peyrere. *Antropología Biológica, 3*, 1–11.

Armelagos, G. J., Carlson, D. S., & Van Gerven, D. P. (1982). The theoretical foundations and development of skeletal biology. In F. Spencer (Ed.), *A history of American physical anthropology, 1930–1980* (pp. 305–328). New York: Academic.

Arroyo-Kalin, M. A., Neves, W. A., Kipnis, R., & Araujo, A. G. M. (2007). *Micromorphological interpretations of the stratigraphy of the Boleiras archaeological site, Minas Gerais, Brazil*. Paper presented at the "Developing International Geoarchaeology Conference", Cambridge, UK.

Atuí, J. P. V. (2005). *Morfologia craniana de ameríndios brasileiros recentes e suas implicações para a questão da ocupação do Novo Mundo: uma análise exploratória*. Master Thesis. Departamento de Genética e Biologia Evolutiva, Instituto de Biociências, Universidade de São Paulo, São Paulo.

Auler, A. S., Piló, L. B., Smart, P. L., Wang, X., Neves, W. A., & Edwards, R. L. (2003). *Cronologia e processos de deposição da megafauna quaternária em cavernas brasileiras*. Paper presented at the "IX Congresso da Associação Brasileira de Estudos do Quaternário", Recife.

Auler, A., Piló, L. B., Smart, P., Wang, X., Hoffmann, D., Richards, D. A., Edwards, L., Neves, W. A., & Cheng, H. (2006). U-series dating and taphonomy of Quaternary vertebrates from Brazilian Caves. *Palaeogeography, Palaeoclimatology, Palaeoecology, 240*, 508–522.

Bernardo, D. V. (2007). *Afinidades morfológicas intra e extra continentais dos paleoíndios de Lagoa Santa: uma nova abordagem*. Master Thesis. Departamento de Genética e Biologia Evolutiva, Instituto de Biociências, Universidade de São Paulo, São Paulo.

Bernardo, D. V. (2011). O nome da tribo. *Revista de História da Biblioteca Nacional, 71*, 34–35.

Bernardo, D. V. (2012). De Luzia ao Taradinho: a "sina" de Lagoa Santa. Informativo do Centro de Arqueologia Annette L. Emperaire, Outubro, p. 3.

Bernardo, D. V., & Neves, W. A. (2009). Diversidade morfocraniana dos remanescentes ósseos humanos da Serra da Capivara: implicações para a origem do homem americano. *FUMDHAMentos, 8*, 94–106.

Bernardo, D. V., Neves, W. A., & Hanihara, T. (2008). *Intra and extra continental morphological affinities of the Paleoindians of Lagoa Santa: A new approach*. Paper presented at the "X Congreso de la Asociación Latinoamericana de Antropología Biológica", La Plata.

Bernardo, D. V., Almeida, T. F., Neves, W. A., & Hanihara, T. (2009). *Global human population structuring seen from craniometric data*. Paper presented at the "78th Annual Meeting of the American Association of Physical Anthropologists", Chicago.

Bernardo, D. V., Almeida, T. F., & Neves, W. A. (2010a). *Human cranial diversity and their consequence to global population structuring*. Paper presented at the "XI Congreso de la Asociación Latinoamericana de Antropología Biológica", Bogotá.

Bernardo, D. V., Neves, W. A., & Guidon, N. (2010b). *Cranial diversity of human skeletal remains from Serra da Capivara, Northeastern Brazil: Implications for the origin of the Native Americans*. Paper presented at the "79th Annual Meeting of the American Association of Physical Anthropologists", Albuquerque.

Bernardo, D. V., Almeida, T. F., & Neves, W. A. (2011a). *Assessing molecular and morphological variation in the Americas: A comparative approach*. Paper presented at the "80th Annual Meeting of the American Association of Physical Anthropologists", Minneapolis.

Bernardo, D. V., Hubbe, M., Neves, W. A., Okumura, M., Almeida, T. F., & Harvati, K. (2011b). *Cranial morphological diversity of Early and Middle Holocene Brazilian groups: Implications for the settlement of South America*. Paper presented at the "XVI World Congress UISPP/XVI Congresso SAB", Florianópolis.

Bernardo, D. V., Strauss, A., Neves, W. A., & Okumura, M. (2011c). Measuring skulls: Getting into the biological realm of the settlement of the New World. In D. Vialou (Ed.), *Peuplements et Préhistoire en Amériques* (pp. 31–42). Paris: Éditions du Comité des Travaux Historiques et Scientifiques.

Bernardo, D. V., Almeida, T. F., & Bernardo, D. V. (2012a). *Detecting modular components on the human skull: Implications for the structuring of global human cranial variation*. Paper presented at the "81st Annual Meeting of the American Association of Physical Anthropologists", Portland.

Bernardo, D., Hubbe, M., Strauss, A., & Neves, W. A. (2012b). *Morfología craneana de los esqueletos humanos tempranos de Lagoa Santa, Brasil Central: Implicaciones para el poblamiento de las Américas*. Paper presented at the "VI Simposio Internacional-El poblamiento temprano de América: Modelos de Poblamiento y Aportes desde las Territorialidades Tropicales", Pereira.

Bernardo, D. V., Neves, W. A., Strauss, A., Almeida, T. F., & Oliveira, R. E. (2012c). *Cranial morphology of early human skeletal remains from Lapa do Santo, Lagoa Santa, Brazil: Implications for the settlement of the New World*. Paper presented at the "81th Annual Meeting of the American Association of Physical Anthropologists", Portland.

Bernardo, D. V., Neves, W. A., & Almeida, T. F. (2013). *Covariance patterns in the human skull: A phylogenetic approach to the structure of human cranial variation*. Paper presented at the "82th Annual Meeting of the American Association of Physical Anthropologists", Knoxville.

Bissaro Jr., M. C. (2008). *Tafonomia como ferramenta zooarqueológica de interpretação: viés de representatividade óssea em sítios arqueológicos, paleontológico e etnográfico*. Master Thesis. Departamento de Genética e Biologia Evolutiva, Instituto de Biociências, Universidade de São Paulo, São Paulo.

Bissaro Jr., M. C., & Kipnis, R. (2007). *Taphonomic signatures of paleontological and ethnographic faunal collections from the Neotropic*. Paper presented at the "XX Congresso Brasileiro de Paleontologia", Búzios.

Blum, M., & Neves, W. A. (2002). O "Modelo dos Dois Componentes Biológicos Principais": sua inserção nos eventos expansionistas do final do Pleistoceno e suas implicações para a origem do *Homo sapiens*. *O Carste, 14*, 42–49.

Blum, M., Neves, W. A., & Lahr, M. M. (2000). *Por que os primeiros sulamericanos se parecem com os africanos e australianos atuais?* Paper presented at the "VI Congreso de La Asociación Latinoamericana de Antropología Biológica", Piriápolis.

Blum, M., Neves, W. A., & Lahr, M. M. (2001). The long and winding road of the first Americans. *Current Research in the Pleistocene, 18*, 69–73.

Boëda, E., Clemente-Conte, I., Fontugne, M., Labaye, C., Pino, M., Felice, G. D., Guidon, N., Hoeltz, S., Lourdeau, A., Pagli, M., Pessis, A., Viana, S., Da Costa, A., & Douville, E. (2014). A new late Pleistocence archaeological sequence in South America: The Vale da Pedra Furada (Piauí, Brazil). *Antiquity, 88*, 927–955.

Brace, C. L., Nelson, A. R., Seguchi, N., Oe, H., Sering, L., Qifeng, P., Yongyi, L., & Tumen, D. (2001). Old World source of the first New World human inhabitants: A comparative craniofacial view. *Proceedings of the National Academy of Sciences of the United States of America, 98*(17), 10017–10022.

Bradley, B., & Stanford, D. (2004). The North American Atlantic ice-edge corridor: A possible Paleolithic route to New World. *World Archaeology, 36*(4), 459–478.

Bryan, A. L. (1973). Paleoenvironments and cultural diversity in Late Pleistocene South America. *Quaternary Research, 3*, 237–256.

Bueno, L. (2010). Tecnologia lítica, cronologia e sequência de ocupação: o estudo de um sítio a céu aberto de Lagoa Santa, MG. *Revista do Museu de Arqueologia e Etnologia, 20*, 91–107.

Bueno, L. (2011). L'occupation initiale du Brésil dans une perspective macro-régionale: les cas des regions de l'Amazonie, du Nordeste et du Centre du Brésil. In D. Vialou (Ed.), *Peuplements et Préhistoire en Amériques* (pp. 209–220). Paris: Éditions du Comité des Travaux Historiques et Scientifiques.

Bueno, L. (2012). Entre abrigos e lagoas: tecnologia lítica e territorialidade em Lagoa Santa, Minas Gerais, Brasil. *Revista de Arqueologia, 25*, 62–83.

Camolez, T., Zaher, H., & Kipnis, R. (2005). *A fauna de squamata do Quaternário Superior encontrada em três sítios arqueológicos no Estado de Minas Gerais*. Paper presented at the "II Congresso Latino-americano de Paleontologia de Vertebrados", Rio de Janeiro.

Cann, R. L. (1994). mtDNA and native Americans: A southern perspective. *American Journal of Human Genetics, 55*, 7–11.

Cruz Jr., F. W., Karmann, I., Viana Jr., O., Burns, S. J., Ferrari, J. A., Vuille, M., Sial, A. N., & Moreira, M. Z. (2005). Stable isotope study of cave percolation water in subtropical Brazil: Implications for paleoclimate inferences from speleothems. *Chemical Geology, 220*(3–4), 245–262.

Da-Gloria, P. J. T. (2012). *Health and Lifestyle in the Paleoamericans: Early Holocene Biocultural Adaptation at Lagoa Santa, Central Brazil*. Ph.D. Dissertation. The Ohio State University, Columbus.

Da-Gloria, P., & Neves, W. A. (2010). Os remanescentes ósseos humanos. In A. G. M. Araujo & W. A. Neves (Eds.), *Lapa das Boleiras: um sítio paleoíndio do carste de Lagoa Santa, MG, Brasil* (pp. 191–196). São Paulo: Editora Annablume/FAPESP.

Da-Gloria, P., Strauss, A., & Neves, W. A. (2011). *Mortuary rituals in the Early Holocene population of Lagoa Santa: The Harold Walter collection*. Paper presented at the "80th Annual Meeting of the American Association of Physical Anthropologists", Minneapolis.

De Oliveira, P. E. (1992). *A palynological record of late Quaternary vegetation and climatic change in Southeastern Brazil*. Doctoral Thesis. The Ohio State University, Columbus, Ohio.

De Oliveira, P. E., Barreto, A. M. F., & Suguio, K. (1999). Late Pleistocene/Holocene climatic and vegetational history of the Brazilian caatinga: The fóssil dunes of the middle São Francisco River. *Palaeogeography, Palaeoclimatology, Palaeoecology, 152*, 319–337.

Dias, A., & Bueno, L. (2013). The initial colonization of South America Eastern Lowlands: Brazilian archaeology contributions to settlement of America models. In K. Graf, C. V. Ketron, & M. R. Waters (Eds.), *The Paleoamerican Odyssey* (pp. 339–359). College Station: Center for the Study of the First Americans.

Dillehay, T. D. (1997). *Monte Verde: A late Pleistocene settlement in Chile 2: The archaeological context*. Washington, DC: Smithsonian Institution Press.

Dillehay, T. D. (1999). The late Pleistocene cultures of South America. *Evolutionary Anthropology, 7*(6), 206–216.

Dillehay, T. D. (2000). *The settlement of the Americas*. New York: Basic Books.

Dillehay, T. D. (2003). Tracking the first Americans. *Nature, 425*, 23–24.

Dillehay, T. D., Calderón, G. A., Politis, G., & Beltrão, M. (1992). Earliest hunters and gatherers of South America. *Journal of World Prehistory, 6*, 145–204.

Dixon, E. J. (2001a). Human colonization of the Americas: Timing, technology and process. *Quaternary Science Reviews, 20*, 277–299.

Dixon, E. J. (2001b). *Bones, boats & bison – archaeology and the first colonization of western north America*. Albuquerque: University of New Mexico Press.
Evans Jr., C. (1950). A report on recent archaeological investigations in the Lagoa Santa region of Minas Gerais, Brazil. *American Antiquity, 15*(4), 341–343.
Fagan, B. M. (1987). *The great journey*. London: Thames and Hudson Ltd.
Fagan, B. M. (1999). *World prehistory – A brief introduction*. Menlo Park: Longman.
Faria, F. F. d. A. (2008). Peter Lund (1801–1880) e o questionamento do catastrofismo. *Filosofia e História da Biologia, 3*, 139–156.
Feathers, J., Kipnis, R., Piló, L. B., Arroyo-Kalin, M., & Coblentz, D. (2010). How Old is Luzia? Luminescence dating and stratigraphic integrity at Lapa Vermelha, Lagoa Santa, Brazil. *Geoarchaeology, 25*, 395–436.
Figgins, J. D. (1927). The antiquity of man in America. *Natural History, 27*, 229–239.
Fontugne, M. (2013). New radiocarbon ages of Luzia woman, Lapa Vermelha IV site, Lagoa Santa, Minas Gerais, Brazil. *Radiocarbon, 55*(2–3), 1187–1190.
Freire, G. de Q. (2011). *Madeiras fósseis holocênicas de Ribeirão da Mata: anatomia ecológica, relações florísticas e interpretação paleoambiental da região arqueológica de Lagoa Santa, MG*. Ph.D. Dissertation. Departamento de Botânica, Instituto de Biociências, Universidade de São Paulo, São Paulo.
González-José, R., González-Martin, A., Hernández, M., Pucciarelli, H. M., Sardi, M., Rosales, A. S., & van de Molen, S. (2003). Craniometric evidence for paleoamerican survival in Baja California. *Nature, 452*, 62–65.
González-José, R., Neves, W. A., Lahr, M. M., González, S., Pucciarelli, H., Martínez, M. H., & Correal, G. (2005). Late Pleistocene/Holocene craniofacial morphology in Mesoamerica Paleoindians: Implications for the peopling of the New World. *American Journal of Physical Anthropology, 128*, 772–780.
González-José, R., Martínez-Abadías, N., González-Martín, A., Neves, W. A., Pucciarelli, H. M., & Hernández, M. (2006). Análisis morfogeométrico de cuatro cráneos antiguos del Valle de Mexico: Peñón III, Metro Balderas, Chimalhuacán y Cueva del Tecolote. *Cuicuilco, 13*, 105–127.
González-José, R., Bortolini, M. C., Santos, F. R., & Bonatto, S. L. (2008). The peopling of America: Craniofacial shape variation on a continental scale and its interpretation from an interdisciplinary view. *American Journal of Physical Anthropology, 137*(2), 175–187.
Greenberg, J. H., Turner II, C. G., & Zegura, S. L. (1986). The settlement of the Americas: A comparison of linguistic, dental and genetic evidence. *Current Anthropology, 27*, 477–497.
Gruhn, R., & Bryan, A. (2011). A current view of the initial peopling of the Americas. In D. Vialou (Ed.), *Peuplements et préhistoire em Amériques* (pp. 17–30). Paris: Éditions du Comité des travaux historiques et scientifiques.
Guidon, N., & Delibrias, G. (1986). Carbon-14 dates point to man in the Americas 32,000 years ago. *Nature, 321*, 269–271.
Hanihara, T. (1993a). Population prehistory of East Asia and the Pacific as viewed from craniofacial morphology: The basic populations in East Asia, VII. *American Journal of Physical Anthropology, 91*, 173–187.
Hanihara, T. (1993b). Craniofacial features of Southeast Asians and Jomonese: A reconsideration of their microevolution since the late Pleistocene. *Anthropological Science, 101*, 25–46.
Hanihara, T. (1996). Comparison of craniofacial features of major human groups. *American Journal of Physical Anthropology, 99*, 389–412.
Hanihara, T. (2000). Frontal and facial flatness of major human populations. *American Journal of Physical Anthropology, 111*, 105–134.
Hansen, S. (1888). Lagoa Santa Racen. *E Museo Lundii, 1*, 1–37.
Harvati, K., Hubbe, M., Bernardo, D. V., & Hanihara, T. (2009). *Population history and cranial morphology in a large human skeletal database*. Paper presented at the "78th Annual Meeting of the American Association of Physical Anthropologists", Chicago.
Howells, W. W. (1973). *Cranial variation in man. A study by multivariate analysis of patterns of difference among recent human populations, Papers of Peabody Museum of archaeology and ethnology*. Cambridge: Harvard University Press.

Howells, W. W. (1989). *Skull shapes and the map. Craniometric analysis in the dispersion of modern Homo, Papers of Peabody Museum of archaeology and ethnology*. Cambridge: Harvard University Press.

Howells, W. W. (1999). *Who's who in skulls, Papers of Peabody Museum of archaeology and ethnology* (Vol. 82). Cambridge: Harvard University Press.

Hrdlička, A. (1912). *Early man in South America, Bureau of American ethnology bulletin* (Vol. 52). Washington, DC: Smithsonian Institution.

Hrdlička, A. (1932). The coming of man from Asia in the light of recent discoveries. *Proceedings of the American Philosophical Society, 71*, 393–402.

Hubbe, A. (2008). *Contextualização taxonômica, tafonômica e morfométrica dos remanescentes ósseos da megamastofauna da gruta Cuvieri (MG), um sítio paleontológico do Pleistoceno tardio*. Master Thesis. Departamento de Genética e Biologia Evolutiva, Instituto de Biociências, Universidade de São Paulo, São Paulo.

Hubbe, A. (2010). *Functional analysis of extinct Brazilian Xenarthra's forelimb: A preliminary 3-D morphometric geometric analysis*. Paper presented at the "9th International Congress of Vertebrate Morphology", Punta del Leste.

Hubbe, M. (2012). *Una breve guía para el poblamiento de las Américas: todo lo que siempre quisiste saber y que todavía no sabemos*. Paper presented at the "XII Congreso de la Asociación Latinoamericana de Antropología Biológica", San José.

Hubbe, M. (2013). *Paleoamerican morphology in European and East Asian Late Pleistocene variation: Implications for human dispersion into the New World*. Paper presented at the "First Peopling of the Americas and the World Heritage Convention", Puebla.

Hubbe, A., & Auler, A. S. (2012). A large cervidae Holocene accumulation in eastern Brazil: An example of extreme taphonomical control in a cave environment. *International Journal of Speleology, 41*, 297–305.

Hubbe, M., & Strauss, A. (2012a). *Diachronic cranial morphological diversity in South America: A functional approach*. Paper presented at the "81st Annual Meeting of the American Association of Physical Anthropologists", Portland.

Hubbe, M., & Strauss, A. (2012b). *The initial Settlement of South America: Perspectives from bioanthropology*. Paper presented at the "77th Annual Meeting of the Society for American Archaeology", Memphis.

Hubbe, M., Azevedo, E. T. M., Atui, J. P. V., & Neves, W. A. (2003a). *A Primeira Descoberta da América*. Ribeirão Preto: Editora da Sociedade Brasileira de Genética.

Hubbe, M., Neves, W. A., Schmitz, P. I., & Bernardo, D. V. (2003b). *Afinidades morfológicas do grupo humano de Içara (SC-IÇ-01) com outras séries esqueletais do sul do Brasil: primeiros resultados*. Paper presented at the "XII Congresso da Sociedade de Arqueologia Brasileira", São Paulo.

Hubbe, M., Mentz-Ribeiro, P. A., Bernardo, D. V., & Neves, W. A. (2004a). Afinidades morfológicas de três crânios associados à Tradição Umbu: uma análise exploratória multivariada. *Revista do CEPA, 28*, 159–185.

Hubbe, M., Neves, W. A., Atui, J. P., Cartelle, C., & Da Silva, M. P. (2004b). A new early human skeleton from Brazil: Support for the "Two Main Biological Components Model" for the settlement of the Americas. *Current Research in the Pleistocene, 21*, 77–81.

Hubbe, A., Hubbe, M., & Cruz, M. V. (2007a). Considerações metodológicas à inferência de dieta de animais extintos através da análise de microdesgaste dentário. In I. S. Carvalho, R. C. T. Cassab, C. Schwanke, M. A. Carvalho, A. C. S. Fernandes, M. A. C. Rodrigues, M. S. S. Carvalho, M. Arai, & M. E. Q. Oliveira (Eds.), *Paleontologia: Cenários de Vida* (pp. 743–752). Rio de Janeiro: Editora Interciência.

Hubbe, A., Hubbe, M., & Neves, W. A. (2007b). Early Holocene survival of megafauna in South America. *Journal of Biogeography, 34*, 1642–1646.

Hubbe, A., Karmann, I., & Neves, W. A. (2008a). *Idades radiocarbônicas inéditas da megafauna pleistocênica da região de Iporanga, Apiaí (Sul do Estado de São Paulo), corroboram ampla dispersão cronológica dos remanescentes fósseis da megafauna extinta*. Paper presented at the "VI Simpósio Brasileiro de Paleontologia de Vertebrados", Ribeirão Preto.

Hubbe, M., Neves, W. A., & Harvati, K. (2008b). *Paleoamerican and Amerindian morphological variation: Implications for the settlement of the New World*. Paper presented at the "25th Southern Illinois University Center for Archaeological Investigations Visiting Scholar Conference: Biological and Archaeological Variation in the New World", Carbondale.

Hubbe, M., Strauss, A., & Neves, W. A. (2008c). *Supervivencia y dispersión geográfica de la morfología Paleoamericana en el Holoceno de Sudamérica*. Paper presented at the "X Congreso de la Asociación Latinoamericana de Antropología Biológica", La Plata.

Hubbe, M., Harvati, K., & Neves, W. A. (2009a). *Paleoamericans in a Late Pleistocene context: Assessing morphological affinities*. Paper presented at the "78th Annual Meeting of the American Association of Physical Anthropologists", Chicago.

Hubbe, A., Hubbe, M., & Neves, W. A. (2009b). New Late-Pleistocene dates for the extinct megafauna of Lagoa Santa, Brazil. *Current Research in the Pleistocene, 26*, 154–156.

Hubbe, M., Neves, W. A., De Oliveira, E. C., & Strauss, A. (2009c). Post-marital residence practices in Southern Brazilian Coastal Groups: Continuity and change. *Latin American Antiquity, 20*, 267–278.

Hubbe, M., Bernardo, D. V., & Neves, W. A. (2010a). *Tasa de diferenciación morfológica entre poblaciones tempranas y tardías de América: implicaciones para los modelos de poblamiento del continente*. Paper presented at the "XI Congreso de la Asociación Latinoamericana de Antropología Biológica", Bogotá.

Hubbe, A., Mayer, E. L., & Haddad-Martim, P. N. (2010b). *O papel modulador da morfologia de ambientes cársticos na proporção de mamíferos carnívoros e necrófagos versus herbívoros e onívoros representados em conjuntos faunísticos decorrentes da queda de animais em abismos*. Paper presented at the "VII Simpósio Brasileiro de Paleontologia de Vertebrados", Rio de Janeiro.

Hubbe, M., Neves, W. A., & Harvati, K. (2010c). Testing evolutionary and dispersion scenarios for the settlement of the New World. *PloS One, 5*, e11105.

Hubbe, M., Neves, W. A., Okumura, M. M., & Bernardo, D. V. (2010d). *La diversidad morfológica craneana de grupos del Holoceno Temprano y Medio de Brasil: implicaciones para la ocupación humana de Sudamérica*. Paper presented at the "V Simposio Internacional-El poblamiento temprano de América", La Plata.

Hubbe, A., Haddad-Martim, P., Hubbe, M., Mayer, E., Strauss, A., Auler, A., Piló, L., & Neves, W. A. (2011a). Identification and importance of critical depositional gaps in pitfall cave environments: The fossiliferous deposit of Cuvieri Cave, eastern Brazil. *Palaeogeography, Palaeoclimatology, Palaeoecology, 312*, 66–78.

Hubbe, M., Harvati, K., & Neves, W. A. (2011b). Paleoamerican morphology in the context of European and East Asian late Pleistocene variation: Implications for human dispersion into the New World. *American Journal of Physical Anthropology, 144*, 442–453.

Hubbe, M., Neves, W. A., Bernardo, D., & Harvati, K. (2011c). *Testing evolutionary and dispersion scenarios for the settlement of the New World*. Paper presented at the "80th Annual Meeting of the American Association of Physical Anthropologists", Minneapolis.

Hubbe, M., Neves, W. A., & Harvati, K. (2011d). *Cranial morphological variation in the Americas: Testing evolutionary scenarios for the settlement of the New World*. Paper presented at the "Second International Conference on the Great Migrations: Asia to America", New York.

Hubbe, A., Strauss, A., Hubbe, M., & Bernardo, D. (2011e). *Understanding Fst and Discriminant Analysis classifications by means of resampled datasets: The influence of different variables and extent to which they actually reflect sub-population divergence*. Paper presented at the "80th Annual Meeting of the American Association of Physical Anthropologists", Minneapolis.

Hubbe, A., Vasconcelos, A. G., Vilaboim, L., Karmann, I., & Neves, W. A. (2011f). Chronological distribution of Brazilian *Glyptodon* sp. remains: A direct 14C date for a specimen from Iporanga, São Paulo, Brazil. *Radiocarbon, 53*, 13–19.

Hubbe, A., Haddad-Martim, P. M., Hubbe, M., & Neves, W. A. (2012a). Comments on: "An anthropogenic modification in an Eremotherium tooth from Northeastern Brazil". *Quaternary International, 269*, 94–96.

Hubbe, M., Strauss, A., Bernardo, D., & Neves, W. A. (2012b). *Diferenciación morfológica diacrónica en Sudamérica*. Paper presented at the "XII Congreso de la Asociación Latinoamericana de Antropología Biológica", San José.

Hubbe, M., Bernardo, D. V., Almeida, T. F., Hanihara, T., & Harvati, K. (2013a). *Modern human phenotypic variation: Exploring patterns of differentiation within and between continents*. Paper presented at the "82th Annual Meeting of the American Association of Physical Anthropologists", Knoxville.

Hubbe, A., Hubbe, M., Karmann, I., Cruz, F. W., & Neves, W. A. (2013b). Insights into Holocene megafauna survival and extinction in Southeastern Brazil from new AMS C-14 dates. *Quaternary Research, 79*, 152–157.

Hubbe, A., Hubbe, M., & Neves, W. A. (2013c). The Brazilian megamastofauna of the Pleistocene/Holocene transition and its relationship with the early human settlement of the continent. *Earth-Science Reviews, 118*, 1–10.

Hubbe, M., Neves, W. A., Bernardo, D., Strauss, A., Araujo, A., & Kipnis, R. (2013d). *Early human occupation of Lagoa Santa, Eastern Central Brazil: Implications for the dispersion and adaptation of the initial settlers of South America*. Paper presented at the "The Paleoamerican Odyssey Conference", Santa Fe.

Hurt, W., & Blasi, O. (1969). O projeto arqueológico "Lagoa Santa", Minas Gerais, Brasil. *Arquivos do Museu Paranaense, 4*, 1–63.

Jangelme, G. M., Porsani, J. L., & Kipnis, R. (2007). *Investigações GPR no sítio aqueológico Lapa do Santo, MG: resultados preliminares*. Paper presented at the "10th International Congress of the Brazilian Geophysical Society", Rio de Janeiro.

Kelner, A. W. A., & Campos, D. d. A. (1999). Vertebrate paleontology in Brazil – a review. *Episodes, 22*(3), 238–251.

Kilby, J. D. (2011). Les caches Clovis dans le cadre du Paléoindien ancien en Amérique du Nord. In D. Vialou (Ed.), *Peuplements et préhistoire em Amériques* (pp. 71–84). Paris: Éditions du Comité des travaux historiques et scientifiques.

Kipnis, R. (1998). Early hunter-gatherers in the Americas: Perspective from central Brazil. *Antiquity, 72*, 581–592.

Kipnis, R. (2003). *The taphonomy of archaeological and paleontological assemblages from Late Pleistocene and Early Holocene sites in Central Brazil*. Paper presented at the "68th Annual Meeting of the Society for American Archaeology", Milwaukee.

Kipnis, R. (2008). *Middle Holocene archaeological gaps and their explanations*. Paper presented at the "73rd Annual Meeting of the Society for American Archaeology", Vancouver.

Kipnis, R., & Da-Gloria, P. (2013). *Paleodietary patterns among early foragers in lowland South America*. Paper presented at the "78th Annual Meeting of the Society for American Archaeology", Honolulu.

Kipnis, R., & Silva, D. G. R. (2005). *Plant collecting and Terminal Pleistocene subsistence economy of foraging societies in the Neotropics*. Paper presented at the "70th Annual Meeting of the Society for American Archaeology", Salt Lake City.

Kipnis, R., Bissaro Junior, M. C., & Prado, H. M. (2010a). Os restos faunísticos. In A. G. M. Araujo & W. A. Neves (Eds.), *Lapa das Boleiras: um sítio paleoíndio do carste de Lagoa Santa, MG, Brasil* (pp. 121–147). São Paulo: Editora Annablume/FAPESP.

Kipnis, R., Santos, R. O., & Ernani, M. (2010b). Indústria óssea. In A. G. M. Araujo & W. A. Neves (Eds.), *Lapa das Boleiras: um sítio paleoíndio do carste de Lagoa Santa, MG, Brasil* (pp. 111–119). São Paulo: Editora Annablume/FAPESP.

Kollmann, J. (1884). Schädeln von Lagoa Santa. *Zeitschrift für Ethnologie, 16*, 194–199.

Lacerda, J. B., & Peixoto, R. (1876). Contribuições para o estudo anthropológico das raças indígenas do Brazil. *Archivos do Museu Nacional do Rio de Janeiro, 1*, 47–75.

Laming-Emperaire, A. (1979). Missions archéologique franco-brésiliennes en Lagoa Santa, Minas Gerais, Brésil: Le grand abri de Lapa Vermelha. *Revista de Pré-história, 1*, 54–89.

Lima, T. A. (2006). O povoamento inicial do continente americano: migrações, contextos, datações. In H. P. Silva & C. Rodrigues-Carvalho (Eds.), *Nossa origem. O povoamento das Américas – visões multidisciplinares* (pp. 77–103). Rio de Janeiro: Viera & Lentz.

Lopes, M. M. (2008). “Cenas de tempos profundos”: ossos, viagens, memórias nas culturas da natureza do Brasil. *História, Ciências, Saúde – Manguinhos, 15*(3), 615–634.

Luna Filho, P. E. (2007). *Peter Wilhelm Lund: o auge de suas investigações científicas e a razão para o término de suas pesquisas*. Ph.D. Dissertation. Departamento de História, Faculdade de Filosofia, Letras e Ciências Humanas, Universidade de São Paulo, São Paulo.

Lund, P. W. (1844). Carta do Dr. Lund, escripta da Lagôa Santa (Minas Geraes) a 21 de abril de 1844 – Lida na sessão do Instituto de 20 de Junho deste mesmo anno. *Revista do Instituto Histórico e Geográfico do Brasil, 6*, 334–342.

Lund, P. W. (1846). Notice sur des Ossements Humains Fossiles Trouvés dans une Caverne du Brésil. *Mémoires de la Societé Royale des Antiquaires du Nord, 3*, 49–77.

Mayer, E. L. (2011). *Processos de formação de um depósito fossilífero em abismo na gruta Cuvieri (MG): taxonomia, tafonomia e distribuição espacial de fósseis de mamíferos do Pleistoceno*. Master Thesis. Departamento de Genética e Biologia Evolutiva, Instituto de Biociências, Universidade de São Paulo, São Paulo.

Mayer, E. L., & Hubbe, A. (2010). *Caracterização preliminar das diferenças de tamanho entre a paca vivente, Cuniculus paca, Linnaeus, 1766 e a extinta, Cuniculus major, Lund 1837 (Rodentia, Cuniculidae)*. Paper presented at the “Reunião Anual da Sociedade Brasileira de Paleontologia”, Rio Claro.

Mayer, E. L., Hubbe, A., & Neves, W. A. (2009a). *Novo registro de pacas (Rodentia, Cuniculidae) extintas no Pleistoceno terminal de Lagoa Santa, MG*. Paper presented at the “XXI Congresso Brasileiro de Paleontologia”, Belém.

Mayer, E. L., Hubbe, A., & Neves, W. A. (2009b). *Processos de acumulação óssea de mamíferos do Pleistoceno terminal da gruta Cuvieri, Lagoa Santa, MG*. Paper presented at the “XXI Congresso Brasileiro de Paleontologia”, Belém.

Mayer, E. L., Hubbe, A., & Haddad-Martim, P. M. (2012). *Variações na suscetibilidade da megafauna ao aprisionamento em abismos no interior de cavernas associadas a dolinas*. Paper presented at the “VIII Simpósio Brasileiro de Paleontologia de Vertebrados”, Recife.

Meggers, B. J. (1985). Advances in Brazilian Archaeology, 1935–1985. *American Antiquity, 50*(2), 364–373.

Meltzer, D. J. (2011). La chasse au bison à Folsom: nouveau regard sur um site de référence. In D. Vialou (Ed.), *Peuplements et préhistoire em Amériques* (pp. 85–96). Paris: Éditions du Comité des travaux historiques et scientifiques.

Meltzer, D. J., Grayson, D. K., Ardilla, G., Barker, A. W., Dincauze, D. F., Haynes, C. V., Mena, F., Núñez, L., & Stanford, D. J. (1997). On the Pleistocene antiquity of Monte Verde, Southern Chile. *American Antiquity, 62*, 659–663.

Mendonça de Souza, S. M. F., Rodrigues-Carvalho, C., Silva, H. P., & Locks, M. (2006). Revisitando a discussão sobre o Quaternário de Lagoa Santa e o povoamento das Américas: 160 anos de debates científicos. In H. P. Silva & C. Rodrigues-Carvalho (Eds.), *Nossa origem. O povoamento das Américas – visões multidisciplinares* (pp. 19–43). Rio de Janeiro: Viera & Lentz.

Morais, J. L. (1999). A arqueologia da região sudeste. *Revista USP, 44*, 194–217.

Munford, D. (1999). *Estudo Comparado da Morfologia Craniana de Populações Pré-Históricas da América do Sul: Implicações para a Questão do Povoamento do Novo Mundo*. Master Thesis. Departamento de Biologia, Instituto de Biociências, Universidade de São Paulo.

Munford, D., Zanini, M. C., & Neves, W. A. (1995). Human cranial variation in South America: Implications for the settlement of the New World. *Brazilian Journal of Genetics, 18*, 673–688.

Nakamura, C. (2011). *Folhas fósseis do Ribeirão da Mata: uma abordagem florística e anatômica para caracterização do paleoambiente do Holoceno Médio de Lagoa Santa, MG*. Master Thesis. Departamento de Botânica, Instituto de Biociências, Universidade de São Paulo, São Paulo.

Neves, W. A. (2006). Origens do homem nas Américas: fósseis *versus* moléculas? In H. P. Silva & C. Rodrigues-Carvalho (Eds.), *Nossa Origem. O Povoamento das Américas: Visões Multidisciplinares* (pp. 45–76). Rio de Janeiro: Vieira & Lent.

Neves, W. A. (2013). *Um esqueleto incomoda muita gente...* Editora UNICAMP, Campinas.
Neves, W. A., & Araujo, A. G. M. (2006). *O Projeto "Origens e microevolução do homem na América": um balanço de seis anos de atividade.* Paper presented at the "1° Simpósio de Arqueologia e Patrimônio de Minas Gerais", Juiz de Fora.
Neves, W. A., & Atui, J. P. (2004). O mito da homogeneidade biológica da população paleoíndia de Lagoa Santa: implicações antropológicas. *Revista de Antropologia, 47*, 158–206.
Neves, W. A., & Blum, M. (2000). The Buhl burial: A comment on Green et al. *American Antiquity, 65*, 191–193.
Neves, W. A., & Blum, M. (2001). "Luzia" is not alone: Further evidence of a non-mongoloid settlement of the New World. *Current Research in the Pleistocene, 18*, 73–77.
Neves, W. A., & Blum, M. (2002). Morfologia craniana da população paleoíndia de Lagoa Santa: implicações para a questão do povoamento da América. *Chronos, 6*, 21–37.
Neves, W. A., & Hubbe, M. (2003). Luzia e a saga dos Primeiros Americanos. *Scientific American Brasil, 15*, 24–31.
Neves, W. A., & Hubbe, M. (2004). Luzia und die Geschichte der ersten Amerikaner. *Abenteuer Archäologie, 1*, 56–60.
Neves, W. A., & Hubbe, M. (2005a). Cranial morphology of early Americans from Lagoa Santa, Brazil – Implications for the settlement of the New World. *Proceedings of the National Academy of Science, 102*, 18309–18314.
Neves, W. A., & Hubbe, M. (2005b). *Cranial morphology of the first Americans – Implications for the settlement of the New World.* Paper presented at the "1st Paleopathology Association Meeting in South America", Rio de Janeiro.
Neves, W. A., & Hubbe, M. (2005c). Os pioneiros das Américas. *Nossa História, 22*, 16–19.
Neves, W. A., & Hubbe, M. (2008). *Morphological affinities of South and Mesoamerican early skeletons: evidence of a widespread Paleoamerican morphology in the New World.* Paper presented at the "77th Annual Meeting of the American Association of Physical Anthropologists", Columbus.
Neves, W. A., & Kipnis, R. (2004). Further evidence of a highly cariogenic diet among Late Paleoindians of Central Brazil. *Current Research in the Pleistocene, 21*, 81–83.
Neves, W. A., & Piló, L. B. (2003). Solving Lund's dilemma: New AMS dates confirm that humans and megafauna coexisted at Lagoa Santa. *Current Research in the Pleistocene, 20*, 57–60.
Neves, W. A., & Piló, L. B. (2005). E Museo Lundii. *O Carste, 17*, 2–7.
Neves, W. A., & Piló, L. B. (2008). *O Povo de Luzia. Em busca dos primeiros americanos.* São Paulo: Editora Globo.
Neves, W. A., & Pucciarelli, H. M. (1989). Extra-continental biological relationships of early South American human remains: A multivariate analysis. *Ciência e Cultura, 41*, 566–575.
Neves, W. A., & Pucciarelli, H. M. (1990). The origin of the first Americans: An analysis based on the cranial morphology of early South American human remains. *American Journal of Physical Anthropology, 81*, 274.
Neves, W. A., & Pucciarelli, H. M. (1991). Morphological affinities of the first Americans: An exploratory analysis based on early South American human remains. *Journal of Human Evolution, 21*, 261–273.
Neves, W. A., & Pucciarelli, H. M. (1998). The Zhoukoudian Upper Cave skull 101 as seen from the Americas. *Journal of Human Evolution, 34*, 219–222.
Neves, W. A., Powell, J. F., & Ozolins, E. G. (1999a). Extra-continental morphological affinities of Lapa Vermelha IV, hominid I: A multivariate analysis with progressive numbers of variables. *Homo, 50*, 263–282.
Neves, W. A., Powell, J. F., & Ozolins, E. G. (1999b). Modern human origins as seen from the peripheries. *Journal of Human Evolution, 37*, 129–133.
Neves, W. A., Blum, M., & Kozameh, L. (2001a). Fuegian cranial morphology: The Haush. *Ciência e Cultura, 53*, 69–71.
Neves, W. A., Blum, M., Prous, A., & Powell, J. (2001b). *Paleoindian skeletal remains from Santana do Riacho I, Minas Gerais, Brazil: Archaeological background, chronological context,*

and comparative cranial morphology. Paper presented at the "70th Annual Meeting of the American Association of Physical Anthropologists", Kansas City.

Neves, W. A., Hubbe, M., & Araujo, A. G. M. (2002). A late paleoindian secondary ritual burial from Lagoa Santa, Minas Gerais, Brazil. *Current Research in the Pleistocene, 19*, 83–85.

Neves, W. A., Prous, A., Gonzalez-José, R., Kipnis, R., & Powell, J. (2003). Early Holocene human skeletal remains from Santana do Riacho, Brazil: Implications for the settlement of the New World. *Journal of Human Evolution, 45*, 19–42.

Neves, W. A., González-José, R., Hubbe, M., Kipnis, R., Araujo, A. G. M., & Blasi, O. (2004a). Early Holocene human skeletal remains from Cerca Grande, Lagoa Santa, Central Brazil, and the origins of the first Americans. *World Archaeology, 36*, 479–501.

Neves, W. A., Kipnis, R., Araujo, A. G. M., & Piló, L. B. (2004b). *Origens e Microevolução do Homem na América: Uma Abordagem Paleoantropológica. Relatório científico final*. FAPESP, São Paulo. Unpublished material.

Neves, W. A., Hubbe, M., Okumura, M. M. M., González-José, R., Figuti, L., Eggers, S., & De Blasis, P. A. D. (2005a). A new Early Holocene human skeleton from Brazil: Implications for the settlement of the New World. *Journal of Human Evolution, 48*, 403–414.

Neves, W. A., Hubbe, M., & Piló, L. B. (2005b). *Early Holocene human skeletal remains from Sumidouro Cave, Lagoa Santa, Brazil: History of discovery, geological and chronological context, and comparative cranial morphology*. Paper presented at the "70th Annual Meeting of the Society for American Archaeology", Salt Lake City.

Neves, W. A., Hubbe, M., & Correal, G. (2006). *Afinidades biológicas de crânios do final do Pleistoceno/início do Holoceno da Savana de Bogotá, Colombia: inferências sobre a dispersão da morfologia Paleoamericana na América do Sul*. Paper presented at the "IX Congreso de la Asociación Latinoamericana de Antropología Biológica", Ouro Preto.

Neves, W. A., Araujo, A. G. M., Kipnis, R., & Piló, L. B. (2007a). O Projeto 'Origens e Microevolução do Homem na América: Uma Abordagem Paleoantropológica'. In A. P. Loures de Oliveira (Ed.), *Arqueologia e Patrimônio de Minas Gerais* (pp. 73–90). Juiz de Fora: Ed. Editar.

Neves, W. A., Bernardo, D. V., & Okumura, M. (2007b). A origem do homem americano vista a partir da América do Sul: uma ou duas migrações? *Revista de Antropologia, 50*, 9–44.

Neves, W. A., Hubbe, M., & Correal, G. (2007c). Human skeletal remains from Sabana de Bogotá, Colombia: A case of Paleoamerican morphology late survival in South America? *American Journal of Physical Anthropology, 133*, 1080–1098.

Neves, W. A., Hubbe, A., & Karmann, I. (2007d). New accelerator mass spectrometry (AMS) ages suggest a revision of the electron spin resonance (ESR) Middle Holocene dates obtained for a *Toxodon platensis* (Toxodontidae, Mammalia) from southeast Brazil. *Radiocarbon, 49*, 1411–1412.

Neves, W. A., Hubbe, M., & Piló, L. B. (2007e). Early Holocene human skeletal remains from Sumidouro Cave, Lagoa Santa, Brazil: History of discoveries, geological and chronological context, and comparative cranial morphology. *Journal of Human Evolution, 52*, 16–30.

Neves, W. A., Araujo, A. G. M., Ceccantini, G. C. T., de Bueno, L. M. R., Oliveira, P. E. de, Kipnis, R., & Bernardo, D. V. (2008). Origens e Microevolução do Homem na América: Uma Abordagem Paleoantropológica II. Relatório científico final. FAPESP, São Paulo. Unpublished material.

Neves, W. A., Bernardo, D. V., & Guidon, N. (2010). *Morphological affinities of human skeletal remains from Serra da Capivara, Northeastern Brazil and their implications for the origins of Native Americans*. Paper presented at the "XI Congreso de la Asociación Latinoamericana de Antropología Biológica", Bogotá.

Neves, W. A., Bernardo, D. V., Okumura, M., Almeida, T. F., & Strauss, A. (2011a). Origem e dispersão dos Tupiguaranis: o que diz a morfologia craniana? *Boletim do Museu Paraense Emílio Goeldi, Ciências Humanas, 6*, 95–122.

Neves, W. A., Bernardo, D. V., Okumura, M., Almeida, T. F., & Strauss, A. (2011b). *The origin and dispersion of the Tupiguarani Indians: A comparative cranial multivariate analysis*. Paper presented at the "80th Annual Meeting of the American Association of Physical Anthropologists", Minneapolis.

Neves, W. A., Araujo, A. G. M., Bernardo, D. V., Kipnis, R., & Feathers, J. (2012a). Rock art at the Pleistocene/Holocene boundary in Eastern South America. *PloS One, 7*(e), 32228.

Neves, W. A., Mayer, E. L., Hubbe, A., Martin, P. M. H., & Hubbe, M. (2012b). Devagar... Se vai ao Pleistoceno. *Ciência Hoje, 48*, 36–42.

Neves, W. A., Strauss, A., Bernardo, D. V., Araujo, A. G. M., & Kipnis, R. (2012c). As promessas de Lagoa Santa. *Scientific American Brasil, 127*, 78–81.

Neves, W. A., Hubbe, M., Bernardo, D., Strauss, A., Araujo, A., & Kipnis, R. (2013). Early human occupation of Lagoa Santa, Eastern Central Brazil: Implications for the dispersion and adaptation of the initial settlers of South America. In K. Graf, C. V. Ketron, & M. R. Waters (Eds.), *The Paleoamerican Odyssey* (pp. 397–414). College Station: Center for the Study of the First Americans.

Nunes, T. (2010). *Afinidades biológicas extra-continentais dos peloíndios tardios de Lagoa Santa e suas implicações para a questão do povoamento do Novo Mundo: uma abordagem odontométrica.* Master Thesis. Departamento de Genética e Biologia Evolutiva, Instituto de Biociências, Universidade de São Paulo, São Paulo.

Okumura, M. M. M., & Araujo, A. G. M. (2011). *Pontas bifaciais do Sudeste e Sul do Brasil: há diferenças regionais importantes?* Paper presented at the "XVI Congresso Mundial da UISPP/XVI Congresso da Sociedade de Arqueologia Brasileira", Florianópolis.

Oliveira, R. E. (2011). *O sepultamento 21 da Lapa do Santo (8500 BP): primeiro caso de manipulação peri-mortem do corpo no Holoceno Inicial na paisagem mortuária de Lagoa Santa.* Paper presented at the "XVI Congresso Mundial da UISPP/XVI Congresso da Sociedade de Arqueologia Brasileira", Florianópolis.

Oliveira, R. E., Strauss, A., Da-Gloria, P., Bernardo, D. V., Kipnis, R., Araujo, A., & Neves, W. A. (2010). *Practicas funerárias paleoamericanas en la región de Lagoa Santa: el uso del cuerpo como símbolo.* Paper presented at the "V Simpósio Internacional-El Hombre Temprano en América, La Plata.

Oliveira, R. E., Strauss, A., Da-Gloria, P., Bernardo, D. V., Kipnis, R., & Neves, W. A. (2012a). *Secondary ritual or peri-mortem body manipulation during Early Holocene in South America: The case of Burial 21 from the site of Lapa do Santo, Lagoa Santa region, Brazil.* Paper presented at the "81st Annual Meeting of the American Association of Physical Anthropologists", Oregon.

Oliveira, R. E., Strauss, A., Da-Gloria, P., & Inglez, M. (2012b). *New burials from Lapa do Santo and its implications for reevaluating Lagoa Santa's Early Holocene archaeological record.* Paper presented at the "XII Congreso de la Asociación Latinoamericana de Antropología Biológica", San José.

Pedler, D. R., & Adovasio, J. M. (2011). The peopling of the Americas. In D. Vialou (Ed.), *Peuplements et préhistoire em Amériques* (pp. 55–67). Paris: Éditions du Comité des travaux historiques et scientifiques.

Perez, C. P. (2009). *Paleoecologia de mamíferos viventes como ferramenta na caracterização do ambiente Holocênico de Lagoa Santa, MG.* Master Thesis. Departamento de Genética e Biologia Evolutiva, Instituto de Biociências, Universidade de São Paulo, São Paulo.

Perez, C. P., & Kipnis, R. (2007). *Reconstituição paleoambiental dos últimos 10.000 anos através do registro fossilífero de Lagoa Santa, M.G.* Paper presented at the "XX Congresso Brasileiro de Paleontologia", Búzios.

Perez, C. P., Bissaro Jr., M. C., & Kipnis, R. (2007). *O uso do registro faunístico fóssil na reconstituição paleoambiental da região de Lagoa Santa, Minas Gerais.* Paper presented at the "II Encontro Brasileiro de Estudos do Carste", São Paulo.

Piló, L. B. (2002). P.W. Lund e a geomorfología cárstica de Lagoa Santa. *O Carste, 14*, 12–17.

Piló, L. B., & Neves, W. A. (2002). A coexistência do homem com os grandes mamíferos extintos e a origem do homem no Novo Mundo, segundo Peter Wilhelm Lund (1801–1880). *Chronos, 5*, 29–41.

Piló, L. B., & Neves, W. A. (2003). *Novas datações de 14C (AMS) confirmam a tese da coexistência do homem com a megamastofauna pleistocênica na região cárstica de Lagoa Santa, MG*. Paper presented at the "IX Congresso da Associação Brasileira de Estudos do Quaternário", Recife.

Piló, L. B., Auler, A. S., Neves, W. A., Wang, X., Cheng, H., & Edwards, R. L. (2004a). Revisitando a Lapa do Sumidouro: marco paleoantropológico do Quaternário americano. *Revista Brasileira de Paleontologia, 7*, 337–348.

Piló, L. B., Neves, W. A., Araujo, A. G. M., & Kipnis, R. (2004b). *Lapa Vermelha IV e Lapa das Boleiras: dois exemplos de registros sedimentares na formação de sítios arqueológicos no Carste de Lagoa Santa, M.G.* Paper presented at the "I Encontro Brasileiro de Estudos do Carste", Belo Horizonte.

Piló, L. B., Auler, A. S., Neves, W. A., Wang, X., Cheng, H., & Edwards, R. L. (2005). Geochronology, sediment provenance, and fossil emplacement at Sumidouro Cave, a classic Late Pleistocene/Early Holocene paleoanthropological site in Eastern Brazil. *Geoarchaeology, 20*, 751–764.

Porsani, J. L., Jangelme, G. M., & Kipnis, R. (2010). GPR survey at Lapa do Santo archaeological site, Lagoa Santa Karstic region, Minas Gerais State, Brazil. *Journal of Archaeological Science, 37*, 1141–1148.

Powell, J. F. (2005). *The First Americans: Race, evolution, and the origin of Native Americans*. Cambridge: Cambridge University Press.

Powell, J. F., & Steele, D. G. (1992). A multivariate craniometric analysis of North American paleo-indian remains. *Current Research in the Pleistocene, 9*, 59–62.

Prous, A. (1991). *Arqueologia Brasileira*. Brasília: Editora UnB.

Prous, A. (1997). O povoamento da América visto do Brasil: uma perspectiva crítica. *Revista USP, 14*, 8–21.

Prous, A. (2006). *O Brasil antes dos brasileiros – a pré-história do nosso país*. Rio de Janeiro: Jorge Zahar Editor.

Prous, A., & Fogaça, E. (1999). Archaeology of the Pleistocence-Holocene boundary in Brazil. *Quaternary International, 53*(54), 21–41.

Pucciarelli, H. M., Neves, W. A., González-José, R., Sardi, M. L., Rozzi, F. R., Struck, A., & Bonilla, M. Y. (2006). East-west cranial differentiation in pre-Columbian human populations of South America. *Homo, 57*, 133–150.

Pucciarelli, H. M., González-José, R., Neves, W. A., Sardi, M. L., & Rozzi, F. R. (2008). East-west cranial differentiation in pre-Columbian populations from Central and North America. *Journal of Human Evolution, 54*, 296–308.

Pugliese Jr., F. A. (2008). *Os líticos de Lagoa Santa: um estudo sobre organização tecnológica de caçadores-coletores do Brasil Central*. Master Thesis. Museu de Arqueologia e Etnologia, Universidade de São Paulo, São Paulo.

Quatrefages, A. (1887). *Histoire générale des races humaines. Introduction a l'étude des races humaines*. Paris: Hennuyer.

Rackza, M. F., De Oliveira, P. E., Bush, M., & McMichael, C. H. (2013). Two paleoecological histories spanning the period of human settlement in southeastern Brazil. *Journal of Quaternary Science, 28*, 144–151.

Reis, M. I., Bernardo, D. V., Almeida, T. F., & Neves, W. A. (2010). *Avaliação de métodos métricos e não-métricos para determinação de sexo em esqueletos humanos*. Paper presented at the "XI Congreso de la Asociación Latinoamericana de Antropología Biológica", Bogotá.

Roosevelt, A. C., Douglas, J., & Brown, L. (2002). The migrations and adaptations of the first Americans: Clovis and pre-Clovis viewed from South America. In N. G. Jablonski (Ed.), *The First Americans: The Pleistocene colonization of the New World*. San Francisco: California University Press.

Seguchi, N., Nelson, A. R., Austin, S., & Brace, C. L. (2005). Early South Americans in craniofacial metric perspective: Lagoa Santa. *American Journal of Physical Anthropology, 40*(Suppl), 187.

Seguchi, N., McKeown, A., Schmidt, R., Umeda, H., & Brace, C. L. (2010). An alternative view of the peopling of South America: Lagoa Santa in craniometric perspective. *Anthropological Science*. doi:10.1537/ase090921.

Sellet, F. (2011). Fallen giants: The story of Paleoindian point types on the North American great plains. In D. Vialou (Ed.), *Peuplements et préhistoire em Amériques* (pp. 98–105). Paris: Éditions du Comité des travaux historiques et scientifiques.

Silva, D. G. R., & Kipnis, R. (2004). *A arqueobotânica em Minas Gerais*. Paper presented at the "I Encontro Brasileiro de Estudo do Carste", Belo Horizonte.

Silva, H. P., & Rodrigues-Carvalho, C. (2006). A busca pelos primeiros americanos. In H. P. Silva & C. Rodrigues-Carvalho (Eds.), *Nossa origem. O povoamento das Américas – visões multidisciplinares* (pp. 11–18). Rio de Janeiro: Viera & Lentz.

Steele, D. G., & Powell, J. F. (1992). Peopling of the Americas: Paleobiological evidence. *Human Biology, 64*, 303–336.

Steele, D. G., & Powell, J. F. (1993). Paleobiology of the first Americans. *Evolutionary Anthropology, 2*, 138–146.

Strauss, A. M. (2010). *As práticas mortuárias dos caçadores-coletores pré-históricos da região de Lagoa Santa (MG): um estudo de caso do sítio arqueológico "Lapa do Santo"*. Master Thesis. Departamento de Genética e Biologia Evolutiva, Instituto de Biociências, Universidade de São Paulo, São Paulo.

Strauss, A. (2011). As práticas mortuárias dos caçadores-coletores pré-históricos da região de Lagoa Santa (MG): um estudo de caso do sítio arqueológico "Lapa do Santo". *Revista Brasileira de Arqueologia, 24*, 136–139.

Strauss, A. (2012). Possibilidades e limitações interpretativas da hipótese Saxe/Goldstein. *Boletim do Museu Paraense Emilio Goeldi-Ciências Humanas, 7*, 525–546.

Strauss, A., & Araujo, A. G. M. (2009). *Geoarqueologia na lagoa: um estudo exploratório do sítio arqueológico do Sumidouro (Lagoa Santa – MG)*. Paper presented at the "XV Congresso da Sociedade de Arqueologia Brasileira", Belém.

Strauss, A., & Araujo, A. G. M. (2012). *The use of sedimentary depositional rates as an interpretative tool in Geoarchaeology: A case study of the archaeological region of Lagoa Santa, Central Brazil*. Paper presented at the "AK Geoarchaeologie", Leipzig.

Strauss, A., & Da-Gloria, P. (2012). *Paradigm shift in Lagoa Santa's research tradition*. Paper presented at the "XII Congreso de la Asociación Lationoamericana de Antropología Biológica", San José.

Strauss, A., & Hubbe, M. (2008). *Estruturación de la variabilidad morfológica mundial: comparación com los datos moleculares*. Paper presented at the "X Congreso de la Asociación Lationamericana de Antropología Biológica", La Plata.

Strauss, A., & Hubbe, M. (2009). *Dissimilarity fraction for metric traits of human skull: Comparison with genetic studies*. Paper presented at the "78th Annual Meeting of the American Association of Physical Anthropologists", Chicago.

Strauss, A., & Hubbe, M. (2010). Craniometric similarities within and between human populations in comparison with neutral genetic data. *Journal of Human Biology, 82*, 315–330.

Strauss, A., Nunes, T., & Neves, W. A. (2009). *Partindo do principio: impacto de distintos métodos de correção de dados em análise de afinidade morfocraniana*. Paper presented at the "XV Congresso da Sociedade Brasileira de Arqueologia", Belém.

Strauss, A., Da-Gloria, P. T., Bernardo, D. V., Kipnis, R., & Neves, W. A. (2010a). *High variability of Paleoamerican mortuary practices in Lagoa Santa region, Central Brazil*. Paper presented at the "75th Anniversary Meeting of the Society for American Archaeology", Saint Louis.

Strauss, A., Hubbe, M., Neves, W. A., & Nunes, T. (2010b). *Morphological diversity of Early South Americans: Testing hypotheses for the origin of the high regional diversity of recent Americans*. Paper presented at the "79th Annual Meeting of the American Association of Physical Anthropologists", Albuquerque.

Strauss, A., Hubbe, M., Nunes, T., & Neves, W. A. (2010c). *Was the high morphological diversity of recent Native Americans present among the first inhabitants of the continent?* Paper presented at the "79th Annual Meeting of the American Association of Physical Anthropologists", Albuquerque.

Strauss, A., Da-Gloria, P. T., Oliveira, R. E., Bernardo, D. V., Araujo, A. G. M., Kipnis, R., & Neves, W. N. (2011a). *Lapa do Santo rockshelter: New evidence of perimortem body manipula-*

tion in Early Holocene South America. Paper presented at the "80th Annual Meeting of the American Association of Physical Anthropologists", Minneapolis.

Strauss, A., Kipnis, R., Oliveira, R. E., Da-Gloria, P., Bernardo, D. V., Araujo, A., & Neves, W. A. (2011b). *The mortuary practices in Lagoa Santa region: implications of the new findings at Lapa do Santo rockshelter*. Paper presented at the "XVI Congresso Mundial da UISPP/XVI Congresso da Sociedade de Arqueologia Brasileira", Florianópolis.

Strauss, A., Kipnis, R., Oliveira, R. E., Da-Gloria, P., Bernardo, D. V., Araujo, A., & Neves, W. A. (2011c). *Lapa do Santo Rockshelter: New evidence of body manipulation in Early Holocene South America*. Paper presented at the "XVI Congresso Mundial da UISPP/XVI Congresso da Sociedade de Arqueologia Brasileira", Florianópolis.

Strauss, A., Koole, E., Oliveira, R. E., Inglez, M., Nunes, T., Da-Gloria, P. T., Robazzini, A., Costa, F., & Neves, W. A. (2011d). Two directly dated Early-Holocene archaic burials from Pains, State of Minas Gerais, Brazil. *Current Research in the Pleistocene, 28*, 123–125.

Strauss, A., Da-Gloria, P., & Neves, W. A. (2012). Beyond Lund's dilemma: New perspectives from Lagoa Santa region, Central Brazil. Paper presented at the "*77th Annual Meeting of the Society for American Archaeology*", Memphis.

Strauss, A., Oliveira, R., Da-Gloria, P., Bernardo, D., Salazar Garcia, D. C., Hubbe, M., Talamo, S., Pugliese, F., Bissaro Jr., M. C., Dos Santos, R., Wilkinson, C., Black, S., Atencio, D., Correa, L., Gunz, P., Richards, M., Araujo, A., Kipnis, R., Hublin, J. J., & Neves, W. A. (2013). *The oldest case of decapitation in the New World*. Paper presented at the "The Paleoamerican Odyssey Conference", Santa Fe.

Ten Kate, H. (1885). Sur les Crânes de Lagoa-Santa. *Bulletins de la Société d'Anthropologie de Paris, 8*, 240–244.

Van Vark, G. N., Kuizenga, D., & Williams, F. L. E. (2003). Kennewick and Luzia: Lessons from the European Upper Paleolithic. *American Journal of Physical Anthropology, 121*(2), 181–184.

Virchow, R. C. (1892). *Crania Ethnica Americana. Sammlung auserlesener amerikanischer Schädeltypen*. Berlin: A. Asher & Company.

Walter, H. V. (1958). *Arqueologia da região de Lagoa Santa*. Rio de Janeiro: Sodegra.

Waters, M. R., & Stafford Jr., T. W. (2007). Redefining the age of Clovis: Implications for the peopling of the Americas. *Science, 315*, 1122–1126.

Part II
Research Topics

Chapter 10
The Repercussions of the Human Skeletons from Lagoa Santa in the International Scenario

Mark Hubbe and Walter A. Neves

Abstract The region of Lagoa Santa has been a centerpiece in the discussion of the settlement of the Americas since the mid of the nineteenth century. The Lagoa Santa Karst generated the largest collection to date of human skeleton remains dating to early Holocene and, as such, called the attention of different researchers interested in the discussions about the origins and dispersion of the first human groups of the New World. In particular, the study of the Lagoa Santa skulls has called the attention of generations of researchers due to its unique morphological characteristics when compared to most of the Native American groups studied so far. In this chapter, we review briefly the most important phases of study of the Lagoa Santa skeletal remains, which begins with the typological descriptions of the material during the nineteenth century and the role it played in discussing the antiquity of humans in the Americas and ends with modern studies that investigate the origin of the biological diversity of early Americans in the context of evolutionary theories. This review illustrates the difficulty of reconstructing the initial millennia of human presence in the Americas and shows how there is still no consensus about the origins of the high morphological diversity that is observed in the continent during the Holocene.

M. Hubbe (✉)
Department of Anthropology, The Ohio State University,
174 West 18th Avenue, Columbus, OH, USA

Instituto de Arqueologia y Antropología, Universdiad Católica del Norte, Antofagasta, Chile
e-mail: hubbe.1@osu.edu

W.A. Neves
Laboratory for Human Evolutionary and Ecological Studies, Department of Evolutionary Biology and Genetics in the Biosciences Institute, University of São Paulo,
R. do Matão 277, sala 218, São Paulo, SP, Brazil
e-mail: waneves@ib.usp.br

P. Da-Gloria et al. (eds.), *Archaeological and Paleontological Research in Lagoa Santa*, DOI 10.1007/978-3-319-57466-0_10

Introduction

This book clearly attests how important the paleontological, archaeological, and speleological record of Lagoa Santa are for studies of the Brazil's remote past. Other chapters describe in detail how Lagoa Santa has been established as a key region for paleontological and archaeological studies ever since the pioneering work of Peter Lund, and how it has become the region with the longest history of that kind of research in Brazil and one of the oldest on the South American continent. During the last 185 years, the Lagoa Santa material has been at the center of major discussions about the settlement of the Americas, and even today, it is the reference collection for any study interested in the biological characteristics of the first Americans. Among the tons of material recovered from the region, the human skeletons have always been particularly important in that discussion (e.g., de Azevedo et al. 2011; González-José et al. 2008; Hubbe et al. 2010, 2011; Neves et al. 2003, 2007a, 2013; Neves and Hubbe 2005; Powell 2005; Powell and Neves 1999; Seguchi et al. 2011). This chapter aims to make a succinct review of the impact the human skeletons of Lagoa Santa have had on the international literature over the last two centuries. The long rich story about the origin of the morphological diversity of the Lagoa Santa skeletons has been present in the international literature ever since Peter Lund published his first descriptions in the nineteenth century, and this chapter makes a historical review of such works particularly interesting in the context of the present volume. Basically, the publications concerning the osteological material from Lagoa Santa can be divided into four stages: the first two comprise descriptive and typological works produced in parts of the nineteenth and twentieth centuries. The last two appeared in the later decades of the twentieth century and consisted largely of analyses of the material using modern quantitative techniques and the inclusion of the Lagoa Santa material in a worldwide comparative framework. In this chapter, we will address the works and ideas proposed in each of those periods.

The First Descriptive Works and the Antiquity of the Human Skeletons

Between 1840 and 1844, Lund described the findings of human remains from the fossiliferous deposits in the caves of Lagoa Santa in a series of published letters (Lund 1840, 1842, 1844). Those publications initiated a long trajectory of studies investigating the earliest groups of Americans that were to make Lagoa Santa one of the pillars of studies on the settlement of the New World. The human remains Lund discovered only represent a small part of the total amount of the material he excavated in the Lagoa Santa caves, recovered from just a few of the hundreds of caves that the Danish naturalist explored. The human skeletons caught his attention immediately because they were found in deposits that also contained the remains of extinct megafauna species, suggesting that the material was of great antiquity. Given the context at the time, when the discussion on the antiquity of the human species among western naturalists was still strongly influenced by the idea that our species had been created by supernatural forces, the possibility that human beings might have coexisted with extinct mammalian species caused a tremendous impact on the international scientific community. The Lagoa Santa remains not only suggested the great antiquity of the human species but also undermined the ideas put forward in Europe by some of the most important naturalists of the day (the most important to Lund being Georges Cuvier) that the extinct species were entirely separated from the living species (humans among them) by cycles of distinct catastrophic events that had obliterated life on the planet (Huxley 2007). Indeed, a similar observation of remains of extinct mammal species that coexisted with mollusk species that are still living today in Argentina was one of the pieces of evidence collected by Darwin during his voyage around the world on board the Beagle and was one that would later influence the elaboration of his theory of the evolution of species through very gradual changes (Darwin 1859).

Against that historical background, it is not surprising that in his first reports, Lund disregarded the possibility that the human remains shared their antiquity with the remains of the extinct megafauna (Lund 1840). What is even less surprising, however, is that when Lund later changed his opinion and instead defended the idea of the coexistence of human beings and the megafauna in Lagoa Santa (Lund 1842, 1844), the Lagoa Santa skeletons became objects of tremendous attention to those North American and European researchers interested in the origin of our species. Lund's argument about the coexistence of humans and megafauna species was based on the finds at the Sumidouro Cave, where the stratigraphic association of the excavated remains, the degree of mineralization of the human remains, and the supposedly primitive morphological features of the cranial remains all led him to conclude that:

> There can be no doubt that the existence of Man on this continent dates back to times earlier than those in which the last races of gigantic animals, whose remains are abundant in the caves of this country, ceased to exist, or in other words earlier than historical times. (Lund 1844:329–330)

Lund went even further in his observations and came to the sweeping conclusion that:

> […] America was already inhabited at a time when the first rays of History had not come over the horizon of the Old World, and the people that inhabited it in those remote times were of the same race as those that were inhabiting it at the time of the Discovery. (Lund 1844:330)

Despite the similarities that Lund observed between the Lagoa Santa skulls and other ancient Old World skulls, he proposed that the Lagoa Santa material was older, representing samples of the "most primitive" human race, and he ended the respective communication by defending the idea that the antiquity of the Lagoa Santa remains was evidence that Brazil was part of the most ancient continent on the planet.

Lund's findings in Lagoa Santa and the implications of his conclusions regarding the origins of the human species caught the attention of a series of researchers in the decades that followed his publications. In addition to Lund's works, specimens of the Lagoa Santa skeletons ended up in various institutions in Europe and Brazil, facilitating access for researchers interested in the material. Most of the material recovered from Lagoa Santa was sent to the Zoological Museum in Copenhagen (Holten and Sterll 2000), but the British museum also acquired a cranium and a series of human bone fragments (Blake 1864), and Lund sent one cranium to the Brazilian Historical and Geographic Institute (*Instituto Histórico e Geográfico Brasileiro* (IHGB)) in Rio de Janeiro. Descriptions and discussions of that material appeared in various publications during the nineteenth century and the early decades of the twentieth century, and the main point of discussion was around the definition of the age of the human remains and the possibility that they could have been contemporary with the extinct megafauna.

Among those authors that defended the antiquity of the Lagoa Santa specimens, Reinhardt (1888) was the first to actually study the collection sent to Copenhagen. In Reinhardt's opinion, the human remains from Lagoa Santa presented the same degree of mineralization as the remains of the extinct megafauna in the region, and, accordingly, they were of a similar age. Quatrefages (1879, 1887) also adhered to the idea of the coexistence of human beings and extinct animals in Lagoa Santa, but he never analyzed the material himself and merely based his conclusions on Lund (1844) and Lacerda and Peixoto's (1876) descriptions.

However, most of the authors who studied the Lagoa Santa collection were more cautious about accepting the antiquity proposed by Lund. Blake (1864), who described the material acquired by the British Museum, argued that the specimens were undoubtedly very old given the extent of their fossilization, but they were probably not as old as the extinct megafauna. Lacerda (1875) drew attention to the fact that some of the crania taken from shell middens in the state of Pará showed the same degree of fossilization as those from Lagoa Santa, even though they were known to be more recent. That refuted the logic that the degree of fossilization in the Lagoa Santa material was a good indicator of its antiquity. Lütken (1883), who was responsible for the description of the findings history of the Copenhagen collection, was also reluctant to accept the antiquity of the Lagoa Santa human skeletons. Considering that there were no marks of any human action on the megafauna remains or any evidence in the human bones that they had been attacked by large mammals (one of Lund's hypotheses to explain how the material got into the Sumidouro Cave), Lütken concluded that, albeit it might be possible, the coexistence of human beings and the extinct megafauna in Lagoa Santa was far from having been proved. Ten Kate (1885), who studied all the human remains in the Copenhagen collection, was of the same opinion as Lütken. In turn, the person responsible for the analysis of the collection Lund sent to Copenhagen, Sören Hansen (1888), argued that it was impossible to reliably determine the geological age of the human specimens due to the high degree of disturbance evident in the Sumidouro deposits when Lund excavated them. At the beginning of the twentieth century, Hrdlička (1912), in his great review on the antiquity of human presence in

South America, also rejected the idea of any association between the megafauna and humans, even though his conclusions, like those of Quatrefages, were not based on any actual analysis of the material from Lagoa Santa but merely derived from the literature.

Typological Studies of the Origin of Paleoamerican Morphology

Given the possibility that the human skeletons from Lagoa Santa might be extremely ancient, the material was promptly included in the discussions of the morphological characteristics of the first human beings in the New World, and how those characteristics were related to the history of our species in general. Various authors in that initial period described the Lagoa Santa individuals as being representatives of a primitive and unique human race, the so-called Lagoa Santa Man, who represented the prototype of the American morphological type. Even though such conclusions were based on the possibility of the great antiquity of the Lagoa Santa material (whether contemporary with the megafauna or not), descriptions of the cranial morphology were recurrently presented to support that idea. As mentioned above, Lund (1844) himself described the morphology of the Lagoa Santa material as being primitive, particularly in the aspect of the slope of the forehead. Even though Lund's description was eventually shown to be incorrect (Hrdlička 1912), most researchers who analyzed the material considered the Lagoa Santa specimens to belong to a singular human race. Lacerda and Peixoto (1876), for example, compared the Lagoa Santa skulls to those of Botocudo Indians from central-eastern Brazil deposited in the IHGB and also to other skeletal series and concluded that "the present day indigenous races represent a mixture of two different types" (p. 74) of which the Lagoa Santa type was the more ancient and more closely resembled that of the Botocudo Indians. Hansen (1888) took a similar view and concluded that the great uniformity of the Sumidouro crania indicated the presence of a very ancient race that occupied most of South America and later mixed with other racial elements, especially groups with brachycephalic crania.

Strongly influenced by the hierarchic and progressive model for human races that prevailed in their day, according to which the populations of Asia and the Native Americans occupied the lower rungs of the ladder in terms of intellectual and moral capacity, most authors who studied the material from Lagoa Santa, including Reinhardt (1888), Quatrefages (1879, 1887), and Kollman (1884), were of the opinion that the Lagoa Santa groups had the same morphological features as the more recent groups of Native Americans, that is, "the same racial features as the Indians still living today" (Kollman 1884:198). However, even that opinion was not entirely generalized. After conducting a complete examination of all the human skulls deposited in Copenhagen, Ten Kate (1885) criticized Kollman's conclusions, which had been based on the study of only part of the collection of 15 crania that were available at the time. Ten Kate concluded that at least one of the skulls studied did not have the same morphological features as the others but showed a brachycephalic tendency.[1] Based on that observation, the author argued that even among the ancient Lagoa Santa crania, it was possible that the mixture of "racial types" was already in progress, and he, therefore, rejected the idea that the Lagoa Santa skulls represented the typical shape of the American aboriginals. That opinion of Ten Kate's was shared by Virchow (1892), who also studied the collection in Denmark, and concluded that during the prehistoric period represented by the Lagoa Santa collection, the continent was inhabited by dolichocephalic and brachycephalic races whose regions of settlement overlapped, thereby allowing for their miscegenation.

During the first decades of the twentieth century, the discussion of the morphological diversity of the groups that inhabited Lagoa Santa began to systematically include the diversity of American groups as a whole. Rivet (1908), for example, described the occurrence of the Lagoa Santa race in other parts of the continent, especially in Ecuador, and he argued that the continent had been occupied by various races during its prehistory. The concept of great morphological diversity among the Native American groups was strongly criticized by Hrdlička, who defended his former point of view (Hrdlička 1907), whereby the American continent had been occupied by a single group, the American homotype. In his work dedicated to South America (Hrdlička 1912), he recognized the existence of morphological diversity, but he did not attribute it with sufficient importance to refute his idea about the American homotype. "The fact is that the American stem or homotype is not homogeneous; it presents in different tribes and localities the extremes of head forms and also numerous other pronounced differences. Yet, the living Indian, as well as his skeletal remains, are characterized throughout America [...] by certain fundamental traits that indicate unity in a more general sense of the word" (Hrdlička 1912:183).

Rivet and Hrdlička's divergent opinions defined the direction of the ensuing debate, which went on for most of the twentieth century. Poech (1938), Imbelloni (1938), and Bastos de Ávila (1950) appeared in that context as authors defending the position of Rivet and Ten Kate, that the Americas had been occupied by more than one race. Among those three, Imbelloni deserves special attention because of the geographic outreach of his studies. The morphological comparison of human crania from various regions of South America enabled him to conclude that the Native American populations were divided into different human races of which the remains found in Lagoa Santa represented the most primitive.

Starting from the middle of the twentieth century, new excavations were conducted in the region but at this time focused on the archaeological record, and details of them can be found in other chapters of this book. Padberg-Drenkpol, sponsored by the National Museum of Rio de Janeiro, excavated different sites, being the Lapa Mortuária of Confins the most important of them (see Chaps. 4 and 5 of this vol-

[1] During the 2000s, Luis Beethoven Piló and one of us (WAN) confirmed that in the Sumidouro collection in Denmark, one of the skulls had different morphological characteristics. The same skull, however, had sediment stuck to the inside of the calotte that was different from the sediment observed in the other craniums, and based on this, it is possible that it comes from a more recent sedimentary stratum. Thus, it is probable that the skull that Ten Kate described was a more recent skull got mixed up with the others before or during the excavation process.

ume). Hurt and Blasi (1969) excavated the Cerca Grande complex (see Chap. 7 of this volume). The French mission excavated the Lapa Vermelha IV rock-shelter (Laming-Emperaire 1979; see Chap. 7 of this volume). The Minas Gerais Academy of Sciences (Walter et al. 1937; see Chap. 6 of this volume) excavated various other sites, among them was a supposed Pleistocene stratum at the Lapa Mortuária, which was the same site previously excavated by Padberg-Drenkpol. Those new excavations generated a large number of human skeletons, considerably increasing the Lagoa Santa human material collection.[2]

The new Lagoa Santa collections established that region as the focus for any study interested in investigating the characteristics of the earliest American groups. Mello e Alvim (1977) and Mello e Alvim et al. (1983–1984) proposed that the Lagoa Santa populations were extremely homogeneous, in contrast with what had been observed among the Brazilian coastal populations. Soto-Heim (1994), who made yet another study of the Copenhagen collection, shared Mello e Alvim's opinion that the Lagoa Santa crania showed great homogeneity, which supported the conclusion arrived at by de Hansen (1888) and Kollman (1884). Their idea of great homogeneity was later criticized by Neves and Atui (2004) based on the work that used modern multivariate statistical analyses to show just the contrary: the skulls from Lagoa Santa represent a considerably diverse population. The results of that analysis served to debunk the myth of Lagoa Santa homogeneity that had formerly been widely adopted by Brazilian archaeologists based on Mello e Alvim's analysis (e.g., Prous 1993; Prous e Fogaça 1999; Schmitz 1984). The question of the within-population morphological diversity of Lagoa Santa was the object of a recent new study (Hubbe et al. 2015) that compared the Lagoa Santa variability with the values obtained for other modern populations around the world, concluding that the Lagoa Santa biological diversity is not different from that found in modern human populations across the planet. In other words, albeit showing a greater degree of heterogeneity than defended by Mello e Alvim, the Lagoa Santa diversity is not sufficiently accentuated to refute the hypothesis that they stem from a single biological population, despite the fact that they represent almost 3000 years of human settlement of the region.

The Chronological and Morphological Contextualization of Lagoa Santa

Despite the fact that they address an essential aspect of the biological characteristic of Lagoa Santa, especially for determining the validity of using the Lagoa Santa collections as representatives of a real biological unit (population), the studies

[2] In fact, adding the skeletons excavated in the most recent decades by a team led by one of us (WAN), more than 200 skeletons dating to the early Holocene are known from Lagoa Santa. As a point of comparison to show the importance of this collection, in the entire North America, fewer than 30 skeletons dating to the same period have been found, and they are highly dispersed across the continent.

dedicated to intrapopulation analyses are only a small part of the studies in the last century that included the Lagoa Santa skeletons. The vast majority of works published at the end of the twentieth century and the beginning of the twenty-first century refer to the morphological affinities of the Lagoa Santa series and other series around the world and to their implications for our understanding of the processes that led to the colonization of the Americas. During the last decades of the twentieth century, Neves and Pucciarelli (1989, 1990, 1991) published a series of studies comparing Lagoa Santa skulls with those of populations representing the morphological diversity of modern human groups on the planet. The results presented in their publications, based on multivariate statistical analyses, clearly show that the Lagoa Santa skulls are not similar to those of modern Native Americans or East Asians but show far greater affinity with African and Australian-Melanesian groups. Neves and Pucciarelli's work has brought the discussion maintained in the last century into the context of modern physical anthropology and has triggered a heated debate about the biological relations of the Lagoa Santa groups and present-day Native Americans.

A large number of publications appeared in the wake of Neves and Pucciarelli's initial analyses, revising and refining both the knowledge of the biological affinities of the Lagoa Santa and the chronological context of the collections (de Azevedo et al. 2011; González-José et al. 2005, 2008; Hubbe et al. 2010, 2011; Neves et al. 1999a, b, 2003, 2005, 2007a, b, 2013; Powell and Neves 1999; Pucciarelli et al. 2006; Seguchi et al. 2011). Regarding the chronology of Lagoa Santa's settlement, dates generated in the last two decades have shown how the great majority of the human remains from Lagoa Santa date to the early millennia of the Holocene (see Neves and Hubbe 2005 and Chap. 9 of this volume for further details). With the exception of one skeleton – Luzia – exhumed by the French mission at the Lapa Vermelha IV rock-shelter (Feathers et al. 2010; Lamming-Emperaire 1979; Chap. 7 of this volume), not a single Pleistocene skeleton was recovered from the region, even though the archaeological record in Lagoa Santa dates back to at least 10.5 thousand years BP (uncalibrated; Araujo et al. 2008, 2012).[3]

Thus, the Lagoa Santa human remains represent human groups that buried their dead in the period from 9.5 to 7.0 thousand years BP (uncalibrated, see Araujo et al. 2013; Neves et al. 2013 for reviews of more recent archaeological work in the region) or even a little more recent than that (Araújo et al. 2012). This is the period in which the limestone rock-shelters in the region were used as cemeteries. The rare individuals with greater antiquity (Luzia and possibly the Confins Man) represent individuals who died or were deposited inside caves that were not systematically occupied by human beings (see Chap. 13 of this volume for further details).

Despite the rarity of Pleistocene human skeletons in the Lagoa Santa context, in recent decades, it has become possible to elucidate Lund's dilemma regarding the

[3] The Confins Man excavated by the Minas Gerais Academy of Sciences at the Lapa Mortuária was also originally proposed as having come from a Pleistocene sedimentary stratum (Walter et al. 1937). However, recent attempts to obtain a date for the material were unable to generate sufficient collagen for an absolute date.

coexistence of human beings and the extinct megafauna in Lagoa Santa (Neves and Piló 2003). The absolute dates of the sedimentary strata obtained in recent decades (see Chaps. 9 and 17 of this volume) have been complemented by evidence that some megafauna species did survive into the Holocene epoch (Hubbe et al. 2013; Neves and Piló 2003), which means that the human groups coexisted for a time with extinct animal species in Lagoa Santa, even though, up until now, no evidence of interaction between humans and the megafauna in the region has come to light (Hubbe et al. 2013). The absence of any signs of interaction is not really surprising, considering that South America as a whole shows very little evidence of the hunting of big animals in spite of the chronological coexistence of humans and extinct mammal species (Borrero 2009; Hubbe et al. 2013).

On the subject of morphological diversity, Neves and Pucciarelli's (1989, 1990, 1991) analyses, as already stated, opened the way for a long discussion that is still going on regarding the origin of the morphological and biological diversity of the American groups. The main focus of the new studies has been on elucidating the origin of Paleoamerican[4] morphology by means of comparative studies aimed at gaining an understanding as to whether morphological pattern is exclusive to the Americas or can be explained in the evolutionary context that marked modern human groups' diversification and expansion over the planet during the Pleistocene. Although the Paleoamerican morphology may be best characterized by the Lagoa Santa skulls, given the size of the collection, this morphological pattern is by no means restricted to the region. In fact, all the human skulls dated to more than 7000 years (uncalibrated) that have been analyzed up until now have shown great affinity with the Lagoa Santa group. In South America, the Paleoamerican morphology has been described in samples from Central Brazil (Lagoa Santa; Neves et al. 2007a; Neves and Hubbe 2005) and Colombia (Savana de Bogotá; Neves et al. 2007b), as well as in isolated specimens in the Brazilian southeast and northeast (Capelinha; Neves et al. 2005; Toca dos Coqueiros; Hubbe et al. 2007) and in the south of Chile (Palli Aike; Neves et al. 1999b). In North America, groups with cranial morphology similar to that of the Lagoa Santa group have been observed in Mexico (Chatters et al. 2014; Gonzalez-José et al. 2003) and the United States (Jantz and Owsley 2001; Powell 2005). Thus, up until now, there is a relative consensus in the literature that the totality of the continent was occupied by groups that shared a similar morphological pattern during the Pleistocene/Holocene transition and, furthermore, that this morphology is distinct from the morphology observed in most of the present-day Native American groups.

---

[4]The term Paleoamerican is frequently used to describe the morphology that typifies the Americas' most ancient groups (see, however, Gonzaléz-José et al. 2008 for a critique of the use made of the term). Paleoamerican morphology must not be confused with the term Paleo-Indian, which is used to describe the cultural or chronological context of the earliest American groups. When correctly used, the term Paleoamerican refers exclusively to the morphological characteristics of a given population.

Exploring the Origin of American Morphological Diversity

Given the context described above, the question of the origin of Paleoamerican morphology became one of great importance, particularly because it is distinct from the morphology observed today in the groups inhabiting eastern Asia, which is the region where the first Americans must have come from. Recent analyses have suggested that the Paleoamerican morphology is a retention of the morphological patterns present among populations around the planet at the end of the Pleistocene (Hubbe et al. 2011; Neves et al. 2003). Hubbe et al. (2011) have shown that there is a high degree of morphological affinity between the Paleoamerican groups (Lagoa Santa and Colombia) and the specimens from the end of the Pleistocene in Europe and eastern Asia, which in turn leads to a proposed human dispersion scenario whereby the modern human groups that migrated out of Africa 70 to 50 thousand years BP and occupied southern and southeast Asia (Lahr 1995; Mellars 2006) did not undergo great changes in their cranial morphological patterns. From Southeast Asia, groups dispersed toward the south, eventually colonizing Melanesia and Australia, and to the north, colonizing Northeast Asia and eventually the Americas. The results obtained by Hubbe et al. (2011) suggest that in the course of that process, there was no accentuated morphological change between the Paleoamerican groups, the Pleistocene populations of Asia and Europe, and the actual groups of Africa and Australo-Melanesia. After the colonization of the Americas, more accentuated morphological differentiation processes may have occurred, giving rise to the morphological diversity that typifies the modern groups in Asia, Europe, and Americas.

Although that explanatory model for the origin of Paleoamerican morphology has not been contested by new analyses up to the moment, the same cannot be said of the discussion about the origin of the morphological pattern that represents the majority of the modern Native American groups. Many different models have been proposed to explain the increase in morphological diversity during the Holocene and to explain the transition from the Paleoamerican pattern to the morphology that is the characteristic of the present-day Native American groups, who strongly resemble present-day Asian group in terms of their morphology. The studies addressing this question tend to fall into one of three categories:

- Those that argue that the morphological diversity observed on the continent during the Holocene was due to multiple waves of human dispersion coming out of Asia (e.g., Hubbe et al. 2010; Neves et al. 2007a, b; Neves and Hubbe 2005)
- Those that defend the idea the morphological transition was due to the adaptation processes that occurred on the continent (Perez et al. 2009, 2011; Powell 2005)
- Those that consider that there was a continuous gene flow with the Asian northeast during the Holocene (de Azevedo et al. 2011; González-José et al. 2008)

Those who defend the first scenario argue that the diachronic morphological differences observed on the continent are too large to be explained merely by local microevolutionary changes. The Two Main Biological Components Model, as it was first presented by Neves and colleagues (Munford et al. 1995), explains the settlement of the continent as occurring through two waves of dispersion: the first brought in the Paleoamerican morphology that characterizes the most ancient groups on the continent and that was present in Asia toward the end of the Pleistocene, and the second introduced the morphology that characterizes most of today's Native American groups, which is similar to the morphology of present-day eastern Asian groups. The Two Main Biological Components Model proposes that the majority of the Paleoamerican groups was replaced by groups with present-day morphology, given that no groups with intermediary morphology have been found on the continent. Nevertheless, in some regions (Baja California, Savana de Bogotá, Patagonia, and Central Brazil), groups can be observed that retained high affinities with the Paleoamerican morphology until more recent times (González-José et al. 2003; Hubbe et al. 2014; Lahr 1995; Neves et al. 2007b; Strauss et al. 2015).

The second scenario suggested to explain morphological diversity on the continent assumes that the morphological changes detected as having occurred during the Holocene can be explained by local microevolutionary processes. That scenario was supported by the results of some dental studies (Powell 2005) and other studies comparing morphological series and molecular data (mitochondrial DNA) in Argentinean human groups (Perez et al. 2009, 2011). However, studies contrasting the different models on a continental scale (de Azevedo et al. 2011; Hubbe et al. 2010) have considered such models to be less parsimonious than the one that assume that extra-continental morphological diversity entered during the Holocene.

The third model proposed to explain morphological diversity on the continent assumes that the first populations to enter the Americas already showed great morphological diversity and that a genetic flow was maintained between the Americas and the Asian northeast throughout the entire period of human settlement of the New World (de Azevedo et al. 2011; González-José et al. 2008). This model defends that that the combination of those two factors would be sufficient to explain the morphological diversity encountered on the continent. This last scenario, like the first one, assumes that morphological diversity on the continent required extra genetic influx during the Holocene, but unlike the Two Main Biological Components Model, it assumed that a greater degree of morphological diversity already existed among the ancient groups. Recently, Hubbe et al. (2015) have shown that there is no great morphological diversity observable among the Paleoamerican groups in South America, which supports the hypothesis that there was an increase in morphological diversity on the continent during the Holocene.

The lack of consensus on the process of human settlement of the continent is not restricted to the question of morphology; in recent decades, genetic studies have become a prolific area in discussions addressing this question. The vast majority of molecular studies is restricted to modern populations, and in the past, they have been used to defend the ideas of one to three migrations to the Americas (Bonatto and Salzano 1997; Fagundes et al. 2008; Perego et al. 2009; Reich et al. 2012; Tamm et al. 2007; Wang et al. 2007; Zegura et al. 2004) and, in some cases, to defend the idea that there was a continuous gene flow with the Asian northeast (González-José et al. 2008; Tamm et al. 2007) or even the hypothesis of a period of isolation of proto-American groups in Beringia prior to the arrival of human groups on the Americas itself (Kitchen et al. 2008; Tamm et al. 2007). Most DNA studies of prehistoric remains are limited to material dating from the end of the Holocene[5] (Raff et al. 2011), but some studies of human remains dating from the end of the Pleistocene (Chatters et al. 2014; Gilbert et al. 2008; Kemp et al. 2007; Rasmussen et al. 2014) have not detected any lineages different from those observed among Native American groups. A detailed review of the molecular studies is beyond the scope of this chapter, but it must be remarked that the lack of consensus among the molecular studies makes any attempt to reconcile them with the morphological ones very difficult and subject to widespread criticism (e.g., González-José et al. 2008; Greenberg et al. 1986).

Regardless of the process, for the purposes of this chapter, it is only left to say that Lagoa Santa plays and will continue to play a central role in the discussions about the initial settlement of the Americas.

Final Remarks

In conclusion, in the last two centuries, the karstic region of Lagoa Santa generated the Americas' greatest collection of human skeletons from early Holocene. That material has attracted the attention of researchers around the world since the nineteenth century and still today is a central piece in the studies of the biological aspects of the first Americans. During its 3000 years of human settlement, Lagoa Santa was occupied by a population with different morphological characteristics from those of most present-day Native American groups – the so-called Paleoamerican morphology – which was dispersed throughout the continent during the first millennia of human presence in the Americas, and one that probably represents the morphology that characterized modern human groups all over the planet up until the end of the Pleistocene. Accordingly, the colonization of the continent must have taken place prior to the morphological differentiation process that gave rise to the current cranial diversity on the planet. Even though there is no consensus regarding the processes associated to the arrival and development of the morphological pattern that characterizes most of the present-day Native American groups, most of the studies investigating the issue suggest that the existence of a genetic flow from outside the continent during the Holocene must be postulated to explain the process. The nature of such a flow, whether it was continuous or limited to discrete dispersion processes cannot yet be determined.

However, despite the importance of its impact on studies investigating the processes of colonization of the Americas, the Lagoa Santa collection does have some limitations. The biggest one is the absence of preserved DNA in the bone remains,

[5] Laboratories in Brazil and overseas have made many attempts to extract DNA from the Lagoa Santa human bone remains, but, up until now, they have been unable to find any DNA in the Lagoa Santa sample, making it impossible to compare molecular data with morphological data for that collection.

which makes direct comparison of morphological data with genetic data very difficult. A second limiting factor is the very uniqueness of the Lagoa Santa collection in the continent. Practically all studies interested in the biological characteristics of the American continent's most ancient human groups depend on Lagoa Santa as the only series that represents such populations. Although Lagoa Santa makes such studies feasible, it represents just a single dot on a continent whose colonization process has increasingly shown itself to be complex and difficult to understand, and there are very few regions on the continent capable of producing archaeological and paleontological records as dense as those of Lagoa Santa. Therefore, until future generations gain access to new collections of ancient material on the Americas, Lagoa Santa will continue to be the central pillar of studies about the biological characteristics of the first Americans.

References

Araujo, A. G. M., Feathers, J. K., Arroyo-Kalin, M., & Tizuka, M. M. (2008). Lapa das Boleiras rockshelter: Stratigraphy and formation processes at a Paleoamerican site in Central Brazil. *Journal of Archaeological Science, 35*, 3186–3202.

Araujo, A. G. M., Neves, W. A., & Kipnis, R. (2012). Lagoa Santa revisited: An overview of the chronology, subsistence, and material culture of Paleoindian sites in eastern Central Brazil. *Latin American Antiquity, 23*, 533–550.

Araujo, A. G. M., Strauss, A., Feathers, J. K., Paisani, J. C., & Schrage, T. (2013). Paleoindian open-air sites in tropical settings: A case study in formation processes, dating methods and paleoenvironmental models. *Geoarchaeology, 28*, 195–220.

Ávila, J. B. (1950). Anthropometry of the indians of Brazil. In J. H. Steward (Ed.), *Handbook of South American Indians* (pp. 71–84). Washington, DC: Smithsonian Institution, Washington DC.

Blake, C. C. (1864). On human remains from a bone Cave in Brazil. *Journal of the Anthropological Society of London, 2*, 265–267.

Bonatto, S. L., & Salzano, F. M. (1997). Diversity and age of the four major mtDNA haplogroups, and their implications for the peopling of the new world. *American Journal of Human Genetics, 61*, 1413–1423.

Borrero, L. (2009). The elusive evidence: The archaeological record of the south American extinct megafauna. In G. Haynes (Ed.), *American megafaunal extinctions at the end of the Pleistocene* (pp. 145–168). New York: Springer.

Chatters, J. C., Kennett, D. J., Asmeron, Y., Kemp, B. M., Polyak, V., Blank, A. N., Beddows, P. A., Reinhardt, E., Arroyo-Cabrales, J., Bolnick, D. A., Malhi, R. S., Culleton, B. J., Erreguerena, P. L., Rissolo, D., Morell-Hart, S., & Stafford Jr., T. W. (2014). Late Pleistocene human skeleton and mtDNA link Paleoamericans and modern Native Americans. *Science, 344*, 750–754.

Darwin, C. R. (1859). *On the origin of species by means of natural selection or the preservation of favored races in the struggle for life*. London: Penguin Books.

de Azevedo, S., Nocera, A., Paschetta, C., Castillo, L., & González-José, R. (2011). Evaluating microevolutionary models for the early settlement of the new world: The importance of recurrent gene flow with Asia. *American Journal of Physical Anthropology, 146*, 539–552.

Fagundes, N. J. R., Kanitz, R., Eckert, R., Valls, A. C. S., Bogo, M. R., Salzano, F. M., Smith, D. G., Silva Jr., W. A., Zago, M. A., Ribeiro-dos-Santos, A. K., Santos, S. E. B., Petzl-Erler, M. L., & Bonatto, S. L. (2008). Mitochondrial population genomics supports a single pre-Clovis origin with a coastal route for the peopling of the Americas. *American Journal of Human Genetics, 82*, 583–592.

Feathers, J., Kipnis, R., Piló, L., Arroyo-Kalin, M., & Coblentz, D. (2010). How old is Luzia? Luminescence dating and stratigraphic integrity at Lapa Vermelha, Lagoa Santa, Brazil. *Geoarchaeology, 25*, 395–436.

Gilbert, M. T. P., Jenkins, D. L., Götherstrom, A., Naveran, N., Sanchez, J. J., Hofreiter, M., Thomsen, P. F., Binladen, J., Higham, T. F. G., Yohe, R. M., Parr, R., Cummings, L. S., & Willerslev, E. (2008). DNA from pre-Clovis human coprolites in Oregon, North America. *Science, 320*, 786–789.

González-José, R., Bortolini, M. C., Santos, F. R., & Bonatto, S. L. (2008). The peopling of America: Craniofacial shape variation on a continental scale and its interpretation from an interdisciplinary view. *American Journal of Physical Anthropology, 137*, 175–187.

González-José, R., González-Martín, A., Hernández, M., Pucciarelli, H. M., Sardi, M., Rosales, A., & Van der Molen, S. (2003). Craniometric evidence for Palaeoamerican survival in Baja California. *Nature, 425*, 62–65.

González-José, R., Neves, W. A., Lahr, M. M., González, S., Pucciarelli, H., Hernández, M., & Correa, G. (2005). Late Pleistocene/Holocene craniofacial morphology in Mesoamerican Paleoindians: Implications for the peopling of the new world. *American Journal of Physical Anthropology, 128*, 772–780.

Greenberg, J. H., Turner II, C. G., & Zegura, S. L. (1986). The settlement of the Americas: A comparison of the linguistic, dental and genetic evidence. *Current Anthropology, 27*, 477–495.

Hansen, S. (1888). Lagoa Santa Racen. En anthropologisk Undersögelse af Jordfundne Menneskelevninger fra brasilianske Huler. Med et Tillaeg om det jordfundne Menneske fra Pontimelo, Rio de Arrecifes, La Plata. *E Museo Lundii, 1*, 1–34.

Holten, B., & Sterll, M. (2000). The Danish naturalist Peter Wilhekm Lund (1801-80): Research on early man in Minas Gerais. *Luso-Brazilian Review, 37*, 33–45.

Hrdlička, A. (1907). *Skeletal remains suggesting or attributed to early man in North America, Bureau of American ethnology bulletin 33*. Washington, DC: Smithsonian Institution.

Hrdlička, A. (1912). *Early man in South America, Bureau of American ethnology bulletin 52*. Washington, DC: Smithsonian Institution.

Hubbe, M., Harvati, K., & Neves, W. A. (2011). Paleoamerican morphology in the context of European and East Asian late Pleistocene variation: Implications for human dispersion into the new world. *American Journal of Physical Anthropology, 144*, 442–453.

Hubbe, A., Hubbe, M., & Neves, W. A. (2013). The Brazilian megamastofauna of the Pleistocene/Holocene transition and its relationship with the early human settlement of the continent. *Earth Science Reviews, 118*, 1–10.

Hubbe, M., Neves, W. A., Amaral, H. L., & Guidon, N. (2007). "Zuzu" strikes again- morphological affinities of the early Holocene human skeleton from Toca dos Coqueiros. Piaui, Brazil. *American Journal of Physical Anthropology, 134*, 285–291.

Hubbe, M., Neves, W. A., & Harvati, K. (2010). Testing evolutionary and dispersion scenarios for the settlement of the new world. *PLoS ONE, 5*, e11105. doi:10.1371/journal.pone.0011105.

Hubbe, M., Okumura, M., Bernardo, D. V., & Neves, W. A. (2014). Cranial morphological diversity of early, middle, and late Holocene Brazilian groups: Implications for human dispersion in Brazil. *American Journal of Physical Anthropology, 155*, 546–558.

Hubbe, M., Strauss, A., & Neves, W. A. (2015). Early south Americans cranial morphological variation and the origin of American biological diversity. *PLoS ONE, 10*(10), e0138090. doi:10.1371/journal.pone.0138090.

Hurt, W. R., & Blasi, O. (1969). O projeto arqueológico "Lagoa Santa", Minas Gerais, Brasil (nota final). Arquivos do Museu Paranaense. *Arqueologia, 4*, 1–63.

Huxley, R. (2007). *The great naturalists*. London: Thames & Hudson.

Imbelloni, J. (1938). Tabla classificatoria de los indios: regiones biológicas y grupos raciales humanos de América. *Physis, 12*, 229–249.

Jantz, R. L., & Owsley, D. W. (2001). Variation among early North America crania. *American Journal of Physical Anthropology, 114*, 146–155.

Kemp, B. M., Malhi, R. S., McDonough, J., Bolnick, D. A., Eshleman, J. A., Rickards, O., Martinez-Labarga, O., Johnson, J. R., Lorenz, J. G., Dixon, E. J., Fifield, T. E., Heaton, T. H.,

Worl, R., & Smith, D. G. (2007). Genetic analysis of early Holocene skeletal remains from Alaska and its implications for the settlement of the Americas. *American Journal of Physical Anthropology, 132*, 605–621.
Kitchen, A., Miyamoto, M. M., & Mulligan, C. J. (2008). A three-stage colonization model for the peopling of the Americas. *PLoS ONE, 3*, e1596. doi:10.1371/journal.pone.0001596.
Kollman, J. (1884). Schädeln von Lagoa Santa. *Zeitschrift für Ethnologie, 16*, 194–199.
Lacerda, J. B. (1875). Documents pour servir à l'histoire de l'homme fossile du Brésil. *Mémoires de la Societé d'Anthropologie de Paris, 2*, 517–542.
Lacerda, J. B., & Peixoto, R. (1876). Contribuições para o estudo anthropologico das raças indígenas do Brazil. *Archivos do Museu Nacional do Rio de Janeiro, 1*, 47–79.
Lahr, M. M. (1995). Patterns of modern human diversification: Implications for Amerindians origins. *Yearbook of Physical Anthropology, 38*, 163–198.
Laming-Emperaire, A. (1979). Missions archéologiques franco-brésiliennes de Lagoa Santa, Minas Gerais, Brésil: Le grand abri de Lapa Vermelha (P.L.) *Revista de Pré-História, 1*, 53–89.
Lund, P. W. (1840). Nouvelles Recherches sur la faune fossile du Brésil (Extraits d'une letter addressee aux rédacteurs, et date de Lagoa Santa, 1er. Avril 1840). *Annales des Sciences Naturelles, 8*, 310–319.
Lund, P. W. (1842). Carta escripta da Lagôa Santa (Minas Geraes), ao Sr. 1.o Secertario do Instituto, pelo socio honorario Sr. Dr. Lund. *Revista Trimensal de História e Geografia, 4*, 80–87.
Lund, P. W. (1844). Carta escripta de Lagôa Santa a 21 de abril de 1844. *Revista do Instituto Histórico e Geográfico do Brasil, 6*, 334–342.
Lütken, C. F. (1883). Exposition de quelques-uns des cranes et des autres ossements humains de Minas-Geraés dans le Brésil central découverts et déterrés par le feu Professeur P. W. Lund. C. R. Congrès international des Americanistes 40.
Mellars, P. (2006). Going East: New genetic and archaeological perspectives on the modern human colonization of Eurasia. *Science, 313*, 796–800.
Mello e Alvim, M. C. (1977). Os antigos habitantes da área arqueológica de Lagoa Santa (MG-Brasil) – Estudo Morfológico. *Arquivos do Museu Nacional de História Natural da Universidade Federal de Minas Gerais, 2*, 119–174.
Mello e Alvim, M. C., Soares, M. C., & Cunha, P. S. P. (1983–1984). Traços não métricos cranianos e distância biológica em grupos indígenas interioranos e do litoral do Brasil. "Homem de Lagoa Santa", índios Botocudos e construtores de sambaquis. *Arquivos do Museu Nacional de História Natural da Universidade Federal de Minas Gerais, 8/9*, 323–338.
Munford, D., Zanini, M. C., & Neves, W. A. (1995). Human cranial variation in South America: Implications for the settlement of the new world. *Brazilian Journal of Genetics, 18*, 673–688.
Neves, W. A., & Atui, J. P. V. (2004). O mito da homogeneidade biológica na população paleoíndia de Lagoa Santa: implicações antropológicas. *Revista de Antropologia USP, 47*, 159–206.
Neves, W. A., & Hubbe, M. (2005). Cranial morphology of early Americans from Lagoa Santa. Brazil: Implications for the settlement of the new world. *Proceedings of the National Academy of Science, 102*, 18309–18314.
Neves, W. A., Hubbe, M., Bernardo, D., Strauss, A., Araujo, A., & Kipnis, R. (2013). Early human occupation of Lagoa Santa, eastern Central Brazil: Implications for the dispersion and adaptation of the initial settlers of South America. In K. Graf, C. V. Ketron, & M. R. Waters (Eds.), *The Paleoamerican Odyssey* (pp. 397–414). Santa Fe: Center for the Study of the First Americans, Santa Fe.
Neves, W. A., Hubbe, M., & Correal, G. (2007b). Human skeletal remains from Sabana de Bogotá, Colombia: A case of Paleoamerican morphology late survival in South America? *American Journal of Physical Anthropology, 133*, 1080–1098.
Neves, W. A., Hubbe, M., Okumura, M. M., González-José, R., Figuti, L., Eggers, S., & De Blasis, P. A. D. (2005). A new early Holocene human skeleton from Brazil: Implications for the settlement of the new world. *Journal of Human Evolution, 48*, 403–414.
Neves, W. A., Hubbe, M., & Piló, L. B. (2007a). Early Holocene human skeletal remains from Sumidouro Cave, Lagoa Santa, Brazil: History of discoveries, geological and chronological context, and comparative cranial morphology. *Journal of Human Evolution, 52*, 16–30.

Neves, W. A., & Piló, L. B. (2003). Solving Lund's dilemma: New AMS dates confirm that humans and megafauna coexisted at Lagoa Santa. *Current Research in the Pleistocene, 20*, 57–60.

Neves, W. A., Powell, J. F., & Ozolins, E. G. (1999a). Modern human origins as seen from the peripheries. *Journal of Human Evolution, 34*, 219–222.

Neves, W. A., Powell, J. F., & Ozolins, E. G. (1999b). Extra-continental morphological affinities of Palli Aike, southern Chile. *Interciencia, 24*, 258–263.

Neves, W. A., Prous, A., González-José, R., Kipnis, R., & Powell, J. F. (2003). Early Holocene human skeletal remains from Santana do Riacho, Brazil: Implications for the settlement of the new world. *Journal of Human Evolution, 45*, 19–42.

Neves, W. A., & Pucciarelli, H. M. (1989). Extra continental biological relationships of early south American human remains: A multivariate analysis. *Ciência e Cultura, 41*, 566–575.

Neves, W. A., & Pucciarelli, H. M. (1990). The origin of the first Americans: An analysis based on the cranial morphology of early south American human remains. *American Journal of Physical Anthropology, 81*, 274.

Neves, W. A., & Pucciarelli, H. M. (1991). The origin of the first Americans: An analysis based on the cranial morphology of early south American human remains. *Journal of Human Evolution, 21*, 261–273.

Perego, U. A., Achilli, A., Angerhofer, N., Accetturo, M., Pala, M., Olivieri, A., Kashani, B. H., Ritchie, K. H., Scozzari, R., Kong, Q. P., Myres, N. M., Salas, A., Semino, O., Bandelt, H. J., Woodward, S. R., & Torroni, A. (2009). Distinctive paleo-Indian migration routes from Beringia marked by two rare mtDNA haplogroups. *Current Biology, 19*, 1–8.

Perez, S. I., Bernal, V., Gonzalez, P. N., Sardi, M., & Politis, G. G. (2009). Discrepancy between cranial and DNA data of early Americans: Implications for American peopling. *PLoS ONE, 4*(5), e5746. doi:10.1371/journal.pone.0005746.

Perez, S. I., Lema, V., Diniz-Filho, J. A. F., Bernal, V., Gonzalez, P. N., Gobbo, D., & Pucciarelli, H. M. (2011). The role of diet and temperature in shaping cranial diversification of south American human populations: An approach based on spatial regression and divergence rate tests. *Journal of Biogeography, 38*, 148–163.

Poech, H. (1938). Bietrag zur kenntnis von den fossilen meschlichen Funden von Lagoa Santa (Brasilien) und Fontezuelas (Argentinien). *Mitteilugen der Anthropologischen Gesellschaft, 98*, 310–335.

Powell, J. F. (2005). *The first Americans: Race, evolution and the origin of Native Americans*. Cambridge: Cambridge University Press.

Powell, J. F., & Neves, W. A. (1999). Craniofacial morphology of the first Americans: Pattern and process in the peopling of the new world. *Yearbook of Physical Anthropology, 42*, 153–188.

Prous, A. (1993). *Arqueologia Brasileira*. Brasília: Editora da Universidade de Brasília.

Prous, A., & Fogaça, E. (1999). Archaeology of the Pleistocene-Holocene boundary in Brazil. *Quaternary International, 53*(54), 21–41.

Pucciarelli, H. M., Neves, W. A., González-José, R., Sardi, M. L., Ramírez-Rozzi, F., Struck, A., & Bonilla, M. Y. (2006). East–west cranial differentiation in pre-Columbian human populations of South America. *Homo, 57*, 133–150.

Quatrefages, A. (1879). *The human species*. New York: Appleton.

Quatrefages, A. (1887). *Histoire générale des races humaines*. Paris: Introduction a l'étude des races humaines. A. Hennuyer, Paris.

Raff, J., Bolnick, D. A., Tackney, J., & O'Rourke, D. (2011). Ancient DNA perspectives on American colonization and population history. *American Journal of Physical Anthropology, 146*, 503–514.

Rasmussen, M., Anzick, S. L., Waters, M. R., Skoglund, P., DeGiorgio, M., Stafford, T. W., Rasmussen, S., Moltke, I., Albrechtsen, A., Doyle, S. M., Poznik, G. D., Gudmundsdottir, V., Yadav, R., Malaspinas, A. S., White, S. S., Allentoft, M. E., Cornejo, O. E., Tambets, K., Eriksson, A., Heintzman, P. D., Karmin, M., Korneliussen, T. S., Meltzer, D. J., Pierre, T. L., Stenderup, J., Saag, L., Warmuth, V. M., Lopes, M. C., Malhi, R. S., Brunak, S., Sicheritz-Ponten, T., Barnes, I., Collins, M., Orlando, L., Balloux, F., Manica, A., Grupta, R., Metspalu,

M., Bustamante, C. D., Jakobsson, M., Nielsen, R., & Willerslev, E. (2014). The genome of a late Pleistocene human from a Clovis burial site in western Montana. *Nature, 506*, 225–229.

Reich, D., Patterson, N., Campbell, D., Tandon, A., Mazieres, S., Ray, N., Parra, M. V., Rojas, W., Duque, C., Mesa, N., García, L. F., Triana, O., Blair, S., Maestre, A., Dib, J. C., Bravi, C. M., Bailliet, G., Corach, D., Hünemeier, T., Bortolini, M. C., Salzano, F. M., Petzl-Erler, M. L., Acuña-Alonzo, V., Aguilar-Salinas, C., Canizales-Quinteros, S., Tusié-Luna, T., Riba, L., Rodríguez-Cruz, M., Lopez-Alarcón, M., Coral-Vazquez, R., Canto-Cetina, T., Silva-Zolezzi, I., Fernandez-Lopez, J. C., Contreras, A. V., Jimenez-Sanchez, G., Gómez-Vázquez, M. J., Molina, J., Carracedo, A., Salas, A., Gallo, C., Poletti, G., Witonsky, D. B., Alkorta-Aranburu, G., Sukernik, R. I., Osipova, L. S., Fedorova, A., Vasquez, R., Villena, M., Moreau, C., Barrantes, R., Pauls, D., Excoffier, L., Bedoya, G., Rothhammer, F., Dugoujon, J. M., Larrouy, G., Klitz, W., Labuda, D., Kidd, J., Kidd, K., Di Rienzo, A., Freimer, N. B., Price, A. L., & Ruiz-Linares, A. (2012). Reconstructing Native American population history. *Nature, 488*, 370–374.

Reinhardt, J. T. (1888). De Brazilianske Knoglehuler og de I dem Forekommende Dyrelevninger. *E Museo Lundii, 1*, 1–56.

Rivet, P. (1908). La race de Lagoa-Santa chez le populations précolombiennes de L'Equateur. *Bulletins et Mémoires de la Société D'Anthropologie de Paris, 19*, 209–275.

Schmitz, P. I. (1984). *Caçadores e coletores da pré-história do Brasil*. São Leopoldo: Instituto Anchietano de Pesquisas.

Seguchi, N., Mckeown, A., Schmidt, R., Umeda, H., & Brace, C. L. (2011). An alternative view of the peopling of South America: Lagoa Santa in craniometric perspective. *Anthropological Science, 119*, 21–38.

Soto-Heim, P. (1994). Les Hommes de Lagoa Santa (Brésil). Caractères Anthropologiques et position parmi d'autres populations paléoindiennes d'Amérique. *L'Anthropologie, 98*, 81–109.

Strauss, A., Hubbe, M., Neves, W. A., Bernardo, D. V., & Atui, J. P. V. (2015). The cranial morphology of the Botocudo Indians, Brazil. *American Journal of Physical Anthropology, 157*, 202–216.

Tamm, E., Kivisild, T., Reidla, M., Metspalu, M., Smith, D. G., Mulligan, C. J., Bravi, C. M., Rickards, O., Martinez-Labarga, C., Khusnutdinova, E. K., Fedorova, S. A., Golubenko, M. V., Stepanov, V. A., Gubina, M. A., Zhadanov, S. I., Ossipova, L. P., Damba, L., Voevoda, M. I., Dipierri, J. E., Villems, R., & Malhi, R. S. (2007). Beringian standstill and spread of Native American founders. *PLoS ONE, 2*, e829. doi:10.1371/journal.pone.0000829.

Ten Kate, H. (1885). Sur les Crânes de Lagoa-Santa. *Bulletins et Mémoires de la Société d'Anthropologie de Paris, 8*, 240–244.

Virchow, R. (1892). *Crania Ethnica Americana*. Berlin: A. Asher & Co..

Walter, H. V., Cathoud, A., & Matos, A. (1937). The Confins man – a contribution to the study of early man in South America. In G. G. Maccurdy (Ed.), *Early man – international symposium* (pp. 341–348). Philadelphia: The Academy of Natural Sciences.

Wang, S., Lewis Jr., C. M., Jakobsson, M., Ramachandran, S., Ray, N., Bedoya, G., Hurtado, A. M., Labuda, D., Klitz, W., Barrantes, R., Bortolini, M. C., Salzano, F. M., Petzl-Erler, M. L., Tsuneto, L. T., Llop, E., Rothhammer, F., Excoffier, L., Feldman, M. W., Rosenberg, N. A., & Ruiz-Linares, A. (2007). Genetic variation and population structure in Native Americans. *PLoS ONE, 3*, e185. doi:10.1371/journal.pgen.0030185.

Zegura, S. L., Karafet, T. M., Zhivotovsky, L. A., & Hammer, M. F. (2004). High-resolution SNPs and microsatellite haplotypes point to a single, recent entry of Native American Y chromosomes into the Americas. *Molecular Biology and Evolution, 21*, 164–175.

Chapter 11
The Lagoa Santa Skeletons and the Cranial Morphology of the First Americans

Danilo Vicensotto Bernardo and Walter A. Neves

Abstract The physical similarities between Native Americans and Asian populations began to draw attention of early naturalists, who have speculated in the earliest chronicles of a possible relation of descent among these peoples. Based on a typological-racialist approach, physical anthropologists suggested, at the beginning of the twentieth century, that Native Americans presented a high degree of morphological homogeneity, consisting of exclusively Asian biological origin. Debates since the first half of the twentieth century, enriched by the development and relative popularization of protein polymorphisms studies, have relegated anthropometric research to a marginal position in the debate over the origins of New World Man. This scenario was only changed in the 1970s, when W. Howells began his extensive production of comparative craniometric studies. In this context, the skeletal material originated from Lagoa Santa started to occupy, since the 1990s, a prominent position in the debate about the first Americans, fomenting a new interpretive model for the origins of man in the Americas. This model is based upon the close cranial morphological similarity of the Lagoa Santa specimens with recent African and Australo-Melanesian populations rather than with other groups like recent Asian or Native Americans.

D.V. Bernardo (✉)
Laboratory for Studies of Biological Anthropology, Bioarchaeology and Human Evolution, Human Sciences and Information Institute, Federal University of Rio Grande, Av. Itália, Km 8, Carreiros, Rio Grande, 96201-900 RS, Brazil
e-mail: danilobernardo@furg.br

W.A. Neves
Laboratory for Human Evolutionary and Ecological Studies, Department of Evolutionary Biology and Genetics in the Biosciences Institute, University of São Paulo, R. do Matão 277, sala 218, São Paulo, SP, Brazil
e-mail: waneves@ib.usp.br

P. Da-Gloria et al. (eds.), *Archaeological and Paleontological Research in Lagoa Santa*, DOI 10.1007/978-3-319-57466-0_11

Introduction: New People in the New World

The arrival of the Europeans in the New World was marked by a fascinating fact: the presence of human beings on the American continent. Right after the first contacts, those men, completely "devoid" of European faith, culture, or technology, evoked a series of questions and ideas among the scholars of the Old World. Who were those human beings with such exotic behavior? What was their faith, their view of the world? What were their origins (Diamond 2003; Fagan 1987, 1999)?

The physical similarities between the Native Americans and Asian populations caught the attention of the first naturalists, and they began to speculate on possible relations of descent and ancestry among those peoples. In 1648, in a very incipient way and with little scientific rigor, Thomas Gage (1597–1656; Gage 1648) noticed those physical similarities and proposed that Asia (Tejera 1996) had been the point of departure for the Amerindian populations and that they had arrived in the New World via the Bering Straits (Fagan 1987, 1999).

From that point on, the exploratory aspect of anthropology and its derivations, such as physical anthropology and archaeology, began to play determinant roles in elucidating the history of the settlement of the Americas in an endeavor to establish biological and cultural relations between Amerindians and Asian populations (Prous 2006). In that same vein, works addressing the Amerindians' material culture as well as their physical characteristics began to constitute the main sources of evidence that generated ideas about the colonization of the New World (Fagan 1987; Harper and Laughlin 1982; Munford 1999).

Derived from the first speculations regarding the physical similarities between Amerindians and native Asians, the incipient physical anthropology of eighteenth century began to contribute heatedly to the debates on the emergence of humans in the New World. Studies of the cranial features of Native Americans conducted by John Winslow (1703–1774), in 1722; George W. Steller (1709–1746), from 1728 to 1739; and Johann F. Blumenbach (1752–1840), in 1795, underscored how closely related the Amerindians and the peoples with mongoloid morphology were (Harper and Laughlin 1982).

In that context, in 1839, Samuel G. Morton (1799–1851) published his work *Crania America* (Morton 1839) which presented a morphological description of a large number of skulls from Native American and other populations in the world, thereby disseminating craniology among the first generation of North American physical anthropologists (Fagan 1987). The study of cranial morphology was fully developed in works like those of the eminent French scientist Paul Broca (1824–1880), who made use of metric relations of the cranium such as the cranial index, for example, to establish a characterization of two morphological types, the dolichocephalic and the brachycephalic (Brace 1982; Brace and Hunt 1990; Gould 1999; Stocking 1988, 2004; Trinkaus 1982). In the results set out in his book and those in his following work *Crania Aegyptiaca* (Morton 1844), Morton used the measurements of cerebral volume indexes to try and demonstrate the existence of a racial hierarchy and even went so far as to subdivide the native populations into barbarians and civilized populations (Fagan 1987; Gould 1999; Newmann 1951).

The extremely typological-racialist nature of the first human cranial morphology studies was not opposed in any way by early twentieth-century physical anthropologists. Ăles Hrdlička (1869–1943) and Earnest A. Hooton (1887–1954), the leading scientists of North American physical anthropology at the time, played an incisive role in the debates that mobilized the scientific community concerning the origins of humans in the New World and the emergence of modern humans as such. Hrdlička defended the idea of linearity in human evolution, and he was a fierce adept of racialist theories like those of Broca. He definitively disseminated the adoption of a typological-classificatory approach in the physical anthropology studies in the USA (Brace 1982; Fagan 1987; Howells 1999; Trinkaus 1982). That approach assumes the existence of morphological homogeneity among the Native Americans in comparison with populations of other regions of the planet (Hrdlička 1912, 1932).

Against that background, the studies of the cranial morphology that typified the specimens from the region of Lagoa Santa, in the state of Minas Gerais, were conducted in Brazil in the same way as they were being undertaken in Europe and North America, that is, adopting a typological approach to address the anthropometric features of the various indigenous races in Brazil and South America at large (Hansen 1888; Hrdlička 1912; Lacerda and Peixoto 1876; Ten Kate 1885). At the time in Brazil, the kind of physical anthropology being practiced at the National Museum of Rio de Janeiro was predominant and headed by figures such as João Batista de Lacerda (1846–1915) and Edgard Roquette-Pinto (1884–1954), who addressed issues that lay on the borders between the anatomy of the human "races" and the cultural specificities of the "barbarian" humans found in Brazilian territory (Castro-Faria 1952; Silva e Sá et al. 2007).

The importance of the skeletal material found in Lagoa Santa (for a review of it, see Chap. 10 of this volume), particularly in relation to the discussion of its chronology and consequently its importance for the study of the earliest Americans (see also Chap. 9 of this volume for a synthesis of information on the history of the research of those materials), caught the attention of the community interested in the subject and guaranteed the international repercussion of the human bone remains discovered in the region. Important work conducted by scientists like Julius Kollmann (1834–1918), Christian Frederik Lütken (1827–1901), Johannes Theodor Reinhardt (1816–1882), Herman ten Kate (1858–1931), and Jean Louis Armand de Quatrefages (1810–1892) revealed singular features of the Lagoa Santa skeletal remains which, according to their interpretation, suggested that they were representative of a primitive human race to which they attributed the denomination of "Lagoa Santa Man (or race)" (Hansen 1888; Kollmann 1884; Lacerda and Peixoto 1876; Quatrefages 1879, 1887; Reinhardt 1888).

In that scenario, in 1908, Paul Rivet (1876–1958) presented a synthesis of the main results observed to date concerning the Lagoa Santa race as compared to other pre-Columbian New World populations. In his work, Rivet corroborated the idea of a "Lagoa Santa Man" (Rivet 1908) distinct from the morphological features associated to the "shell-midden (*sambaqui*) man" (Lacerda 1882), reinforcing the notion of the great antiquity of the Lagoa Santa specimens and of their morphological affinities with other indigenous groups considered to be ancient and primitive, like the Botocudos (Lacerda and Peixoto 1876; Lund 1844, 1849; Ten Kate 1885). That conclusion favored Rivet's own interpretation that the Australian and Melanesian populations must also have participated in the genesis of New World Man (Munford 1999; Rivet 1908, 1942; Smith 1984).

Although it was widely disseminated and adopted by many scholars investigating the subject (e.g., Mendes Corrêa 1948), Rivet's proposal met with fierce opposition from the North American physical anthropologists. Even though at the time it was firmly anchored in the typological-racial nature of their anthropometric studies, the dominant idea among physical anthropologists in the USA was based on the concept of the "American homotype" (Hrdlička 1907), suggesting that all Amerindians made up a single common homogeneous type delineated by a set of distinctive physical characteristics (such as skin color, eye shape, hair type and color, and the distribution of hair on the body), including skeletal features (like height, cranial capacity, and robustness) and physiological functions (like heartrate and breathing rate). Hrdlička associated that "American homotype" to the Asian populations, establishing the theory of an exclusively Asian origin for the Amerindians (Hrdlička 1932).

Such racial-typological thinking reached a peak in the 1940s. It served to support totalitarian philosophies in Europe such as Nazi-fascism, and the anthropometric typological classification was coopted by theories that sought to eliminate or subjugate human groups or those that were considered undesirable from either the ethnic or socioeconomic point of view (Gould 1999). That improper use of anthropometric techniques evoked protests from the cultural anthropologists who, mainly supported by Franz Boas's (1858–1942) work addressing the question of morphological plasticity in response to environmental influence, defended the idea of the anti-archetypical nature of human morphological variability and consequently questioned the validity of cranial morphology as an anthropometric marker for the reconstruction of the histories of populations (Boas 1912).

Starting in the 1960s, with the emergence and popularization of techniques for manipulating protein polymorphism, such as those developed in the work of James N. Spuhler (1917–1992), cranial morphology was practically abandoned as an anthropological marker. The very few works that still went ahead with it were of an adaptationist nature and attempted to associate skeletal morphological features to the geographic distribution of the populations being investigated (Barbujani et al. 1997; Lewontin 1972). Among anthropologists at the time, especially in the wake of physical anthropology's downfall, the idea that skeletal remains had nothing to contribute to the elucidation of the history of recent human evolution, including the question of human settlement of the Americas, became a widely held belief (Armelagos et al. 1982; Genovés 1967; Gould 1999; Harper and Laughlin 1982; Newmann 1962).

In the global academic sphere of human cranial morphology studies, that state of affairs dramatically affected the research investigating the material from Lagoa Santa, despite the progress achieved by the archaeological sciences. Accordingly, there was very little progress in increasing knowledge about the rich osteological material produced by the excavations in Minas Gerais in the sense of gaining a better understanding of the first Americans' biological characteristics. Research such as that undertaken by Messias and Mello (1962), Mello e Alvim (1963, 1977, 1992–1993), and Mello e Alvim et al. (1983–1984) was limited to superficial discussions of topics such as the association between biological diversity and population identity, based on unsophisticated and analytical-typological methods of little importance for the understanding of the origins of humans on the Americas. Even so, those anachronistic studies enjoyed surprisingly great prestige in the Brazilian archaeological community and constituted the biological basis for interpreting certain aspects of the biology of the first Americans (for a review, see Neves and Atuí 2004).

Modern Craniometrics and the Rebirth of an Anthropometric Marker

Anthropometric protocols consist of comparisons of the outlines and proportions of anatomical organs, points, or regions (Bass 1987). In terms of morphometric analyses, those comparisons must be translated into objective measurements so that the observations can be analyzed together in large data series or groups (Slice 2005). The first critical stage in developing and applying a craniometric protocol is to make an efficient selection of the landmarks to be used as reference points; these are bone parts or structures that measurements will be taken from. The measurements themselves are the linear distances between the landmarks, and they need to cover all the morphological variation observed in the skull under study (Slice 2005). Figure 11.1 shows the coverage of a cranium by linear measurements.

All the tradition and development of the various measurement protocols date back to the conventions initially proposed at the Munich Craniometrics Congress of 1877 and consolidated by the International Congress of Anthropologists held in Frankfurt in 1884. Those meetings established the primordial references for anthro-

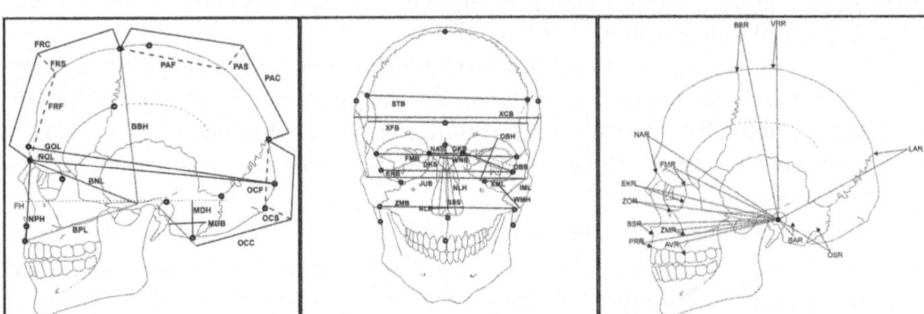

Fig. 11.1 Representation of some of the craniometric linear measurements defined by Howells (1973). The measurements displayed on the cranium at the far right are the measurements of radius projections. Note that the measurements made of the base of the cranium have been omitted

pometrics and defined the fundamental concepts subjacent to the techniques, such as the Frankfurt horizontal plane, that are in use up to the present day. They also established two classes of anatomical landmarks, namely, paired landmarks and non-paired landmarks.

The measurements set out in the Howells (1973) protocol are taken from 21 landmarks. The definition of those points and the subsequent measurements were based on the work of Broca (1875), Buxton and Morant (1933), Martin (1928), Kherumian (1949), and Vallois (1965).

Thus, the measurement protocol that Howells (1973) developed and published stems from the earlier protocols classically used by physical anthropologists at that time. As Howells (1973) himself explained, the definitions of each measurement, its instructions, and the landmarks employed are very similar to the techniques formerly proposed by Broca (1875), Martin (1928), and Kherumian (1949) and those stipulated in the 1906 Monaco Convention (Howells 1973). Given the advantages offered by his new methodology (uniformity in measuring process, reduction of subjectivity in determining the landmarks, and the adoption of logical nomenclature for cranial measurements), Howells' protocol came to dominate the craniometric studies that followed. His protocol also included a technique which, albeit subjective, was a considerable contribution to cranial measuring techniques: the proposal of a logical sequence for taking the measurements.

Lastly, the adoption of a suitable theoretical framework for the exploratory analysis of the observed cranial measurements, one that united statistical and quantitative approaches and allowed for a biological investigation of the specimens being studied, was of the greatest importance in establishing cranial variation as an anthropometric marker.

Very early on in the history of genetics, it became apparent that there is no direct, one-to-one connection between genes and phenotype features (Futuyma 1998). Instead, some genes have pleiotropic effects, while other characteristics can be derived from polygenic inheritance, that is, the selective advantage of a gene may depend on another gene or on other genes, thereby complicating the analysis of their interactions and quite often making it impossible to visualize any simple correlations among different variables (Ridley 2006).

That multifactor aspect of biological phenomena constantly disqualifies analyses simply based on univariate or bivariate statistical analysis and obliges the researcher to model methods and algorithms in a way that makes it feasible to observe microevolutionary forces at work in quantitative variations among populations (Relethford 1982).

Univariate techniques are only useful for simple analyses and descriptive inferences insofar as they can only handle one variable at a time, making it impossible to integrate different measurements of the same phenomenon (Mingoti 2005). Bivariate techniques serve for analyses involving two variables and may establish whether there are cause and effect relationships between them. They can be detected either by Pearson or Spearman's coefficients of linear correlation or by the use of the classic model of simple linear regression (Reis 2001).

However, it must be stated that from the biological-evolutionary and the statistical and quantitative points of view, inferring human populations' genetic structures through studies of cranial morphology, for example, must necessarily be based on multivariate statistical techniques (Relethford 1982, 1994, 2002). They are the only techniques that make it feasible to examine and obtain a complete, realistic understanding of phenomena that show multiple, natural correlations (Hair et al. 1998). "Multivariate statistics" is not an easy term to define. In a way, it refers to all those statistical methods that analyze, simultaneously, various measurements of an object or individual being investigated (Hair et al. 1998). It includes methods that analyze multiple dependent variables and/or multiple independent variables, whether they establish causal relationships among them or not. They can be applied to the analysis of individual or group levels, that is, to analyses of intragroup and intergroup relationships (Reis 2001).

Usually when measurements are made of the characteristics of two or more sample groups, the characteristics are correlated. When there is proven correlation among the variables, it is necessary to "disentangle" the information they display and make the structure they form more evident (Reis 2001). Consequently, one of the objectives of multivariate analysis is to simplify the data, describing the information by reducing the number of dimensions, without, however, distorting the greater part of the variance present in the set of variables adopted (Mingoti 2005). As such, multivariate techniques make it possible to describe sample being studied and to infer relationships and generalize the sample results to the population from which it was taken with a reasonable degree of certainty (Reis 2001).

In biological anthropology, or to be precise, in the studies of cranial morphological variability, the use of multivariate analyses has become widespread as can be seen by the overwhelming predominance of its use reported in the specialized literature produced since the time of Howells's seminal work. Among the various techniques, the derivatives of principal component analysis and Mahalanobis distance (van Vark and Schaafsma 1992) are the most commonly used.

Principal component analysis (PCA) was introduced by Karl Pearson (Mingoti 2005) in a paper in which he discussed the mathematical significance and the physical, statistical, and biological implications of the configurations of points and lines in n-dimensional systems (Pearson 1901). The transition from the theoretical basis to the practical application of the technique was developed based on a paper published by Hotelling (1933), addressing the analysis of variance and covariance in complex statistical data (Reis 2001).

PCA is an interdependence technique and the most popular multivariate approach in biological anthropology because of its conceptual facility and practicability in terms of establishing morphological (dis)similarities among human populations (van Vark and Schaafsman 1992). It is a reductive multivariate statistical technique, which means that it uses matrix data of correlation or covariance to construct vectors of complex variables that provide information on more than one variable or original dimension (Reis 2001; Valentin 2000). Basically, the reduction consists in condensing the information contained in a group of correlated variables into orthogonal axes that express the cumulative percentages of variance in a matrix of original data. In other words, each of the axes or principal components summarizes the information contained in various variables according to the correlation or covariance among them (Gower 1966).

The reduction makes it possible to execute the analyses made on the basis of the orthogonal relationship between two principal components, even when there is a high degree of original variance in the sample, something that would be impossible to achieve using bivariate techniques (Madrigal 1998; Hair et al. 1998).

The generalized Mahalanobis distance (D^2) is a powerful tool for measuring the dissimilarity that exists between sets of samples, making it possible to identify the extent to which two or more subsets of a sample may be associated (Clark et al. 1993). For that reason, it is commonly used to measure biological distances (Hubbe 2006; Reis 2001; Valentin 2000). The D^2 between two cases is obtained by summing the differences between vectors of measurements, while weighting them by the matrix of covariance between the variables. That means that the Mahalanobis distance is a generalized distance measurement that takes into account the covariance between the variables; the smaller the degree of association between variables, the greater the weight attributed to it in the overall distance (Mahalanobis 1936). That property is what enables the algorithm to convert and explain the conversion of a matrix of distances between studied cases into a matrix of biological dissimilarity. With that analytical framework and starting with his work in the mid-1970s, Howells reestablished the importance of cranial morphological variation as an anthropometric marker that could be applied to investigations of the demographic-evolutionary trajectories of populations.

Regarding the autochthone populations of the New World, Howells noted the marked morphological similarities among the three populations of the Americas that he studied and the strong morphological association of those populations with those of Northeast Asia, thereby confirming, from the anthropometric point of view, the geographic ancestry of modern Amerindians (Howells 1973, 1989, 1999).

Despite the methodological improvements that Howells proposed and used himself, that field of anthropology continued to be absent from the discussions of the origins of the Native American until the end of the 1980s. The characterization of Native American populations continued to be done on the basis of molecular biology, and in that context Hrdlička's model continued to predominate in physical anthropological studies underscoring homogeneity, a common geographic origin, and the proposal that diversity only emerged after the settlement of the continent (Armelagos et al. 1982; Fagan 1987; Munford et al. 1995; Szathmary 1993; for a more expanded model of New World settlement according to the hypothesis of biological homogeneity proposed by Hrdlička, see Greenberg et al. 1986). That context ensured that little or no material from Lagoa Santa was studied using the more advanced craniometric techniques of the day.

Lagoa Santa Strikes Again

The application of modern comparative craniometrics to the biological characterization of the Brazilian and South American prehistoric populations was destined to change at the beginning of the 1980s, when one of us (WAN) presented the first exploratory results of craniometric variations among the shell-midden populations of the south of Brazil (Neves 1982). After the end of the 1980s, the comparatively modern techniques mentioned above began to be used for the morphological characterization of the first Americans when one of us (WAN) together with Hector Pucciarelli presented new contributions to that debate, with the publication of the first results showing the metric characteristics of crania recovered from the Lagoa Santa archaeological record (accepted as belonging to the chronological period corresponding to the Pleistocene/Holocene transition period in South America) compared to data on recent and ancient populations with a broad global distribution. In their work, the authors noted a strong morphological association between the first Americans and Australo-Melanesian and African populations (Neves and Pucciarelli 1989, 1990, 1991).

Neves and Pucciarelli's results placed the evolutionary history of the first American populations in disagreement with the claims of the extant model at the time. To justify their results, those authors proposed a new model for New World settlement: the four migrations model, which was later renamed to two main biological components model (2MBCM) (Neves et al. 1996b, 1999a; Neves and Pucciarelli 1989, 1990, 1991) (Fig. 11.2).

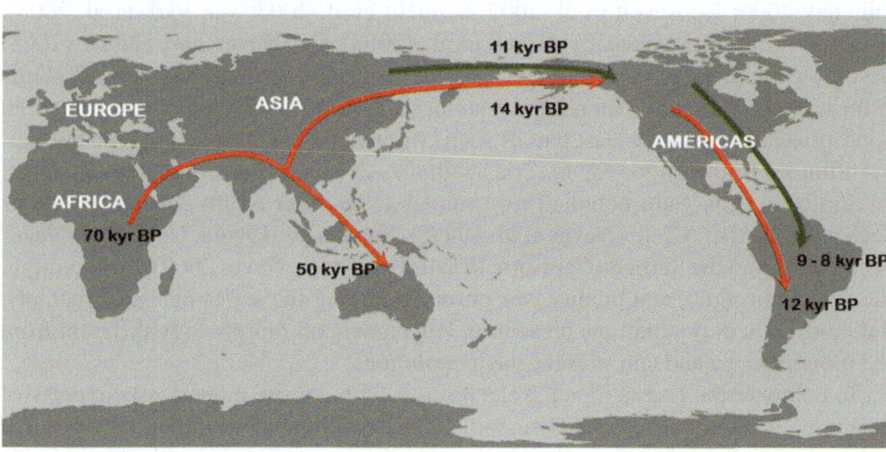

Fig. 11.2 Schematic representation of the two main biological components model. The *red arrows* represent the generalized morphology corresponding to the Paleoamerican morphology. The *green arrows* represent the biological component of specialized morphology corresponding to the Amerindians. Observe their chronological succession, that is, Paleoamerican morphology entering the Americas before the Amerindians. Observe, also, that the route of entry for both biological components (morphologies) is via Beringia (Adapted from Neves et al. 1997 and Piló et al. 2003. kyr = thousand years ago)

In this scenario, the disclosure of the results obtained from a single skeleton, popularly known as Luzia, discovered in the Lapa Vermelha IV, and exhumed in the mid-1970s by the French-Brazilian mission caused a considerable impact on the international community, because it is the most ancient human bone remains ever found in the Americas to date. The comparative analyses conducted of the Luzia cranium clearly reveal the relationship of this specimen with African and Australo-Melanesian populations, as the earlier studies had suggested (Neves et al. 1996a, 1999b; Neves and Pucciarelli 1998).

In order to broaden their research of cranial morphology variation among the first Americans, Powell and Neves (1999) compared the morphology of the most ancient skeleton of South America (Lagoa Santa – Lapa Vermelha IV) with what was, at the time, the most ancient North American specimen (Warm Mineral Springs). The results were convergent with those of other research results, insofar as the early skulls presented a strong association with one another and with Australo-Melanesian and African populations. Similar results were obtained independently by other researchers for a series of skeletons from other regions of the American continent (González-José et al. 2003, Neves et al. 2007; Powell and Steele 1992 for some examples), thereby strengthening the hypothesis put forward by the South American researchers.

Despite the considerable scientific production associated to the 2MBCM, the model was the target of a series of criticisms from the specialized scientific community, especially the North Americans, mostly based on the argument that the results obtained by Neves and associates were the fruit of anomalous crania, of small samples of outlier crania, or based on isolated skulls (Brace et al. 2001; Dillehay 2000; Roosevelt et al. 2002; Seguchi et al. 2005; van Vark et al. 2003). However, results obtained by Neves et al. (2003) and Neves and Hubbe (2005) based on much larger collections of archaeologically well-contextualized skulls from Lagoa Santa, corresponding to almost a hundred individuals, constituted an appropriate rebuttal to at least part of such criticisms. They ratified the high biological affinities of the Lagoa Santa series with the African and Australian populations.

Similar results were obtained for samples from other South American regions such as Palli Aike, Chile (Neves et al. 1999c), Sabana de Bogotá, Colombia (Neves et al. 2007), and the Serra da Capivara, Brazil (Bernardo e Neves 2009). Furthermore, the same morphological duality was observed among the series analyzed, namely, the more ancient populations presenting Paleoamerican morphology different from the more modern and sub-present-day populations.

To compare the Lagoa Santa skeletal series with a more numerically expressive database, Bernardo (2007) reassessed the morphological affinities between the Lagoa Santa skeletal series and other autochthone populations with a worldwide distribution, based on the database organized by T. Hanihara (1993a, b, 1996, 2000). Making use of modern, multivariate statistical analysis techniques commonly applied to morphometric analyses (Manly 1994), the author showed that when analyzed in an intracontinental perspective, the Lagoa Santa specimens only show an association with the other Paleoamerican specimens in the sample, represented by series originating from the Savana de Bogotá, Colombia (Neves et al. 2007). In the same work, Bernardo (2007) observed that, in comparisons on an extra-continental scale, the Lagoa Santa Paleoamericans are associated primarily to Africans and Australo-Melanesians as the 2MBCM proposes (Neves and Hubbe 2005). Figure 11.3 represents the results of the comparative analysis of the cranial morphology of Lagoa Santa specimens and series from other regions of the planet.

To conclude, the extensive production regarding the cranial morphology of the skeletal series originating from Lagoa Santa (see Bernardo 2007; Neves et al. 2004, 2005, 2007; Neves and Hubbe 2005, for just some of the examples) demonstrates the peculiar nature of the Paleoamerican cranial morphology in the New World context typified by long and narrow crania, low cheeks, prognathism, and relatively low orbits and noses. That morphology is considered to be generalized and similar to what can be found today among sub-Saharan African and Australo-Melanesian peoples and is diametrically distinct from the specialized morphology that is found in the majority of East Asian populations and in the subrecent Native Americans, referred to in the traditional literature as "mongoloid" or "specialized" (Lahr 1995).

The consistency of those observations led to the formulation of two distinct hypotheses: either the Paleoamerican morphology is indeed indicative of a migration prior to the entrance of mongoloid morphology in the Americas, as the model proposed by Neves and colleagues originally suggested, or mongoloid morphology in the Old and the New World represents a case of evolutionary convergence (Powell and Neves 1999). Succinctly stated, the groups associated to Paleoamerican cranial morphology from different regions of the American continent all present the same morphological features and are more closely associated to the characteristics of African and Australo-Melanesian populations when they are compared to the morphological variation that exists worldwide. This observation does not agree with the interpretation that the morphology of modern Native Americans would be the result of a local microevolutionary process, convergent with what occurred in Asia. Therefore, the better explanation for Paleoamerican morphology is the one proposed by Neves and Pucciarelli (1991). However, it must be underscored that this model does not coincide with models proposed on the basis of molecular biology studies of Native Americans, and even in the sphere of comparative morphological studies, alternative scenarios cannot be easily ruled out (Azevedo et al. 2011; Bonatto and Salzano 1997; Fagundes et al. 2008a,b; Goebel et al. 2008; Gonzalez-José et al. 2008; Perego et al. 2009; Reich et al. 2012; Tamm et al. 2007; Wang et al. 2007).

New Models and the Future of Craniometric Investigations in Lagoa Santa

If we take both kinds of markers to be legitimate, that is, cranial morphology and molecular data, and if we consider the different interpretations usually presented in the specialized literature, then we can imagine the following evolutionary scenarios,

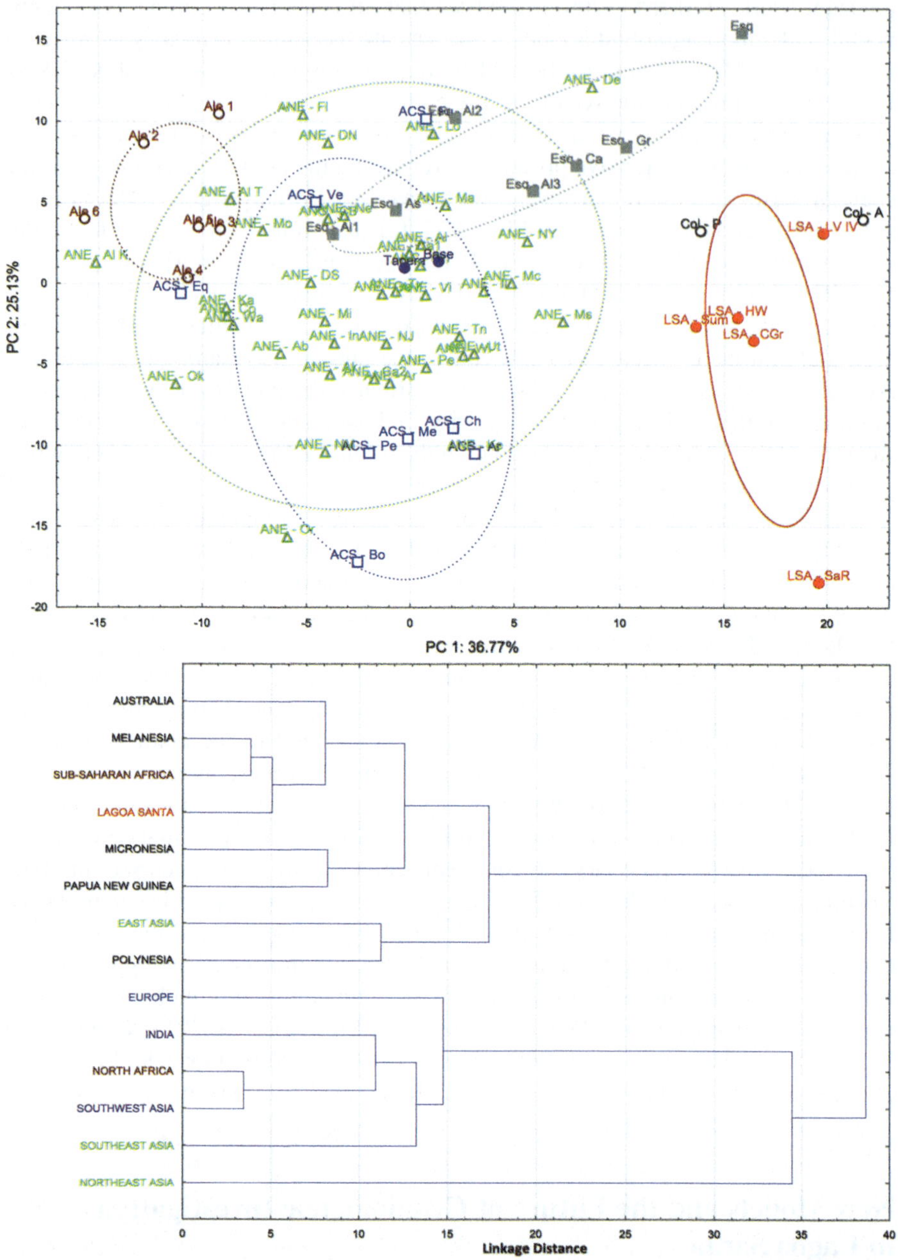

Fig. 11.3 Results of analyses of intra- and extra-continental morphological affinities of Lagoa Santa specimens. The graph at the *top* represents the intra-continental affinities obtained by principal component analysis (PCA). The analysis covers 61.9% of the variability contained in the data, represented by the sum of the two principal components (PC1 and PC2). The Lagoa Santa

in all of which the morphology observed in the Lagoa Santa material invariably occupies an outstanding position: (1) the populations with the generalized morphology, that is, the Paleoamericans who first arrived in the New World, were partly or totally replaced by the population with the specialized morphology that succeeded them; (2) it could be possible that the molecular markers and the morphological markers are reflecting distinct evolutionary stories, that is, the same set of genes could accommodate more than one morphological pattern; and (3) the second population contingent may have absorbed the individuals of the first one, who, accordingly, had their genes diluted in the genetic pool that exists today. All those scenarios, however, require extensive skeletal series and large samples of genetic material (in the case of Lagoa Santa, despite the great efforts made in the last two decades, it has been not yet possible to obtain any DNA material) in order to be properly tested. This situation, with the exception of Lagoa Santa, is extremely rare in the main archaeological areas of the Americas.

New studies have been undertaken in that direction, generating new contributions to the debate on the first Americans. Hubbe et al. (2010) presented a comparison of the American cranial morphology, including the Lagoa Santa Paleoamericans, with other series worldwide and adopted a genetic-quantitative approach to test the evolutionary scenarios for the dispersion of human groups in the New World. The authors concluded that when models that assume that the diachronic morphological duality observed in South America is the result of different population dispersion waves are confronted with models proposing that the duality has been the result of in situ microevolutionary processes, the former models stand out as being more parsimonious, which suggests that the morphological diversity observed in the Americas is, indeed, of an extra-continental origin.

Similarly, Hubbe et al. (2014) analyzed the dispersion patterns of cranial morphological diversity in recent and subrecent groups distributed in the Brazilian territory. Comparing D^2 matrixes resulting from the craniometric variation observed in those series with geographic distance matrixes serving as proxies for the time of divergence of the population groups, the authors observed that, for the series analyzed, the model that assumed an extra-continental origin of two biological dispersion waves with a late survival of the more ancient (Paleoamerican) one was the best

Fig. 11.3 (continued) series are represented on the *right side* by *solid circles* (*red*), while the *open circles* (*black*) at the *top right corner* represent the specimens from Sabana de Bogotá. The other series are represented by *open circles* (*brown*) for the Aleuts, *open triangles* for North America (*green*), *open squares* for Central and South America (*blue*), *solid circles* (*purple*) for the shell middens, and *solid squares* for the Eskimos (*gray*). Observe that Lagoa Santa is associated to the Sabana de Bogotá series, but in complete opposition to all the other representatives of the American continent. The second graph displays the extra-continental affinities analyzed in the form of a dendrogram obtained by cluster analysis of a Mahalanobis distance matrix. The Lagoa Santa series forms a cohesive group together with the Australo-Melanesian and sub-Saharan African series. The morphology observed among the East Asian series is only slightly aggregated in that cluster and even then only through the association with the morphology observed among the Polynesian series. Other Asian series appear in a group that is in complete opposition to the one that contains the Lagoa Santa series (Adapted from Bernardo 2007)

adjusted compared with the model that proposed a common extraterritorial origin for all the variation observed. Again, this model was better adjusted than the one that proposed the entrance of different population waves, but with the survival of only the most recent one.

In short, despite the extensive production addressing the morphological characteristics of the Paleoamericans, the specimens coming from Lagoa Santa have an outstanding role in the study of the first Americans' biological origins (González-José et al. 2005, 2006; Hubbe et al. 2004; Jantz and Owsley 2001; Munford et al. 1995; Neves et al. 1999a, b, c, 2003, 2004, 2005; Neves and Hubbe 2005; Neves and Pucciarelli 1989, 1990, 1991; Powell and Neves,1999; Powell and Steele 1992; Steele and Powell 1992, 1993, for just a few examples). The results obtained here make it possible to structure conclusions with different degrees of likelihood, from highly likely, like the intra-continental distinction between Paleoamerican populations and later Native American populations, to other hardly likely, such as the supposition that at some moment in the Holocene, a rapid and significant change in morphological variation occurred locally among Native Americans. The next studies need to address the new techniques and quantitative models in the hopes that new generations can produce and enjoy the benefits of collections as rich, dense, and generous as those produced in the archaeological region of Lagoa Santa.

Acknowledgements The authors wish to thank all those who have dedicated their efforts to the study of the origins of Native Americans. The list would be immense and impossible to reproduce here, but it is only fair to say that their contributions have been, are, and always will be an inspiration of fundamental importance for new investigations into the history of the first Americans. The research undertaken in Lagoa Santa was only possible because of the expressive support given by the *Fundação de Amparo à Pesquisa do Estado de São Paulo* (FAPESP – São Paulo Research Foundation), which benefited both authors (grant process 04/01321-6 for WAN and 05/52273-4 for DVB). During the elaboration of this manuscript, the authors were supported by the *Conselho Nacional de Desenvolvimento Científico e Tecnológico* (National Council for Scientific and Technological Development – CNPq) through grant processes 300917/2010-4 (WAN) and 461122/2014-6 (DVB).

References

Armelagos, G. J., Carlson, D. S., & Van Gerven, D. P. (1982). The theoretical foundations and development of skeletal biology. In F. Spencer (Ed.), *A history of American physical anthropology, 1930–1980* (pp. 305–328). New York: Academic.

Azevedo, S., Nocera, A., Paschetta, C., Castillo, L., González, M., & González-José, R. (2011). Evaluating microevolutionary models for the early settlement of the New World: The importance of recurrent gene flow with Asia. *American Journal of Physical Anthropology, 146*, 539–552.

Barbujani, G., Magagni, A., Minch, E., & Cavalli-Sforza, L. L. (1997). An apportionment of human DNA diversity. *Proceedings of the National Academy of Sciences of the USA, 94*, 4516–4519.

Bass, W. M. (1987). *Human osteology: A laboratory and field manual* (3rd ed.). Columbia: Missouri Archaeological Society.

Bernardo, D. V. (2007). *Afinidades morfológicas intra e extra continentais dos paleoíndios de Lagoa Santa: uma nova abordagem*. Master Thesis. Departamento de Genética e Biologia Evolutiva, Instituto de Biociências, Universidade de São Paulo, São Paulo, Brasil.

Bernardo, D. V., & Neves, W. A. (2009). Diversidade morfocraniana dos remanescentes ósseos humanos da Serra da Capivara: implicações para a origem do homem americano. *FUMDHAMentos, 8*, 94–106.

Boas, F. (1912). New evidence in regard to the instability of human types. *American Antiquity, 14*, 530–562.

Bonatto, S. L., & Salzano, F. M. (1997). A single and early migration for the peopling of the Americas supported by mitochondrial DNA sequence data. *Proceedings of the National Academy of Science of USA, 94*, 1966–1971.

Brace, C. L. (1982). The roots of the race concept in American physical anthropology. In F. Spencer (Ed.), *A history of American physical anthropology, 1930–1980* (pp. 11–29). New York: Academic.

Brace, C. L., & Hunt, K. D. (1990). A nonracial craniofacial perspective on human variation: A(ustralia) to Z(uni). *American Journal of Physical Anthropology, 82*, 341–360.

Brace, C. L., Nelson, A. R., Seguchi, N., Oe, H., Sering, L., Qifeng, P., Yongyi, L., & Tumen, D. (2001). Old world source of the first New World human inhabitants: A comparative craniofacial view. *Proceedings of the National Academy of Science of USA, 98*(17), 10017–10022.

Broca, P. (1875). Instructions craniologiques et craniométriques. *Mem de la Soc d'Anthrop de Paris, 2*, 1–203.

Buxton, L. H. D., & Morant, G. M. (1933). The essential craniological technique. *J Roy Anthrop Inst, 63*, 19–47.

Castro-Faria, L. (1952). Pesquisas de antropologia física no Brasil. *Boletim do Museu Nacional, 13*, 1–106.

Clark, J. D., Dunn, J. E., & Smith, K. G. (1993). A multivariate model of female black bear habitat use for a geographic information system. *Journal of Wildlife Management, 57*, 519–526.

de Quatrefages, A. (1879). *The human species*. New York: Appleton.

de Quatrefages, A. (1887). *Histoire générale des races humaines: Introduction a l'étude des races humaines*. Paris: A. Hennuyer.

Diamond, J. (2003). *Armas, Germes e Aço: o Destino das Sociedades Humanas*. Rio de Janeiro: Record.

Dillehay, T. D. (2000). *The settlement of the Americas*. New York: Basic Books.

Fagan, B. M. (1987). *The great journey*. London: Thames and Hudson Ltda.

Fagan, B. M. (1999). *World prehistory – A brief introduction* (p. 327). Menlo Park: Longman.

Fagundes, N. J. R., Kanitz, R., & Bonatto, S. L. (2008a). A reevaluation of the native American mtDNA genome diversity and its bearing on the models of early colonization of Beringia. *PloS One, 3*(9), e3157.

Fagundes, N. J. R., Kanitz, R., Eckert, R., Valls, A. C. S., Bogo, M. R., Salzano, F. M., Smith, D. G., Silva Jr., W. A., Zago, M. A., Ribeiro-dos-Santos, A. K., Santos, S. E. B., Petzl-Erler, M. L., & Bonatto, S. (2008b). Mitichondrial population genomics supports a single pre-Clovis origin with a coastal route for the peopling of the Americas. *American Journal of Human Genetics, 82*(3), 583–592.

Futuyma, D. J. (1998). *Biologia Evolutiva* (3rd ed.). Ribeirão Preto: Funpec Editora.

Gage, T. (1648). *The English American or a new survey*. London: G. Routledge & Sons, Limited.

Genovés, S. T. (1967). Some problems in the physical anthropological study of the peopling of America. *Current Anthropology, 3*, 297–312.

Goebel, T., Waters, M. R., & O'Rourke, D. H. (2008). The late Pleistocene of modern humans in the Americas. *Science, 319*(5869), 1497–1502.

González-José, R., Bortolini, M. C., Santos, F. R., & Bonatto, S. L. (2008). The peopling of America: Craniofacial shape variation on a continental scale and its interpretation from an interdisciplinary view. *American Journal of Physical Anthropology, 137*(2), 175–187.

González-José, R., González-Martin, A., Hernández, M., Pucciarelli, H. M., Sardi, M., Rosales, A. S., & van de Molen, S. (2003). Craniometric evidence for paleoamerican survival in Baja California. *Nature, 452*, 62–65.

González-José, R., Martínez-Abadías, N., González-Martín, A., Neves, W. A., Pucciarelli, H. M., & Hernández, M. (2006). Análisis morfogeométrico de cuatro cráneos antiguos del Valle de Mexico: Peñón III, Metro Balderas, Chimalhuacán y Cueva del Tecolote. *Cuicuilco, 13*, 105–127.

González-José, R., Neves, W. A., Lahr, M. M., González, S., Pucciarelli, H., Martínez, M. H., & Correal, G. (2005). Late Pleistocene/Holocene craniofacial morphology in Mesoamerica Paleoindians: Implications for the peopling of the New World. *American Journal of Physical Anthropology, 128*, 772–780.

Gould, S. J. (1999). *A falsa medida do Homem*. São Paulo: Martins Fontes.

Gower, J. C. (1966). Some distance properties of latent roots and vector methods used in multivariate analysis. *Biometrika, 53*, 325–338.

Greenberg, J. H., Turner II, C. G., & Zegura, S. L. (1986). The settlement of the Americas: A comparison of linguistic, dental and genetic evidence. *Current Anthropology, 27*, 477–497.

Hair Jr., J. F., Anderson, R. E., Tatham, R. L., & Black, W. C. (1998). *Análise Multivariada de Dados* (5rd ed.). São Paulo: Bookman.

Hanihara, T. (1993a). Population prehistory of East Asia and the Pacific as viewed from craniofacial morphology: The basic populations in East Asia, VII. *American Journal of Physical Anthropology, 91*, 173–187.

Hanihara, T. (1993b). Craniofacial features of southeast Asians and Jomonese: A reconsideration of their microevolution since the late Pleistocene. *Anthropological Science, 101*, 25–46.

Hanihara, T. (1996). Comparison of craniofacial features of major human groups. *American Journal of Physical Anthropology, 99*, 389–412.

Hanihara, T. (2000). Frontal and facial flatness of major human populations. *American Journal of Physical Anthropology, 111*, 105–134.

Hansen, S. (1888). Lagoa Santa Racen. *E Museo Lundii, 1*, 1–37.

Harper, A. B., & Laughlin, W. S. (1982). Inquiries into the peopling of the New World: Development of ideas and recent advances. In F. Spencer (Ed.), *A history of American physical anthropology, 1930–1980* (pp. 281–304). New York: Academic.

Hotelling, H. (1933). Analysis of a complex of statistical variables into principal components. *Journal of Educational Psychology, 24*, 417–441.

Howells, W. W. (1973). *Cranial variation in man. A study by multivariate analysis of patterns of difference among recent human populations. Papers of Peabody Museum of archaeology and Ethnology*. Cambridge: Harvard University Press.

Howells, W. W. (1989). *Skull shapes and the map. Craniometric analysis in the dispersion of modern Homo. Papers of Peabody Museum of Archaeology and Ethnology*. Cambridge: Harvard University Press.

Howells, W. W. (1999). *Who's who in skulls. Papers of Peabody museum of archaeology and ethnology* (Vol. 82). Cambridge: Harvard University Press.

Hrdlička, A. (1907). *Skeletal remains suggesting or attributed to early man in North America, Bureau of American Ethnology Bulletin* (Vol. 33). Washington, DC: Smithsonian Institution.

Hrdlička, A. (1912). *Early man in South America, Bureau of American Ethnology Bulletin* (Vol. 52). Washington, DC: Smithsonian Institution.

Hrdlička, A. (1932). The coming of man from Asia in light of recent discoveries. *Proceedings of the American Philosophical Society, 71*, 393–402.

Hubbe, M. (2006). *Análise biocultural dos remanescentes ósseos humanos de Porto do Rio Vermelho 02 (SC-PRV-02) e suas implicações para a colonização da costa brasileira*. Ph.D. Dissertation. Departamento de Genética e Biologia Evolutiva. Instituto de Biociências. Universidade de São Paulo, São Paulo, Brasil.

Hubbe, M., Neves, W. A., Atui, J. P., Cartelle, C., & Da Silva, M. P. (2004). A new early human skeleton from Brazil: Support for the "Two Main Biological Components Model" for the settlement of the Americas. *Current Research in the Pleistocene, 21*, 77–81.

Hubbe, M., Neves, W. A., & Harvati, K. (2010). Testing evolutionary and dispersion scenarios for the settlement of the New World. *PloS One, 5*, e11105.

Hubbe, M., Okumura, M., Bernardo, D. V., & Neves, W. A. (2014). Cranial morphological diversity of early, middle and late Holocene Brazilian groups: Implications for human dispersion in Brazil. *American Journal of Physical Anthropology, 155*, 546–558.

Jantz, R. L., & Owsley, D. W. (2001). Variation among early North America crania. *American Journal of Physical Anthropology, 114*, 146–155.

Kherumian, R. (1949). Répertoire des points craniométriques et anthropométriques. *Rev de Morpho-physiologie humaine, 2*, 1–22.

Kollmann, J. (1884). Schädeln von Lagoa Santa. *Zeitschrift für Ethnologie, 16*, 194–199.

Lacerda, J. B. (1882). A morfologia craniana do homem dos sambaquis. *Revista da Exposição Antropológica Brasileira* 22–23.

Lacerda, J. B., & Peixoto, R. (1876). Contribuições para o Estudo Anthropológico das Raças Indígenas do Brazil. *Archivos do Museu Nacional do Rio de Janeiro, 1*, 47–75.

Lahr, M. M. (1995). Patterns of modern human diversification: Implications for Amerindian origins. *Yearbook of Physical Anthropology, 38*, 163–198.

Lewontin, R. C. (1972). The apportionment of human diversity. *Evolutionary Biology, 6*, 381–398.

Lund, P. W. (1844). Carta do Dr. Lund, escripta da Lagôa Santa (Minas Geraes) a 21 de abril de 1844 – Lida na sessão do Instituto de 20 de Junho deste mesmo anno. *Revista do Instituto Histórico e Geográfico do Brasil, 6*, 334–342.

Lund, P. W. (1849). Notice sur des Ossements Humains Fossiles Trouvés dans une Caverne du Brésil. *Mémoires de la Societé Royale des Antiquaires du Nord, 3*, 49–77.

Madrigal, L. (1998). *Statistics for anthropology*. Cambridge: Cambridge University Press.

Mahalanobis, P. C. (1936). On the generalized distance in statistics. *Proceedings of the National Institute of Science of India, 12*, 49–55.

Manly, B. F. J. (1994). *Multivariate statistical methods: A primer*. London: Chapman and Hall.

Martin, R. (1928). *Lehrbuch der Anthropologie in systematischer darstellung, Kraniologie, Osteologie* (Vol. 2). Jena: Gustav Fischer.

Mello e Alvim, M. C. (1963). Diversidade morfológica entre os índios "Botocudos" do leste brasileiro (séc. XIX) e o homem de Lagoa Santa. *Boletim do Museu Nacional, N.S., 23*, 1–70.

Mello e Alvim, M. C. (1977). Os antigos habitantes da área arqueológica de Lagoa Santa (MG-Brasil) – Estudo Morfológico. *Arquivos do Museu Nacional de História Natural da Universidade Federal de Minas Gerais, 2*, 119–174.

Mello e Alvim, M. C. (1992–1993). Os antigos habitantes da Serra do Cipó (MG-Brasil): estudo morfológico preliminar. *Arquivos do Museu de História Natural da UFMG, 13/14*,107–128.

Mello e Alvim, M. C., Soares, M. C., & Cunha, P. S. P. (1983–1984). Traços não métricos cranianos e distância biológica em grupos indígenas interioranos e do litoral do Brasil. "Homem de Lagoa Santa", índios Botocudos e construtores de sambaquis. *Arquivos do Museu Nacional de História Natural da Universidade Federal de Minas Gerais, 8/9*, 323–338.

Mendes Corrêa, A. A. (1948). *Ameríndios*. Porto: Instituto de Antropologia da Universidade do Porto.

Messias, T. T., & Mello e Alvim, M. C. (1962). Contribuições ao estudo do Homem de Lagoa Santa. *Boletim do Museu Nacional, N.S., 20*, 1–55.

Mingoti, S. A. (2005). *Análise de dados através de métodos de estatística multivariada – uma abordagem aplicada*. Belo Horizonte: Editora UFMG.

Morton, S. G. (1839). *Crania Americana: Or a comparative view of the skulls of various aboriginal nations of America*. London: J. Dobson.

Morton, S. G. (1844). *Crania Aegyptiaca: Or, observations on Egyptian ethnography, derived from anatomy, history and the monuments*. Philadelphia: J. Penington.

Munford, D. (1999). *Estudo Comparado da Morfologia Craniana de Populações Pré-Históricas da América do Sul: Implicações para a Questão do Povoamento do Novo Mundo*. Master Thesis. Departamento de Biologia, Instituto de Biociências, Universidade de São Paulo, São Paulo, Brasil.

Munford, D., Zanini, M. C., & Neves, W. A. (1995). Human cranial variation in South America: Implications for the settlement of the New World. *Brazilian Journal of Genetics, 18*, 673–688.

Neves, W. A. (1982). Variação métrica nos construtores de sambaquis do sul do Brasil: primeira aproximação multivariada. *Revista de Pré-História, 3*(4), 83–108.

Neves, W. A., & Atuí, J. P. (2004). O mito da homogeneidade biológica da população paleoíndia de Lagoa Santa: implicações antropológicas. *Revista de Antropologia, 47*, 158–206.

Neves, W. A., do Zanini, M. C., Munford, D., & Pucciarelli, H. M. (1997). O povoamento da América à luz da morfologia craniana. *Revista USP. Dossiê surgimento do homem na América, 34*, 97–105.

Neves, W. A., González-José, R., Hubbe, M., Kipnis, R., Araujo, A. G. M., & Blasi, O. (2004). Early Holocene human skeletal remains from Cerca Grande, Lagoa Santa, Central Brazil, and the origins of the first Americans. *World Archaeology, 36*, 479–501.

Neves, W. A., & Hubbe, M. (2005). Cranial morphology of early Americans from Lagoa Santa, Brazil – implications for the settlement of the New World. *Proceedings of the National Academy of Science of USA, 102*, 18309–18314.

Neves, W. A., Hubbe, M., & Correal, G. (2007). Human skeletal remains from Sabana de Bogotá, Colombia: A case of Paleoamerican morphology late survival in South America? *American Journal of Physical Anthropology, 133*, 1080–1098.

Neves, W. A., Hubbe, M., Okumura, M. M. M., González-José, R., Figuti, L., Eggers, S., & De Blasis, P. A. D. (2005). A new early Holocene human skeleton from Brazil: Implications for the settlement of the New World. *Journal of Human Evolution, 48*, 403–414.

Neves, W. A., Meyer, D., & Pucciarelli, H. M. (1996a). Early skeletal remains and the peopling of the Americas. *Revista de Antropologia, 39*, 121–139.

Neves, W. A., Munford, D., & Zanini, M. C. (1996b). Cranial morphological variation and the colonization of the New World: Towards a four-migration model. *American Journal of Physical Anthropology (Suppl.), 26*, 169.

Neves, W. A., Powell, J. F., & Ozolins, E. G. (1999a). Modern human origins as seen from the peripheries. *Journal of Human Evolution, 37*, 129–133.

Neves, W. A., Powell, J. F., & Ozolins, E. G. (1999b). Extra-continental morphological affinities of Lapa Vermelha IV, hominid I: A multivariate analysis with progressive numbers of variables. *Homo, 50*, 263–282.

Neves, W. A., Powell, J. F., & Ozolins, E. G. (1999c). Extra-continental morphological affinities of Palli Aike, southern Chile. *Interciencia, 24*, 258–263.

Neves, W. A., Prous, A., Gonzalez-José, R., Kipnis, R., & Powell, J. (2003). Early Holocene human skeletal remains from Santana do Riacho, Brazil: Implications for the settlement of the New World. *Journal of Human Evolution, 45*, 19–42.

Neves, W. A., & Pucciarelli, H. M. (1989). Extra-continental biological relationships of early south American human remains: A multivariate analysis. *Ciência e Cultura, 41*, 566–575.

Neves, W. A., & Pucciarelli, H. M. (1990). The origin of the first Americans: An analysis based on the cranial morphology of early south American human remains. *American Journal of Physical Anthropology, 81*, 274.

Neves, W. A., & Pucciarelli, H. M. (1991). Morphological affinities of the first Americans: An exploratory analysis based on early south American human remains. *Journal of Human Evolution, 21*, 261–273.

Neves, W. A., & Pucciarelli, H. M. (1998). The Zhoukoudian Upper Cave skull 101 as seen from the Americas. *Journal of Human Evolution, 34*, 219–222.

Newmann, M. T. (1951). An historical resume of the concept of differences in Indian type. *American Anthropologist, 53*, 19–36.

Newmann, M. T. (1962). Evolutionary changes in body size and head form in American Indians. *American Anthropologist, 64*, 237–257.

Pearson, K. (1901). On lines and planes of closest fit to systems of points in the space. *Philosophical Magazine, 2*, 559–572.

Perego, U. A., Achilli, A., Angerhofer, N., Accetturo, M., Pala, M., Olivieri, A., Kashani, B. H., Ritchie, K. H., Scozzari, R., Kong, Q. P., Myres, N. M., Salas, A., Semino, O., Bandelt, H. J., Woodward, S. R., & Torroni, A. (2009). Distinctive Paleo-Indian migration routes from Beringia marked by two rare mtDNA haplogroups. *Current Biology, 19*, 1–8.

Piló, L. B., Brito, M., Araujo, A. G. M., Costa, F. A., Bárbara, J., Freitas, J. D. R., Bastos, M. C. A. G., Kipnis, R., Albano, R., Navarro, T., & Neves, W. A. (2003). *Mapa Arqueológico da APA Carste de Lagoa Santa, MG*. São Paulo: Fapesp.

Powell, J. F., & Neves, W. A. (1999). Craniofacial morphology of the first Americans: Pattern and process in the peopling of the New World. *Yearbook of Physical Anthropology, 42*, 153–188.

Powell, J. F., & Steele, D. G. (1992). A multivariate craniometric analysis of North American Paleo-indian remains. *Current Research in the Pleistocene, 9*, 59–62.

Prous, A. (2006). O Brasil antes dos brasileiros. A pré-história do nosso país. Rio de Janeiro: Zahar.

Reich, D., Patterson, N., Campbell, D., Tandon, A., Mazieres, S., Ray, N., Parra, M. V., Rojas, W., Duque, C., Mesa, N., García, L. F., Triana, O., Blair, S., Maestre, A., Dib, J. C., Bravi, C. M., Bailliet, G., Corach, D., Hünemeier, T., Bortolini, M. C., Salzano, F. M., Petzl-Erler, M. L., Acuña-Alonzo, V., Aguilar-Salinas, C., Canizales-Quinteros, S., Tusié-Luna, T., Riba, L., Rodríguez-Cruz, M., Lopez-Alarcón, M., Coral-Vazquez, R., Canto-Cetina, T., Silva-Zolezzi, I., Fernandez-Lopez, J. C., Contreras, A. V., Jimenez-Sanchez, G., Gómez-Vázquez, M. J., Molina, J., Carracedo, A., Salas, A., Gallo, C., Poletti, G., Witonsky, D. B., Alkorta-Aranburu, G., Sukernik, R. I., Osipova, L. S., Fedorova, A., Vasquez, R., Villena, M., Moreau, C., Barrantes, R., Pauls, D., Excoffier, L., Bedoya, G., Rothhammer, F., Dugoujon, J. M., Larrouy, G., Klitz, W., Labuda, D., Kidd, J., Kidd, K., Di Rienzo, A., Freimer, N. B., Price, A. L., & Ruiz-Linares, A. (2012). Reconstructing native American population history. *Nature, 488*, 370–374.

Reinhardt, J. T. (1888). De Brazilianske Knoglehuler og de I dem Forekommende Dyrelevninger. *E Museo Lundii, 1*, 1–56.

Reis, E. (2001). *Estatística Multivariada Aplicada*. Lisboa: Edições Sílabo.

Relethford, J. H. (1982). The use of quantitative traits in the study of human population structure. *Yearbook of Physical Anthropology, 25*, 113–132.

Relethford, J. H. (1994). Craniometric variation among modern human populations. *American Journal of Physical Anthropology, 95*, 53–62.

Relethford, J. H. (2002). Apportionment of global human genetic diversity based on craniometrics and skin color. *American Journal of Physical Anthropology, 118*, 393–398.

Ridley, M. (2006). *Evolução* (3rd ed.). Porto Alegre: Artmed.

Rivet, P. (1908). La race de Lagoa Santa chez les populations précolombiennes de l'Equateur. *Bulletins et Mémoires de la Société d'anthropologie de Paris, IX*(2), 209–274.

Rivet, P. (1942). *Les Origines de l'Homme Américain*. Montreal: Les Éditions l'Arbre.

Roosevelt, A. C., Douglas, J., & Brown, L. (2002). The migrations and adaptations of the first Americans: Clovis and pre-Clovis viewed from South America. In N. G. Jablonski (Ed.), *The first Americans: The Pleistocene colonization of the New World*. San Francisco: California University Press.

Seguchi, N., Nelson, A. R., Austin, S., & Brace, C. L. (2005). Early south Americans in cranifacial metric perspective: Lagoa Santa. *American Journal of Physical Anthropology, 40*(Suppl), 187.

Silva e Sá, G. J., Santos, R. V., Rodrigues-Carvalho, C., & Silva, E. C. (2007). Crânios, corpos e medidas: a constituição do acervo de instrumentos antropométricos do Museu Nacional na passagem do século XIX para o XX. *História, Ciências, Saúde, 15*(1), 197–208.

Slice, D. E. (2005). Modern morphometrics. In D. E. Slice (Ed.), *Modern morphometrics in physical anthropology* (pp. 1–46). New York: Springer.

Smith, F. H. (1984). Fossil hominids from the upper Pleistocene of Central Europe and the origino f modern Europeans. In F. H. Smith & F. Spencer (Eds.), *The origins of modern humans: A world survey of the fossil evidence* (pp. 273–299). New York: Allan Liss.

Steele, D. G., & Powell, J. F. (1992). Peopling of the Americas: Paleobiological evidence. *Human Biology, 64*, 303–336.

Steele, D. G., & Powell, J. F. (1993). Paleobiology of the first Americans. *Evolutionary Anthropology, 2*, 138–146.
Stocking Jr., G. W. (Ed.). (1988). *Bones, bodies, behavior: Essays in biological anthropology*. Madison: The University of Wisconsin Press.
Stocking Jr, G. W. (Org.). (2004). Franz Boas. A formação da Antropologia Americana. Rio de Janeiro: Contraponto Editora.
Szathmary, E. J. E. (1993). mtDNA and the peopling of the Americas. *American Journal of Human Genetics, 53*, 793–799.
Tamm, E., Kivisild, T., Reidla, M., Metspalu, M., Smith, D. G., Mulligan, C. J., Bravi, C. M., Rickards, O., Martinez-Labarga, C., Khusnutdinova, E. K., Fedorova, S. A., Golubenko, M. V., Stepanov, V. A., Gubina, M. A., Zhadanov, S. I., Ossipova, L. P., Damba, L., Voevoda, M. I., Dipierri, J. E., Villems, R., & Malhi, R. S. (2007). Beringian standstill and spread of native American founders. *PloS One, 2*, e829. doi:10.1371/journal.pone.0000829.
Tejera, D. (1996). Spanish-English relationship in the work of Thomas Gage, al. Tomás de Santa María. *Sederi, 6*, 203–208.
Ten Kate, H. (1885). Sur les Crânes de Lagoa-Santa. *Bulletins de la Société d'Anthropologie de Paris, 8*, 240–244.
Trinkaus, E. (1982). A history of *Homo sapiens* paleontology in America. In F. Spencer (Ed.), *A history of American physical anthropology, 1930–1980* (pp. 261–280). New York: Academic.
Valentin, J. L. (2000). *Ecologia numérica – uma introdução à analise multivariada de dados ecológicos*. Rio de Janeiro: Editora Interscienca.
Vallois, H. V. (1965). Anthropometric techiniques. *Current Anthropology, 6*(2), 127–143.
van Vark, G. N., Kuizenga, D., & Williams, F. L. (2003). Kennewick and Luzia: Lessons from European upper Paleolithic. *American Journal of Physical Anthropology, 121*, 181–184.
van Vark, G. N., & Schaafsma, W. (1992). Advances in the quantitative analysis of skeletal morphology. In S. R. Saunders & M. A. Katzenberg (Eds.), *Skeletal biology of past peoples: Research methods* (pp. 225–257). New York: Wiley Liss.
Wang, S., Lewis Jr., C. M., Jakobsson, M., Ramachandran, S., Ray, N., Bedoya, G., Hurtado, A. M., Labuda, D., Klitz, W., Barrantes, R., Bortolini, M. C., Salzano, F. M., Petzl-Erler, M. L., Tsuneto, L. T., Llop, E., Rothhammer, F., Excoffier, L., Feldman, M. W., Rosenberg, N. A., & Ruiz-Linares, A. (2007). Genetic variation and population structure in native Americans. *PloS One, 3*, e185. doi:10.1371/journal.pgen.0030185.

Chapter 12
History of the Research into Health and Lifestyle in Lagoa Santa

Pedro Da-Gloria and Rodrigo Elias Oliveira

Abstract Lagoa Santa region was one of the first places where systematic paleontological and archaeological studies were conducted in Brazil. Along more than 180 years, hundreds of human skeletons were exhumed in the rock-shelters of the region. Despite this long history of research, there has been little progress in understanding health and lifestyle of the New World's first inhabitants. In this chapter, we first describe the history of research in health with Lagoa Santa material, starting from the pioneer holistic studies of Peter Lund, passing through the biomedical and clinical-pathological view, and finishing with more recent anthropological approaches using skeletal markers. In the second part, we propose some approaches that could contribute toward a more detailed understanding of the health of the Lagoa Santa prehistoric population. We divide them into three parts: skeletal material, archaeological context, and living populations. We effectively propose that the association of studies of human skeletons with the archaeological context, the essence of the field of bioarchaeology, is the way to achieve these goals. The task of revealing the health of prehistoric populations involves a much broader effort than the mere registration of the prevalence of pathologies.

P. Da-Gloria (✉)
Laboratory for Human Evolutionary and Ecological Studies, Department of Evolutionary Biology and Genetics in the Biosciences Institute, University of São Paulo, R. do Matão 277, sala 218, Cidade Universitária, São Paulo, SP, Brazil
e-mail: da-gloria@ib.usp.br

R.E. Oliveira
Laboratory for Human Evolutionary and Ecological Studies, Department of Evolutionary Biology and Genetics in the Biosciences Institute, University of São Paulo, R. do Matão 277, sala 218, Cidade Universitária, São Paulo, SP, Brazil

Stomatology Department in the Odontology Faculty, University of São Paulo, Av. Professor Lineu Prestes, 2227 São Paulo, SP, Brazil
e-mail: eliaso@usp.br

P. Da-Gloria et al. (eds.), *Archaeological and Paleontological Research in Lagoa Santa*, DOI 10.1007/978-3-319-57466-0_12

Introduction

The Lagoa Santa region was one of the first places where systematic paleontological and archaeological studies were conducted in Brazil. Since Peter Lund's interventions in the first half of the nineteenth century (Holten and Sterll 2011), a large quantity of archaeological material older than 7000 years has been excavated. The good state of preservation of the human skeletal material is the most notable feature of the region, because it allows for a variety of studies of population history and lifestyle of those first inhabitants, as can be seen in the various chapters of this book. In the long history of research in the region, there has been a notable emphasis on the cranial morphology typology of those early inhabitants, whether in an attempt to characterize "Lagoa Santa Man" (e.g., Ávila 1950) or in an effort to understand the first migrations to the Americas (e.g., Neves and Hubbe 2005; Chaps. 10 and 11 of this volume). Those lines of research had a notorious international impact and made Lagoa Santa an obligatory reference in any studies of the first Americans (Neves et al. 2013; Chaps. 10 and 11 of this volume). On the other hand, despite the long history of research in the region, there has been little or no progress in understanding health and lifestyle of the New World's first inhabitants. With that in mind, we will endeavor, in this chapter, to narrate the history of health research in the region as a way of casting light on a theme that has been marginalized in the Lagoa Santa literature. After the historical review, we propose lines of investigation that could contribute toward a deeper understanding of the health and lifestyle of those earliest inhabitants.

The History of Research

The pioneer of paleontological and archaeological studies in Brazil was Peter Lund (1801–1880). In the decade of 1840, he discovered about 30 human craniums together with the remains of megafauna at the Gruta do Sumidouro in the Lagoa Santa region (for details, see Chaps. 2 and 3 of this volume). Those finds were of the greatest importance in establishing subsequent research, investigating the antiquity of the human skeletons of Lagoa Santa, their cranial morphology, and the possible coexistence of humans and the megafauna (see Chaps. 10 and 11 of this volume). Lund described the human craniums he found with details of their anatomical and pathological aspects. He noticed that in regard to buccal pathologies, the craniums showed evidence of loss of teeth in vivo due to alveolar reabsorption. Lund's explanation for the presence of such pathologies was that they were associated to the advanced age of most of the skeletons, relating the precarious state of the teeth of some individuals with their having been sick. In addition, he observed holes in the temporal bones of many of the craniums he studied,[1] suggesting that violent attacks

[1] Something that has never been confirmed by subsequent research.

may have been the cause of death of those individuals (Lund 1844). He also noticed that the incisors were heavily worn down. To understand the reason, Lund used ethnographic analogies and comparison with other archaeological populations to discard the hypothesis that the inhabitants of Lagoa Santa ate earth or chewed some kind of narcotic substance. Later, in a letter written in 1858, Lund related incisors wear with the chewing of roots or grains to ferment in pottery containers, as was the habit among tribes on the east coast of Brazil (Holten and Sterll 2011:217–222). Regardless of the veracity of his conclusions,[2] Lund can also be considered a pioneer in the use of ethnography to understand the behavior of past populations.

During the nineteenth century, various researchers studied the material Lund had collected. Those early studies of the collection were of a predominantly anatomical-descriptive nature. Reinhardt (1866) characterized the Lagoa Santa individuals as being tall with a light bone structure, whereas Kollmann (1884) concluded that the craniums had belonged to individuals that were strong but without many visible muscular marks. Hansen (1888) concluded that those inhabitants had enjoyed good health, because he did not identify many cases of caries or abundant evidence of trauma or pathologies in general. He estimated that the Lagoa Santa individuals had heights of 1570 mm and called attention to the robustness of the long bones. Lacerda and Peixoto (1876) observed a cranial fracture caused by a cutting instrument in one cranium from the Sumidouro cave that was deposited in Rio de Janeiro. It is very difficult to conciliate all those visions of the Lagoa Santa inhabitants' biology and health, because each author interpreted the markers he studied in a different way. In all their descriptions of the Lagoa Santa material, however, the authors agreed that the dentition showed a high degree of wear.

The first half of the twentieth century was marked by expeditions to collect fossils in the Lagoa Santa region. Lanari (1909) reported having found at least three human skeletons at Lapa do Caetano, which looked to him as if they had been thrown down an opening in the roof of the cave. His report on the bone material, however, is highly superficial, mentioning only missing incisors in the skeletons found. That expedition was followed by others, organized by the National Museum of Rio de Janeiro and by the Minas Gerais Academy of Sciences (see Chaps. 4, 5, and 6 of this volume). Despite the large quantity of human skeleton remains found, the National Museum expedition reports are very scanty and barely offer descriptions of the geological context and the archaeological finds (e.g., Ávila 1937; Padberg-Drenkpohl 1926). A central discussion topic in that period was the possible coexistence of humans with the extinct megafauna of the region. In that discussion, the discovery of a human cranium (Confins Man) associated to extinct horse cranium and a teeth and part of a long bone of a Mastodon in the Confins cave aroused an intense debate (Walter et al. 1937; see Chaps. 5 and 6). None of those authors mentioned anything about the lifestyle or health of the Lagoa Santa population.

[2]Today, it is an established fact that the human skeletons Lund found in the Gruta do Sumidouro date back to a time thousands of years before the appearance of pottery or the domestication of plant species in Central Brazil (Neves et al. 2007).

Regarding the academic production of the members of the Minas Gerais Academy of Sciences, two synthetic works about Lagoa Santa deserve to be highlighted. They were the works that made most progress toward obtaining an understanding of aspects of the Lagoa Santa inhabitants' behavioral and cultural aspects, albeit the analyses were strongly biased by a typological and racialist approach. The first, entitled *A Raça de Lagoa Santa* (The Lagoa Santa Race), was written by Anibal Mattos (1941), and it presented a discussion of the stature, dentition, and traumas in the Lagoa Santa skeletons. Of special interest to the present chapter is the way Mattos underscored the importance of the combination of racial (genetic) and environmental factors in determining the stature of the individuals. Comparing a sample of long bones from Lagoa Santa, which he recognized as being small, with those of indigenous populations, he concluded that the inhabitants of Lagoa Santa were of medium to low stature. Mattos also presented a discussion of diet and dentition in Lagoa Santa. He reported that the teeth of the Lagoa Santa skeletons were considerably worn with "partial destruction of the alveoli" (p. 223) and attributed the fact to a diet of roots, insects, freshwater mollusks, and ground seeds, frequently ingested with earth. Using a theoretical reference of unilineal, progressive evolution, Mattos suggested that the Lagoa Santa inhabitants were still at the primitive stage of vegetarianism insofar as they only consumed plant foods and that they had not yet reached the stage of facultative carnivores in which consumption is mainly of meat. He assessed the Lagoa Santa dwellers' diet and health as having been good and similar to that observable in Brazilian indigenous people, with few physical deformities and low incidence of caries.

Mattos believed that the low incidence of caries was the result of abundant sunlight and the long period of suckling of their young, as occurs among the Brazilian Indians, guaranteeing a good ingestion of calcium. Lastly, Mattos described the occurrence of a trauma in the basal part of a female Lagoa Santa cranium similar to traumas reported by Lutken (1888) in craniums collected by Lund from the Sumidouro cave. Based on ethnographic analogies, Mattos suggested that the trauma could have been inflicted by the sacrificing of women when their husbands had been killed in conflicts. Although Mattos made an important contribution to the understanding of the behavior of Lagoa Santa inhabitants, his book is mainly dedicated to anatomical descriptions of the teeth and craniums with the underlying intention of typifying a Lagoa Santa race. Curiously enough, the chapters on stature, dentition, and trauma are absent from the second edition of his book entitled *O Homem das Cavernas de Minas Gerais* (Mattos 1961) (The Caveman of Minas Gerais).

The second work of interest is the book *Arqueologia da Região de Lagoa Santa, Minas Gerais: Índios Pré-Colombianos dos Abrigos-Rochedos* (Archaeology of the Lagoa Santa Region, Minas Gerais: Pre-Columbian Indians of the Rockshelters) by Harold Walter, published in 1958 (Walter 1958). The book presents archaeological evidence to support a proposed model of unilineal cultural evolution in Lagoa Santa, beginning from the simplest culture and evolving toward more complex one, inspired by the cultural sequences of European prehistory. That model attributes overriding importance to the typological classification of the material and the use of ethnographic data as the means to establish the chronological sequences. As an example, Walter observed two large traumas in the craniums uncovered at the Lapa de Limeira and Lapa do Galinheiro, and, using ethnographic examples, he suggested that they might be cases of the sacrifice of war prisoners. Although the author admitted that only a small number of craniums had such fractures, he inferred that violence was an important aspect of Lagoa Santa inhabitants' lives. Walter also drew inferences about Lagoa Santa demography. He reported that 50% of the skeletons exhumed were of subadults, suggesting that the populations living in times prior to the discovery of antibiotics had high infant mortality rates. Regarding buccal pathologies, Walter described the dentition of the Lagoa Santa skeletons as being free from caries and infections. On the other hand, he observed high numbers of teeth lost in vivo and severe wear of the teeth. According to Walter, chewing hard foods such as roots and raw palm kernels demanded a lot of the teeth, and as a result, most of the Lagoa Santa individuals had lost their incisors and premolars by the time they were 40 years old. He also observed that the upper incisors were the first to drop out, probably because of their use in biting and grinding. Walter also referred to the effect of earth grains in the food, which further exacerbated the wear caused by grinding. He suggested that the inhabitants used to eat large slugs that they roasted as well as deer, armadillos, alligators, small mammals, birds, wild fruits, and honey. Despite his praiseworthy efforts to characterize the cultural behavior of the Lagoa Santa population, Walter did not describe his osteological analysis methodology nor report the numerical results of his analyses, thereby impeding any posterior verification or validation of his results or even their comparison with other studies.

In addition to the archaeological community, other professionals became interested in the prehistoric material from Lagoa Santa. The contributions of dentistry professionals in Lagoa Santa began with the work of Suelyo Santos Oliveira, who showed that the dental arcades of the Lagoa Santa skeletons deposited in the National Museum of Rio de Janeiro were wide and the teeth were heavily worn (Oliveira 1954). That fist study was quite short and added little to what was already known about Lagoa Santa dentition. It was only when the professor of pathology at the Fluminense Faculty of Dentistry and at the Faculty of Pharmacy and Odontology of the State of Rio de Janeiro, Ernesto Salles Cunha, began his studies that there was any significant contribution to the understanding of oral health in Lagoa Santa. He made a detailed study of the alveolar-dental infections in the Lagoa Santa skeletons deposited in the Federal University of Minas Gerais. Examining a skeletal collection of 50–62 individuals, he encountered evidence of oral pathologies such as enamel hypoplasias, dental caries, loss of teeth in vivo, granulomas and cysts, alveolar atrophy resulting from periodontal infections, and heavy tooth wear (Salles Cunha 1961). In the Lagoa Santa sample, 32.2% (20/62) to 40.0% (20/50) of the individuals had dental caries, 11.2% (3/62) to 14.0% (3/50) had granulomas, and 4.8% (3/62) to 6.0% (3/50) had cysts. All in all, Salles Cunha concluded that the oral health of Lagoa Santa Man was poor, with the presence of many deep caries and destruction of the pulp, loss of teeth in vivo early in life, frequent periodontal inflammations, and a high degree of tooth wear. That situation was different from the one found among the skeletons of the Sambaqui Man (shell-mound builders of the Brazilian coast) of which had very few cavities and a lot of dental calculus. Similar to Lagoa Santa, Sambaquieiros showed considerable tooth wear. However, according to the author, the pattern of tooth wear in Lagoa Santa that lead to the drop shape of the teeth was the result of anteroposterior movements, unlike the Sambaqui Man where the teeth are shaped like the mouthpiece of a flute and are the result of lateral movements. According to that study, which made comparisons with the Sambaqui group, the overall pattern of oral health in Lagoa Santa may be the result of a low-protein diet and the absence of sunlight in the caves of the region or even to general climatic conditions. That was the most detailed study investigating the oral health of the Lagoa Santa skeletons up until that time and included information on prevalence and comparisons with other skeletal populations, albeit little information was offered regarding the methodology used to identify the pathologies.

The studies of the Lagoa Santa collection at the National Museum of Rio de Janeiro in the second half of the twentieth century were led by Marília Carvalho de Mello e Alvim. Making use of a traditional approach based on the anatomical description of the skeletons, the author used various indexes to characterize biological affinities and morphological diversity. In 1977, Mello e Alvim published a great synthesis of her morphological studies of the Lagoa Santa collection that included around 200 skeletons deposited in the museums of Rio de Janeiro and Belo Horizonte (Mello e Alvim 1977). It was by far the most complete examination of the collection up until the time it was published, expanding an earlier study based on a much smaller skeletal sample (Messias and Mello e Alvim 1962). As regards the osteological markers of health and lifestyle, she emphasized the presence of oral pathologies and reported the prevalence of caries and dental abscesses. Alvim observed the presence of caries in 40% of a sample of around 130 individuals. She also found that 16.9% of the individuals had granulomas and 9.1% had cysts. The rate of loss of teeth in vivo was also considered to have been very high and mainly affected the incisors and premolars. On the topic of tooth wear, she observed a moderate degree of wear as being the most common condition. Although this study embraced a much larger skeletal sample, it was not as detailed as the oral pathologies analysis conducted by Salles Cunha (1961). In regard to bone pathologies, the study does contain a brief report of an infection in the auditory meatus in one cranium from the Sumidouro cave, problems of deformations in the temporal-mandibular articulation, and one case of spina bifida. She calculated the average male height to be 1628 mm and female height 1518 mm. The work is predominantly descriptive, and little attempt was made to understand aspects of the Lagoa Santa population's behavior. In her final remarks, Alvim concludes that the Lagoa Santa inhabitants' oral health was poor except for some of the skeletons from Lapa do Eucalipto, which were relatively free from caries (see above). She explained that poor health as being due to a diet that was mainly dependent on plants. In general terms, Alvim described the Lagoa Santa population as having a rudimentary culture and low population density and practicing seasonal nomadism. She also concluded that the skeletons at Lagoa Santa are small statured, delicate, and showing a high degree of sexual dimorphism.

Jorge Ferigolo was the first researcher to make a systematic analysis of the paleopathological osteological markers in Lagoa Santa, calculating prevalence values and comparing them to the skeletons from the Sambaqui de Cabeçuda in Rio de Janeiro (Ferigolo 1987). His work was conducted in a comparative paleopathological perspective and included analyses of human skeletons and the remains of Pleistocene mammals. Ferigolo addressed a wide range of lesions including traumas, degenerative lesions of the joints, and nutritional, dental, and infectious diseases. In addition to describing the pathological cases, that author made an extensive review of the biological-evolutionary hypotheses for explaining the occurrence of those lesions, addressing aspects such as biological development and aging. As an example, Ferigolo raised the possibility that bony growth along the edges of articulations, that is, the formation of osteophytes, is in fact an adaptation to increase the articulations' sustaining capacity. Another example is a discussion of Schmorl nodes (depressions in the body of the vertebrae) which the author relates to neoteny[3] of the superior primate vertebral column, such as retardation of the process of notochord involution. Regarding his analysis of the Lagoa Santa skeletal material, Ferigolo found very few osteomas in the external auditory tract, which suggested that the Lagoa Santa inhabitants were not in the habit of diving as the Sambaquieiros (shell midden groups) were. Markers of stress during growth (enamel hypoplasia and Park-Harris lines) showed that in Lagoa Santa, there were more episodes of nutritional shortages than at the Sambaqui de Cabeçuda, suggesting a protein deficiency in the diet of the Lagoa Santa group. The author identified more osteoarthritis and traumas in the Lagoa Santa skeletons, whereas those from Sambaqui de Cabeçuda showed a greater number of articulation and thoracic vertebrae lesions. Joint lesions among the Lagoa Santa skeletons were more frequently found in the cervical vertebrae, wrist, and knee joints. Bone infections were not frequent in either population, but one case of syphilis was detected in the Lagoa Santa group. Regarding dental diseases, Lagoa Santa presented a high incidence of tooth loss in vivo, little evidence of periodontal lesions, a high incidence of pulp exposure from tooth wear, and very few temporomandibular lesions. There were few dental caries in either of the skeletal collections. Ferigolo concluded that on the whole, the Lagoa Santa inhabitants showed that they lived in worse nutritional conditions than the shell midden group mainly due to a lack of protein. Despite the wealth of data generated and the hypotheses it led to, very little was concluded about the behavior of those prehistoric populations. We believe that it was partly due to the lack of an archaeological or ethnographic data to support the paleopathological analyses. Furthermore, it is very difficult to test the biological hypotheses proposed in very ancient bones and more so when there is no rigid control of the individuals' age at death. Anyway, that study was the first systematic, in-depth study focusing on health in Lagoa Santa.

The second set of significant studies focused on the paleopathology of the Lagoa Santa inhabitants was conducted by Sheila Mendonça de Souza who analyzed 43 skeletons excavated in the 1970s at the Santana do Riacho I site.[4] She

[3] Neoteny means the retention of juvenile features in the adult form.

[4] The site is located in the Serra do Cipó adjacent to the Lagoa Santa region.

produced a qualitative description of the pathologies detected, adopting an osteobiographic approach (Mendonça de Souza 1992/1993a). Mendonça de Souza noticed that the skeletons of Santana do Riacho I presented many accidental traumas, many bone infections, few dental caries, a high degree of tooth wear, a high number of Harris lines, very little osteoarthritis, and not very pronounced muscular marks. Her research, however, does not include prevalence information of the pathologies detected, making it impossible to compare the results with those for other skeletal samples. In another article, Mendonça de Souza (1992/1993b) reported the results of a paleodemographic analysis of the Santana do Riacho skeletal collection in which she had found a large proportion of subadults in the skeletal sample. From that she inferred a low fertility rate and a short life expectancy for the Lagoa Santa population, suggesting that the prehistoric group lived under considerable stress. This second study, however, assumed that the indexes derived from life tables are sensitive to variations in mortality, a principle that was contested in subsequent publications, which attribute greater influence on such indexes to fertility rates (e.g., Bocquet-Appel and Naji 2006).

Lastly, Ronaldo Radicchi (1992/1993) investigated the order of dental eruption in Santana do Riacho. He concluded that the Lagoa Santa inhabitants' anterior teeth erupted even earlier than the posterior ones when compared to the eruption pattern observed in present-day populations. Unfortunately, the causes of that pattern were not defined by the study. Despite that all the above studies focused on the health of the Lagoa Santa inhabitants, the way of life of the population was only addressed in a superficial way.

As regards the impacts in the international sphere, the first studies that led to international publications on the health of the Lagoa Santa inhabitants were those of Neves and Cornero (1997) and Neves and Kipnis (2004). Those authors found a high prevalence of caries in the teeth of Santana do Riacho (9.0%; 8/88) and Sumidouro (8.3%; 12/145) as compared to the prevalence found in other skeletal samples of hunter-gatherers from around the world. The authors, therefore, suggested that the inhabitants of Lagoa Santa fed mainly on carbohydrates of plant origin. Although those studies were short and succinct, they had the merit of inserting the oral health results in the international sphere. Indeed, the hypothesis of a subsistence diet rich in carbohydrates at Lagoa Santa was opposed to the dominant model at the time (Clovis-first), whereby hunting animals was the primary mode of subsistence of the first Americans. However, the sample of Lagoa Santa teeth investigated in those studies was too small to be controlled by sex and by age, thereby impeding the establishment of a solid foundation for the conclusions.

An in-depth study of osteological markers of health and lifestyle in the Santana do Riacho skeletal collection was finally conducted by Silvia Cornero (2005). Her work used quantitative methods such as prevalence calculation and statistical tests to gain an understanding of population aspects and behavior of Lagoa Santa inhabitants, thereby making it possible to compare them with other skeletal samples. The study results revealed a moderate prevalence of porotic hyperostosis, high prevalence of dental caries, a high incidence of accidental traumas, very little osteophytosis in the vertebral column, the absence of interpersonal violence, and low rate of occurrence of infectious diseases in the bones. Cornero also found a high degree of sexual dimorphism, expressed in the stature, robustness, and osteoarthritis. In her final remarks, she concluded that the Santana do Riacho population had good health and represented a case of adaptive success. Despite the statistical rigor observed in the analysis, the sample size was too small to allow for control of the impact of sex and age on the pathologies prevalence observed. Furthermore, the comparison of different skeletal samples was made without using a standardized methodology. Nevertheless, the work was of fundamental importance insofar as it showed the informational potential of osteological health markers in enabling an understanding of behavioral and adaptive aspects of the Brazilian prehistoric populations.

In 1999, José Ossian de Lima defended his doctoral thesis introducing an isotopic approach to gain an understanding of subsistence based on the prehistoric skeletons of Minas Gerais, including the skeletons of the Santana do Riacho 1 and 3 sites, and Piauí states. He used methods involving the detection of stable isotopes of carbon ($\delta^{13}C$), nitrogen ($\delta^{15}N$), and some trace elements (Sr, Ca, Ba, Zn, among others), which are commonly used in bioarchaeology to reconstruct paleodiets (Lima 1999). However, chemical degradation, contamination, and the absence of an isotopic ecology approach made it impossible to draw solid conclusions about the past diet in the region. Tiago Hermenegildo, in his Master's thesis, was the first one to measure isotopic values of animal bones found in the archaeological record of Minas Gerais to establish the panorama of local isotopic ecology. He integrated these data with stable carbon ($\delta^{13}C$) and nitrogen ($\delta^{15}N$) isotope values of 12 Lagoa Santa human skeletons (Hermenegildo 2009). He found that the $\delta^{15}N$ values were rather low in the human bones, suggesting a diet based on plants and primary consumers, like deer and Brazilian guinea pigs. Furthermore, individuals with an age of around 4 years at death presented high values for $\delta^{15}N$, indicative of prolonged breast feeding. Regarding $\delta^{13}C$ values, he found that 10–15% of the diet was derived from C4 plants but probably obtained indirectly from the consumption of rodents like the Brazilian guinea pig (*Cavia aperea*). Those studies opened the way for an important line of research in Lagoa Santa about isotopic markers of lifestyle.

Finally, the most comprehensive health and lifestyle analysis of the Lagoa Santa collection was conducted by Pedro Da-Gloria (2012). The work included human skeletons deposited in Rio de Janeiro and Belo Horizonte and the most recently excavated skeletons from the Lapa do Santo and Lapa das Boleiras sites deposited in the University of São Paulo. Radiocarbon dating of bones was used to select a skeletal sample corresponding to around 195 Lagoa Santa individuals. Dating ranged from 10,000 to 7000 BP years. To overcome difficulties in making comparisons, a standardized methodology was used, and the results were compared with a continental database composed of more than 6733 skeletons derived from 36 prehistoric populations. The methodology also made it possible to control effects associated to age and sex. The results of the study showed that the Lagoa Santa sample, when compared to other samples of hunter-gatherer skeletons, presented evidence of poor oral health (caries and dental abscesses; Fig. 12.1), great physical activity of the limbs (osteoarthritis: Fig. 12.2), little mobility (external dimension of the femur), few accidental traumas in the post-cranial skeleton, many bone infections (periosteal

Fig. 12.1 Cavity resulting from a cariogenic process in a Lagoa Santa tooth

Fig. 12.2 Osteoarthrosis in the humerus distal epiphysis of a Lagoa Santa skeleton

Fig. 12.3 Periosteal reaction in the fibula distal extremity of a Lagoa Santa skeleton

reactions; Fig. 12.3), and high stress levels during growth and development (stature and cribra orbitalia; Fig. 12.4). The lesions in the vertebral column (Fig. 12.5) and the traumas in the head from interpersonal violence (Fig. 12.6), albeit with high prevalence, were similar to the prevalence encountered in other hunter-gatherers of

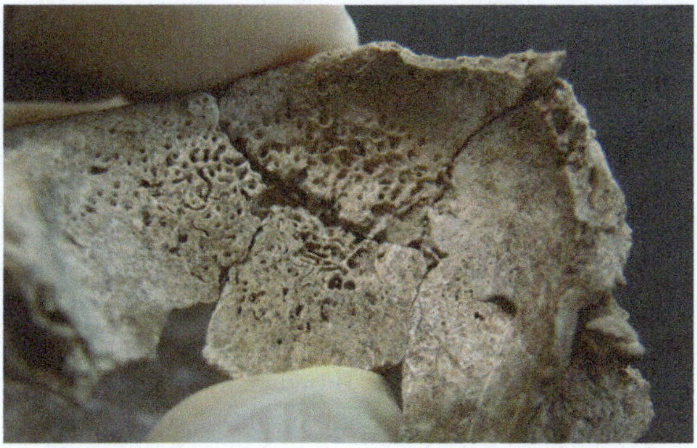

Fig. 12.4 Cribra orbitalia on the orbit roof of a Lagoa Santa skeleton

Fig. 12.5 Osteophytosis on the vertebral body margins of a Lagoa Santa skeleton

Fig. 12.6 Fracture of the parietal eminence of the cranium of a Lagoa Santa skeleton

the Americas. The overall conclusion was that the presence of lesions in the Lagoa Santa skeletons was high. The oral pathologies were particularly surprising because the high prevalence in Lagoa Santa, especially among the women, was not to be expected for a population of hunter-gatherers. The combination of tubers with native fruits that have high cariogenic potential was put forward as a possible explanation for those results (Da-Gloria and Larsen 2014). Indeed, Da-Gloria addressed the theme of health in an anthropological and bioarchaeological perspective, seeking to use the skeletal markers to gain an understanding of the behavior, way of life, health, and adaptation of the prehistoric population of Lagoa Santa. Despite its broad scope, this study did not make any more profound analyses of each one of the health and lifestyle aspects addressed. Based on the conclusions generated by these more recent studies, we propose some approaches that could contribute toward a more detailed understanding of the health of the Lagoa Santa prehistoric population.

New Prospects

Skeletal Collection

Considerable efforts have been made to characterize classic osteological markers of lifestyle associated to the skeletal collection of Lagoa Santa. Among them, we can mention oral pathologies, traumas, bone infections, and stature. Those markers are easy to observe, and there is already an extensive database of comparative data. On the other hand, new osteological markers with great informative potential have gained increasing importance in the bioarchaeological studies undertaken since the 1990s. The first of them consists of the muscular stress marks (MSM) on the bones. Muscular activities with strong impacts can be detected at the points of insertion of tendons and muscles by observing the marks of bone remodeling and hypertrophy in the external layer of the cortex. Some methods have been elaborated to register such marks like the pioneering method developed by Hawkey and Merbs and, more recently, Villote and his collaborators' (2010) method for registering fibrocartilaginous muscular insertions. Furthermore, controls for body mass and age have been introduced in MSM analyses of skeletal collections (Weiss 2003). An example is the prominent MSMs detected in the shoulders of Hudson Bay Eskimos probably due to their intense activity of paddling kayaks (Hawkey and Merbs 1995). The second method is the study of geometric sections of the femur, which is also directly associated to the intensity of an individual's physical activity during his life. The technique consists of measuring the cortical area of the bone in a transverse section and then calculating the second moment indexes of the polar area and module of polar section. Those indexes reflect the forces applied to the bone during the lifetime and must be controlled for the individual's body mass. Studies of that kind have managed to detect a substantial reduction in femur force parameters from 2 million years ago to the present day, reflecting a reduction in the mechanical load on the skeleton in the course of our evolution (Ruff 2008). Lastly, the third marker is the micro-wear marks on teeth. The investigation of the microscopic marks made on the occlusal surface of the enamel as a result of chewing food can provide important elements for inferences about diet and the extra-masticatory use of teeth. Very hard foodstuff, for example, can lead to greater numbers of marks on the dental enamel surface (Teaford 1988). The classic technique for registering such marks used a scanning electron microscope to count the scratches and depressions in the enamel. Recently, however, a more objective technique has been developed based on measuring enamel texture using a confocal microscope (Scott et al. 2006). Those new skeletal markers are important complements of the classic markers already studied to enable a better understanding of health and lifestyle in Lagoa Santa. The application of these methods to the Lagoa Santa collection is currently in progress under the scope of the project "Health and Lifestyle of the Lagoa Santa Paleoamericans: an ethnobioarchaeological approach" based at the Laboratory for Human Ecological and Evolutionary Studies at the University of São Paulo (LEEEH-USP).

A second category for skeletal analysis is formed by the isotopic markers. Hermenegildo (2009) measured carbon ($\delta^{13}C$) and nitrogen ($\delta^{15}N$) in the Lagoa Santa collection. However, the number of samples was very small ($n = 12$) compared to the estimated number of 200 skeletons in the Lagoa Santa collection. Furthermore, the isotopic analysis of animals and plants present in the region has not yet been sufficiently extensive to achieve a characterization of the region's isotopic ecology. Other chemical elements could potentially be used to characterize the lifestyle of past populations. As an example, the proportions of strontium isotopes ($^{87}Sr/^{86}Sr$) vary according to the composition of the rocks that formed the soil in the region, while $\delta^{18}O$ varies according to the climate and the source of water (Katzenberg 2008). By obtaining those chemical elements from the bones and teeth, it is possible to make inferences regarding migrations and mobility during a lifetime (e.g., Knudson and Torres-Rouff 2014). Strontium analyses were begun in the Lagoa Santa region thanks to the collaboration of the Max Plank Institute for Evolutionary Anthropology and the LEEEH-USP (Strauss et al. 2015) and have obtained results related to mobility and funeral rites. Another analysis that could add knowledge about the lifestyle of the early inhabitants of Lagoa Santa is the investigation of molecular biology of the bony tissues of individuals affected by pathologies, something that has already been done in the Americas in the case of leishmaniosis (Costa et al. 2009) and syphilis (Melo et al. 2010).

Forming yet another class of promising studies targeting the Lagoa Santa collection are the microscopic analyses of dental calculus. In expansion since the year 2000, the techniques for extracting information about diet of prehistoric populations from dental calculus (also known as tartar) have contributed toward a better understanding not only of the health of the individuals studied but also indirectly of the environment in which they lived (Boyadjian 2007; Scheel-Ybert et al. 2003). Microfossils of plant material such as phytoliths (microscopic plant structures present in leaves, stalks, and fruits) and starch grains trapped in the calculus make it an effective reservoir of information on the food ingested by the individual (Boyadjian 2012; Wesolowski et al. 2010). More recently, techniques for extracting proteins that can also be found in calculus have been added as sources of raw material for molecular analyses (Bastos 2009; Hardy et al. 2012). In the Lagoa Santa region, based on reports collected from longtime residents in the karst (see below) and on prospections in the neighborhood of the archaeological sites, researcher Célia Helena Boyadjian and one of us (REO) have been constructing a database containing information on the phytoliths and starch grains of plants of the Minas Gerais savannahs (*cerrado*) that are probably edible, with the aim of obtaining a comparative collection for the future analysis of the microfossils found in the dental calculi.

Lastly, paleoparasitology studies could provide valuable information on the health of prehistoric populations even when no marks are produced on the skeletons. The parasitic infections of the early Lagoa Santa inhabitants have been under analysis by a Fiocruz team coordinated by researcher Alena Iñiguez, thanks to the meticulous archaeological excavation work carried out at the Lapa do Santo site. The presence of parasite eggs or even the identification of molecular markers of those pathogens is being sought for in the burial sediments and in the anatomical cavities of the skeletons exhumed at that site, thereby helping to expand knowledge about the environment and the health conditions in which those ancient inhabitants of the eastern-central Brazil lived (Iñiguez et al. 2003, 2006; Jaeger et al. 2012; Leles et al. 2010).

Archaeological Context

The study of the health of the prehistoric population of Lagoa Santa has traditionally been addressed in the form of a description of bone and dental pathologies. We, however, defend a much broader understanding of health that considers the lifestyle of the population being studied. In this sense, we highlight two types of archaeological information of fundamental importance for an understanding of, for example, the diet of those past populations. First, the remains of animals' bones can provide crucial data for understanding the selection and preparation of food and the paleoenvironment. Kipnis (1998, 2002) applied evolutionary ecology models to the set of fauna remains excavated in the sites of Minas Gerais state and showed that the diet at the end of the Pleistocene and beginning of the Holocene was based on small animals such as the Brazilian guinea pigs (*Cavia aperea*), rock cavies (*Kerodon rupestris*), and armadillos (*Dasypus novemcinctus*), while medium-sized animals like deer (*Mazuma* sp.) and peccaries (*Tayassu* sp.) were only found occasionally. Basing himself on the optimal foraging theory, Kipnis suggested that the diet of that time was diversified and included a significant amount of vegetable food and an absence of megafauna hunting. Preliminary analysis of the zooarchaeological record from the recently excavated sites of Lapa das Boleiras and Lapa do Santo, both in the Lagoa Santa region, supported the results obtained by Kipnis. Small reptiles and mammals, for example, were abundant at both sites during the Early Holocene (Kipnis et al. 2010; Perez 2009). Future analyses of the same materials should provide even more detailed data regarding the food choices of those earliest inhabitants.

The importance of plants in the diet of the prehistoric population of Lagoa Santa leads us to another class of archaeological evidence, the paleobotanical remains. Recent excavations in Lagoa Santa have been careful to collect the remains of fruits and seeds from the sediment using the flotation technique. The Early Holocene strata in Santana do Riacho, for example, present large amounts of macroremains of pequi (*Caryocar brasiliense*; Resende and Prous 1991). At the Lapa do Santo site, preliminary analyses carried out by Silva (2006) identified 39 kinds of vegetable macroremains. At Lapa das Boleiras, remains of a palm species (*Syagrus flexuosa*), jatobá (*Hymenaea* sp.), pequi (*Caryocar brasiliense*), araticum (*Annona classiflora*), and chicha (*Sterculia chicha*) were found in Early Holocene strata. Considering that most of them were found in a carbonized condition, they must have been used as food by the inhabitants (Nakamura et al. 2010).

The high prevalence of oral pathologies in the Lagoa Santa skeletons (Da-Gloria and Larsen 2014) can only be fully understood in the light of the archaeological data set out above. The suggestion derived from the archaeology of a high consumption of plants in the region is corroborated by the high prevalence of dental caries, which is an indicator for high consumption of carbohydrates (Larsen 2015). That interaction between skeletal analysis and archaeological context, which lies in the field of bioarchaeology, opens up various pathways toward an understanding of other health-related aspects such as physical activity, stress during growth, and violence. Actually, the use of modern techniques in the archaeological excavations, such as spatial registration and meticulous collection of the archaeological materials, has opened up new possibilities for the health studies.

Living Populations

Trying to delineate the lifestyle of populations in the distant past runs up against the lack of information about edible plant material available in the local environment. In an attempt to fill in the gap, in the period from 2009 to 2011, Walter Neves and one of us (REO) administered a questionnaire to the oldest living residents in the Lagoa Santa region, seeking to gain a vision, through in the memories of those older volunteers, of the local ecosystem when it was less perturbed by agriculture and ranching. The responses obtained orientated a collection of fruits and tubers consumed by the Lagoa Santa population, many of which were unknown to the Minas Gerais botanists. Those plant materials, as mentioned above, are being used to form a database of phytoliths and starch grains to support the dental calculus analyses. In addition, with the collaboration of the Nutrition Laboratory coordinated by Beatriz Cordenunsi, in the Faculty of Pharmaceutical Sciences at the University of São Paulo, the centesimal compositions (nutrients) of the main food plants mentioned in the interviews are being analyzed to help to elucidate the possible cause of the poor state of the oral health detected in the human skeletons of Lagoa Santa (Da-Gloria and Larsen 2014).

With the paleobotanical data complete and using the interviews of the local residents, it should be possible to conduct controlled food ingestion experiments. Those experiments would be to observe the cariogenic potential of the various foods that the prehistoric populations probably consumed and to observe the interaction of those foods with the mouth environment. After the experimental subjects have ingested the food, Stephan curves, which measure pH variations in the biofilm for a period of 60 min, can be calculated (Lingstrom et al. 1993). Those foods that generate a drop in pH values for a long period of time would be strong candidates for explaining the high prevalence of caries in the skeletal samples. That methodology, however, does have some severe limitations insofar as it does not take into account the frequency of ingestion or the combinations of food materials consumed by the prehistoric population. Furthermore, the experiments do not simulate the saliva composition or the original dental structure of the prehistoric population in question. Even so, such experiments can contribute, together with other sources of archaeological information, to generate a broad reconstruction of subsistence in the past. The proposed experiments are being planned in the scope of the project "Health and Lifestyle of the Lagoa Santa Paleoamericans: an ethnobioarchaeological approach," based at the LEEEH-USP.

Last but not least, the study of living populations can contribute toward the formulation of interpretive models for understanding health of past populations. The use of ethnographic analogies in archaeology is a long-standing practice (Stahl 1993). Beginning in the 1970s, the occurrence of ethnographic expeditions for observing patterns of material culture deposition to assist the interpretation of the archaeological record started to take place more frequently. Indeed, ethnoarchaeological studies, as such approaches are called, seek to understand the ethnographic context and the relation between the material culture and the people that produce it (David and Kramer 2001). The use of interpretative models originating from the field of ethnography is essential for archaeology, because it is only possible to visualize the dynamic interaction between people and their material culture in living populations. In the field of bioarchaeology, some studies have already used that approach, albeit in a very incipient manner, because it is not common for surveys in the ethnographic field to be attempting to find answers to bioarchaeological questions. Nevertheless, Harrod and collaborators (2012), for example, investigated trauma patterns among pastoral people in Tanzania, registering the location and accumulation of the fractures as well as gathering information on the connection between life histories and interpersonal violence. That data can make a significant contribution to the interpretation of markers of violence in skeletal populations. Other works exploring that relation studied enamel hypoplasia (Dobney and Goodman 1991), oral health (Walker et al. 1998; Walker and Hewlett 1990), stature (Vercellotti et al. 2014), and porotic hyperostosis (Piperata et al. 2014). In the case of oral health, the studies sought to establish a connection between the living populations and the skeletal samples by focusing on the oral pathologies in an ethnographic context (Walker et al. 1998; Walker and Hewlett 1990). There has been little progress, however, in the formulation of models that include physiological and nutritional aspects, such as the composition of the biofilm and the saliva of the

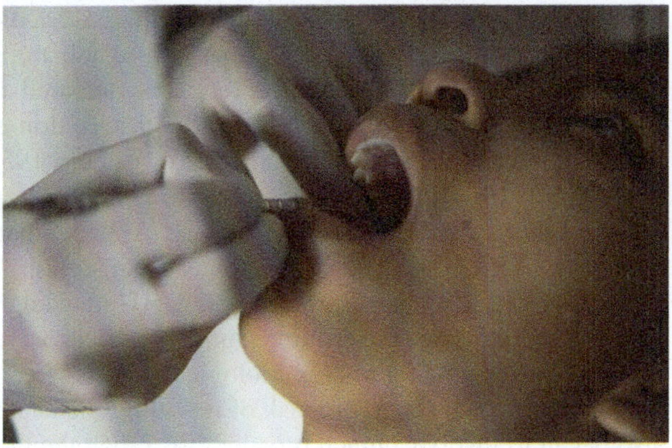

Fig. 12.7 Collecting microbiological material from a woman resident of a rural riverside community – middle course of the Solimões River (Photo: Rodrigo Erib)

investigated individuals. One project in course at the LEEEH-USP is researching the dynamics of oral health factors involved in relatively isolated populations in riverside communities along the middle course of the Solimões River (Fig. 12.7). The study is attempting to integrate socioeconomic, ecological, nutritional, and physiological aspects with oral pathologies to make it feasible to test specific hypotheses about the relative importance of diet and physiology in dental caries etiology (Lukacs 2011; Lukacs and Largaespada 2006). We believe that approach can notably enrich the interpretation of health and lifestyle indicators in skeletal samples.

Final Remarks

The historical trajectory of health research in Lagoa Santa reveals tendencies associated to the theories of the day in archaeology and anthropology. Lund's pioneering studies reveal a more holistic view of prehistorical issues, bringing together disciplines which at the time were not as clearly separated as they are today. Subsequent studies in the region, toward the end of the nineteenth and at the beginning of the twentieth century, however, relegated health to a marginal position, focusing work on the human craniums in a racialist perspective. Any mention of health in those studies was based on a biomedical and clinical-pathological view that used a merely descriptive approach. Health itself, in that context of the examination of the osteological markers, was conceived as merely being the absence of sickness; a healthy individual was one with no signs of disease (Almeida-Filho 2011). That kind of approach in Lagoa Santa went against broader views of what constituted health, like the one the World Health Organization formulated in 1946, whereby "Health is a state of complete physical, mental and social well-being and not merely the absence of disease or infirmity" (WHO 1946). The holistic nature of this definition shows us the importance of broadening the biomedical view extant in the earlier studies.

Other theoretical contributions suggested that archaeology would be greatly benefited by connecting itself with other more holistic sciences like anthropology (Binford 1962), while other theoreticians declared that physical anthropology should be less concerned with technical description and more population based when formulating scientific hypotheses (Washburn 1951). It was only in the 1990s and 2000s, however, that the approach to health in Lagoa Santa began to change, and even then, it was a very timid process. The association of the studies of human skeletons with the archaeological context, constituting the field of bioarchaeology (Larsen 2015), was decisive in that respect. Understanding the health of prehistoric populations then began to require a more comprehensive view of the environment in which people lived, not only in biological but also in cultural terms (Goodman and Leatherman 1998). Thus, the task of revealing the health of prehistoric populations also began to involve a much broader effort than the mere registration of the prevalence of pathologies. In this chapter, we have endeavored to present some proposals to better understand health of the Lagoa Santa prehistoric population and to show that there is still a long way to go to complete that task. Nevertheless, in our view, the situation is very promising. Studies of skeletal collections, archaeological material, and living populations should provide a far more consistent scenario regarding the lifestyle of Lagoa Santa's first inhabitants.

References

Almeida-Filho, N. (2011). *O que é saúde?* Rio de Janeiro: Editora Fiocruz.

Ávila, J. B. (1937). *Excursão às grutas e cavernas "Carrancas", Nova Granja, M.G – 1937.* Rio de Janeiro: Relatório, Museu Nacional da UFRJ.

Ávila, J. B. (1950). Anthropometry of the Indians of Brazil. In J. H. Steward (Ed.), *Handbook of South American Indians* (Vol. 6, pp. 71–84). Washington, DC: Smithsonian Institution.

Bastos, M. Q. R. (2009). *Mobilidade Humana no Litoral Brasileiro: análise de isótopos de estrôncio no sambaqui do Forte Marechal Luz.* Master Thesis. Fundação Oswaldo Cruz.

Binford, L. R. (1962). Archaeology as anthropology. *American Antiquity, 28*, 217–225.

Bocquet-Appel, J., & Naji, S. (2006). Testing the hypothesis of a worldwide Neolithic demographic transition. *Current Anthropology, 47*, 341–365.

Boyadjian, C. H. C. (2007). *Microfósseis contidos no cálculo dentário como evidência do uso de recursos vegetais nos Sambaquis Jaboticabeira II (SC) e Moraes (SP).* Masters Degree Dissertation, Universidade de São Paulo.

Boyadjian, C. H. C. (2012). *Análise e identificação de microvestígios vegetais de cálculo dentário para a reconstrução de dieta sambaquieira: estudo de caso de Jaboticabeira II, SC.* Ph.D. Dissertation, Universidade de São Paulo.

Cornero, S. E. (2005). *Biologia esqueletal de los primeros americanos: un caso de estudio en Santana do Riacho I, Brasil Central.* Ph.D. Dissertation, Universidad Nacional de La Plata.

Costa, M. A., Matheson, C., Iachetta, L., Llagostera, A., & Appenzeller, O. (2009). Ancient leishmaniosis in a highland desert of Northern Chile. *PloS One, 4*, e6983.

Da-Gloria, P. (2012). *Health and lifestyle in the Paleoamericans: early Holocene biocultural adaptation at Lagoa Santa, central Brazil.* Ph.D. Dissertation, The Ohio State University.

Da-Gloria, P., & Larsen, C. S. (2014). Oral health of the Paleoamericans of Lagoa Santa, central Brazil. *American Journal of Physical Anthropology, 154*, 11–26.

David, N., & Kramer, C. (2001). *Ethnoarchaeology in action*. Cambridge: Cambridge University Press.

Dobney, K., & Goodman, A. (1991). Epidemiological studies of dental enamel hypoplasias in Mexico and Bradford: Their relevance to archaeological skeletal studies. In H. Busch & M. Zvelebil (Eds.), *Health in past societies: Biocultural interpretations of human skeletal remains in archaeological contexts* (pp. 81–89). Oxford: TEMPVS REPARATVM.

Ferigolo, J. (1987). *Paleopatologia comparada de vertebrados. Homem de Lagoa Santa, Homem do sambaqui de Cabecuda e mamiferos Pleistocênicos*. Ph.D. Dissertation, Universidade Federal do Rio Grande do Sul.

Goodman, A. H., & Leatherman, T. L. (1998). Traversing the chasm between biology and culture: An introduction. In A. H. Goodman & T. L. Leatherman (Eds.), *Building a new biocultural synthesis: Political-economic perspectives on human biology* (pp. 3–41). Ann Arbor: University of Michigan Press.

Hansen, S. (1888). Lagoa Santa racen. En Anthropologisk Undersogelse af Jordfundne Menneskelevninger fra Brasilianske Huler. Med et Tillaeg om det Jordfundne Menneske fra Pontimelo, Rio de arrecifes, La Plata. En Samling af Afhandlinger. *E Museo Lundii*, I, 5, Kjobenhavn, pp. 1–37.

Hardy, K., Buckley, S., Collins, M. J., Estalrrich, A., Brothwell, D., Copeland, L., García-Tabernero, A., García-Vargas, S., de la Rasilla, M., Lalueza-Fox, C., Huguet, R., Bastir, M., Santamaría, D., Madella, M., Wilson, J., Cortés, A. F., & Rosas, A. (2012). Neanderthal medics? Evidence for food, cooking, and medicinal plants entrapped in dental calculus. *Naturwissenschaften, 99*, 617–626.

Harrod, R. P., Liénard, P., & Martin, D. L. (2012). Deciphering violence in past societies: Ethnography and the interpretation of archaeological populations. In D. L. Martin, R. P. Harrod, & V. R. Pérez (Eds.), *The bioarchaeology of violence* (pp. 63–80). Gainesville: University Press of Florida.

Hawkey, D. E., & Merbs, C. F. (1995). Activity-induced musculoskeletal stress markers (MSM) and subsistence strategy changes among ancient Hudson Bay Eskimos. *International Journal of Osteoarchaeology, 5*, 324–338.

Hermenegildo, T. (2009). *Reconstituição da dieta e dos padrões de subsistência das populações pré-históricas de caçadores-coletores do Brasil Central através da ecologia isotópica*. Master Thesis, Escola Superior de Agricultura "Luiz de Queiroz".

Holten, B., & Sterll, M. (2011). *Peter Lund e as grutas com ossos em Lagoa Santa*. Belo Horizonte: Editora UFMG.

Iñiguez, A. M., Reinhard, K. J., Araújo, A., Ferreira, L. F., & Vicente, A. C. P. (2003). *Enterobius vermicularis*: Ancient DNA from north and South American human coprolites. *Memórias do Instituto Oswaldo Cruz, 98*(Suppl. 1), 67–69.

Iñiguez, A. M., Reinhard, K., Gonçalves, M. L. C., Ferreira, L. F., Araújo, A., & Paulo Vicente, A. C. (2006). SL1 RNA gene recovery from *Enterobius vermicularis* ancient DNA in pre-Columbian human coprolites. *International Journal of Parasitology, 36*, 1419–1425.

Jaeger, L. H., Leles, D., Lima, V. D. S., da Silva, L. D. P., Dias, O., & Iñiguez, A. M. (2012). Mycobacterium tuberculosis complex detection in human remains: Tuberculosis spread since the 17th century in Rio de Janeiro, Brazil. *Infection, Genetics and Evolution, 12*, 642–648.

Katzenberg, M. A. (2008). Stable isotope analysis: A tool for studying past diet, demography, and life history. In M. A. Katzenberg & S. R. Saunders (Eds.), *Biological anthropology of the human skeleton* (pp. 413–441). New York: Wiley Liss.

Kipnis, R. (1998). Early hunter-gatherers in the Americas: Perspectives from Central Brazil. *Antiquity, 72*, 581–592.

Kipnis, R. (2002). *Foraging societies of Eastern Central Brazil: An evolutionary ecological study of subsistence strategy*. Ph.D. Dissertation, The University of Michigan.

Kipnis, R., Bissaro, M. C., & Prado, H. M. (2010). Os restos faunísticos. In A. G. M. Araujo & W. A. Neves (Eds.), *Lapa das Boleiras: Um sítio paleoíndio do carste de Lagoa Santa, MG, Brasil* (pp. 121–147). São Paulo: Annablume.

Knudson, K. J., & Torres-Rouff, C. (2014). Cultural diversity and paleomobility in the Andean middle horizon: Radiogenic strontium isotope analysis in the San Pedro de Atacama oases of Northern Chile. *Latin American Antiquity, 25*, 170–188.

Kollmann, J. (1884). Shadeln von Lagoa Santa. *Zeitschrift fur Ethnologie, XVI*, 194–199.

Lacerda, J. B., & Peixoto, J. R. (1876). Contribuições para o estudo anthropológico das raças indígenas do Brasil. *Archivos do Museu Nacional, I*, 47–76.

Lanari, C. U. (1909). Ossadas humanas fósseis encontradas numa caverna calcárea das vizinhanças do Mocambo. *Annaes da Escola de Minas de Ouro Preto, 11*, 15–35.

Larsen, C. S. (2015). *Bioarchaeology: Interpreting behavior from the human skeleton* (2nd ed.). Cambridge: Cambridge University Press.

Leles, D., Araújo, A., Vicente, A. C. P., & Iñiguez, A. M. (2010). ITS1 intra-individual variability of Ascaris isolates from Brazil. *Parasitology International, 59*, 93–96.

Lima, J. O. G. (1999). *Estudo químico comparativo de ossadas humanas pré-históricas*. Ph.D. Dissertation, Universidade Federal de Minas Gerais.

Lingstrom, P., Imfeld, T., & Birkhed, D. (1993). Comparison of three different methods for measurement of plaque pH in humans after consumption of soft bread and potato chips. *Journal of Dental Research, 72*, 865–870.

Lukacs, J. R. (2011). Sex differences in dental caries experience: Clinical evidence and complex etiology. *Clinical Oral Investigations, 15*, 649–656.

Lukacs, J. R., & Largaespada, L. (2006). Explaining sex differences in dental caries prevalence: Saliva, hormones, and life-history etiologies. *American Journal of Human Biology, 18*, 540–555.

Lund, P. W. (1844 [1950]). Notícia sobre ossadas humanas fósseis achadas numa caverna no Brasil. In: Couto, C. P. (Org.) *Memórias sobre a paleontologia brasileira*. Instituto Nacional do Livro, Rio de Janeiro, pp. 465–484.

Lutken, C. F. (1888). Indledende Bemaerkninger om Menneskelevninger i Brasiliens Huler og i de Lundske Samlinger. En Samling af Afhandlinger. *E Museo Lundii, I*(4), 1–29.

Mattos, A. (1941). *A raça de Lagôa Santa: Velhos e novos estudos sobre o homem fóssil americano*. São Paulo: Companhia Editora Nacional.

Mattos, A. (1961). *O Homem das cavernas de Minas Gerais*. Belo Horizonte: Editora Itatiaia.

Mello e Alvim, M. C. (1977). Os antigos habitantes da área arqueológica de Lagoa Santa, Minas Gerais, Brasil – Estudo morfologico. *Arquivos do Museu de Historia Natural da UFMG, 2*, 119–174.

Melo, F. L., Mello, J. C. M., Fraga, A. M., Nunes, K., & Eggers, S. (2010). Syphilis at the crossroad of phylogenetics and paleopathology. *PLoS Neglected Tropical Diseases, 4*(1), e575.

Mendonça de Souza, S. (1992a/1993). Paleopatologia humana de Santana do Riacho. *Arquivos do Museu de História Natural da UFMG, 13/14*, 129–160.

Mendonça de Souza, S. (1992b/1993). Paleodemografia da população do Grande Abrigo de Santana do Riacho, Minas Gerais: Uma hipotese para verificação. *Arquivos do Museu de História Natural da UFMG. 13/14*, 161–171.

Messias, T. T., & Mello e Alvim, M. C. (1962). Contribuições ao estudo do Homem de Lagoa Santa. *Boletim do Museu Nacional, N.S. Antropologia, 20*, 1–55.

Nakamura, C., Melo Jr., J. C., & Ceccantini, G. (2010). Macro-restos vegetais: uma abordagem paleoetnobotânica e paleoambiental. In A. G. M. Araujo & W. A. Neves (Eds.), *Lapa das Boleiras: Um sítio paleoíndio do carste de Lagoa Santa, MG, Brasil* (pp. 163–190). São Paulo: Annablume.

Neves, W. A., & Cornero, S. (1997). What did South American Paleoindians eat? *Current Research in the Pleistocene, 14*, 93–96.

Neves, W., & Hubbe, M. (2005). Cranial morphology of early Americans from Lagoa Santa, Brazil: Implications for the settlement of the New World. *Proceedings of the National Academy of Sciences of the United States of America, 102*, 18309–18314.

Neves, W. A., Hubbe, M., Bernardo, D., Strauss, A., Araujo, A., & Kipnis, R. (2013). Early human occupation of Lagoa Santa, eastern Central Brazil: Craniometric variation of the initial settlers of South America. In K. E. Graf, C. V. Ketron, & M. R. Waters (Eds.), *Paleoamerican odyssey* (pp. 397–412). College Station: Texas A&M University Press.

Neves, W. A., Hubbe, M., & Piló, L. B. (2007). Early Holocene human skeletal remains from Sumidouro Cave, Lagoa Santa, Brazil: History of discoveries, geological and chronological context, and comparative cranial morphology. *Journal of Human Evolution, 52*, 16–30.

Neves, W. A., & Kipnis, R. (2004). Further evidence of a highly cariogenic diet among late Paleoindians of Central Brazil. *Current Research in the Pleistocene, 21*, 81–83.

Oliveira, S. S. (1954). Nota prévia sôbre um estudo das arcadas dentária do Homem de Lagoa Santa. *Anais da Faculdade Nacional de Odontologia, 7*, 99–107.

Padberg-Drenkpohl, J. H. (1926). *Relatório de duas excursões à região calcárea de Lagoa Santa em 1926*. Rio de Janeiro: Museu Nacional da RFRJ.

Perez, C. P. (2009). *Paleoecologia de mamíferos viventes como ferramenta na caracterização do ambiente holocênico de Lagoa Santa, MG*. Master Thesis, Universidade de São Paulo.

Piperata, B. A., Hubbe, M., & Schmeer, K. K. (2014). Intra-population variation in anemia status and its relationship to economic status and self-perceived health in the Mexican family life survey: Implications for bioarchaeology. *American Journal of Physical Anthropology, 155*, 210–220.

Radicchi, R. (1992/1993). Estudo dos dentes, maxila e mandíbula dos restos esqueletais do sítio arqueológico Santana do Riacho (escavação n°1). Determinação da idade pelo método radiológico. *Arquivos do Museu de História Natural da UFMG, 13/14*, 173–187.

Reinhardt, J. (1866 [1888]). De brasilianske Knoglehuler og de I dem forekommende Dyrelevninger. En Samling af Afhandlinger. *E Museo Lundii, I*(4), 1–56.

Resende, E. T., & Prous, A. (1991). Os vestigios vegetais do Grande Abrigo de Santana do Riacho. *Arquivos do Museu de História Natural da UFMG, 12*, 87–112.

Ruff, C. B. (2008). Biomechanical analyses of archaeological human skeletons. In M. A. Katzenberg & S. R. Saunders (Eds.), *Biological anthropology of the human skeleton* (pp. 183–206). New York: Wiley Liss.

Salles Cunha, E. (1961). Afecções alvéolo-dentárias no Homem de Lagoa Santa. *Revista da Associação Paulista de Cirurgiões Dentistas, 15*, 1–10.

Scheel-Ybert, R., Eggers, S., Petronilho, C. C., Boyadjian, C. H., & Gaspar, M. D. (2003). Novas perspectivas na reconstituição do modo de vida dos sambaquieiros: uma abordagem multidisciplinar. *Revista de Arqueologia do MAE, 16*, 109–137.

Scott, R. S., Ungar, P. S., Bergstrom, T. S., Brown, C. A., Childs, B. E., Teaford, M. F., & Walker, A. (2006). Dental microwear texture analysis: Technical considerations. *Journal of Human Evolution, 51*, 339–349.

Silva, D. G. R. (2006). *Estratégias de subsitência de sociedades forrageiras pré-históricas da região de Lagoa Santa, MG: uma abordagem paleoetnobotânica*. Universidade de São Paulo: Manuscrito não publicado.

Stahl, A. B. (1993). Archaeology concepts of time and approaches to analogical reasoning in historical perspective. *American Antiquity, 58*, 235–260.

Strauss, A., De-Oliveira, R., Bernardo, D., Garcia, D. C. S., Talamo, S., Jaouen, K., Hubbe, M., Black, S., Wilkinson, C., Richards, M., Araujo, A., Kipnis, R., & Neves, W. (2015). The oldest case of decapitation in the New World (Lapa do Santo, East-Central Brazil). *PloS One, 10*(9), e0137456. doi:10.1371/journal.pone.0137456.

Teaford, M. F. (1988). A review of dental microwear and diet in modern mammals. *Scanning Microscopy, 2*, 1149–1166.

Vercellotti, G., Piperata, B. A., Agnew, A. M., Wilson, W. M., Dufour, D. L., Reina, J. C., Boano, R., Justus, H. M., Larsen, C. S., Stout, S. D., & Sciulli, P. W. (2014). Exploring the multidimensionality of stature variation in the past through comparisons of archaeological and living populations. *American Journal of Physical Anthropology, 155*, 229–242.

Villotte, S., Castex, D., Couallier, V., Dutour, O., Knüsel, C. J., & Henry-Gambier, D. (2010). Enthesopathies as occupational stress markers: Evidence from the upper limb. *American Journal of Physical Anthropology, 142*, 224–234.

Walker, P. L., & Hewlett, B. S. (1990). Dental health diet and social status among central African foragers and farmers. *American Anthropologist, 92*, 383–398.

Walker, P. L., Sugiyama, L. S., & Chacon, R. J. (1998). Diet, dental health, and cultural change among recently contacted South American Indian hunter-horticulturists. In J. R. Luckacs (Ed.), *Human dental development, morphology, and pathology: A tribute to Albert A Dahlberg* (pp. 355–386). Eugene: University of Oregon Press.

Walter, H. V. (1958). *Arqueologia da região de Lagoa Santa, Minas Gerais: índios pré-colombianos dos abrigos-rochedos*. Rio de Janeiro: Sedegra.

Walter, H. V., Cathoud, A., & Mattos, A. (1937). The Confins man: A contribution to the study of early man in South America. In G. G. Maccurdy (Ed.), *Early man* (pp. 341–348). Philadelphia: J. B. Lippincott.

Washburn, S. L. (1951). The new physical anthropology. *Transactions of the New York Academy of Science Series, II*(13), 298–304.

Weiss, E. (2003). Understanding muscle markers: Aggregation and construct validity. *American Journal of Physical Anthropology, 121*, 230–240.

Wesolowski, V., Mendonça de Souza, S. M. F., Reinhard, K. J., & Ceccantini, G. (2010). Evaluating microfossil content of dental calculus from Brazilian sambaquis. *Journal of Archaeological Science, 37*, 1326–1338.

WHO (World Health Organization). (1946). *Constitution of the World Health Organization. Basic documents*. Geneva: WHO.

Chapter 13
Burial Practices in the Lagoa Santa Region

André Strauss

Abstract In reconstructing the life of past populations, human burials are highly informative of symbolic and ritual behavior. In eastern South America, however, skeletal remains dating to the early Holocene are rare, precluding the proper study of their ritual dimensions. Lagoa Santa region in central Brazil is an important exception, as hundreds of well-preserved early Holocene human skeletons were recovered throughout 170 years of research. Here, I present a critical review of the history of discoveries of human remains in the region starting with the first interventions of Peter Lund in the first half of the nineteenth century. New excavations in Lapa do Santo starting in 2001 have revealed an elaborated setting of mortuary behavior in Lagoa Santa. Between 9.4 and 9.6 cal kyBP, the reduction of the body by means of mutilation, defleshing, tooth removal, exposure to fire, and possibly cannibalism, followed by the secondary burial of the remains according to strict rules, became a central element in the treatment of the dead. In the absence of monumental architecture or grave goods, these groups were using parts of fresh corpses to elaborate their rituals, showing this practice was not restricted to the Andean region at the beginning of the Holocene. Between 8.2 and 8.6 cal kyBP, another change occurred, whereby pits were instead filled with disarticulated bones of a single individual without signs of body manipulation, showing that during the early Archaic, Lagoa Santa was a region inhabited by dynamic groups that were in constant transformation over a period of centuries.

The Lagoa Santa region is renowned worldwide for its rich fossil collection and for the conditions that led to an excellent degree of preservation of human skeletal remains such as is seldom observed in other ancient archeological sites in the Americas. Furthermore, the presence of humans in the region dates back to the end of the Pleistocene, endowing it with an outstanding role in the studies investigating the first Americans.

A. Strauss (✉)
Palaeoanthropology Senckenberg Centre for Human Evolution and Palaeoenvironment, Eberhard Karls Universität Tübingen, Rumelinstrasse 23, 72074, Tübingen, Germany
e-mail: andre.strauss@uni-tuebingen.de

P. Da-Gloria et al. (eds.), *Archaeological and Paleontological Research in Lagoa Santa*, DOI 10.1007/978-3-319-57466-0_13

Ever since Danish naturalist Peter Lund first discovered human skeletons in the first half of the nineteenth century, the study of the skeletal remains in the region concentrated on the morphology of the craniums that were discovered and their relation to the morphology that typifies Native American groups and their implications for the antiquity of the first Lagoa Santa inhabitants (Neves et al. 2007b; see Chaps. 10 and 11 of this volume).

On the other hand, despite the abundance of well-preserved human skeletal remains in the Lagoa Santa, little has been discussed about the mortuary practices of those populations. The reasons for that involve from lack of interest in the subject on the part of archeologists to the use of field registration techniques fairly unsuited for documenting the subtleties inherent to funeral structures in the region.

This chapter retrieves the importance of the Lagoa Santa mortuary records by conducting a review of the history of research in the region, specifically in the perspective of the archeology of mortuary practices. The Map in Fig. 13.1 indicates the location of the sites discussed below.

Fig. 13.1 Map of the Lagoa Santa region. *Red small circles* indicate position of sites where human skeletal remains were found (Credit: Strauss 2010)

In the period from 1835 to 1843, Peter Lund visited more than 800 caves and identified paleontological material in at least 70 of them. In six, he also found the remains of human skeletons. Among all the caves he explored, however, none was as important as the one located at the base of the Sumidouro lake massif. Most of the time, that cave was flooded, making any kind of exploration inside it unfeasible. However, during the periods of severe drought that occur every 30 years, the water table level is so low that it is possible to enter the cave. During one such period, Lund excavated the subterranean deposits of the Gruta de Sumidouro, which he suspected to be extremely old (Lund 1844). In them, he found the bones of human individuals mixed with those of extinct animal species, which convinced him that New World Man was very ancient indeed.

Thus, the first human skeletal remains found in Lagoa Santa were not in a burial context. How the human bones came to be accumulated in the Sumidouro cave is still unknown. One possible explanation is that they were thrown into the cave through a lateral entrance. The truth is that in the face of such a fascinating discovery, Lund was not immediately concerned to find out how the bones of those groups came to be buried there. After all, it was in that mixture of extinct and still living species that he found the enigmatic remains of the horse and of man, all at the same stage of fossilization, so there was the possibility in his mind about the coexistence of those beings (Lund 1844; Neves et al. 2007a).

Even though he was in an isolated location in the interior of Brazil, Lund took pains to disclose his revolutionary ideas, and in doing so, he put Lagoa Santa on the world stage. His main interlocutor in Brazil was the Brazilian Historical and Geographic Institute (*Instituto Histórico e Geográfico Brasileiro* – IHGB) with which he corresponded constantly and to which he gave the only cranium that would remain in Brazil (Mendonça de Souza 1991). Internationally, his closest relations were with his country of origin to which he shipped practically all the material he found, including the collection of human craniums recovered from the Sumidouro cave. The study of that material by European and North American physical anthropologists back in the nineteenth century (e.g., Hansen 1888; Kollman 1884; Ten Kate 1885) was an important factor in making the Lagoa Santa region known worldwide (Mendonça de Souza et al. 2006; see Chap. 10 of this volume). In North America, Lund's works and the Lagoa Santa region became known mostly because they were the target of fierce criticism on the part of one of the most influential scholars of the day, physical anthropologist Alês Hrdlíčka Hrdlíčka (1912). In any event, none of those eminent scientists were particularly interested in the mortuary aspects of the Lagoa Santa people.

Despite all the international recognition that Lagoa Santa received, 60 years went by before there were any new investigations in the region. The first person to continue the work that Lund had begun was not a trained academic at all but the eldest son of the proprietor of the Mocambo farm, a large tract of land on which many of the caves that Lund himself had explored were located (Lanari 1909). During the first decade of the twentieth century, Cassio Umberto Lanari "explored with rare enthusiasm and daring and even risking his life, various labyrinthine caves on the paternal property and managed to find interesting remains, mainly of the human bones of two or three human individuals, about which he left a curious study published posthumously in the *Annaes da Escola de Minas* (Annals of the Minas School)" (Padberg-Drenkpol 1926:3). After his death in 1910, those human skeletal remains he had examined remained in the family's possession. Much later, in 1926, when an institutionalized expedition finally came to the region, the material was donated to the National Museum of Rio de Janeiro (Padberg-Drenkpol 1926).

In June and July of 1926, Jorge Henrique Padberg-Drenkpohl carried out the first twentieth century official excursion to the Lagoa Santa region. He was a researcher attached to the National Museum and interested in taking up Lund's work once more, particularly in the aspect of investigating the possible coexistence of prehistoric man and the region's extinct megafauna species. Some of the caves he visited had already been explored by Lund himself such as the Cerca Grande and especially "the celebrated Gruta de Sumidouro" (Padberg-Drenkpol 1926:2). However, when he realized that the former cave had been exhausted in archeological terms and the latter was inaccessible due to the high level of water in the lake, he redirected his attention to the caves that Lanari had explored.

Lanari had been dead for over a decade, but Padberg-Drenkpohl managed to establish good relations with his family who showed the researcher where Lanari had found the bones. So, Padberg-Drenkpohl then excavated the very same site, known as the Lapa do Caetano, in which Lanari had found human skeletons, and the researcher managed to increase the number of individuals whose remains were found in that location. He also found human skeletons in another site that the Lanari family identified for him, the Lapa do Cássio (Padberg-Drenkpol 1926). In both cases the skeletons not only presented a highly fossilized appearance, but they were recovered from the midst and underneath of a thick layer of stalagmitic deposit which led him to attribute great antiquity to them. The National Museum naturalist believed that the individuals were part of what could already be called at the time, the famous "Lagoa Santa race," with craniums that were characteristically dolichocephalic (Padberg-Drenkpol 1926).

Encouraged by the results of his first incursion, Padberg-Drenkpohl organized a second one in the same year from September 18 to December 8. After unfruitful investigations in the Zumby, Carrancas, Vermelha de Vespasiano, and Lagoa dos Mares caves, he directed his efforts to the vicinity of Confins where he was to find "the most fertile mine of prehistoric and paleontological remains" (Padberg-Drenkpol 1926:3). The Lapa Mortuária, as the rock-shelter came to be called, contained the skeletal remains of at least 80 individuals and the bones of extinct animals like the mastodon and an extinct horse species. Even though Padberg-Drenkpohl categorically denied that the material found in Confins indicated the coexistence of man and the megafauna (Padberg-Drenkpohl 1939), the possibility of such an association endowed the Lapa Mortuária with notoriety among the scientists of that time. Unfortunately, although the Lapa Mortuária was the site with the greatest number of individuals exhumed in the entire region, no description of the burials was ever published to the knowledge of the author of this present contribution.

Padberg-Drenkpohl never excavated in Lagoa Santa again after those two incursions, but he did revisit the area 3 years later when he dedicated himself to elaborating a detailed map. He felt that the spatial distribution of the sites and rock-shelters was very important and declared that only a reliable map would make it possible to "locate the innumerable caves visited by Lund or by ourselves, giving an idea of their distribution and the geographic and geological features of such an interesting and important zone" (Padberg-Drenkpol 1926:5).

After that intervention sponsored by the National Museum of Rio de Janeiro, the archeological excavations in the area were only taken up again in the period from the 1930s to the 1950s by the Minas Gerais Academy of Sciences (Walter 1958; see Chap. 6 of this volume). The person responsible for them was the Englishman Harold Walter with the strong support of his Brazilian colleagues Aníbal Mattos and Arnaldo Cathoud. The group's first effort was carried out from 1933 to 1935 in the same Lapa Mortuária that Padberg-Drenkpohl had excavated in 1926. That researcher had concentrated his attention on the sheltered area external to the cave itself, but the Minas Gerais team was much more interested in the paleontological material and removed tons of earth sediments from the deepest part of the cave. At a depth of more than 2 m beneath the surface, together with the bones of a horse and a mastodon, they found the skeleton of what came to be known as the "Confins Man." The supposed association of the skeleton with the Pleistocene megafauna sparked heated debates between the Rio de Janeiro team, which contested the legitimacy of that association, and the Minas Gerais team responsible for the discovery. The relations between them were far from cordial.

Because the Minas Gerais team, at that moment, was more interested in paleontology than archeology, 10 years passed before it redirected its efforts to the latter field. From 1948 to 1954, they excavated the Eucalipto, Mãe Rosa, Limeira, Marciano, and Samambaia sites. In addition to polished and flaked stones, pottery, and bone artifacts, the excavations in the rock-shelters uncovered a considerable number of human bones in the Lagoa Santa region (Walter 1958). Eucalipto was excavated in 1948, 1949, and 1954. In it, "various human skeletons were found, some of which were still articulated due to the limestone incrustations adhering to the bones. Of these last, 14 were in a flexed position with their hands above their heads and small blocks of stone or pebbles rolled on top of them, which were most certainly placed there at the time of burial" (Walter 1958:43).

In 1949, Harold Walter worked at the Mãe Rosa site, whose inhabitants he believed to have been the most primitive of the aboriginal people that ever lived in the region (Walter 1958). That interpretation was merely based on the extreme simplicity of the artifacts found in the rock-shelter, but it would hardly be sustained by any present-day archeologist. The human bones were found at a depth of 2 m, but because of the blocks of stone placed over them at the time of burial, they were in a very poor state of conservation. There is no description of the burials available.

The years 1952 and 1953 were dedicated to excavating the Limeira rock-shelter. The descriptions available of some of the burials that were found there merely indicate that the bones were found at a depth of 1 and 1.5 m and all completely broken. That must have been at least partly due to the great stones that had been placed on top of them or to later trampling of the graves. In addition, a well-preserved cranium was found some way off from the majority of the finds, in the northern portion of the rock-shelter. According to Walter (1958), the cranium presented a huge fracture which most certainly caused the death of the individual, probably killed by a blow from a club.

Like Lund and Padberg-Drenkpohl before him, Harold Walter investigated the Sumidouro site. However, whereas the first two excavated the subterranean part of the limestone massif, Walter concentrated on the terrace formed by the western slope of the limestone massif on which there were various rock paintings. The usual remains of prehistoric material were found, such as charcoal, fragments of pots, quartz flakes, and stone artifacts. In addition, several human skeletons were exhumed, all highly fragmented due to the stones that had been placed on top of them, indicating that the place had been used for funeral purposes. Most of the bones were buried at a depth of 1 and 2 m, in a flexed position and in the midst of blackened earth (Walter 1958:83).

Finally, in 1956, the Minas team went back to the Lapa Mortuária de Confins and carried out new excavations, this time in the front part of the site in a sedimentary deposit that was plainly anthropogenic and which had already been excavated by Padberg-Drenkpohl. The main goal of the intervention was precisely to establish, in a reliable manner, the relation between the cranium of the "Confins Man" and the material discovered by Padberg-Drenkpohl. After very careful excavation, Walter concluded that despite the presence in the rock-shelter of an archeological stratum consisting of earth mixed with charcoal, bone artifacts, and lithic instruments, the "Confins Man" was not part of it and was located at a depth 70 cm lower (Walter 1958). Not surprisingly, that work in 1956 confirmed the original hypotheses of the Minas Gerais team that the "Confins Man" was associated to the megafauna and not to the other human skeletons uncovered in 1926. Even though no burials had been found in this latter excavation of the Minas team, they did recover a fragment of a human femur which, according to Walter, had been manipulated to serve as a pipe (Walter 1958). Regardless of that interpretation of its function, the piece had indeed been deliberately modified.

In that same year of 1956 that the Minas team went back to the Lapa Mortuária de Confins, a research team formed on the basis of a collaboration between the University of South Dakota (USA) and the National Museum of Rio de Janeiro carried out an archeological project that led to the excavation of eight archaeological sites in the Lagoa Santa region (Hurt and Blasi 1969; see Chap. 7 of this volume). The group's mentor was Wesley Hurt, a North American archeologist who had begun to take an interest in the settlement of the Americas (Plew 1998). In the decades that followed, he was to become one of the main international Latin American archeologists. Thus, he collaborated with the Federal Universities of Paraná and Santa Catarina in the studies of Brazilian shell middens (Sambaquis), and in a partnership with the National University of Colombia, he excavated the El Abra site, one of the most important Paleo-Indian settlements in South America (Hurt et al. 1977).

The fieldwork at Lagoa Santa went on for about 3 months, and almost all the effort was dedicated to the seven rock-shelters in Cerca Grande and the Lapa das Boleiras site. In them, many human burials were found as well as projectile points and bone artifacts. The charcoal collected in Cerca Grande provided the first radiocarbon dates for the Lagoa Santa region, namely, 10,500–9764[1] cal BP and 11,597–10,679cal BP (Hurt and Blasi 1969).

No skeletons were found in Cerca Grande caves 1, 3, or 4. Four burials were identified in Cerca Grande 2. Burial 1 presented a series of stones set up in a vertical position around it. The skeleton at a depth of 34 cm and highly fragmented was flexed with the knees up against the face. The anatomical position revealed that the trunk had been twisted and the arms crossed over the chest. A quartz crystal scraper was found with the skeleton, but the excavators believed that was merely a fortuitous association. Burial 2 was just under the surface and had been partially excavated by earlier teams so that it was impossible to obtain a detailed description. Burial 3 was 44 cm below the surface. Just like burial 1, the skeleton was flexed with the knees up near the cranium and the face pointing down. The grave also presented a covering of stones. Burial 4 was at a depth of 55 cm and flexed like the others. However, instead of a covering of stones, the stones were at the base of the grave (Hurt and Blasi 1969).

Shelter number 5 is the biggest one at Cerca Grande complex, and five human burials were found in it. Burial 1 had been partially excavated by earlier teams so that no description could be made based on the scanty material that remained. Burial 2 was at a depth of from 2 to 10 cm. The skeleton was flexed with the cranium and knees together on the left side. The bones were highly mineralized and had red stains on them. At the bottom of the grave, there was a hardened bed of ashes and a fireplace near to the face. Part of the grave was encircled by stones. Burial 3 was 10 cm below the surface with the skeleton in a flexed sitting position (head between the knees). Burial 4 was found at a depth of 20 cm. None of the bones of this child's skeleton were in an anatomical position. Burial 5 was an adult buried in a flexed position with the knees very close to the head. It was at a depth of 13 cm (Hurt and Blasi 1969).

Although Cerca Grande shelter 5 provides the greatest sheltered area, the rock-shelter with the most extensive human occupation is Cerca Grande 6. The sedimentary stratum was divided into nine levels dating from 10,500–9764 cal BP to 11,597–10,679 cal BP. Most of the bones of burial 1 were broken, away from their anatomical positions, and at a depth of 25–50 cm. Hurt and Blasi (1969) suggested that the body was originally in a flexed position with the knees right up against the cranium. Burials 2 and 3 were both at a depth of 40–50 cm. The bodies were in a flexed position with the back curved, the head between the knees, and the arms crossed over the pelvis. The grave was covered by rocks. In burial 4 the skeleton was in a sitting position with the knees next to the head. The cranium and the elbows were resting on the pelvis with the forearms extended forward. There was a limestone slab over the grave. Fragments of bark were found there which Hurt and Blasi (1969) interpreted as possibly being some kind of offering. The burial was at a depth of 50–75 cm.

[1] All the dates presented in this chapter were calibrated using the IntCal 2013 curve, and the corresponding confidence interval is 95.4 %.

Burial 5 was at a depth of 30 cm. The skeleton was incomplete and not in an anatomical position, and bones were broken or missing. The grave was covered and circled by stone slabs. Burial 6 was also represented by few bones and in a flexed position. Burial 7 was found at a depth of 40–50 cm, and even though the bones were crushed, it was described as being in a flexed position. There are no descriptions available for burials 8 and 9. Burial 10 was at the same level as burials 6 and 7 and also flexed. Burial 11 was found at a depth of 50 cm and "had been disturbed with few bones remaining in their anatomical positions. It was impossible to say what its original position had been" (Hurt and Blasi 1969:34). The last burial found at Cerca Grande shelter 6 was partly covered by a large rock which meant that it could only be partly exhumed.

Lastly, from Cerca Grande 7, only a single burial was recovered. It was found at a depth of 40 cm. The body was bent with the knees up next to the head and the arms crossed above the waist. A layer of stones covered the grave and the sides were lined with big rocks.

In 2004, Neves and his collaborators presented new dates for Cerca Grande. One of them dated 10,385–10,227 cal BP (BETA 84446) refers to a charcoal sample found close to the burial in Cerca Grande 7. The other two were obtained based on bone samples from burials 2 to 3 of the Cerca Grande shelter 6, and they were, respectively, 9400–9321 cal BP (BETA 161666) and 9400–9033 cal BP (BETA 161668).

The North American team found no burials at the Lapa do Chapéu site, but two were found at the Lapa das Boleiras. Burial 1 was just 10 cm below the surface, and many of the bones were missing. Burial 2 was found underneath 1.35 m of sediment. The skeleton of an adolescent was laid on its side completely flexed with the head and knees together (Hurt and Blasi 1969:41).

Attracted by the antiquity of the dates obtained for the Cerca Grande 6 site, French archeologist Annette Laming-Emperaire set up a French-Brazilian mission which studied the caves and rock-shelters of the Lagoa Santa region in the periods 1971–1976 (see Chap. 7 of this volume). The work of her team focused on the Lapa Vermelha IV rock-shelter which was intensely excavated all the way down its 13-m depth. Radiocarbon dating indicated ages of as much as 29,265–28,800 cal BP for the sedimentary strata, and lithic artifacts were recovered for levels from which charcoals were dated from 24,231–23,887 cal BP to 18,357–18,079 cal BP. At a depth of 11 m, a bone and a coprolith of a terrestrial sloth were found in stratum that dated 11,591–10,268 cal BP (Laming-Emperaire 1979), and 2 m deeper in an archeological level dated at 12,055–11,827 cal BP and another at 13,480–13,405 cal BP, the skeletal remains of an adult female human being[2] were identified. Some authors suggested that the skeleton had been discovered in situ (Laming-Emperaire 1979; Prous 1986), while others pointed out the disarticulated state of the remains, suggesting that there had been postdepositional forces at work (Cunha and Guimarães 1978). Despite the many attempts made, no date was obtained directly from those bones. Even so, considering the stratigraphic position and a minimum dating of 10,703–10.299 cal BP obtained on the basis of acid residues extracted from the bone, the age of the skeleton found in the Lapa Vermelha IV site is presumed to be

[2] Nicknamed "Luzia" by Walter Neves.

between 12,935–12,730 cal BP and 13,399–13,283 cal BP (see also Feathers et al. 2010). According to Neves et al. (1999), this individual was not formally buried but simply deposited at the bottom of a cleft.

Mihály Bányai (1997) excavated two rock-shelters in which human burials were found. In 1970, he explored the Lapa de Miguel Fernandes site, also known as the Lapa da Samambaia. As far as it is possible to follow the limited descriptions, it seems that he excavated two adjacent chambers in the cave. In the first, four individuals were found "laid on their sides." In the second, which could only be accessed after the sediment had been removed from the first chamber, there was a small mound which was excavated and proved to be "nothing less than two craniums belonging to two entwined skeletons." On top was the cranium of 6–7-year-old child and underneath the skeleton of an elderly woman. At the side of the skeletons, there were vestiges of a hearth with fragments of animal bones. "In the corridor in front of the entrance, on its back, with its legs sticking out, was the very well preserved skeleton of an elderly man" (Bányai 1997:45). Twenty-six skeletons of various ages and both sexes were recovered altogether. Twenty-five of them had a "pillow of black stone" underneath the cranium. For that reason, the place was baptized as "The Black Pillows Cemetery." In 1987, Bányai found another cemetery, this time at the Lapa do Acácio site. Four individuals were found at a depth of 20 and 30 cm. From the photograph, it can be inferred that they were articulated (Bányai 1997). Beside the cranium of one of them, possibly an old man, an arrowhead was found made of a bluish material.

Thus, after more than a century of archeological excavations in the region, a great amount of information regarding the human skeletons in the Lagoa Santa has been accumulated. Outstandingly, it has been consensually established in the literature that the burial pattern in the Lagoa Santa region was characterized by its simplicity and homogeneity. The first person to offer a regional view regarding funeral patterns was Walter (1958:118):

> All the bodies, without exception, were buried in a flexed position with the knees up next to the chin and the arms stretched up in such a way that the hands were near the head or on it. That was a common practice among the tribes of South America and it is a custom that has been identified in many parts of the world. The lack of proper instruments to open up deeper graves must have been the cause of their squeezing the bodies into such small spaces. In Lagoa Santa, the dead were placed in shallow holes with a maximum depth of one meter and usually covered with big stones before being filled in with earth. Sometimes stones were placed inside the hole around the body. That method of inhumation resulted in the deformation of many of the craniums and the disintegration of the bones of the limbs.

Four decades later, the academic world's view of the burial patterns in the Lagoa Santa region remained unaltered (Neves et al. 2004:481):

> All the burials were similar: the graves were very shallow (maximum 60 cm in depth), and the bodies were hyper-flexed. Few funerary goods were present, and small stone slabs surrounded and covered the graves (Hurt and Blasi 1969). As Prous (1991) has emphasized, this is the regular funerary pattern found in the late Paleoindian sites of Lagoa Santa.

Neves and Hubbe (2005) asserted that the mortuary practices in the Lagoa Santa region during the early Holocene were so homogeneous that it could even be used as a chronological marker (Neves and Hubbe 2005:18311):

> Because of its homogeneity, the burial pattern of the early settling of Lagoa Santa can also be used as a complementary indicator of the chronology of the human remains uncovered when neither stratigraphic control nor direct methods of dating were available. In general, the corpses were deposited in hyperflexed position in very shallow graves topped by small blocks of limestone or quartz. Sometimes, these blocks were also used to cover the lateral walls of the grave pits. A small hearth was always established adjacent to the top of the pit, and burning charcoals were thrown into the pit, before sealing. Secondary burials were also popular in ancient days in the region. Red pigment was vastly used in these cases. Because most naturalists and amateurs described the burial conditions of the human skeletons they recovered in Lagoa Santa, these descriptions can also be used, as a last resource, to help to contextualize the old material in time.

In that scenario, the disarticulated burials with cut marks found at the Lapa das Boleiras site and described by Neves et al. (2002) were an important exception.[3] After all, the procedures involved in burial 3 cannot by any means be described as simple or common to the region. Quite the contrary, they include complex secondary burial procedures involving manipulation of the body, removing bones, applying ochre, and the action of fire. Thus, assuming that the descriptions of all the other authors are correct, then Boleiras stands out as a unique case in the burial landscape of Lagoa Santa. As such it requires an explanation to understand that site in relation to all the other rock-shelters in the region, where skeletons have been found. However, as Neves et al. (2002) have pointed out, there are various reasons to suppose that the view whereby the burial practices of Lagoa Santa were simple and homogeneous is a simplification that does not correspond to the reality. Instead, there are indications that the complex funeral procedures involving the manipulation of the body that was documented at the Boleiras site were a common practice at other sites in the region.

One piece of evidence in that direction has already been presented: the femur diaphysis that Walter believed to have been a pipe. More likely it is a yet another case in which the diaphyses of the long bones were separated from the extremities as was seen at the Boleiras rock-shelter. Another piece of evidence was identified by Messias and Mello e Alvim (1961) who described cut marks on bones that Hurt and Blasi (1969) originally described as primary burials in the Lapa de Cerca Grande. Analyzing the same material in 2009 as part of a doctoral thesis project, Pedro Da-Gloria also observed the cut marks on the Cerca Grande bones. During the process of curating the Harold Walter collection, Walter Neves (personal communication) identified various cases of long bone extremities that had been removed from the diaphyses.

In addition to that already known evidence, the descriptions set out in this chapter on the Lagoa Santa burials also seem to indicate that the practice of primary burials was far from ubiquitous in the region. At the Mãe Rosa rock-shelter, for example, skeletons were found in an extremely poor condition due to the blocks that had been placed over them. Although the poor standards of the documental records of the time make it impossible to take a definitive stance, it is quite feasible that the

[3] In addition to the excavations mentioned in the text the Boleiras site was excavated recently as part of the Origins project (see below).

"very bad state of conservation" (Walter 1958) was not due to the fragmentation of the bones under the weight of the blocks at all but instead resulted from secondary mortuary practices. The same can be said of the burials "fragmented by the overlying stones" that Harold Walter found in the Limeira and Sumidouro site and that Hurt and Blasi (1969) found at the Cerca Grande 5 and Cerca Grande 6 rock-shelters. In the Limeira site, in addition to the fragmented bones, an isolated cranium was found (Walter 1958). Once again it is impossible to be sure, but it may have been intentionally isolated as part of a mortuary ritual. In Cerca Grande 5, ochre was found on the bones (Walter 1958).

However, even though the indications mentioned above may raise doubts regarding the classical portrayal of funeral patterns in the region, they lack sufficient density, so to speak, to permit the confection of a new scenario. Until the collections can be revisited with the specific objective of reassessing the original characterizations of the burials, the truth is that it is impossible to discuss the mortuary practices in the region of Lagoa Santa in a regional perspective.

In the year 2000, with the formation of a new team to study the Lagoa Santa region, the situation began to change. Coordinated by Walter Neves, the project "Origins and micro-evolution of Man in America: a paleoanthropological approach" brought together experts from different fields of knowledge such as paleontology, palynology, geoarcheology, zooarcheology, bioanthropology, geology, and geomorphology. Generously financed by the FAPESP[4] and institutionally attached to the Laboratory of Human Evolutionary and Ecological Studies (*Laboratório de Estudos Evolutivos e Ecológicos Humanos* – LEEEH),[5] the project was able to explore the region intensely, not only in prospecting new sites but also in the excavation of archeological and paleontological sites (see Chap. 9 of this volume). The Origins project finalized in 2009. Human skeletal remains were found in three of the sites excavated under its scope, namely, the Lapa Grande de Taquaraçu (also known as Niáctor Rock-shelter), the Lapa das Boleiras (excavated by Hurt and Blasi in 1956), and the Lapa do Santo (also known as the Lapa Cinzenta [grayish rock-shelter]). The Taquaraçu site is located outside of the Lagoa Santa Karst and will not be discussed here.

As has been mentioned, the Lapa das Boleiras was excavated by the North American mission in 1956. In the new excavations, however, three new human burials were uncovered (Neves et al. 2002). None of them was articulated, but in the case of burial 3, it was possible to confirm that it was a case of secondary burial and not postdepositional perturbation. Burial 3 consisted of various long bones of the arms and legs of a subadult individual. The bones were laid out in parallel forming a bundle. On top of them was a cranial calotte. It was as if the bundle of bones had been placed inside the calotte. There were the remains of a fireplace near the cranium. The skeleton had been painted with red ochre. The proximal epiphysis of the left ulna had been completely removed, and the femurs looked as if they had been intentionally broken as part of the ritual (Fig. 13.2).

[4] Fundação de Amparo à Pesquisa do Estado de São Paulo (São Paulo Research Foundation).

[5] Founded in 1994, the LEEEH is part of the Biosciences Institute of the University of São Paulo.

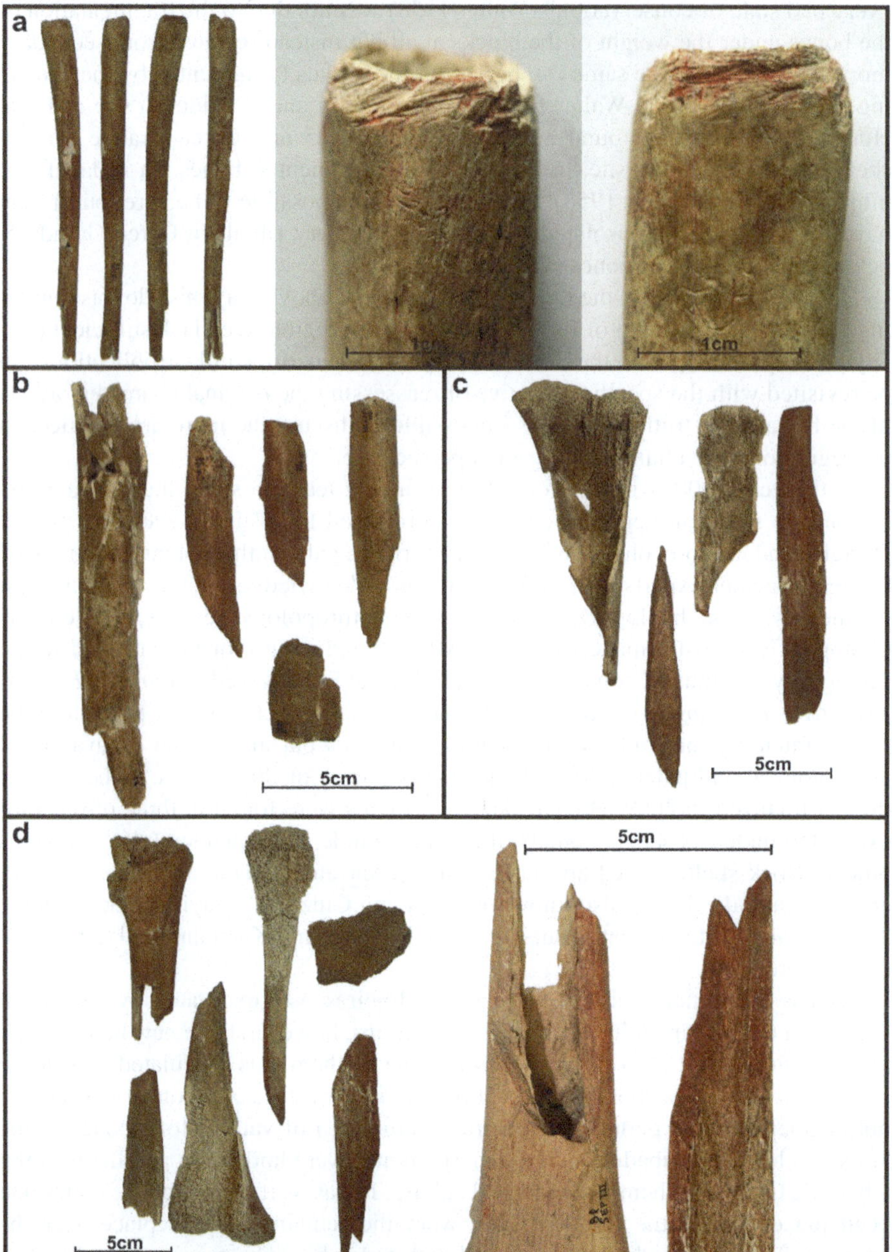

Fig. 13.2 Burial 3 at the Lapa das Boleiras. (**a**) Anterior, lateral, posterior, and medial view of the left ulna whose epiphysis was intentionally removed. In detail, the cut region, note the *cut marks* and the *red coloring*: the right tibia (**b**), the *left* femur, (**c**) and the right femur (**d**) show postmortem fractures possibly associated to the funeral ritual (Credit: Strauss 2010)

Fig. 13.3 Lapa do Santo Burial 1 (Credit: LEEEH-USP)

The other site where human burials were found is the Lapa do Santo. This is an archeological site located in the Lagoa Santa region for which the earliest evidence of human occupation has been dated as 12,700–11,700 cal BP. Excavations began there in 2001 and went on through to 2009. During that period, 26 human burials were exhumed and attributed to the early Holocene based on direct radiocarbon dating and stratigraphic interpretation of the site (Villagran et al. 2017). The human burials were classified into six mortuary patterns based on their chronology and shared features (Strauss 2010, 2011, 2012, 2016; Strauss et al. 2016). The Lapa do Santo Mortuary Patterns (LSMP) 1, 2, and 3 will be described succinctly in this chapter. LSMP-1 is dated to 9700–10,600 cal BP, and it is typified by primary burials without any signs of body manipulation. It includes burials 1 and 27 (Fig. 13.3). The dating for LSMP-2 is from 9400 to 9600 cal BP, and it can be subdivided into

Fig. 13.4 Lapa do Santo Burial 14. Photographs taken in the field show successive stages of the exhumation of the bones that made up the bundle and the exposure of the individualized cranium

three subcategories: LSMP-2a (burials 21 and 26), LSMP-2b (burials 9, 14, 17, 18, and 23), and LSMP-2c (isolated bones).

LSMP-2a consists of skeletons that are incomplete but fully articulated with the presence of cut marks (Fig.13.4). The diaphyses of both tibias and fibulas of burial 21 were removed when the soft tissues were still present. Burial 26 is a case of decapitation with the first six cervical vertebrae in their anatomical position, but the hyoid bone is missing (see Strauss et al. 2015a for details). The two hands were amputated at the wrist and placed on top of the cranium. Analyses of the cut marks using electronic and confocal microscopy show that the probable cutting agents were stone flakes.

Fig. 13.5 Lapa do Santo Burial 15. The bones of a single adult individual were deposited in a shallow circular grave. Overall, the arrangement of the skeleton was chaotic, although some of the bones remained anatomically connected

LSMP-2b is noted for graves filled with completely disarticulated bones of up to five individuals and showing evidence of a rigorous selection of anatomical parts. Some bones had the marks of exposure to fire, the application of red pigment, removal of flesh, cuts, and intentional removal of the teeth. Burials 14, 17, and 18 consisted of a bundle of long bones of one or two individuals deposited together with the cranium and/or lower maxilla of another individual (Fig. 13.5). Bundles of bones of subadult individuals were associated with the craniums of adults and vice versa. The long bones in the bundle were often cut in such a way as to separate their proximal and distal extremities. Black stains were evidence of burning restricted to the anterior portion of the alveolar border of the maxilla, indicating possible exposure to fire in the presence of soft tissues. The occurrence of such marks, in association with signs of flesh removal, suggests that in LSMP-2, some form of cannibalism may have occurred. In burials 17 and 18, all the teeth had been deliberately removed, and in the case of the latter, both coronoid processes of the mandibles had been perforated. Red pigment was widely applied to the bones in burials 14 and 18. Burial 23 consisted of the cranial calotte of a subadult individual inside which 54 permanent teeth and 30 deciduous teeth had been deposited. Some of the teeth among them actually belonged to the cranium in burial 17. Burial 9 consisted of the isolated cranium of a child. The teeth had been removed, and an arrangement of bones and teeth was found 15 cm away from the cranium. LSMP-2c is typified by isolated bones with cut marks and signs of burning, and none of them are found near to formal burial spots. In some cases, there were gnawing marks suggesting that they had probably been exposed on the surface before being buried.

Overall, the set of LSMP-2 burials can be interpreted as a ritual based on secondary burial principles (sensu Hertz 1907) in which LSMP-2a represents the initial stage, LSMP-2b the final stage, and LSMP-2c the remains of the dismembering process. The careful organization and arrangement of the body parts is compatible with an intermediary stage in which the custody and public exhibition would be elements to support the visual communication of the ceremony. The strict observation of such specific and technically elaborate procedures could indicate the presence of funerary agents specializing in those tasks.

LSMP-2 shows that, in the absence of a sophisticated architecture or rich funerary accompaniments, the elaboration of mortuary rites involved the use of the actual body of the dead person as a symbol. Regarding the archeological record, it is expressed in the form of disarticulated burials, consisting of individualized craniums, bundles of bones (of two or more individuals), cut marks, chamfers, tooth extraction, selection of anatomical parts, exposure to fire, and the application of ochre. At the same time, the presence of articulated skeletons, among them the most ancient case of decapitation ever recorded for the Americas, shows that the selection of anatomical parts was made right after death, while the soft tissues were still present. Later, the bones were reallocated and arranged in accordance with a series of very well-defined principles.

In many human societies, organic remains constitute a powerful cultural resource. The manipulation and organization of human bones and body parts has commonly been used to reify cosmological principles (Brown 2010). Notably, it is those procedures and arrangements that are the means to reifying the logical directives that possibly reflect aspects of those groups actual world vision. The disarticulated skeletons of the PFLS-2 type burials show an underlying dichotomy in the logic that prescribes the way the bones should be organized. That logic can be inferred from the presence of a double dichotomy between "adult" and "subadult" and between the "cranium" and "postcranium", the dichotomy between the "diaphysis" and "extremity," and the dichotomy between the "tooth" and "empty tooth socket."

That system, emphasizing pairs of opposition, inevitably imparts a Lévi-Straussian (structuralist) tone to the adopted logic (Lévi-Strauss 1966, 1969). Héretier (1982:158–159, *apud* Viveiros de Castro) has stated that such logic is based on the initial supposition that the "elementary symbolism of the identical and the different" is a basic tool that each society finds with which to generate its self-representation. More specifically, it is worth noting that among two of Brazil's main macrolinguistic groups, that is, the Jê and the Tupi, even though such symbolism is present in both, there is a profound difference in the way in which that binary logic affects their social morphologies.

Among the Tupis, albeit that dialectic expresses itself magnificently in the strictly cosmological plane (Viveiros de Castro 1992), it makes very little impression on social organization as such. On the other hand, among the Jê, that logic is reified in the social fabric in a far more explicit and intense manner. In the words of Viveiros de Castro (1992:5), among the Jê groups, "we find the maximum development of complementary oppositions in the social categories and cosmological values; they are oppositions that unfold, intercept and echo in one another in a dizzying Baroque progression." Thus, it is a case of the classic contrast between "metaphoric societies and metonymic societies" between "totemic societies and sacrificial societies" and, finally, between "readable societies and imperceptible societies" (Viveiros de Castro 1992:11).

Acknowledging, as one must, the limitations that interpretations with a more subjective content are liable to when they address the boundary between the Pleistocene and the Holocene, I propose that what we see in the LSMP-2 burials are precisely those elements that are part of Viveiros de Castro's "dizzying Baroque progression." Without ever losing sight of the fact that we are entering the plane of the purest conjecture and that possibly we will never be able to test the validity of such hypotheses, I nevertheless propose, that in a similar way to the present-day Jê groups, the groups responsible for the LSMP-2 also had an organization that was crystalized and readable and based on explicit structuralist dialectics (Maybury-Lewis 1979; *apud* Viveiros de Castro 1992). That situation stands out in contrast with the more amorphous possibility represented by many of the Tupi groups (Viveiros de Castro 1992).

Another similarity between the groups that settled the Lagoa Santa region during the early Holocene and the Jê groups, more specifically the Botocudos, is their cranial morphology (Strauss et al. 2015a, b). A possible interpretation of that observation is that the Jê groups, or at least some of them, have an ancestral relationship with the inhabitants of Lagoa Santa.

It is worth noting that those two groups, unlike the Tupis, are also similar insofar as they have subsistence strategies strictly based on hunting and gathering with a smaller participation of the cultivation that is so very common among the Tupi groups. Given the huge time interval involved, it is difficult to be sure whether the similarities identified here are not merely a coincidence, but one thing is sure, they are suggestive and deserve to be the object of future research.

The dating for LSMP-3 is from 8200 to 8600 cal BP, and it includes nine burials, namely, 6, 7, 10, 11, 12, 13, 15, 19, and 22. They are characteristically shallow circular graves completely filled by the bones of a single individual, almost always completely disarticulated. Adult and subadult individuals of both sexes were identified. Some of the graves are covered by a circular structure of blocks of limestone, but such structures were also found without being directly associated to graves. There was no evidence of anatomical selection, and, except for some small bones, most of the skeletal elements were present. In some cases, the long bones had been broken in the intermediary part of the diaphyses to make them fit into the tiny grave (Fig. 13.6). The LSMP-3 burials are very similar to one another, in contrast to the great variation observed among those of the LSMP-2. Furthermore, some of the features typifying LSMP-2 burials, such as cut marks, tooth removal, red pigment, and signs of burning, were not detected among the LSMP-3 burials.

LSMP-3 is interpreted as being the result of the "postponed burials" practice. The bones would have been taken to the Lapa do Santo rock-shelter for their final deposition in circular graves. The facultative presence of a circle of stones suggests that they were not merely used as weights to press down the bones inside the grave but may have served some other demarcation or symbolic purpose.

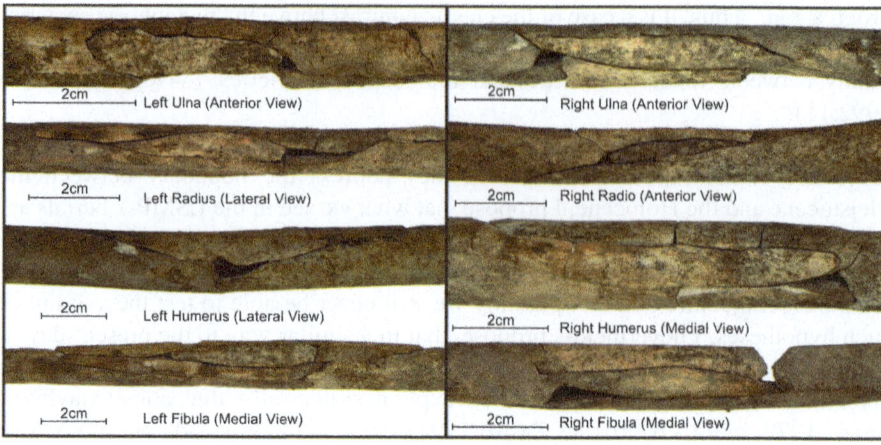

Fig. 13.6 Lapa do Santo Mortuary Pattern 3. The long bones usually present postmortem fractures as shown in these images of the bones of burial 22Credits: Strauss 2010

When analyzed as a whole, the archeological context of the Lapa do Santo reveals dynamic groups that were in constant transformation over the centuries, contradicting the view that hunter-gatherer groups were temporally static. The Lapa do Santo was not used exclusively for burying the dead, and so it cannot be classified as a cemetery. In that regard, it would more appropriate to use the concept of a "persistent place" (Schlanger 1992): one that was repeatedly occupied due to its unique features in the regional landscape. Given the fact that non-articulated burials rarely reflect aspects of social organization (Schroeder 2001), the funerary record of the Lapa do Santo should be considered as reflecting cosmological aspects of the groups that occupied the rock-shelter (Brown 2010; Strauss 2012).

By way of conclusion, it must be underscored that the critical review of the literature and of the new burials found in Lagoa Santa shows that the prevailing idea, whereby mortuary practices in the region were marked by a simple pattern of primary burials in flexed positions and covered with slabs, is actually the result of a strong bias. On the contrary, the burials in Lagoa Santa are highly complex and that makes their exhumation and subsequent interpretation extremely difficult. Again, it is notable that the documentation and excavation techniques used by the teams that worked in the region were incapable of addressing that great complexity. Thus, the supposed simplicity and homogeneity attributed to the burial patterns were actually more attributable to the archeological methods and theoretical frameworks themselves. Once more, the latent complexity of the archeological registration can only be revealed based on the most meticulous investigation.

The Origins project finalized in 2009 and with it the excavations at Lagoa Santa. The excavated area was totally filled up. In 2011, however, a new research project was set up coordinated by myself and Rodrigo de Oliveira. Entitled as "The Mortuary Practices of the First Americans," it aims to continue with the excavations in Lagoa Santa. In 2011, 2012, 2014, and 2016, a total of 20 weeks of excavation was carried out, and a new area was opened, separate from those previously excavated. Up to now, 14 new burials have been exhumed. The curating of this material and the excavations at the site are still in full swing, and the new finds will shortly be presented to the archeological community.

Acknowledgments I wish to thank Pedro Da-Gloria, Mark Hubbe, and Walter Neves for the invitation to participate in this book. This chapter owes a lot to the fruitful intellectual environment of the LEEEH-USP and, outstandingly, to the following people attached to it: Astolfo Araujo, Danilo Bernardo, Marcos César Bissaro Júnior, Pedro Da-Gloria, Tiago Hermenegildo, Max Hernani, Alex Hubbe, Mark Hubbe, Mariana Inglez, Renato Kipnis, Elver Mayer, Rui Sergio Murrieta, Walter Neves, Mercedes Okomura, Rodrigo de Oliveira, and Francisco Pugliese. For casting a critical external eye on the Lagoa Santa material that was of great assistance in establishing the theoretical basis for this work, I thank Camila Jacome, Gustavo Martinez, Alejandra Ortiz, André Prous, and Igor Rodrigues. I also thank all those who participated in the project Origins in general, especially those who carried out the work at Lagoa Santa. The work of IPHAN and IBAMA in analyzing and processing the authorization requests also deserves recognition. Special thanks go to Jean-Jacques Hublin, not only for the financial support for the new excavations in Lagoa Santa but also for having believed in the site's importance in an international scenario. In this new stage of the research, the support of the IEF in the persons of Rogério Tavares and Luisa Cota has been of fundamental importance. My eternal gratitude goes to João Bárbara Filho for his reception and for having always going on ahead to make sure everything went well. Finally, many thanks go to José Hein for his enthusiasm and the unconditional support given to our research endeavors on his property. The research in Lagoa Santa was financed by the FAPESP, CNPq, and the Max Planck Institute for Evolutionary Anthropology.

References

Bányai, M. (1997). *Minhas pesquisas arqueológicas na região de Lagoa Santa*. Budapest: Symbiose.

Brown, J. A. (2010). Cosmological layouts of secondary burials as political instrument. In L. P. Sullivan & R. C. Mainfort (Eds.), *Mississippian mortuary practices – beyond hierarchy and the representationist perspective*. Florida: The University Press of Florida.

Cunha, F. L. S., & Guimarães, M. L. (1978) Posição geológica do homem de Lagoa Santa no grande abrigo de Lapa Vermelha Emperaire Pedro Leopoldo–Estado de Minas Gerais. In: *Coletânea de Estudos em Homenagem a Annette Laming-Emperaire*. São Paulo: Museu Paulista da Universidade de São Paulo.

Feathers J., Kipnis R., Piló L. B., Arroyo-Kalin M., & Coblentz D. (2010). How old is Luzia? Luminescence dating and stratigraphic integrity at Lapa Vermelha, Lagoa Santa, Brazil. *Geoarchaeology* 25:395–436.

Hansen, S. (1888). Lagoa Santa racen. Anthropologisk Undersogelse af Jordfundne Menneskelevninger fra Brasilianske Huler. Med et Tillaeg om det Jordfundne Menneske fra Pontimelo, Rio de arrecifes, La Plata. *E Museo Lundii, 1*, 1–34.

Héretier, F. (1982). The symbolics of incest and its prohibition. In M. Izard & P. Smith (Eds.), *Between belief and transgression: Structuralist essays in religion, history and myth*. Chicago: University Press.

Hertz, R. (1907). Contribution à une étude sur la représentation collective de la mort. *Année Sociologique, 10*, 48–137.

Hrdlička, A. (1912). *Early man in South America, Bureau of American Ethnology Bulletin* (Vol. 52). Washington, DC: Government Printing Office.

Hurt, W., & Blasi, O. (1969). O projeto arqueológico Lagoa Santa–Minas Gerais, Brasil (nota final). *Arquivos do Museu Paranaense–Arqueologia, 4*, 1–63.

Hurt, W., Van Der Hammen, T., & Correal, U. G. (1977). *The El Abra Rockshelters, Sabana De Bogotá, Colombia, South America*. Bloomington: Indiana University Museum.

Kollman, J. (1884). Schadeln von Lagoa Santa. Z. *Zeitschrift für Ethnologie, 16*, 194–199.

Laming-Emperaire, A. (1979). Missions archéologiques franco-brésiliennes de Lagoa Santa, Minas Gerais, Brésil–Le Grand abri de Lapa Vermelha. *Revista de Pré-história, 1*, 53–89.

Lanari, C. U. (1909). Ossadas humanas fósseis encontradas numa caverna calcárea das vizinhanças do Mocambo. *Annaes da Escola de Minas de Ouro Preto, 11*, 15–35.

Lévi-Strauss, C. (1966). *The savage mind*. Chicago: The University of Chicago Press.

Lévi-Strauss, C. (1969). *The raw and the cooked: Introduction to a science of mythology*. Chicago: The University of Chicago Press.

Lund, P. W. (1844). Notícia sobre ossadas humanas fósseis achadas numa caverna do Brasil. In C. Paula Couto (Ed.), *Memórias sobre a paleontologia brasileira*. Rio de Janeiro: Instituto Nacional do Livro.

Maybury-Lewis, D. (1979). *Dialectical societies: The Gê and Bororo of Central Brazil*. Cambridge, MA: Harvard University Press.

Mendonça de Souza, A. A. C. (1991). História da arqueologia brasileira. *Pesquisas, 46*, 1–157.

Mendonça de Souza, S. M. F., Rodrigues-Carvalho, C., Silva, H. P., & Locks, M. (2006). Revisitando a discussão sobre o Quaternário de Lagoa Santa e o povoamento das Américas: 160 anos de debates científicos. In H. P. Silva & C. Rodrigues-Carvalho (Eds.), *Nossa origem – o povoamento das Américas visões multidisciplinares*. Rio de Janeiro: Vieira & Lent Casa Editorial.

Messias, T. T., & Mello e Alvim, M. C. (1961). Material das escavações arqueológicas na região de Lagoa Santa, Minas Gerais, Brasil (1956). *Boletim do Museu Nacional, Nova Série, Antropologia, 20*, 1–55.

Neves, W. A., & Hubbe, M. (2005). Cranial morphology of early American from Lagoa Santa, Brazil: Implications for the settlement of the New World. *Proceedings of the National Academy of Sciences of the United States, 102*, 18309–18314.

Neves, W. A., Powell, J. F., Prous, A., Ozolins, E. G., & Blum, M. (1999). Lapa Vermelha IV Hominid 1: Morphological affinities of the earliest known American. *Genetics and Molecular Biology, 22*, 1–5.

Neves, W. A., Hubbe, M., & Araujo, A. G. M. (2002). A late-paleoindian secondary ritual burial from Lagoa Santa, Minas Gerais, Brazil. *Current Research in the Pleistocene, 19*, 83–85.

Neves, W. A., González-José, R., Hubbe, M., Kipnis, R., Araujo, A. G. M., & Blasi, O. (2004). Early Holocene human skeletal remains from Cerca Grande, Lagoa Santa, Central Brazil, and the origins of the first Americans. *World Archaeology, 36*, 479–501.

Neves, W. A., Hubbe, A., & Karmann, I. (2007a). New accelerator mass spectrometry (AMS) ages suggest a revision of the electron spin resonance (ESR) Middle Holocene dates obtained for a *Toxodon platensis* (Toxodontidae, Mammalia) from southeast Brazil. *Radiocarbon, 49*, 1411–1412.

Neves, W. A., Hubbe, M., & Piló, L. B. (2007b). Early Holocene human skeletal remains from Sumidouro Cave, Lagoa Santa, Brazil: History of discoveries, geological and chronological context, and comparative cranial morphology. *Journal of Human Evolution, 52*, 16–30.

Padberg-Drenkpohl, J. H. (1939). *Carta de Padberg-Drenkpohl a Heloísa Torres, diretora do Museu Nacional*. Rio de Janeiro: Museu Nacional da UFRJ.

Padberg-Drenkpol, J. H. (1926). *Relatório interno do Museu Nacional*. Rio de Janeiro: Museu Nacional da UFRJ.

Plew M.G. (1998). Wesley R. Hurt 1917-1997. SAA Bulletin 16 (http://www.saa.org/portals/0/saa/publications/saabulletin/16-5/SAA18.html).

Prous, A. (1986). L'Archéologie au Brésil. 300 Siécles d'occupation humaine. *L'Anthropologie, 90*, 257–306.

Schlanger, S. (1992). Recognizing persistent places in Anasazi settlement systems. In J. Rossignol & L. Wandsnider (Eds.), *Space, time and archaeological landscapes* (pp. 91–112). New York: Plenum Press.

Schroeder, S. (2001). Secondary disposal of the dead: Cross-cultural codes. *World Cultures, 12*, 77–93.

Strauss, A. (2010). *As práticas mortuárias dos caçadores-coletores pré-históricos da região de Lagoa Santa (MG): um estudo de caso do sítio arqueológico "Lapa do Santo"*. Master Thesis, Universidade de São Paulo.

Strauss, A. (2011). As práticas mortuárias dos caçadores-coletores pré-históricos da região de Lagoa Santa (MG): um estudo de caso do sítio arqueológico "Lapa do Santo". *Revista Brasileira de Arqueologia, 24*, 136–139.

Strauss, A. (2012). Possibilidades e limitações interpretativas da Hipótese Saxe/Goldstein. *Boletim do Museu Paraense Emílio Goeldi, Ciências Humanas, 7*, 525–546.

Strauss, A. (2016). The burial patterns in the archaeological site of Lapa do Santo (early Holocene, east-central Brazil). *Bulletin of the Emílio Goeldi Museum of Para: Human Sciences, 11*, 243–276. Belém, Brazil. Portuguese.

Strauss, A., Oliveira, R. E., Bernardo, D. V., Salazar-García, D. C., Talamo, S., Jaouen, K., Hubbe, M., Black, S., Wilkinson, C., Richards, M. P., Araujo, A. G. M., Kipnis, R., & Neves, W. A. (2015a). The oldest case of decapitation in the New World (Lapa do Santo east-central Brazil). *PloS One, 10*, e0137456. USA. English.

Strauss, A., Hubbe, M., Neves, W. A., Bernardo, D. V., & Atui, J. P. V. (2015b). The cranial morphology of the Botocudo Indians, Brazil. *American Journal of Physical Anthropology, 157*, 202–216.

Strauss, A., Oliveira, R. E., Villagran, X., Bernardo, D. V., Salazar-Garcia, D. C., Bissaro, M. C., Pugliese, F., Hermenegildo, T., Santos, R., Barioni, A., Oliveira, E. C., Moreno de Souza, J. C., Jaouen, K., Ernani, M., Hubbe, M., Inglez, M., Rockwell, H. M., Machado, M. C., Souza, G. N., Chemale, F., Kawashita, K., O'Connel, T., Israde, I., Black, S., Feathers, J., Castro, C., Richards, M. P., Kipnis, R., Wahl, J., Araujo, A., & Neves, W. A. (2016). Early Holocene ritual complexity in South America: The archaeological record of Lapa do Santo (east-central Brazil). *Antiquity, 90*, 1454–1473. UK, English.

Ten Kate, H. (1885). Sur les Crânes de Lagoa Santa. *Bulletins de la Société d'Anthropologie de Paris, 8*, 240–244.

Villagran, X. S., Strauss, A., Miller, C., Ligouis, B., & Oliveira, R. (2017). Buried in ashes: Site formation processes at Lapa do Santo rockshelter, east-central Brazil. *Journal of Archaeological Science, 77*, 10–34.

Viveiros de Castro, E. (1992). *From the enemy point of view – humanity and deity in an Amazonian society*. Chicago: The University of Chicago Press.

Walter, H. V. (1958). *Arqueologia da região de Lagoa Santa*. Rio de Janeiro: SEDEGRA.

Chapter 14
Constructing the Past. A Look at Lagoa Santa Paleontology

Cástor Cartelle

Abstract The genius of naturalist P. W. Lund is clearly revealed in his masterly publications addressing animal physiology, zoology botany, and especially paleo-mammal zoology. Other contributions include the identification of extant and extinct species. His discoveries have served as the basis and inspiration for many researchers. Outstanding among them is H. Winge and his monumental 1,370-page work *E Museo Lundii*, published in the years from 1888 to 1915. This exceptional work was the result of studies of Lund's discoveries in Lagoa Santa. After Winge, in the mid-twentieth century, there was a new wave of research and publications addressing the area Lund had investigated, most notably the work of Carlos de Paulo Couto, who, in addition to studying a series of extinct species identified by Lund, published in Portuguese the scientific memorials of that most Brazilian of all Danes, making his work accessible to Brazilians. Lund has been the starting point for very many of the subsequent research efforts unfolded in that region. In addition to morphological aspects, they continue to produce phylogenetic, paleoenvironmental, taphonomic, and chronological information and interpretations. P. W. Lund must be credited with being the pioneer of these new efforts and perspectives, including archaeological research.

The Studies after Peter Lund

One aspect of Peter Lund that is worth highlighting, apart from his magnificent intellectual preparation, is the broad scope of the themes he addressed with such competence: physiology, zoology (invertebrates, birds, and living mammals), botany, and lastly paleomammal zoology. Extensive lists of his works can be consulted in Holten and Sterll (2011) and in Paula Couto (1950). His first publication in 1825, when he was only 24 years old, enjoyed great prestige, and his last work was published in 1846 when Lund, after 10 years of fieldwork in the Lagoa Santa region and

C. Cartelle (✉)
Museum of Natural Sciences of the Catholic University of Minas Gerais,
Av. Dom José Gaspar, 500, Coração Eucarístico, Belo Horizonte, MG, Brazil
e-mail: cartelle@pucminas.br

P. Da-Gloria et al. (eds.), *Archaeological and Paleontological Research in Lagoa Santa*, DOI 10.1007/978-3-319-57466-0_14

assiduously publishing numerous accounts of his discoveries, finally put an end to his activities and shipped the paleontological collection he had built up to Denmark, donating it to King Christian VIII. In his letter of donation, he wrote "this collection should be used by science as soon as it arrives, and as completely as possible in view of its great scientific interest and value" (Paula Couto 1950:10). His works were written in Danish, and they did not get the widespread diffusion they deserved. Nevertheless, when he was still alive, some of them were translated into French and English. His work was also referred to by Charles Darwin in *The Origin of Species*. It was only after the collection's arrival in Denmark that it came to be studied again, together with small collections of fossils collected by Claussen, also a Dane, who used to sell fossils collected in caves in the vicinity of the village of Curvelo and the Velhas River to the Natural History Museums of London and Paris.

The first publications appeared in the second half of the nineteenth century, well after the collection had arrived in Copenhagen, but it was as if they had rediscovered material that had just reached Europe. Boas (1881) believed that one of the two equine specimens Lund described, *Hippidion principale* Lund, was a zebra. There were various works (Burmeinster 1885, 1886a, b, 1887; Lütken 1886; Reinhardt 1878) discussing Lund's first discovery of an extinct species, a small terrestrial sloth, *Nothrotherium maquinense*.

The first of those three authors was a German naturalist who visited Lund in Lagoa Santa in the 1860s. Reinhardt, who was destined to become the first curator of the Lund Collection in the Copenhagen Zoology Museum, stayed with Lund on three occasions. He received Lund's diaries and registered the paleontological work Lund carried out from September 2, 1835, right through to the last excavation on June 2, 1844 (Reinhardt 1868, 1888). During that period, Lund used to participate directly in the excavations for 230 days a year. In the dry months, from April to September, he would keep some workers busy collecting, and in the rainy months, he would dedicate himself to preparing and studying the collected material.

In addition to those records, Reinhardt wrote works of his own about the Glyptodontidae (Reinhardt 1875), Dasypodidae, and Tayassuidae (Reinhardt 1880). In this last work, he gave a detailed description and produced an excellent illustration of the cranium of a *Brasiliochoerus stenocephalus* Lund. Gervais (1867–1869) and Liais (1872) made a review of Xenarthra and the Brazilian ungulates based on Lund's discoveries.

In another direction, the material traded by Claussen became well known through the drawings produced by Gervais (1874, 1877) and Lydekker (1887) and others, but they added little to what Lund had already written, except perhaps that Gervais described a new species, *Valgipes deformis* (Gervais 1873). Some of the abovementioned authors can be considered as mere revisers of Lund's nomenclature, which was not always correct. Later in this chapter, there is a reference to a Scelidotheriinae species which Lund named *Scelidotherium bucklandi* (Lund 1839), but the correct name for the genus is *Valgipes*. The species name Lund used, however, has been maintained.

E Museo Lundii

It is very much as if those early works were merely an introduction to Herluf Winge's (1857–1923) monumental work, *E Museo Lundii*. This publication, dedicated to the study of Lund's collection at the Copenhagen Zoological Museum, was published in three volumes with the second and third being divided into two fascicles. The 1,370-page work (1,188 pages written by Winge himself) contains 81 illustrations. It was a publication worthy of the Lund collection and one that displays the admirable knowledge of one of the nineteenth to twentieth centuries' leading zoologists. His book is evidence of his tremendous work capacity and his encyclopedic knowledge based on a bibliography that embraced practically all that was known at the time on the subjects he addressed. That knowledge enabled him to make a solid revision of Lund's work making use of the diaries and notes that accompany the collection and analyzing them with incomparable depth and breadth of vision. In the monographies written for *E Museo Lundii*, he not only studied the extinct fauna but also the current fauna of the Lagoa Santa region, using the collection of modern animals that Lund had also collected. At the same time, he discussed the interrelations of various extinct and living taxa, including other similar forms of life akin to the Brazilian fauna.

Before giving a summary of Winge's work, it is worth to present his opinion about the work undertaken by P. Lund, even though he never knew him personally:

> When Lund began his work on the animal remains in the bone-bearing caves of Brazil, he was not qualified in any way for that activity. He had to work in isolated regions in the remote interior of Brazil without the aid of sufficient material for comparison or books. The material was only extracted from the caves a little at a time, so that it took a long time before it was possible for him to review it all together as a whole. Most of the remains that were found were just incoherent fragments, and the cave animals were very different from those of today. It would have been a miracle if Lund had not frequently made mistakes in interpreting the bones. What is practically a miracle, however, is that it did not happen more often than it actually did, and especially, that generally speaking, Lund himself corrected his own mistakes. From his manuscripts, it becomes very clear that in the end he acquired great skill in determining and interpreting the bones. Anyone who limits themselves to merely reading his printed scientific memoirs can have no idea of the real extent of his knowledge. (Winge 1915:262) [Translated from a Portuguese translation by Couto 1950]

Winge had the broader view that was largely lacking in Lund. His great merit lies in the global analysis he made of the fauna discovered by his fellow compatriot, adopting new systematic determinations to conclude impeccable morphological assessments and phyletic analyses of extraordinary depth for that day and age, many of the latter with a strong influence of Lamarckism. Eventual mistakes in the nomenclature were largely due to the insufficient numbers of specimens available at the time, constituting almost "logical" errors but not affecting the morphological aspects.

He began his review with a study of the fossil birds that Lund had not described. We consider that work (Winge 1888) to have been the very first study of Brazilian fossil birds. He identified 126 species in Lund's collection and also made a comparative study with Reinhardt's (1870) work regarding the ornithological fauna of the "Brazilian fields." We can take that term to mean the *cerrado* (Brazilian savannahs). Winge's work is the starting point for the study of the Brazilian Pleistocene bird fauna and, consequently, of Lagoa Santa. There are very few posterior records of fossils for this class in the Lagoa Santa region (Alvarenga et al. 2008).

Right from the start, in the first volume of *E Museo Lundii*, Winge's (1888) study of the Rodentia is masterly. In the difficult and mutant universe of the Rodentia (due to the constant changes, especially in the nomenclature), he registered each fossil specimen against the cave where it was found (26 caves) and listed and identified both extinct and living species fossils that Lund had collected (51 and 31 species, respectively). The very act of distinguishing or determining extinct and extant species shows the mastery of a researcher whose only knowledge of Brazilian specimens came from collections or the literature, for he had never been in South America. Altogether, he recognized 13 fossil species and 16 living species as they had been identified by Lund. Today those numbers are probably lower due to the presence of synonyms in some of the revisions.

Meticulous descriptions, innumerable measurements, and a compendium of Families are present in this far-reaching work, together with a complete (for that time) bibliography. Ninety-two drawings of specimens distributed in eight plates completed the publication. Later in the chapter, an analysis will be made of the importance of that work insofar as it could serve as an inspiration for a new study of the Order.

The work continued with the production of the second volume of *E Museo Lundii*, in two fascicles dedicated to the orders Chiroptera and Marsupialia, respectively, and including fossil and contemporary species. Winge (1893) identified 27 fossil species of Chiroptera and 26 living species among the specimens Lund collected in Lagoa Santa and shipped to Denmark. According to Winge, among those species only one was new, *Stenoderma humerale* Lund, and it was not represented in the fossils. In fact, it was a species of the *Sturnira* genus. This work displayed 37 illustrations distributed in two plates of very high quality.

Among the Marsupial material, Winge identified 13 fossil species. He emphasized that when comparing them to the living species, which Lund had also collected in the region, there were no morphological differences or differences in composition. Most notable among them are *Didelphis marsupialis*, which Lund determined as *Didelphis albiventris*, and *Grymaeomys griseus*, which was named *Marmosops incanus* (Lund 1840). The work is enriched by 61 illustrations set out in four plates.

The third fascicle was published in 1895–1896 and dealt with the fossil primates and carnivores. Among the primate material, the author identified five fossil species of which four are still living species. Outstanding here is the study and illustration of the first record of a fossil Primate, *Protopithecus brasiliensis* (Lund 1838). This is an extinct species that Lund mistakenly attributed to a different genus. We will comment on this species later in the chapter.

In the part dedicated to carnivores, Winge mentions 21 fossil species, and he attributed the denomination of seven of them to Lund. Thirty-one illustrations are set out on eight plates, completing the magnificent work. Nowadays four of the species determined by Lund are considered to be extinct, and two are still living. Again, this aspect will be expanded on later.

In 1906, the first fascicle of the third volume was published, and in it Winge expanded certain themes that Lund had only addressed succinctly, namely, the Tayassuidae, Cervidae, Lamini, Mammutidae (mastodon), Equidae, and Tapiridae, or did not mention at all, such as the Litopterna and Toxodontia. Winge (1906) considered them all to be ungulates, which was not correct in the case of Litopterna, Toxodontia, and the mastodon. Once again, Winge produced excellent descriptions and a complete panorama of all that was known at the time about those groups. Distributed in nine plates, there are nine figures of *Palaeolama maior* (Lund. In Liais 1840), 12 of *Mazama gouazoubira* (Fisher 1814), four of *Brasiliochoerus stenocephalus*, three of *Equus (Amerhippus) neogeus* (Lund 1840), two of *Tapirus cristatellus* (Winge 1906), and four of *Stegomastodon waringi* (Holland 1920).

The second fascicle of the third volume was finally published in 1915, finalizing with a flourish this monumental publication. It is possibly the most profound volume of the collection. Winge (1915) himself wrote the magnificent texts, and in them, he displayed his mastery of a theme that was unusual for a European zoologist, namely, the Xenarthras. In addition to the meticulous and precise descriptions, there are abundant notes, bibliographic registrations, and observations concerning interrelations in this order. The 42 illustrations are up to the same high standard; 7 of them are of armadillo species, 8 of *Pampatherium humboldti* (Lund 1839), 4 of *Hoplophorus euprhractus* (Lund 1840), 4 of *Glyptodon clavipes* (Owen 1839) (= *Glyptotherium* sp?), 2 of *Nothrotherium maquinense* (Lund), 1 of *Mylodon robustus* (Owen 1842) (actually the figures displayed represent three species: *Ocnotherium giganteum* (Lund 1839), *Glossotherium* sp., and Megalonychidae still at the stage of being determined), 12 of *Catonyx cuvieri* (Lund 1839), and 4 of *Valgipes bucklandi* (Lund 1839).

The Second Half of the Twentieth Century

With the exception of some references made to the species that Winge determined, especially by Argentinean authors like Ameghino and Kraglievich, the next moment of studies targeting Lagoa Santa only occurred in the mid-twentieth century, and they were undertaken by the now extinct Minas Gerais Academy of Sciences. Its members, Harold Walter, Aníbal Mattos, Arnaldo Cathoud, and Josaphat Pena, carried out various excavations in caves and rock-shelters in the Lagoa Santa region (see Chap. 6 of this volume). Some of the notes they published refer to the discovery of the remains of a mastodon and of a bear. The various finds built up a collection which was later acquired by the Federal University of Minas Gerais and is now deposited in its Natural History Museum. Some of the specimens were studied by

Carlos de Paula Couto and to a lesser extent by Fausto L. Souza Cunha, both of the National Museum of Rio de Janeiro (see Chap. 8 of this volume).

Paula Couto is our second Lund. His production, brilliant in quality and quantity, is worthy of his stature as a human being. Regarding Lagoa Santa, he wrote various works over a period of more than 30 years. His paleontological work involving Lagoa Santa began in 1946 when he undertook a meticulous review of the nomenclature that Winge had used. Today, that work is out of date, but, at the time, it was an excellent work tool. In 1947, based on new material discovered by the Minas Gerais Academy of Sciences, he added to the knowledge concerning the glyptodont (*Hoplophorus euphractus*), which Winge (1915) had studied (Paula Couto 1947). Shortly after that he published his first book on the paleontology of Brazilian mammals, and much space in it was dedicated to the discoveries in Lagoa Santa (Paula Couto 1953).

Prior to that, one of the great works we owe to him had been published in the monumental work *Memórias sobre a Paleontologia Brasileira* (Memoirs on Brazilian Paleontology), an indispensable reference for that field in Brazil. It is a version in Portuguese of all of Lund's work, including 56 plates with the figures from the original publications. The introduction presented a summary of the lives of the naturalist Lund and of Winge. In addition to the technical index that facilitates consultation of the 590-page volume, there are comments and footnotes in which he updated nomenclature and concepts. That work alone would justify considering Paula Couto to be one of the leading South American paleontologists.

During the 1950s, Couto published a series of important papers regarding Lagoa Santa paleontology. In 1951, he described a new species for the region, the extinct rodent *Tetrastylus walteri* (Paula Couto 1951). There had been no identification of a new extinct species for Lagoa Santa since Winge had published his work.

Restudying *Pampatherium humboldti*, he proposed a new subfamily: the Pampatheriinae. Later (Paula Couto1955), he went on to study the saber-toothed tiger *Smilodon populator* (Lund 1842) and restudied the glyptodont (*Hoplophorus euphractus*), and, toward the end of the decade, he published a description of a scientific expedition to the Lagoa Santa area (Paula Couto 1970a). In this last work, he made observations about the karst, about the new material that he would be studying later, and included a discussion of the question of the coexistence of man and the extinct fauna species. At that time, he believed there were insufficient elements available to affirm that humans and the extinct fauna had been synchronic.

More works followed based on the material found in Pedro Leopoldo (Minas Gerais state) by members of the Minas Gerais Academy of Sciences, and Couto undertook a restudy of the bear species *Arctotherium brasiliense* (Lund 1839) on two occasions: in 1957 and 1960, with the second study being the more exhaustive of the two (Paula Couto 1957, 1960). There followed a publication summarizing the Paleontology of Lagoa Santa (Paula Couto 1970b, 1971a). Shortly after that, he went back to address the smallest of the Brazilian terrestrial sloth species *Nothrotherium maquinense* (Paula Couto 1971b).

Toward the end of the 1970s, his monumental work *Tratado de Paleomastozoologia* (*Paleomammal Zoology Treatise*) was published. In the opinion of the celebrated paleontologist G. G. Simpson, it is one of the most profound works synthesizing knowledge of fossil mammals ever published. All mammal species that had been identified up to that date were cited in the book, when the order they belonged to came under analysis (Paula Couto 1979).

Shortly after that, he published a paper on *Propraopus punctatus* (Lund 1839), based on the material he received during the abovementioned scientific expedition (Paula Couto 1980). A year before his death, he published and revalidated work on a Tayassuidae species that Lund had named *Brasilochoerus stenocephalus* in his manuscripts. Reinhardt (1880) had previously made a detailed description of that species, as has already been mentioned. It was one of Paula Couto's last works (Paula Couto 1981). A little before Masters' student Silvio Fonseca (1979), whom Paula Couto was tutoring, had produced an excellent dissertation based on a study of the Tayassuidae specimens in the Natural History Museum of the Federal University of Minas Gerais, material that had been collected by the Minas Gerais Academy of Sciences, as recorded earlier in the chapter.

There was a notable increase in publications referring to the fossil fauna of Lagoa Santa in the second half of the twentieth century. In the 1950s, French paleontologist Robert Hoffstetter published his doctoral thesis on the Pleistocene mammals of Ecuador (Hoffstetter 1952). In that work, there are frequent allusions to, and comparisons with, Megatheriidae, Pampatheridae, Equidae, Scelidotheriinae, and Lamini species that Lund and Winge had defined for Lagoa Santa.

It seemed that his doctoral research had aroused Hoffstetter's interest in the Xenarthra Pilosa of Lagoa Santa, and he wrote an article in which he drew a series of conclusions and introduced a set of errors regarding the group (Hoffstetter 1954). However, his conclusions were widely accepted for a considerable period and led some authors to produce wrong analyses of Brazilian species. Cartelle et al. (2009) contested almost all of Hoffstetter's proposals.

Hoffstetter had stated categorically that the cranium and dentition that Winge (1915) had considered to belong to the species *Catonyx giganteus* was actually of the species *Scelidotherium* (*Catonyx*) *cuvieri*, and what Winge (1915) had classified as the postcranial skeleton of *Catonyx giganteus* was actually a member of the Megalonychidae, *Ocnopus gracilis* (Lund). That interpretation is incorrect. In fact, two species were grouped together in the material that Winge (1915) identified as *Catonyx giganteus*, both belonging to the Scelidotheriinae. The figures of the cranium and dentition correspond to a species peculiar to Brazil, *Catonyx cuvieri* (Lund), while the postcranial material belongs to the species *Valgipes bucklandi* (Cartelle et al. 2009). Hoffstetter's proposal to consider *Nothrotherium maquinense* as a subgenus is also not accepted today.

Hoffstetter reassessed *Valgipes deformis* as being an indeterminate member of the Megalonychidae, based on an incomplete calcaneus described by Paul Gervais, as mentioned earlier in the text. Cartelle et al. (2009) identified the specimen as being the holotype of *Valgipes bucklandi,* as Lund had quite rightly determined. Thus, only the name of the genus needs to be corrected, and that attributed by P. Gervais is in fact valid, as we have stated.

Regarding the Mylodontidae, Hoffstetter identified *Glossotherium giganteum* (Owen 1842), attributing that denomination to specimens that Winge (1915) had identified as *Mylodon robustus* (Owen 1842). In fact, the teeth illustrated by Winge belong to the sloth species *Ocnotherium giganteum* (Lund 1839) and a cuboid bone with the same identification belongs to a large Megalonychidae (Winge 1915, plate XXVI, Fig. 6), which is currently being studied by the author of this chapter, as is a juvenile jawbone (Winge 1915, plate XXVI. Fig. 5), belonging to a *Glossotherium.*

When he was a paleontologist attached to the National Museum of Rio de Janeiro, Souza Cunha published two papers. The first approached the remains of *Hippidion principale*, discovered in the Lapa Mortuária of Confins by members of the Minas Gerais Academy of Sciences, which had been denominated as such by Lund (Souza Cunha 1960, 1964). In the second publication, Souza Cunha (1971) reassessed the holotype of another horse species which Lund had also defined, *Equus (Amerhippus) neogaeus*.

Later, Alberdi et al. (2003) took up the topic once more but in a broader perspective that took in not only morphological analyses but also the presence of *Equus (Amerhippus) neogeus* and *Hippidion principale* in other regions of Brazil.

Lund's initial view of the Brazilian fauna, which he was to abandon and replace by the different reality that he encountered inside the caves, was the current view in his day, namely, that the various animal species had suffered global destruction. That was in keeping with his religious belief in God the Creator. Accordingly, it can be understood why that predisposition led him to identify remains of a deer that he recovered in the Maquiné Cave as being those of an antelope, which is an old world animal. In fact, they were bones of a South American species of deer of the *Mazama* genus. That group of animals has never been the object of paleontological studies since Winge (1906).

Souza Cunha and Magalhães (1986) identified new deer material in Cerca Grande (municipality of Matozinhos), and Magalhães (1987) in his dissertation for a Master's degree made a review of the fossil cervids found in various Brazilian regions emphasizing the finds in Lagoa Santa, where he registered species of the following genera: *Antifer*, *Ozotoceros*, *Mazama*, *Blastoceros*, and, somewhat doubtfully, *Morenelaphus.*

In 1836, Lund made a highly important discovery. He hurried to send off a communication to Europe reporting his discovery of *Protopithecus brasiliensis* (Lund 1838). It was the first ever record of a fossil monkey. At the beginning of the nineteenth century, Cuvier had denominated a species that he discovered as *Adapis parisiensis* (Cuvier 1812), but it was only recognized as a primate in 1859. Among the primates, fossils of the *Homo* genus were already known, but they were largely ignored. The oldest discovery was the cranium of Cannstatt, discovered in 1700, and the skeleton of Paviland, recovered from a cave by the Englishman Buckland in 1822. The famous Neanderthal man discovery was much later, in 1856.

After Winge's (1895) publication of illustrations of Lund's discovery, it was practically forgotten. The find was only brought to light again by Hartwig (1995a). The title of his article says exactly what it is about: Protopithecus : rediscovering the first fossil primate. Later, Hartwig (1995b, 2002) made a detailed description of the material curated in the Copenhagen Zoology Museum. Hartwig and Cartelle (1996) heard about an almost complete skeleton of the species that had been recovered from a cave in Bahia and compared it with the material Lund had found. More recently, Halenar and Rosenberg (2013), disagreeing with the denomination, identified the specimen found in Bahia as being a new species, *Cartelles coimbrafilhoi* (Halenar and Rosenberg 2013).

Lund made notable efforts in his discovery of fossils of bats. He took a long time to find them, and from his writings we can deduce that his interest was aroused by the impression those animals made on him when he first entered the caves, where they were frequent and numerous inhabitants. In his homeland, which is entirely flat, he had never had the opportunity of entering a cave. As we have said, Winge (1893) identified 27 species of Chiroptera in the fossil material collected by Lund.

Czaplewski and Cartelle (1998), in their work on fossil species of Brazilian Chiroptera, described the pieces of the various specimens and provided a list of their occurrence in the various Brazilian states. They highlight the Lagoa Santa material where there have been no new finds of fossil species of this order since Lund's time. According to those authors, after a thorough review of the material preserved in Denmark, the specimens collected by Lund proved to contain the remains of 26 species. That was practically the same number of species that Winge (1893) had identified.

Lund dedicated his attention to the carnivores on various occasions and managed to establish several species. The most spectacular one was *Smilodon populator* (Lund 1842), the saber-toothed tiger. Another occurrence without precedent in mammal zoology was Lund's determination of an extinct carnivore species, which he denominated *Speothos pacivorus* (Lund 1839). He had heard from the local people about a rare, mysterious dog that lived in the caves. From what they told him, he suspected that it might be a new species. Finally, he managed to get hold of two specimens, one dead and the other alive, and he kept it in his yard and observed it for over a year. He had been right; the elusive animal was indeed a new species and he gave it the name of *Speothos venaticus* (Lund 1842), the vinegar dog. That living species was correctly included in the same genus as the extinct species he had discovered earlier.

Regarding the carnivore species Lund described, Berta (1984) expanded what was known about the canine species *Speothos pacivorus* (Lund 1839), and Cartelle and Langguth (1993, 1999) prepared a comprehensive description with various illustrations of another canine, *Protocyon troglodytes* (Lund 1840). Cartelle and Abuhid (1989) also registered new occurrences of the feline species *Smilodon populator*.

The large mammal remains that Lund found in the greatest quantities were of the Xenarthra, and he dedicated many pages to studying them, as did Winge (1915). In addition to the publications cited above, various other studies focusing on the discoveries of Lagoa Santa appeared in the second half of the twentieth century.

Cartelle and Fonseca (1983) described a very complete specimen of *Nothrotherium maquinense*. Up until then not enough had been known about that extinct species first discovered by Lund.

In their studies of material from the United States, Central America, and various South American countries, Cartelle and De Iuliis (1995) concluded that the species *Eremotherium laurillardi* (Lund 1842) was Pan-American, and at the end of the Pleistocene era, its territory extended from the US state of New Jersey to the Brazilian state of Rio Grande do Sul. They revalidated the species that Lund had denominated *Megatherium laurillardi*, using, for that purpose, two molariform teeth of a young specimen that Lund had found in the Lagoa Santa region. Guérin and Faure (2000) used the molariform teeth of a young individual that Lund had found to justify their proposal that the denomination of *Eremotherium laurillardi* should be attributed to a dwarf species of *Eremotherium* that lived at the end of Pleistocene epoch, while the larger species which Cartelle and De Iuliis (1995) had designated as *Eremotherium laurillardi* should be called *Eremotherium lundi*. The two denominations proposed by those authors are not sustainable. They are nomina nuda for reasons which will be set out below.

The teeth Lund found are of very young specimens. Cartelle and De Iuliis (2006) have described and illustrated the odontogenic development of the dentition of *E. laurillardi*. To be sure of their position, they examined over 900 teeth of that species coming from the same deposit. There can be no doubt that the molariform teeth Lund found are those of a very young individual of the *Eremotherium laurillardi* species. In addition to the quantity of molariform teeth that were available to them, they had an opportunity to study the crania and mandibles of young individuals with the molars implanted. Those specimens repeated the morphology that Guérin and Faure (2000) had used to justify the existence of a dwarf Megatheriinae species. In their justification, they described and illustrated a portion of the cranium, arguing that its small size was consistent with the existence of an individual of small size. Cartelle and De Iuliis (2006), however, not only demonstrated the differences between young individual's molariform teeth and those of adults but they also illustrated another fragment practically identical with the one Guérin and Faure (2000) used in their work.

The parts of crania described in the two works indubitably belong to very young individuals. The justification for that affirmation is very obvious. All adult mammal craniums, and obviously all Xenarthra, present their sutures closed. In the specimen illustrated by Guérin and Faure, just like the specimen illustrated by Cartelle and De Iuliis, it is easy to visualize the open sutures in the lacrimal bone, in the suture with the zygomatic arch, in the frontal bones, in the palatine suture of the maxillae, and between the maxilla and the palatine bone. In the light of those considerations, it is really inexplicable how specimens belonging to very young individuals of *E. laurillardi* came to be interpreted as representing a dwarf species of *Eremotherium* that never existed in the Brazilian Pleistocene fauna. It is a basic tenet in comparative anatomy that open sutures in a mammalian skeleton indubitably define the non-adult stage of an individual.

In their systematic review of the Brazilian Scelidotheriinae, Cartelle et al. (2009) identified all the mistakes committed by Hoffstetter (1954) in this respect as commented above. That author believed that one of the two species that Lund determined belonged to Megalonychidae, which Winge (1915) had denominated as

Scelidotherium magnum (Owen). Cartelle et al. (2009) revalidated the species that Lund had originally called *Scelidotherium bucklandi* by correcting the name of the genus. According to the rules of the International Code of Zoological Nomenclature, the correct name of the species should be *Valgipes bucklandi*. More information on this has already been supplied in an earlier paragraph.

Lund defined a species of Mylodontidae based on very few elements which were illustrated by Winge (1915, Plate XXVI, Figs. 1, 2, 3, and 4), namely, two caniniform teeth and one molariform tooth. Nowadays, the species is known as *Ocnotherium giganteum* (Lund). Many authors had attributed them to a large *Glossotherium* species. Cartelle and Lessa (1999) were able to confirm the validation of the species because they had access to a reasonably complete skeleton and because the pieces that Winge illustrated were singular and characteristic. Work is currently in course that includes a comparative description and the phylogenetics of the species, showing that, instead of belonging to the *Glossotherium,* it is actually of the *Lestodon* genus.

Lund collected appreciable amounts of the remains of the camelid *Palaeolama maior*. It is apparently a stranger to the Brazilian intertropical fauna of Late Pleistocene. Even stranger is that he did not attribute greater importance to those finds. At first, he believed that such pieces belonged to a camel. Perhaps the reason for his disinterest was that he lived in Lagoa Santa, where scientific news and specialized literature were both rarities. So how could he decide whether it was a camel or a llama species? He eventually opted for the latter due geographic considerations, but he had never had a Lamini skeleton available to him. An error of that nature on the part of someone who lived in South America would never have been understood in European scientific circles. I believe that it was responsible for the few allusions he made in his writings regarding that particular material. Winge's (1906) work, in which he correctly attributed the material to a Lamini species, revealed how Lund had collected many specimens of such species.

In the 1970s, after the research carried out by Laming-Emperaire and André Prous in Lapa Vermelha IV in the Lagoa Santa region, which led to the finding of "Luzia," the earliest Brazilian (see Chap. 7 of this volume), Souza Cunha and Guimarães (1978), Magalhães and Locks (1983), and Locks and Montenegro (1983) identified a series of small mammal species recovered from the same site, including a new rodent species.

For some unknown reason, one volume of *E Museo Lundii* dedicated to Lagoa Santa Osteichthyes, amphibians, and reptiles, in a similar way that preceding volumes had been specifically organized, was not published. Partly filling that gap, Hansen (2012) published an elegant edition of an addendum to *E Museo Lundi.* Earlier, at the very time when *E Museo Lundii* was being published, Lütken (1875) had written a valuable monograph on the fish species that Reinhardt had collected in the Velhas River during his stays in Lagoa Santa.

Most of the fossil fish finds that Lund discovered were from the Sumidouro Cave, and eight species are listed in the Addendum referred to above. Illustrations were produced of part of the cranium of a *Pseudoplatystoma corruscans* (Spix and Agassiz 1829), and the fin rays of *Pimelodus maculatus* (Lacépede 1803).

The only amphibian species registered in the Lund collection is *Ceratophrys aurita* (Raddi 1823), which was present in four caves. In the same addendum, there is a figure showing a well-preserved cranium collected from the Lapa da Escrivânia 5. Regarding extant species collected in Lagoa Santa, 18 species are registered.

Among the fossil reptiles collected by Lund, the author in question referred to two Chelonia specimens, one Crocodilia (1), one Amphisbaenia (2), four Lacertilia (10), and four snakes (29). According to Lund, those fossil species were collected in 13 caves, and there are various pieces that belong to the same species. The numbers in brackets refer to the number of extant species that Lund and Reinhardt collected in the Lagoa Santa region. There were also illustrations of a fragment of the mandible and of several osteoderms of the species *Caiman latirostris* (Daudain 1802).

In addition to those records and descriptions of fossil species, the second part of the work consists of a list of 25 rodents, 3 carnivores, and 1 bird. They were illustrated by excellent images and holotypes of some of those species. It is, without doubt, a publication that revalues the findings from the Lagoa Santa region.

Other Viewpoints

Other approaches, different from those that have been described so far, placed their emphasis on taphonomic aspects. After a long period of work in a cave in the municipality of Matozinhos in the Lagoa Santa region, Perônico and Araújo (2002), using archaeological excavation techniques, managed to recover around 100 pieces of the Late Pleistocene fauna and made an intense study of the taphonomic phenomena observed in the cave.

Taking advantage of the experience gained in excavating the Cuvieri Cave (municipality of Matozinhos) during the project that was unfolded in the course of several years by Walter Neves and his team (see Chap. 9 of this volume), Hubbe (2008) dedicated his Master's dissertation to a taphonomic analysis of the findings from that cave, as well as their taxonomic and osteometric aspects. The fossil findings consisted of specimens of the terrestrial sloth *Catonyx cuvieri*, the tapir *Tapirus terrestris* (Linnaeus 1758), and the cervid species *Mazama gouazoubira* (G. Fisher 1814), among others. Mayer et al. (2009a, b) and Mayer (2011) also studied the sedimentary processes and the findings recorded for the same cave.

More recently, Vasconcelos (2014) described specimens and identified taphonomic processes observed in sediments that he collected in a small cavity in the Escrivânia complex, which Lund had once investigated. Neves and Piló (2003) and Hubbe et al. (2009) have provided the absolute dates for specimens from the region. They were probably the very first direct dates of Lagoa Santa fauna.

The region has also been the object of a series of more general, schematic views of the fauna, which we can refer to as "generalist" works. In that category, we can consider Lund's lists and the rectification that Paula Couto (1946) made of Winge's nomenclature, both mentioned previously in this text. In the same category, we can include the indispensable publication of Lund's work in Portuguese, commented by

Paula Couto (1950). In it can be found innumerable notes that update the systematics in use at the time, as mentioned above. The first of those works that set out to list the Lagoa Santa fauna was undertaken by Marshall et al. (1984), who included the fauna in the broader context of other sets of South American Pleistocene fauna.

Cartelle (1994) identified the fossil mammal fauna of other localities in Minas Gerais, in addition to the species found in the area that Lund researched. A little later, Cartelle et al. (1998), in a far-reaching report produced to justify the creation of the Lagoa Santa Karst Environmental Protected Area (which up until today has not been completely implanted), plotted and listed all the caves and their respective findings. With very few exceptions, they were all caves that Lund had explored.

Another relevant item under this topic was the analysis of the extinct fauna of the Cerrado and the Caatinga biomes carried out by Cartelle (1999), which included the Lagoa Santa region. In it, the author defended the hypothesis that the main cause of extinction at the end of the Pleistocene was climate change, and it especially affected the megafauna of grazing mammal species. Examples of the fauna that Lund had discovered provided strong support for that hypothesis.

A new publication of Cartelle expanded and deepened his 1994 work (Cartelle 2012). It places special emphasis on the Lagoa Santa area, the person of Lund himself, and, obviously, the fauna that he discovered. In that work, Lund's illustrations and descriptions were especially highlighted, and a list of Pleistocene mammal species for the state of Minas Gerais was included in the publication.

Lastly, and understandably, some publications referring to Lund's life, or some aspect of it, and his discoveries must be mentioned. Among them are the works of Marchesotti (2005) and Luna Filho (2007). Recently, Portuguese language versions have been published of two excellent books that were first published in Danish. In the first one, Holten and Sterll (2011), basing themselves on a large number of documents, have reported on their significant research work, portraying the life of Lund and the story of his discoveries. In the second, Holten et al. (2012) set out information on the life and work of Lund's main collaborator, his Norwegian secretary and artist, Peter Andreas Brandt.

Lund's abandonment of his research efforts has been the object of much speculation (Luna Filho 2007). I believe the explanations can be summed up in two underlying principles: the findings began to be repetitive, and the expeditions involved very high costs for Lund, after the assistance from Denmark ended. Furthermore, his concern not to incur in scientific errors must have weighed heavily on him, given his intellectual isolation in Lagoa Santa, and he must also have been worried about the consequences or the criticism that would stem from the irrefutable abandonment of Cuvier's catastrophism that his work had brought about.

Deciding to send his entire collection to Denmark relieved him of his main motivation for carrying on with the research. At the same time, new winds were blowing through Europe in the field of biological interpretations. Lund felt himself to be marginalized from the scientific world by his isolation in Lagoa Santa and that most certainly made him feel insecure. Discovering that his original scientific base, founded on the thinking of Cuvier (Hoch 1982), was bankrupt made him very uncomfortable. His undeniable certainty that he had discovered synchronic fossils of men, certain extinct species, and other similar extant species all in the same Sumidouro Cave came as a great shock to him and obliged him to review and revise almost everything that he had previously published. In that situation, he preferred to withdraw from the field. It was a courageous and understandable act and perhaps the only thing he could have done in the circumstances.

Future Prospects

To my mind, there are two examples of actions that Lund undertook that point to a line of research that has never been followed up on. Based on his second scientific memoir concerning the cave fauna, Lund (1839) began keeping records of mammal species in the Lagoa Santa region, which I consider have not been valued as they should be. It was the starting point for the production of posterior publications of lists, which led to the first *fide digno* survey of the mammal fauna ever made in Brazil. Today it could be considered worthy of imitation.

Those early lists, which he expanded in each subsequent publication, were evidence of a zoologist of rare talent, not only in the aspect of the growth of his acquired knowledge on the extinct fauna but also of his profound mastery of the extant species in the region. In those lists he displayed his admirable capacity for recognizing extinct fossil species and extant species among the pieces he collected in the caves of the region.

Before him, there was nothing known about the living regional fauna species other than sporadic observations made by traveling naturalists. Those lists were surprising because they not only revealed Lund's scientific competence but, as mentioned above, were also the first ever far-reaching survey of a Brazilian regional fauna.

The second example of his actions was the collection Lund made over a period of months in a cave of the Escrivânia complex. The ground was covered with a thick layer of the remains of small animals as a result of the regurgitation of a pair of cave owls that lived there. His observations led him to believe that each cave was inhabited by a single pair of owls and that the cave must have been occupied by a continuous succession of pairs. Lund, accordingly, kept a pair of owls in captivity to observe and calculate their daily consumption. After a painstaking counting of the enormous quantity of tiny pieces collected, he arrived at highly interesting conclusions. He calculated that somewhere around seven million such animals must have been transported into the cave by the succession of owls. Many of their remains had probably not survived the passage of time. The research led him to conclude that, to account for those deposits of bones, around 5,000 years must have passed since the first pair of owls had moved in. He obtained that result at a time when most people believed that the age of the Earth itself was somewhere around 7,000 years.

Neither the undervalued lists of extant mammal species set out in Lund's works with due systematic modifications nor the micro-mammal material stemming from the cave owls' regurgitation, which is deposited in the Copenhagen Zoology

Museum, has been duly explored by the scientific community. The number of tiny pieces of micro-vertebrates preserved in that museum is somewhere around two million specimens. That goes far beyond any other fauna survey that could possibly be made today. I do believe that it is a piece of research which, if it were to use such a complete and broad sample of an area, could hardly be carried out anywhere else. Lagoa Santa alone possesses the necessary elements for conducting such research. Carrying out that proposal would be no easy task because it would require financing for researchers dedicated to the work for a long period, an equation that is hard to balance.

That study could consist of:

(a) The systematization of Lund's lists; they probably could be reinforced by collecting specimens of the regional fauna that Lund and Reinhardt also collected more than 150 years ago, which are deposited in the Copenhagen Zoological Museum (CZM).
(b) The Herculean task of identifying and quantifying the species that occur in the enormous sample of micro-mammals in the CZM collection and which portray the presence and proportion of a type of fauna that probably occurred in the course of thousands of years.
(c) A far-reaching and judicious survey that would make it possible to get an exact idea of the microfauna still present in the region.
(d) An analysis of the entire assemble that could make it possible to draw important conclusions not only from the zoological point of view but also from the environmental one and, in addition, could correct and indicate new directions for the preservation of the fauna in that important region. Such work embraces three historical moments that would be almost impossible to find anywhere else, namely, specimens from the Late Pleistocene, others from 150 years ago, and those of the present day.

We believe that direct paleontological research, that is, by conducting excavations would merely lead to repetitive conclusions in regard to those carried out by Lund. It would be unlikely to come up with anything new regarding species found in the region. Today, the species discovered in the Lagoa Santa region are well known due to finds in other regions, with very few exceptions like *Propraopus sulcatus* (Lund 1842) and *Propraopus punctatus* (Lund 1839). The very limited amount of material found leaves the validity of those two species in doubt. One of them could be synonymous with the other. As mentioned, since Lund's time, only one new extinct mammal species has been identified in material collected from Lagoa Santa (the rodent *Tetrastylus walteri*).

It has been a long time since the last review of the Lund collection, and making one now could add new facts that have previously gone undetected. As an example, we related in previous paragraphs that we identified a cuboid bone that Winge (1915) considered to be of *Glossotherium robustum* (Owen 1842). The bone, however, actually belonged to a very large Megalonoquidius, and it is currently being studied by a team at the Natural Sciences Museum of the Minas Gerais Catholic University. Lund found it in the Escrivânia I Cave. It is quite possible that, in the rest of the material collected from that cave, there may be another specimen belonging to the same species. That fact alone justifies this proposal for a future research into the Lund collection. It must be underscored that the last review of the collection targeting Pleistocene mammals was made by Hoffstetter (1954), who introduced a series of errors, as has already been explained.

Another fact that deserves special attention and specific studies is the extinct mega-mammal material that Lund discovered. There is a certain "air of mystery" associated to the findings of this kind of fauna, and it deserves to be cleared up and interpreted, as will be explained below.

In certain regions of the Central-Northeast Brazil, especially on the borders of the states of Bahia and Minas Gerais and in municipalities in the north of Minas Gerais (e.g., Janaúba, Manga, Pote, and Montes Claros), findings of extinct herbivorous mega-mammals have been more frequent. Among the species are *Stegomastodon waringi*, *Xenorhinotherium bahiens* Cartelle and Lessa, 1988, *Toxodon platensis* Owen, 1837, and *Eremotherium laurillardi*. Of the species *Stegomastodon waringi*, Lund only managed to find fragments of teeth, a juvenile cranium, and some molars. In the case of the Macraucheniidae (*Xenorhinotherium bahiens*?), the findings consisted of two fragments of vertebrae, the distal extremity of a femur, and six dental fragments. For *Eremotherium laurillardi,* as already stated, Lund's findings consisted of just a pair of juvenile molariform teeth.

In the extensive and constant collecting done by Lund in hundreds of caves, the findings of extinct herbivorous mega-mammals corresponded to *Tapirus cristatellus* and *Palaeolama maior*, various species of terrestrial sloths like *Valgipes bucklandi* and *Catonyx cuvieri,* and equine species, mainly *Hippidion principale* and *Equus (Amerhippus) neogeus, Neocheorus sulcidens* (Lund 1839), *Pampatherium humboldti*, and *Hoplophorus euphracthus.* However, the larger species mentioned in the preceding paragraph, with numerous records of occurrence in regions beyond Lagoa Santa, are minimally represented in the lists of the 12,622 pieces that appear in Lund's handwritten catalogue.

Lund only had two teeth of adult *Eremotherium laurillardi*, supplied by Claussen and collected in an area outside of the Lagoa Santa region, and two other molariform teeth of a young specimen as mentioned earlier. Those molariform teeth are so characteristic that they were sufficient for Cartelle and De Iuliis (1995) to revalidate the species as *Eremotherium laurillardi.* Lund had classified the species as *Megatherium laurillardi* based on two molariform teeth that he believed to have belonged to an adult species of *Megatherium*, which his intellectual mentor Georges Cuvier had determined in 1796.

To date, we have no explanation as to why there were such minimal findings of those mega species in the Lagoa Santa region. The scarcity of findings of those species cannot be attributed to restricted collections or to the extension of the area that was researched. In other intertropical regions of Brazil, specimens have been collected in quantities equivalent not only to those of the mega species collected by Lund but also quantities of specimens of those that were practically not collected in Lagoa Santa at all. What could be the reason for that parsimony regarding the species we have referred to? What causes could possibly explain the very small number of records while in other nearby regions the representation of those species is extensive? What environmental circumstances are behind that scarcity?

Another task that needs to be undertaken concerns the collections traded by Peter Claussen with the Natural History Museum of Paris and especially with the British Museum. Claussen was Danish, just like Lund, and he worked with the German paleontologist Friedrich Sellow, who had taken part in the Langsdorff expedition. After Sellow's death by drowning in the Rio Doce, Claussen established himself in Curvelo (Minas Gerais) around the year 1840. Mining saltpeter from the caves in the region around Lagoa Santa, he came across and collected fossils. Many were sold to the abovementioned museums, especially the British Museum. Lydekker (1887) identified some such pieces, but after that record made over 125 years ago practically, none of them has been reidentified or redescribed. The "rediscovery" of those specimens could very well expand the knowledge we have of the Pleistocene fauna of Lagoa Santa.

The proposals that have been put into effect in the past have made Lagoa Santa a fertile and perhaps inexhaustible field for paleontological research of the kind we have referred to. However, there are many other correlated fields to be entered into, such as those related to chronology, taphonomy, analysis of sediments, and speleology.

The question must be raised as to whether there is still time left to actually put into effect those research proposals. The pressures on that fragile area increase daily. The protective measures adopted to date have not restrained the frenetic settlement of the area. Such pressures take the form of deforestation, air and water pollution, mining, cement manufacturing plants, illegal occupation of the banks of rivers and lakes, the highway bypass project, the negative influence of the Confins Airport, inadequate sanitation, various industries, etc. The future seems to be very short, a whole lot shorter than the past.

References

Alberdi, M. T., Cartelle, C., & Prado, J. L. (2003). El registro Pleistoceno de *Equus* (*Amerhippus*) e *Hippidion* (Mammalia, Perissodactyla) del Brasil: consideraciones paleoecologicas y biogeograficas. *Ameghiniana, 40*, 173–196.

Alvarenga, H., Brito, G., Migotto, R. L., Hubbe, A., & Höfling, E. (2008). *Pleistovultur nevesi* gen. et sp. nov. (Aves: Vulturidae) and the diversity of condors and vultures in the South American Pleistocene. *Ameghiniana, 45*, 613–618.

Berta, A. (1984). The pleistocene bush dog *Speothos pacivorus* (Canidae) from the Lagoa Santa caves, Brazil. *Journal of Mammalogy, 65*, 549–559.

Boas, J. E. V. (1881). Om en fossil Zebra-Form fra Brasilienes Campos. Med. et Tillaeg om to Arter af Salaegten Hippidion. *Det Kongelige Danske Videnskabernes Selskabs Skrifte Naturvidenskabelig og Mathematisk Afdeling, 6*, 305–330.

Burmeinster, H. G. (1885). Berichtigung zu Coelodon. *Sitzungsberichte der Königlich Preussischen Akademie der Wissenschaften zu Berlin, 2*, 567–573.

Burmeinster, H. G. (1886a). Weitere Bemerkungen über Coleodon. *Sitzungsberichte der Königlich Preussischen Akademie der Wissenschaften zu Berlin, 1*, 357–358.

Burmeinster, H. G. (1886b). Nochmalige Berichtigung zu Coelodon. *Sitzungsberichte der Königlich Preussischen Akademie der Wissenschaften zu Berlin, 2*, 1127–1132.

Burmeinster, H. G. (1887). Neue Beobachtungen na Coelodon. *Sitzngsberichte der Königlich Preussischen Akademie der Wissenschaften zu Berlin, 2*, 857–862.
Cartelle, C. (1994). *Tempo passado: mamíferos do pleistoceno em Minas Gerais*. Belo Horizonte: Palco/Acesita. 131p.
Cartelle, C. (1999). Pleistocene mammals of the Cerrado and Caatinga of Brazil. In J. F. Eisenberg & K. H. Refford (Eds.), *Mammals of the neotropics* (pp. 27–46). Chicago: The University of Chicago Press.
Cartelle, C. (2012). *Das grutas à luz: os mamíferos pleistocênicos de Minas Gerais*. Belo Horizonte: Editora Bicho do Mato. 236p.
Cartelle, C., & Abuhid, V. (1989). Novos espécimes brasileiros de *Smilodon populator* (Lund, 1842), Carnívora, Machairodontinae: Morfologia e conclusões taxonômicas. *Anais do Congresso Brasileiro de Paleontologia, XI*, 607–620.
Cartelle, C., & De Iuliis, G. (1995). *Eremotherium laurillardi* the panamerican late Pleistocene megatheriid sloth. *Journal of Vertebrate Paleontology, 15*, 830–841.
Cartelle, C., & De Iuliis, G. (2006). *Eremotherium laurillardi* (Lund) (Xenarthra, Megatheriidae), the Panamerican giant ground sloth: Taxonomic aspects of the ontogeny of skull and dentition. *Journal of Systematic Palaeontology, 4*, 199–209.
Cartelle, C., & Fonseca, J. S. (1983). Contribuição ao melhor conhecimento da pequena preguiça terrícola *Nothrotherium maquinense* (Lund) Lydekker, 1889. *Lundiana, 2*, 11–48.
Cartelle, C., & Langguth, A. (1993). *Protocyon troglodytes* (Lund): um canídeo intertropical extinto. *Resumos do Congresso Brasileiro de Paleontologia, XIII*, 141.
Cartelle, C., & Langguth, A. (1999). *Protocyon Troglodytes* (Lund): um canídeo intertropical extinto. *Anais da Academia Brasileira de Ciências, 71*, 371–384.
Cartelle, C., & Lessa, G. (1999). Um Lestodontini intertropical do pleistoceno final brasileiro. XV Jornadas Argentinas de Paleontologia de Vertebrados. La Plata. Resúmenes:6.
Cartelle, C., Abuhid, V., & Ferreira, M. A. (1998). Levantamento paleontológico. *Série APA Carste de Lagoa Santa, 3*, 11–48.
Cartelle, C., De Iuliis, G., & Ferreira, R. L. (2009). Systematic revision of intertropical Scelidotheriinae (Xenarthra, Mylodontidae). *Journal of Vertebrate Paleontology, 29*, 555–566.
Czaplewski, N. J., & Cartelle, C. (1998). New records of Pleistocene bats from cave deposits in Bahia, Brazil. *Journal of Mammalogy, 79*, 748–803.
Fonseca, J. S. (1979). *Taiassuídeos do pleistoceno de cavernas calcárias de Minas Gerais*. Master Thesis, Universidade Federal do Rio Grande do Sul.
Gervais, P. (1867–9). *Zoologie et Paleontologie générales. Nouvelles recherches sur les vertebrés vivants et fossiles*. Première série, Lams 1–50, A. Bertrand, Paris, figs 1–41, pp. VII + 1–263.
Gervais, P. (1874). Lestodon trigonidens et Valgipes deformis. *Journal de Zoologie, 3*, 161–164.
Gervais, P. (1877). Remarques ostéologiques au sujet du pied des édentés. *Journal de Zoologie*, Lam. 2–4, Paris 6:79–82, 198–228.
Guérin, C., & Faure, M. (2000). La veritable nature de *Megatherium laurillardi* Lund, 1842 (Mammalia, Xenarthra) un nain parmi les géants. *Geobios, 33*, 475–488.
Halenar, L. B., & Rosenberg, A. L. (2013). A closer look at the "Protopithecus" fossil assemblage: New genus and species from Bahia, Brazil. *Journal of Human Evolution, 65*, 374–390.
Hansen, L. (2012). *E Museum Lundii addendum*. Copenhagen: Natural History Museum of Denmark.
Hartwig, W. C. (1995). A giant new world monkey from the pleistocene of Brazil. *Journal of Human Evolution, 28*, 189–196.
Hartwig, W. C. (1995). *Protopithecus*: Rediscovering the first fossil primate. *History and Philosophy of Life Sciences, 17*, 447–460.
Hartwig, W. C. (2002). *A possible Giant Muriqui from the pleistocene deposits of Lagoa Santa, Minas Gerais, Brazil*. Chicago: Field Museum Publications/National Zoological Park.
Hartwig, W. C., & Cartelle, C. (1996). A complete skeleton of the giant South American primate Protopithecus. *Nature, 381*, 307–311.
Hoch, E. (1982). The influence of Georges Cuvier on the danish naturalist Peter Lund, "Father of Brazilian Paleontology". *Actes du Symposium Paleontologique Georges Cuvier* (pp. 273–287).

Hoffstetter, R. (1952). Les mammifères pléistocénes de la R'epublique del'Equateur. *M'emoires de la Société Géologique de France, Nouvelle Série, 31*, 1–391.

Hoffstetter, R. (1954). Les gravigrades (Edentés Xénarthres) des cavernes de Lagoa Santa (Minas Gerais, Brasil). *Annales des Sciences Naturelles – Zoologie et Biologie Animale, 16*, 741–764.

Holten, B., & Sterll, M. (2011). *Peter Lund e as Grutas com ossos em Lagoa Santa*. Belo Horizonte: Editora UFMG.

Holten, B., Sterll, M., & Fjeldsa, J. (2012). *O artista desaparecido. P.W. Lund e P. A. Brandt no Brasil*. Belo Horizonte: Editora UFMG.

Hubbe, A. (2008). *Contextualização taxonômica, tafonômica e morfométrica dos remanescentes ósseos da megamastofauna da gruta Cuvieri (MG), um sítio paleontológico do Pleistoceno tardio*. Master Thesis, Universidade de São Paulo.

Hubbe, A., Hubbe, M., & Neves, W. (2009). New late-pleistocene dates for the extinct megafauna of Lagoa Santa, Brazil. *Current Research in the Pleistocene, 26*, 154–156.

Liais, E. (1872). *Climats, Géologie et Geographie botanique au Brèsil*. Paris: Garnier.

Locks, M., & Montenegro, E. (1983). Contribuição à paleontologia do "Grande abrigo da Lapa Vermelha Emperaire", Pedro Leopoldo. Um novo caviíneo quaternário (Mammalia, Rodentia). *Resumos das Comunicações do Congresso Brasileiro de Paleontologia VIII*.

Luna, P. E. (2007). *Peter Wilhelm Lund: o auge das suas investigações científicas e a razão para o término das suas pesquisas*. Ph.D. Dissertation, Universidade de São Paulo.

Lund, P. W. (1839). Blik paa Brasiliens Dyreverden för Sidste Jordomvaeltning. Anden Afhandling: Patterdyrene. *Det Kongelige Danske Videnskabernes Selskbas Naturvidenskabelige og Mathematiske Afhandlinge, 8*, 61–144.

Lütken, C. (1875 [2010]). *Peixes do Rio das Velhas. Passado e Presente*. Projeto Manuelzão, Belo Horizonte.

Lütken, C. (1886). Antikritiske bemaerkninger i anledning af kaempe-sovendyr-alaegten coelodon. *Oversigt over det Kongelige Danske Videnskabernes Selskabs Forhandliger og dets Medlemmers Arbeider i Aarte, 2*, 78–84.

Lydekker, R. (1887). *Catalogue of the fóssil Mammalia in the British Museum Natural History*. London: British Museum of Natural History.

Magalhães, R. M. (1987). *Os cervideos quaternários das Regiões Nordeste, Sudeste e Sul do Brasil (Cervidae, Odocoileinae)*. Master Thesis, Universidade Federal do Rio de Janeiro.

Magalhães, R. M., & Locks, M. (1983). Contribuição à paleontologia do "Grande abrigo da Lapa Vermelha Emperaire", Pedro Leopoldo, MG. Sobre os cervídeos quaternários (Mammalia Artiodactyla). *Resumos das Comunicações. Congresso Brasileiro de Paleontologia VIII*.

Marchesotti, A. A. P. (2005). *Peter Wilhelm Lund (1801–1880): O naturalista, sua rede de relações e sua obra, no seu tempo*. Master Thesis, Universidade Federal de Minas Gerais.

Marshall, L. G., Berta, A., Hoffstetter, R., Pascula, R., Reig, O. A., Bombim, M., & Mones, A. (1984). Mammals and stratigraphy: Geocronology of the continental mammal bearing quaternary of South America. *Paleovertebrata, Mémoire Extraordinaire, 1984*, 1–76.

Mayer, E. L. (2011). *Processo de formação de um depósito fossilífero em abismo na Gruta Cuvieri (MG): taxonomia, tafonomia e distribuição espacial de fósseis de mamíferos do Pleistoceno*. Master Thesis, Universidade de São Paulo.

Mayer, E. L., Hubbe, A., & Neves, W. (2009). Novo registro de pacas extintas (Rodentia, Cuniculidae) do Pleistoceno Terminal de Lagoa Santa, Minas Gerais. *Anais do Congresso Brasileiro de Paleontologia, XXI*, 196.

Mayer, E. L., Hubbe, A., & Neves, W. (2009). Processos de acumulação óssea da mesomastofauna do Pleistoceno Terminal da Gruta Cuvieri, Lagoa Santa, Minas Gerais. *Anais do Congresso Brasileiro de Paleontologia, XXI*, 197.

Neves, W. A., & Piló, L. B. (2003). Solving Lund's dilemma: New AMS confirm that humans and megafauna coexisted at Lagoa Santa. *Current Research in the Pleistocene, 20*, 57–60.

Paula Couto, C. (1946). Atualização da nomenclatura genérica e específica usada por Herluf Winge em "E Museo Lundii". *Estudos Brasileiros de Geologia, 1*, 59–80.

Paula Couto, C. (1947). Contribuição para o estudo de *Hoplophorus euphractus* Lund, 1839. *Summa Brasiliensis Geologiae, 4*, 33–53.

Paula Couto, C. (1950). *Memórias sobre a paleontologia brasileira*. Rio de Janeiro: Instituto Nacional do Livro.

Paula Couto, C. (1951). Uma espécie do gênero *Tetrastylus* Ameghino, 1886, em Lagoa Santa. *Boletim do Museu Nacional: Geologia, 15*, 1–16.

Paula Couto, C. (1953). *Paleontologia brasileira. Mamíferos*. Rio de Janeiro: Instituto Nacional do Livro.

Paula Couto, C. (1955). O "Tigre dente de sabre" no Brasil. *Boletim do Conselho Nacional de Pesquisas, 1*, 1–30.

Paula Couto, C. (1957). Sobre um Gliptodonte do Brasil. *Boletim Divisão de Geologia e Mineralogia, 165*, 1–37.

Paula Couto, C. (1960). Um urso extinto do Brasil. *Boletim da Sociedade Brasileira de Geologia, 9*(1), 5–27.

Paula Couto, C. (1970a). Paleontologia da região de Lagoa Santa, Minas Gerais, Brasil. *Boletim do Museu de História Natural, 1*, 1–21.

Paula Couto, C. (1970b). Lund e a paleolontologia do Brasil. *Boletim Informativo do Centro de Informação Arqueológica, 2*, 5–13.

Paula Couto, C. (1971a). Mamíferos fósseis das cavernas de Minas Gerais. *Espeleologia, 3*, 1–14.

Paula Couto, C. (1971b). On two small Pleistocene ground-sloths. *Anais da Academia Brasileira de Ciências, 43*, 499–513.

Paula Couto, C. (1979). *Tratado de paleomastozoologia*. Rio de Janeiro: Academia Brasileira de Ciências.

Paula Couto, C. (1980). *Propraopus punctatus* (Lund, 1840) no Pleistoceno de Cerca Grande, Minas Gerais. *Anais da Academia Brasileira de Ciências, 52*, 323–325.

Paula Couto, C. (1981). On an extinct peccary from the Pleistocene of Minas Gerais. *Iheringia: Série Geologia, 6*, 75–78.

Perônico C., & Araújo, A.C.S. (2002). Diversidade faunística baseada em escavação tafonomicamente orientada de material subrecente preservado em caverna da região cárstica de Lagoa Santa, Minas Gerais. Arquivos do Museu Nacional. *6*(3), 195–198.

Reinhardt, J. (1868). The bones caves of Brazil and their animal remains. *American Journal of Science and Art, 96*, 264–265.

Reinhardt, J. (1870). Bidrag til Kundskab om Fuglefannaeu i Brasiliens Campos. *Videnskabelige Meddelelser fra deu Naturhistoriske Forening*. Coopenhague.

Reinhardt, J. (1875). De i Brasil ens knoglehuler fundne Glyptodont – Levninger og em ny, til de gravigrade Edentater hörende slaegt. *Videnskabelige Meddelelser fra deu Naturhistoriske Forening, 9–15*, 165–236. Copenhague.

Reinhardt, J. (1878). Kaempedovendyr-Slaegten Coelodon Vidensk: Det kongelige danske Videnskabernes Selkab Skriffter. *Natturvidenskabelig og Mathematisk Afdelingh, 12*, 253–350.

Reinhardt, J. (1880). *Naturforskeren Peter Wilhelm Lund, hans Liv og hans Virksomhed*. Cobenhavn. Separata de *Vindensk. Medd. Naturh. Foren.*

Reinhardt, J. (1888). De brasilianske Knoglehuler og de i dem forekommende Dyrelevninger. *E Museo Lundii, 1*, 1–49.

Souza Cunha, F. L. (1960). *Sobre o Hippidion da Lapa Mortuária de Confins, Lagoa Santa, Minas Gerais*. Rio de Janeiro: Editorial Sul-Americana.

Souza Cunha, F. L. (1964). A Lapa Mortuária de Confins – Lagoa Santa. *Revista Brasileira de Arqueologia, 1*, 30–41.

Souza Cunha, F. L. (1971). Sobre os holótipos de *Equus (Amerhippus) neogaeus* Lund, 1840 e *Equus (Amerhippus) curvidens* Owen, 1844. *Anais da Academia Brasileira de Ciências, 43*(Suplemento), 619–627.

Souza Cunha, F. L., & Guimarães, M. L. (1978). A fauna sub-recente de vertebrados do "Grande abrigo da Lapa Vermelha" de Pedro Leopoldo, Minas Gerais. *Arquivos do Museu de História Natural, 3*, 201–231.

Souza Cunha, F. L., & Magalhães, R. M. (1986). A fauna de vertebrados sub-recentes de Cerca Grande, Matozinhos, MG. *Publicações Avulsas do Museu Nacional, 65*, 119–131.

Vasconcelos, A. G. (2014). *Mamíferos quaternários da cavidade ES-08, município de Prudente de Morais, Minas Gerais: análise tafonômica e taxonômica*. Master Thesis, Universidade Federal de Minas Gerais.
Winge, H. (1888). Jorfundne og nulevende Gnawere (Rodentia) fra Lagoa Santa, Minas Gerais, Brasiliens: med udsigt over gnavares indbyrdes slaegtskab. *E Museo Lundii, 1*(3), 1–200.
Winge, H. (1893). Jorfundne og nulevende Gnawere (Marsupialia) fra Lagoa Santa, Minas Gerais, Brasilien: med udsigt over gnavares pungdyrdes slaegtskab. *E Museo Lundii, 2*(2), 1–149.
Winge, H. (1895). Jorfundne og nulevende Gnawere (Primates) fra Lagoa Santa, Minas Gerais, Brasilien: med udsigt over abernes indyrdes slaegtskab. *E Museo Lundii, 2*(3), 1–57.
Winge, H. (1906). Jorfundne og nulevende Hovdyr (Ungulata) fra Lagoa Santa, Minas Gerais, Brasilien: med udsigt over hovdyrenes indbyrdes slaegtskab. *E Museo Lundii, 3*(1), 1–239.
Winge, H. (1915). Jorfundne og nulevende Gumlere (Edentata) fra Lagoa Santa, Minas Gerais, Brasilien: med udsigt over rovdyrenes indbyrdes slaegtskab. *E Museo Lundii, 3*(2), 1–321.

Chapter 15
The History of the Studies of Prehistoric Rock Paintings in the Lagoa Santa Karst

Alenice Baeta and André Prous

Abstract This chapter addresses the rock art in the Lagoa Santa region. First, it provides an outline of the history of research and the methodologies employed. There is a presentation of general patterns and aspects of the graphisms related to the insertion of rupestrian sites in the karst landscape and of the surfaces selected inside the rockshelters. The chapter also addresses the documentation and diffusion of information concerning rupestrian record in addition to an analysis of the most common and less common stylistic traditions and reflections on the need to conserve and value more highly this kind of archeological heritage.

The Lagoa Santa region became internationally well -known as far back as the first half of the nineteenth century19th century because of the research carried out by the Danish naturalist Peter Wilhelm Lund (see Chaps. 2 and 3 of this volume), who installed himself there permanently, stimulated by the richness of the landscape adorned with innumerable caves and the presence of rare species of plants and animals. For practically a century and a half, archaeological research, there was limited to excavations in search of the remains of either lithic industries or skeletons as ancient as possible. It was only in the 1970s that systematic surveys began to be made of the graphic record (usually referred to as rock art paintings), and 58 sites with pictogramms and/or petroglyphs have been inventoried for the Lagoa Santa Karst to date. It should be noted that rock paintings and engravings made on the natural rock faces are not the only prehistoric graphic records, as there are also other paintings on pottery, clay figures, stone sculptures, and so on. However, in this chapter, our interest will be limited to the graphic manifestations found on the rock walls.

A. Baeta (✉) • A. Prous
Archeology Sector of the Museum of Natural History and Botanical Gardens of the Federal University of Minas Gerais (MHNJB/UFMG), Rua Gustavo da Silveira, 1.035, Santa Inês, Belo Horizonte, MG, Brazil
e-mail: alenicebaeta@yahoo.com.br; aprous80@gmail.com

P. Da-Gloria et al. (eds.), *Archaeological and Paleontological Research in Lagoa Santa*, DOI 10.1007/978-3-319-57466-0_15

Rock Painting Studies in the Lagoa Santa Region: A History

Lund excavated many caves, among them the Escrivânia and Ballet, which have clearly visible rock paintings, but he did not ever describe them:

> curiously enough, Lund did not seem to have set much store on that kind of prehistoric record; he does not even describe the graphisms when he is admiring the spectacle of nature in the location. (Prous 2002:51)

However, Lund did report the rock paintings found at the Rochedo dos Índios which forms part of the rocky outcrops of the Lapa da Cerca Grande, which pannel was illustrated by his friend and collaborator P. A. Brandt and later published in Europe (Neves and Piló 2008:121; Prous et al. 2003:15). Brandt made drawings of a great number of plants, of the entrances to the caves, the landscapes, and the paleontological pieces found in the region.

At the turn of the nineteenth to the twentieth century, mining and civil engineer Álvaro Astolfo da Silveira, working for the Minas Gerais Geographic and Geological Commission (*Comissão Geográfica e Geológica de Minas Gerais*), visited and described caves and landscapes in various places in the state, including the regions that encompass Sete Lagoas, Cordisburgo, Pedro Leopoldo, and Serra do Cipó. His field notes only came to be organized in 1939 in a publication produced by the General Statistics Department (*Departamento Geral de Estatística*), and it contained drawings and sketches made by cartographer Afonso de Guaíra Herbele and by the artist and museologist Raul Tassini, who reproduced some of the figures from the Lapa Maior site in the municipality of Jaboticatubas (DNPM 1939).

Still back in the first decade of the twentieth century, mining and metallurgy engineer Cássio Umberto Lanari, whose family owned a farm property in the region of Mocambeiro in the municipality of Matozinhos, visited the Lapa do Caetano (near to Cerca Grande), from which he removed bones found under a stalagmitic crust, which he concluded were human bones dating back to the Pleistocene (Lanari 1909; Neves and Piló 2008:123). In that same site, on a little ceiling off to one side, there are sets of figures painted on the rock surface.

Lund's original discoveries had caught the attention of two teams from the National Museum of Rio de Janeiro who began excavations at the Lapa Mortuária of Confins. The first excavations were coordinated by Padberg-Drenkpohl and the second incursion by Bastos de Ávila with the participation of Rui Lima and Ney Vidal. They also excavated the caves known as Lapas da Serra de Carrancas in the former farm of Nova Granja. However, they did not find human remains associated to extinct animal species remains, and that defused the stimulus for continuing research in the region. Up to the moment, their reports have not been published, despite the fact that they are an important source for consultation (see Chap. 5 of this volume). There is no mention in their reports of any rock graphisms in the caves they visited or excavated (Ávila et al. 1939; Padberg-Drenkpohl 1926, 1929).

The next group to visit the caves in the region was made up of researchers from Belo Horizonte, members of the Minas Gerais Academy of Sciences, namely, the intellectual Aníbal Mattos, doctor Arnoldo Cathoud, and the British honorary consul

Harold Victor Walter, who, despite the fact that they were all amateurs, continued to publish and disseminate information about the region's prehistory. Generally speaking, they paid workers to do the excavation work and did not concern themselves with the aspects of stratigraphy or with any less visible remains. They built up important collections and publicized the finding of a skeleton at the Lapa Mortuária, possibly associated to the fossils of ancient animal species (Prous et al. 2003:16; Walter et al. 1937). In 1948, H. Walter published a work containing the results of his research in the region, including an illustration of rock paintings at the Lapa de Cerca Grande site (Walter 1948).

In 1940, Aníbal Mattos published a short article in a specialized journal, where he wrote about the archaeological site of Lapa Vermelha in Pedro Leopoldo (nowadays, part of the municipality of Confins), with a special focus on the lithic material, pottery, and the rock paintings there (Mattos 1940). Together with Josaphat Penna and Alfredo Mucci, he produced some reproductions of prehistoric figures and presented them in exhibitions organized in Belo Horizonte. According to H. Walter, "they made many copies of the paintings, helping to identify and catalogue the animals represented, their dimensions and the colors used." In 1958, in São Paulo, J. Penna presented a brief communication on rock painting in Lagoa Santa (Penna 1964; Prous et al. 2003:17). On that occasion, he extended an invitation to A. Laming-Emperaire, who came to Lagoa Santa in 1962 together with researchers from the Prehistory Institute of the University of São Paulo. From that visit, Luciana Pallestrini would take the material for the brochure she was to write some years later (Pallestrini 1969).

In 1958, H. Walter, in his work *Arqueologia da Região de Lagoa Santa* (Archaeology of the Lagoa Santa Region), dedicated one topic of the contents to the rock paintings and presented plates showing the most commonly represented themes found in the Cerca Grande site, which were anthropomorphic figures, geometrical designs, and zoomorphic figures. There was a succinct description of the representations, such as the figures of deer, fish, monkeys, peccaries, birds, and armadillos, and the paintings at other sites were also mentioned, among them, Lapa de Caetano, Porções (or Poções), Lapa Vermelha, and Sumidouro (Walter 1958:104–105).

H. Walter (1958:100) suggested that the rock representations in the region were perhaps: "(...) only a few centuries old. The wish to paint hunted animals and curious symbols may have been part of some magic ritual or custom which developed spontaneously in some countries insofar as Man gradually acquired a certain degree of intelligence and culture."

According to the scholar "the succession of lines" visible high up at the Lapa do Sumidouro must be "a system of linear annotation." However that may be, it was the first attempt at interpreting those records and defining an age for them (Walter 1958:101–103).

Still in the 1950s, a young social sciences student, Marcos Magalhães Rubinger, started visiting the archaeological sites with rock paintings in the Lagoa Santa Karst, in the Serra do Cipó, and in the surrounding areas, registering his reflections on them as well as making sketches, taking photographs, and describing some of the graphisms in his field book notes dated in 1956 and 1959 (Rubinger 1956/1959).

Among his other observations, he showed interest in the alternations of colors and the spatial distribution of bar alignments at the Lapa do Sumidouro. In 1958, he founded the Minas Gerais Society of Anthropology (*Sociedade de Antropologia de Minas Gerais*) and in partnership with Sigefredo Marques Souza and Paulo Apgaua organized courses on the prehistory of Minas Gerais, among other topics related to ethnography, human geography, and bioanthropology. They took some of the students from the Sociology course at the Federal University of Minas Gerais to some of the archaeological sites. Paulo Apgaua also prepared a seminar on the markings at the Lapa de Sucupira, a rockshelter with hundreds of figures on the rock face, located in the Serra do Cipó, around 60 km from Lagoa Santa, which demonstrates just how much attention they were dedicating to that site at the time.

Rubinger wrote a work entitled *Pintura Rupestre – Algo mais do que Arte Pré-Histórica* (Rock Painting: something more than mere prehistoric art), but it was only published posthumously in 1978. In it, he took up the ideas exposed in his earlier notebooks prior to his exile during the military dictatorship in Brazil. He proposed that the rock art functioned as a kind of language within a "concrete-abstract" system and that in fact prehistoric Man had first created "isolated" figures in the rockshelters and on the cave walls, which he called "figurative language," and suggested that it had been produced "primarily as an instrument of magic to ensure the subsistence of the community." As an example, he used a reproduction of the figure of a deer taken from the Pedra Pintada site in Serra da Conceição in the municipality of Barão de Cocais (Rubinger 1979:26).

Following that, he suggested that initially the language had been "symbolic or ideographic," basing his affirmation on the gestaltic representation of the figures and the signs of interrelation among them, and proposed the notion of "ideas." That, he proposed, would have gone on to attain the level of "syllable-ism," which he said would be the representation of "stylized signs." Finally, he considered that the language would attain the level of being "literal" or "written." To illustrate what might be a scene of "symbolic" language, he used the example of a set of figures from the same Pedra Pintada archaeological site (Rubinger 1979). Rubinger considered that the proposed sequence would be merely logical – that is, it would not necessarily provide any means of dating specific works at Lagoa Santa because, as he put it, "the Australians, our primitive contemporaries, are still at the rock painting stage" (Rubinger 1979:26). In addition, he associated rock figurations to possible contemporary indigenous designs and to heavenly bodies and myths, using ethnographic interpretations. In 1977, Rubinger's colleague, Sigefredo Marques Soares, published an article with a similar approach to the rock paintings at Pedra Grande and Pedra Pintada in the region of Serra do Cipó and also in the Serra dos Veados, Rio Piracicaba, and Alto do Rio, all of them far away from the Lagoa Santa Karst (Baeta 2011; Soares 1977:411, 413).

Years earlier, Minas Gerais psychiatrist Clovis de Faria Alvim (1964), who dedicated himself to studying the sexual lives of indigenous peoples, among other things, wrote an essay on the mental level and personality of pre-Columbian Indians in the valley of the Velhas River. Adopting a reductionist and evolutionist stance, and subliminally presuming that the designs on the cave walls were a form of art and that, as such, it was necessarily obliged to be a photographic reproduction of nature, he arrived at the "brilliant" conclusion that the authors of those prehistoric designs had the mental age of children (we do not know what would have been the graphic skills and mental level the eminent doctor would have attributed to himself based on his own scrawls on the blackboards of the university where he taught).

In 1956, Wesley Hurt, a museologist and archaeologist from the University of South Dakota, visited Lagoa Santa together with a team from the National Museum of Rio de Janeiro. Among the team members were Oldemar Blasi and Carlos de Paula Couto. They excavated sites such as Cerca Grande, Lapa do Ballet, and Lapa das Boleiras, but they did not study the pannels of rock paintings and merely mentioned the presence of such graphisms in the first two of the sites mentioned.

In 1966, for the first time, the cave art of the Lagoa Santa region was mentioned in a book on the global history of art in an international collection in which historian Andreas Lommel (1978) presented a representation of a deer from Cerca Grande describing it as a "typical representation of Amazonian [*sic*] rock painting."

In 1973, students at the Federal University of Minas Gerais (UFMG) together with the Director of the Natural History Museum of UFMG, F. de Ávila Pires, created a project for the biological study of the Cerca Grande Massif and included in the activities a survey of the rock art. One of the students, Márcio Rosa (nephew of the famous writer Guimarães Rosa) would later go on to excavate in Santana do Riacho as a member of the first UFMG team.

The first expeditions of the Franco-Brazilian archaeological mission took place from 1971 to 1976 and visited roughly 30 sites in the region. In the years that followed, excavations went ahead at the Grande Abrigo de Lapa Vermelha in Pedro Leopoldo because it presented more precise information on human occupation and the stratigraphy was well preserved.

In Europe, Laming-Emperaire (1962), based on the structuralist ideas of C. Levi-Strauss and M. Rafael, had elaborated a social explanation for Paleolithic art, and she had become one of the most important authorities in that field of research. Unfortunately, she did not manage to study the cave art of Minas Gerais. She had prepared for that task by working with the members of her seminar in Paris to prepare a descriptive vocabulary for it. After her premature death, the research in Minas Gerais was continued by one of us (AP) who later went on to create the archaeology sector of the Natural History Museum at the UFMG.

The Lagoa Santa mission sought to establish the relations among various kinds of archaeological remains excavated and identified, including the figures on the rock surfaces, the landscape, and the climate changes in the course of time. A 1974 publication written by Laming-Emperaire, Prous, Vilhena de Moraes, and Beltrão shows their concern to photograph and to make tracings of the rock art of the Lagoa Santa region, given the ongoing destruction of the massifs for the extraction of limestone.

> The threat of destruction in the Lagoa Santa region has led us to make use of reliefs, tracings, and photographs of all the accessible rock art works. (Laming-Emperaire et al. 1974:7)

In 1974, one of the authors of this chapter (AP) presented a preliminary synthesis of the observations made of the cave art of Lagoa Santa (of which the text of Prous 1977, presents a version in Portuguese) at the International Congress of Americanists held in Mexico City, as well as presenting an analysis of the structure of the rock art records at the Lapa do Ballet site.

In the course of the 1970s, rock painting experts Pierre Colombel and Nadine Orloff with the help of Suzana Monzon and Sydney Anthonioz, all members of the Franco-Brazilian mission, published part of the inventory of rock figurations of the Lagoa Santa Karst. In 1984, microfiches of the rock figurations of the Lapa do Caetano and Lapa de Cerca Grande sites were produced (Anthonioz et al. 1974a, b), as well as published articles (Anthonioz et al. 1978; Anthonioz and Monzon 1977). The last publication of that team was *Importância e Significado da Arte Rupestre* (Importance and Meaning of Rupestrian Art) by P. Colombel and N. Orloff (1981) in the catalogue of an exhibition jointly organized by Museum of Natural History at the UFMG and the Institute of Prehistory of the University of São Paulo, in the art gallery of the USP and then in Belo Horizonte. Parallel to that, the first chronological-stylistic framework was proposed for the region (Prous et al. 1980; Prous and Paula 1979/1980).

In 1982, two artists, Maria Irene Neves and Sílvia Gaia Santana, hired by the Minas Gerais Technology Center (*Centro de Tecnologia de Minas Gerais*) and trained by the mission team, made reproductions of various sets of figures on the cave walls of sites in the Lagoa Santa region. Those works of art are on permanent display at the Tancredo Neves International Airport in Confins (CETEC 1982).

Toward the end of the 1980s, the team of the erstwhile Prehistoric Archaeology Sector of the Natural History Museum at the UFMG, financed by the Minas Gerais Research Foundation (*Fundação de Amparo à Pesquisa de Minas Gerais* – FAPEMIG) and by the new *Mission Archéologique* of Minas Gerais, coordinated this time by one of us (AP), began a new program to map the archaeological sites in the region of the Lagoa Santa Karst. They made copies of sets of rock paintings at sites that up until then had been insufficiently documented by the members of the first Franco-Brazilian archaeological mission or that had not even been the object of any prior registration at all. In the 1990s, the mission continued with some of the activities directed at recording the rupestrian graphisms, especially those at the Serra do Cipó in the Lapa do Gentio and Lapa de Sucupira and nearby sites like the Vargem da Pedra, Lapa do Trevo, Abrigo Rei do Mato, Gruta Rei do Mato, Lapa Vermelha, and others. At that moment, particular attention was devoted to the distribution of the figures and their styles in the internal topographies of the sites (Baeta et al. 1992). Again, at that time, the figurations found at the sites of Altamira, in the municipality of Nova União; Pedra Pintada, in the municipality of Barão de Cocais; and Escrivânia and Capão das Éguas, in the municipality of Prudente de Moraes, all located in the region around the Karst, were properly documented (Baeta 1989, 1991, 1992; Baeta and Prous 1991; Moura 1994; Siqueira et al. 1989a, b). It had also been planned to document the Lapa do Arco site in the municipality of Matozinhos. However, when the team arrived to schedule its entrance on the property to collect the tracings, the rocky outcrop where the art remains had existed had been totally dynamited out of existence by the mining company Calsete, which quarried limestone on the property (Baeta 1998b).

The work of actually removing the markings from the Lapa do Ballet, which was unfolded and coordinated by the restorer Helena David in 2002, was a landmark regarding the conservation of rock drawings, paintings and designs and the depollution of the caves of Minas Gerais state, but, unfortunately, it was the one and only pioneering work for that purpose in the state (David and Moura 2002).

In 2011, A. Baeta made a regional study in which she elaborated a chronological-stylistic review of the sites with rock paintings in the Lagoa Santa Karst and the Serra do Cipó for her thesis. She revisited the sites observing and registering aspects that had previously been neglected, especially vestigial figures (Baeta 2011). The identification of unexplored caves and rockshelters with rupestrian figures on the part of speleology groups[1] that were carrying out projects in the region contributed greatly to the recognition and registration of such sites, as was the case with the rockshelters of Mato Seco, Teto Preto, Lapa Cinzenta,[2] and others.

The research carried out by the University of São Paulo and headed by Walter Neves did not include studies of rock paintings among its priorities. Even so, the discovery of a buried rock engraving at the Lapa Cinzenta, dated to at least 9,000 years old, was an important contribution, insofar as it helped to establish a chronology for such records in the region (Neves et al. 2012).

The Art of Researching Rupestrian Records

Researchers from the Federal University of Minas Gerais have been carrying out research targeting the rupestrian records in Lagoa Santa for some decades now (1977–2013). Among their constant concerns are the aspects of chronology, characterizing the surfaces, the disposition of the figures, and relations among them and their insertion in the landscape.

Observation of Chronological Elements

Some of the buried figures in Lagoa Santa and in the neighboring Serra do Cipó provide absolute (maximum and/or minimum) dating. With exception of the rock engraving at Lapa Cinzenta, all of them were revealed in the excavations of the 1970s.

More recently, researchers from the Physics Department of the Catholic University of Rio de Janeiro (PUC/RJ) and the Brazilian Physics Research Center of Rio de Janeiro (*Centro Brasileiro de Pesquisas Físicas do Rio de Janeiro*), in a part-

[1] Bambuí Speleological Research Group (*Grupo de Pesquisas Espeleológicas* – GBPE) and the Speleological Activities Nucleus (*Núcleo de Atividades Espeleológicas* – NAE)

[2] This site is also known as Lapa do Santo.

nership arrangement with the archaeology sector of the UFMG, have begun studies of dating techniques for rupestrian paintings made with ferruginous pigments that involve measurement of their intrinsic magnetization. Some tests were carried out by a team led by P. Ribeiro at the Grande Abrigo de Santana do Riacho. Preliminary results suggested the presence of a residual magnetism in the cave wall paintings that may make it possible to use the archaeomagnetism as an alternative means of dating (Ribeiro et al. 2007). On the other hand, dating the carbonates deposited over the paintings would be possible in some sites such as the Lapa Vermelha I/II and the Mato Seco rockshelter. Currently, a new program is specifically needed to develop that study further.

Considering that absolute chronological information is extremely rare, it was necessary to find other means of determining the chronological moments at which the graphisms were made. To that end, careful observations in many of the sites were made of the cases of figures that had been superimposed on others, which at least provided relative chronological information. Even more useful, because they separate the productions of ages far removed from one another, was the registration of figures with different patinas in successive layers of peeling (Cerca Grande and Sucupira). It is hoped that they will make it possible to determine moments in time typified by certain themes (that define traditions) or by specific representational conventions (styles). In recent years, there has also been an increasing concern to highlight the way in which older figures have been inserted in more recent graphisms and reappropriated or, on the other hand, have been deliberately deteriorated. In that way, chronological-stylistic studies have been the mainstay of research undertaken by the UFMG's archaeology sector in various places but especially in the river basins of the São Francisco (upper and middle course), Doce, and Jequitinhonha and in the Serra do Cipó region (Prous 1989, 2012).

The Insertion of the Rock Painting Sites in the Landscape and the Choice of Surfaces

The question arises as to why certain places were chosen to be decorated and not others, and the answer may very be different from one population to another. The criteria may have been their visibility in the local landscape, the orientation of the caves and their surfaces, and a standardized average distance between one site and another. At a given site, there may have been different preferences for corridors, vertical or sloping walls, smooth walls, flat roofs or stepped ones, availability of water, the presence of earlier graphisms or there being a virgin surface, and so on. The question is even more complicated because each one of the successive populations in the region may have adopted different criteria. Worse than that, different groups within a given population (clan members, gender groups, or age groups) may have enjoyed spatial privileges. Again, specific occasions may have called for special graphisms to be made in determined places, in the same way that a Catholic group may associate a certain ceremony to a specific church or within the same church prefer certain internal chapels, while other rituals are performed outdoors. The only way to address all those considerations is to search for regularities in the graphisms and try to relate them to the regional, local, and intra-site topography.

Trying to Understand the Wall Devices

It has in fact proved possible, in some situations, to verify the performance of a certain graphic set in relation to its antecessors. The inventorying, detailed description, and analysis of the rupestrian paintings at the Grande Abrigo de Santana do Riacho, in the Serra do Cipó area (Brito et al. 1992/1993), are an example of the use of that approach in the central region of Minas Gerais. First, the way the figures had been superimposed was carefully observed to define the proposition of the pictorial moments and their sequences. Then, the various techniques used to elaborate them were described, including colors, tones, the type of filling of the figures; dimensions, proportion, typologies and anatomical details; and the way the surface had been prepared, as well as the distribution of the themes in the various compartments of the rockshelter (Prous and Baeta 1992/1993). The second part of the study was dedicated to an analysis of the overall set with special attention to relations among the graphisms, identifying similarities and detecting any typological or thematic differences among the layers and any evidence of reuse of the same surface (Baeta and Prous 1992/1993). That work was only possible because of the rigor of the registration process and the reproduction of innumerable sets of rupestrian figures, despite the notorious difficulty involved in documenting certain paintings. A similar approach was adopted by Siqueira et al. (1989a, b) at the sites of Altamira and Pedra Pintada, where they identified various superimposed layers of the Planalto Tradition. Another work done by Baeta et al. (1992) examines the options of surfaces available around the uvala of the Lapa Vermelha in Confins, some of which had been painted, while others remained untouched.

Examples of the recognition of the disposition of various chronological-stylistic assemblages and of the interplay of colors, themes, styles, and topography have been particularly well developed in studies of sites such as the Lapa do Ballet (Prous 1977, 1985/1986), or the Cauaia, Abrigo Rei do Mato, and Vargem da Pedra sites (Baeta et al. 1992) and more recently at the Caieiras I and II, Três Porquinhos, Maquiné, Sumidouro, and Porco Preto sites (Baeta 2011). In 2013, texts about the graphisms at the Lapa do Ballet and Vargem da Pedra (Baeta 2013a, b) were published in the journal *IN SITU*, published by the CAALE[3] for diffusion to the population of the Lagoa Santa region.

Although we have analyzed pigments from the Serra do Cipó, there has been no similar research in the Karst. That was originally to be done by H. David at the Sumidouro site, but the researcher's premature death left the project in abeyance.

---

[3]Annette Laming-Emperaire Center of Archaeology (*Centro de Arqueologia Annette Laming-Emperaire)* coordinated by the archaeologist Rosângela Albano

Filing, Communicating, and Publishing the Documentation

In the files of the UFMG Archaeology Sector, there are thousands of slides produced from 1975 to 1995. At our invitation, C. Andujar in 1976 and M. Consens in 1985 made some partial infrared surveys. Another important collection of digital photos was put together a few years ago in preparation for the publishing of a book (Prous et al. 2003). One of the authors of the present chapter (AB) is currently participating in the "Pesu Rupestre" project,[4] which aims to register the rupestrian art at Lapa do Sumidouro in high-definition images, including figures that are hardly visible and have only recently been detected. The UFMG also maintains thousands of tracings made in the period from 1975 to 1996, and they represent a complete inventory of dozens of sites in the Lagoa Santa region. Based on those tracings, verified in the field on various occasions and duly compared with the photographic records, complete scale models have been made on a scale of 1:5, which have been used for various subsequent studies. The records of some sites have been published in France in the form of microfilm sheets. That kind of record is no longer used, and, accordingly, it is important that the images should be computerized and published in digital form.

The research work has been publicized to the public at large in the region by means of exhibitions in those municipalities that expressed an interest, such as Belo Horizonte (Dean's Office of the UFMG and Atlético FC/Ecological Park, in 1978; Coach station in 1980). In France, an exhibition of the Lagoa Santa rupestrian art went on display at the *Galerie Debret*, in 1978. In São Paulo, there were exhibitions in 1970 and 1980 and in Matozinhos and in the Palace of Culture in Belo Horizonte in the 10-year period that followed. Films and videos to publicize the rupestrian art research have been made by the UFMG (in 1977; see section "Filmography" in this chapter) or by TV channels (Rede Minas/Bem Cultural Program 2012; see section "Filmography" in this chapter).

Scientific publications include articles of synthetic analyses (published in Australia, Bolivia, the United Kingdom, Italy, France, and, of course, Brazil), thematic analyses (in Brazilian journal *Arquivos do Museu de História Natural*, volume 10), and monographies (in microfilm sheets or in *Cahiers d'Archéologie d'Amérique du Sud*) and books for the general public (Jorge et al. 2007; Prous 2007).

State of the Art: The Stylistic Assemblages

We will now present the stylistic units that we have recognized in decreasing order of antiquity, within the limits of present-day knowledge. The possibility that some of them may have been contemporaries or even that their graphisms may have been executed by the same authors cannot be excluded.

[4] "Pesu Rupestre" project implemented by the team of Artefactto Consultoria and Orange Editorial, and financed by Construir.

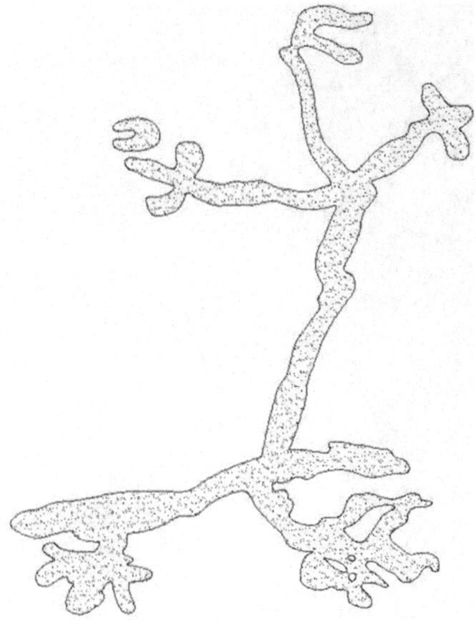

Fig. 15.1 Anthropomorphic figure marked out by fine pecking on a block of stone uncovered in the 2009 excavation. Archaeological site – Lapa do Santo (or Cinzenta), Matozinhos, and Minas Gerais

The Engraving at the Lapa Cinzenta

This isolated engraving was found at the Lapa Cinzenta (name traditionally used by regional residents and used in the Lagoa Santa Karst Environmental Protected Area system). This same site was later registered by the Brazilian Speleology Society as Lapa da Fenda and finally by the USP as "Lapa do Santo." So far, it contains the most ancient dated graphism in the region (Neves et al. 2012). The figure is a small anthropomorphic, itiphallic, filiform design marked out by tiny pecking chipping of the rock (Fig. 15.1). Later in this chapter, we will discuss its possible relations with other assemblages of figures in the region.

Planalto Tradition

Because of the position of its figures on the stratigraphy of the rock walls, this is the most ancient set type of graphic expression to be found in the Lagoa Santa Karst, the Serra do Cipó, and the highlands of central Minas Gerais in general, with the single exception of the Lapa Cinzenta engraving. It also represents the stylistic unit most well represented in the this same geographic area. We believe that it corresponds in time to the Middle Holocene.

It is typified by the visual and usually quantitative predominance of the figures of animals in a variety of sizes (Fig. 15.2). Deer and other quadrupeds are the most common representations, but according to the time and the locality, there may be

Fig. 15.2 A set of red rock paintings at the archaeological site of Cerca Grande. A typical scene of the Planalto Tradition. Municipality of Matozinhos, Minas Gerais (Photo by Ézio Rubbioli)

other zoomorphic forms as well. Although the animals are usually represented in isolation, there are examples of monothematic representations of animals. There are some representations of hunting scenes of an animal surrounded by smaller, schematic, anthropomorphic figures or being shot by a spear (at the sites of Cerca Grande, Vargem da Pedra, Escrivânia, and Capão das Éguas). There are also representations of quadrupeds to which bars have been added (Cerca Grande); another is associated to a circle (loope?) (Vargem da Pedra), and one at Lapa Maior seems to represent fishing with a hook. At the Lapa Pedra Pintada, there is a scene where anthropomorphic figures seem to have cornered their prey.

Generally speaking, the forms of the animals are fairly crude; the body is represented in varying ways, either in outline only or totally filled in. More rarely, at least in the Lagoa Santa Karst, the bodies are outlined or partly filled by parallel lines or dots. This latter style is more characteristic of the regions farther north (Serra do Cipó, Diamantina, Serranópolis de Minas, and Grão Mogol). As a rule, only a single color is used (usually red or yellow), although very rarely a figure may be outlined or decorated with marks of a different color.

The Planalto style sets usually consist of a few dozen figures painted on different parts of the rockshelters, ranging from the point of contact with the floor to heights of as much as 8 m (Fig. 15.3). As a rule, most of them range from 50 cm to 2 m above the floor level, on smooth surfaces, on flat or slightly sloping walls, on high or low ceilings, on ledges, or in niches. The size of the individual zoomorphic figures varies from 10 cm to 1.5 m in length, with the majority being in the range of 20–45 cm. As a rule, the geometrical and anthropomorphic figures are smaller than the zoomorphic representations.

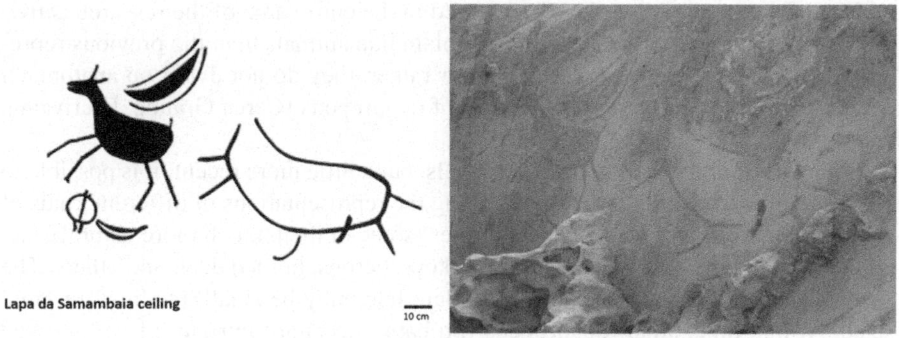

Fig. 15.3 Reproduction and photograph of the ceiling of the uppermost floor of the Lapa da Samambaia showing the rock paintings. Sumidouro State Park – PESU. Municipality of Lagoa Santa, Minas Gerais (Photo: A. Baeta)

The Earliest Period

The period was typified by the zoomorphic figures of quadrupeds such as deer, most of them with the body simply outlined or with plain hue (Lapa do Ballet and Lapa do Sumidouro). At that time, they were made on walls and ceilings of the rockshelters. Some of them were reworked in later times.

Intermediate Moments

It is usual for entire walls or even the entire site to be decorated exclusively with Planalto style representations of animals and filiform anthropomorphic figures (Siqueira et al. 1989a). In other sites, however, certain panels that would seem to be contemporary with a median stage of the tradition were covered by simple geometric figures (Cerca Grande) or by alignments of short rods (Sumidouro and Sucupira). At the Lapa da Pedra Pintada, the Planalto animals have been systematically covered with dots, which are more suggestive of an association than of a succession of events. In the Lapa de Sucupira, the linear geometric figures are chronologically inserted between two distinct moments of animal representation (both classified as being within the Planalto Tradition). The style of animal representation is not the same after the irruption of the bars. That being so, it is possible to postulate two hypotheses to explain the situation. The first is that there must have been a horizon with dots and/or rods that one of us (AP) once has named the Sumidouro Tradition (Prous and Paula 1979/1980). However, the fact that that "horizon" has not been found at other sites does not help to confirm the hypothesis. Another explanation would be that, at some intermediary time, the bearers of the Planalto Tradition had not only painted zoomorphic figures characteristic of the tradition but also sets of dots and bars.

One of the intermediary levels observed in the entire area of the research activities and its vicinities consists of smaller plain hue animals than the previous representations. They too are notably crude, or rather, they do not dwell on anatomical details, especially in the representations of quadrupeds (Cerca Grande, Escrivânia, and Sumidouro).

Still on the subject of intermediate levels, but a little more recent, it is possible to see that there is far greater diversity among the representations of different kinds of geometric figures, anthropomorphic figures (some of them much more natural), and zoomorphic figures, such as jaguars, monkeys, herons, horned deer, and others. The sizes of these figures, all of them with plain hue, may be small, medium, or large sized. At this time, small quadrupeds (probably deer) are introduced with curved bodies and short straight limbs, which were attributed to a "Samambaia facies" in the earliest publications (Prous et al. 1980). Retouching and outlining of some figures were done during this period to some of the earlier zoomorphic and geometric figures (Sumidouro), and sometimes figures were added.

Late Manifestations

In what seems to have been the end of the tradition, there are figures that may have been influenced by units coming from outside, perhaps from the Northeast Brazil. That can be seen, for example, at the Lapa de Cerca Grande, where the traditional deer figures received the addition of anthropomorphic figures strange to the region, all worked in black crayon. New styles appeared, possibly with the entrance in the more recent millennia of new human groups bringing in new forms of representation or graphic expression. The Vargem da Pedra style might have been the result of a combination of the themes and techniques of execution of the Planalto Tradition, and external influences marked by the presence of sets of deer figures filled in with dashes and large birds with elongated bodies worked in crayon (Vargem da Pedra and Mato Seco; Fig. 15.4). At the Lapa do Sumidouro, white spots were added to lines of yellow and red spots, and yellow quadrupeds were designed on top of black crayon figures. At Capão das Éguas, solid anthropomorphic figures and quadrupeds with elongated bodies indicating movement or action seem to have been influenced by the Ballet panels.

The Ballet Figures and the Caieiras Rockshelter

The manifestations of the Ballet stylistic unit usually appear at later dates than those of the Planalto Tradition (Ballet, Pedra Pintada, and Rei do Mato Cave), although some isolated figures may have been contemporary with late representations of deer (Capão das Éguas). The figures that typify this unit are sets of representations of humans forming queues or processions. The clearly characterized sexes and scenes depicting birth suggest an emphasis on sexuality and spatially differentiated treatment for the genders, reproduction, and bodily expression.

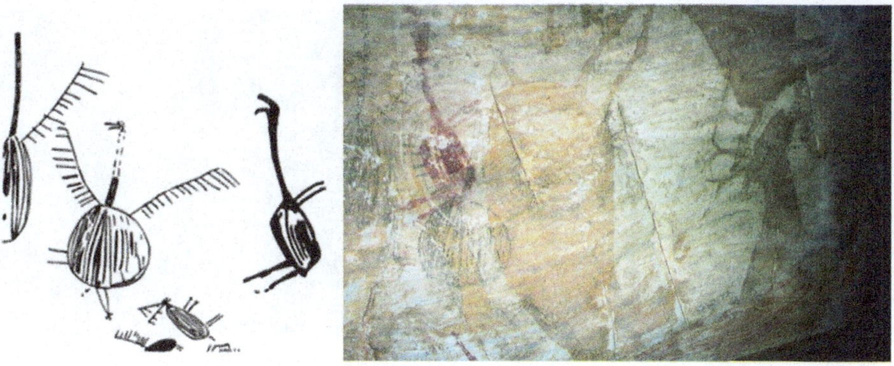

Fig. 15.4 Ceiling of the Mato Seco Cave with rock paintings composed of fine *crayon lines*. Municipality of Prudente de Morais, Minas Gerais (Photo: Ézio Rubbioli)

A typical figure shows a filiform body (1–2 cm wide) with sex identified as male or female, arms held up, and sometimes the mouth is represented in a beak-like form. At the Maquiné and Pedra Grande sites, there are representations of figures upside down as if they were doing somersaults or falling down – a theme that was also identified at some sites in the north of the Minas Gerais state (e.g., at the Abrigo do Janelão in the Peruaçu river valley). In the Lagoa Santa Karst, the Ballet figures are found on the internal walls and platforms of the first chamber of extensive caves or those with a certain horizontal development (Rei do Mato Cave, Maquiné, and Ballet), in small compartments in limestone outcrops (Capão das Éguas), and also in one very large rockshelter or rock face (Campinho). Each assemblage, consisting of 20 figures at the most, seems to have been designed in a single session, and a single color is used for a given site (red, black, yellow, or white). Only rarely are there cases of different sets of figures in the same cave being colored differently (Maquiné, Pedra Pintada, and Ballet).

At Lapa do Ballet, one moment of the Ballet style is superimposed on another. The underlying one is made up of red Ballet figures resembling those in the Rei do Mato Cave and on the ceiling of the Maquiné Cave (Fig. 15.5). The more recent one has filiform figures colored black. Associated to the assemble of Ballet representations, there are compartmentalized concentric circles (Rei do Mato Cave and Maquiné), and what are possibly artifacts such as a rattle next to the hand of the first anthropomorphic figure in the upper line of figures, a basket on the head of another figure in the same line (Lapa do Ballet), and what could be a sack on the body of one of the anthropomorphic figures at the Campinho site.

In the 1990s, we began to consider that the Ballet figures would have some similarities with the Seridó sub-tradition found in the state of Rio Grande do Norte, which, in turn, according to the researchers of the Brazilian Northeast, was a late transformation of the northeast tradition. However, we must state that this is just a hypothesis, because there are differences between the two manifestations. For example, the Seridó paintings do not usually identify the sex of the figures. The figure on the buried stone found at the Lapa Cinzenta is similar to a Ballet figure and

Fig. 15.5 Reproduction and photograph of the figures on the high ceiling of the first chamber at the Lapa da Maquiné, composed by vestiges of typical Ballet style anthropomorphic figures. Peter Lund State Natural Monument (*Monumento Natural Estadual Peter Lund* (MNEPL)). Municipality of Cordisburgo, Minas Gerais (Photo: A. Baeta)

accordingly found in the region. It is, however, unquestionably far older. It should be noted that there are filiform anthropomorphic figures with beaks like birds engraved by pecking in other places too, among them the upper shelter at Caieiras I and Caieiras II (where, in 1971, one of us (AP) excavated a level dated at 9,600 years old, the same age proposed for the anthropomorphic engraving at the Lapa Cinzenta. We cannot be sure, however, that it was associated to the neighboring engravings; Fig. 15.6). That being so, the question arises as to what the relations among all those figures would be. The first possibility that comes to mind is that the similarities are merely a result of convergence. That is indeed possible, but not very convincing because of the degree of similarity in all the details. Another explanation would be that the graphic (and most certainly ideological) model expressed by the anthropomorphic figure found at the Lapa do Santo (and perhaps at Caieiras I/II and Serrinha too) is indeed extremely ancient, and it was maintained for 9 or 10,000 years, including during the period of the Planalto Tradition's hegemony, only to reemerge with renewed force in more recent times (probably about 2,000 years ago) by means of the paintings, and that characterizes the Ballet moment which replaced (or was probably contemporary with) late stages of the Planalto Tradition. Despite their morphological similarities, they present themselves in very different scenographic contexts from that of the typical sets of more recent Ballet figures.

From the chronological point of view, it could be considered that the first Ballet configurations had been produced at a time a little farther back than had been imagined and that it had also influenced the more recent pictorial levels of the Planalto Tradition.

In turn, the sets of graphisms engraved by fine pecking indicate a possible mixing of influences coming from the Northeast Brazil and from the Planalto Tradition (and possibly even the Ballet stylistic unit) in the Lagoa Santa Karst. At the Caieiras I and II site, there are engravings of a variety of themes, among them deer and birds,

Fig. 15.6 Figures engraved by pecking at the Lapa Caieiras I with filiform representations of what are probably human beings, quadrupeds, plants, and others. Municipality of Matozinhos, Minas Gerais (Photo: Ézio Rubbioli)

and phytomorphic, geometric, and anthropomorphic figures. In the upper shelter, there are sets of human figures with their arms raised, suggesting scenes of dance and action. It could be a pecked variant of Ballet, but in this case, it is associated to other types of figuration not identified in the painted version. Again, at the Serrinha rockshelter, there is an exclusive set of anthropomorphic engravings (possibly representing adults and children) with characteristics similar to the Ballet sets of painted figures.

Panels with figures of this stylistic unit, which up until recently were only known in the vicinity of Lagoa Santa, were found by one of us (AB) much farther north (as far as Doresópolis in the upper São Francisco river region), indicating the possibility that engraved Ballet graphisms may exist beyond the Lagoa Santa Karst.

Black Crayon

Until very recently, certain figures sketched in black crayon on more recent surfaces of some sites were considered to be merely modern graffiti, so much so that, although their presence was recorded, they were excluded from the reproductions of the Cerca Grande and Vargem da Pedra made in the 1970s. However, it is possible that they may be an important stylistic expression of late prehistoric times. They are very rough and tiny anthropomorphic or bio-anthropomorphic figures[5] (2–10 cm long) together with lines and combs. They can be seen at the Capão das Éguas, Sumidouro, Pia, Vitrine, and Piquenique (indeed, they are the only kind of graphism at the last-mentioned site).

Pecked Engraving on Stalagmitic Floors or Fallen Blocks of Stone.

Engravings made by fine chipping the rock, very different from the ones described above for the Caieiras I/II rockshelters, were made on the floors of lateral platforms (Lapa Cinzenta) or on blocks located in sheltered zones (Samambaia, Porco Preto, and Serrinha). In most cases, the stone surface had been prepared by polishing in readiness for chipping out the lines of the figures. If there had ever been paintings or any other type of figuration on those horizontal floors, they could have been erased by the activities of preparing the rock surface. There is an anthropomorphic figure engraved on a recent surface on one of the walls of the upper chamber of the Lapa da Samambaia, indicating either the persistence or recurrence of that technique in the region.

In the great majority of the sites where there are engravings, in addition to the human and biomorphic figures, there almost always are circular or elliptical figures accompanying them. The main sites with engravings are the Caieiras I and II, Porco Preto, Cinzenta, cave n°1 at the Lapa Vermelha, and Serrinha. At the latter site, engravings are the only type of graphism found.

Sets of Cupules and Deep Incisions

Small polished cupules of about 2 cm in diameter are found in separate blocks of stone in various rockshelters – some of them in a fairly recent stratigraphic context (Carroção, Caieiras II, and Rei do Mato rockshelter). In some places (Porco Preto and Capão das Éguas) deep-cut lines, short or long, are found radiating outwards from the cupules and giving them the appearance of comets. In other cases, larger depressions are associated with smaller polished ones (Samambaia). Deep incisions in parallel lines or forming tridactyl figures were observed at the base of the Rei do Mato rockshelter, somewhat reminiscent of what we refer to as the meridional geometric tradition (Prous 1989, 1994).

There is a set of larger depressions (7 cm diameter) at the Abrigo da Lagoa in the Lagoa de Santo Antônio. They seem to have been reworked to make them wider,

[5]This term refers to graphisms that mix humanoid features with features of animals and/or plants.

because there is the evidence of a double patina. Recently, a set of very tiny cupules (diameter of around 5 mm) has been identified on the floor of one of the levels of the Lapa do Sumidouro (Baeta et al. 2013).

Narrow/Shallow Incisions and Scrapings

In some rockshelters, very shallow, discreet incisions were identified that had been elaborated over Planalto Tradition figures. They are in the form of fish bones or combs; others are elliptical, grids, or parallel lines. In the Lapa Vermelha IV, the lines cross each other perpendicularly creating a checkerboard effect. They were buried by the sediments around 3,000 years ago (Prous et al. 2003). Therefore, they would be later than the most recent Planalto Tradition manifestations or at the most contemporary with them. The same technique was detected at the Samambaia site in half-buried blocks of stone and at the Abrigo Congonhas. In one of the rockshelters at Lapa de Cerca Grande, there is a series of horizontal alignments of some didactyl and tridactyl scratches that follow the lines of the horizontal limestone strata. There are vestiges of coloring in them, but it is impossible to say whether they actually form figures. Other similar incisions were observed near to the assemble, but with little patina, so that they are apparently just modern depredations.

In the Lapa do Sumidouro, one of us (AB) recently identified unusually small incisions (1–3 cm high) worked on vertical walls below the platforms with the painted panels. They were made on irregular, whitened surfaces (this wall is often washed by the waters of the lake when there is an unusually high flood), and that makes it particularly difficult to visualize them. They are sets of parallel or perpendicular lines, which in some cases have their extremities worked differently, suggesting anthropomorphic or phytomorphic figures, and some even look like representations of palm trees.

In one of the chambers of the Vargem da Pedra site, there is a set of parallel curved lines scratched into a rock surface that had been scraped and painted red, probably to highlight the visibility of the incisions.

The Lapa da Pia site has the greatest number of incisions (Fig. 15.7), and they are concentrated in a single chamber. Pectiform (comb-like) sequences, parallel and anarchic lines, fish, ellipses, and bundles have been elaborated on top of red and yellow figures of the Planalto Tradition. Over those incisions, there are still other designs marked out in black crayon, and they represent the same themes as the incisions.

Incisions at the Limeira site were elaborated on red Planalto Tradition figures, but they seem to have been made with the intention of contributing to the original composition of scratches in parallel and in bundles that have been elaborated on a solid painted circle in this rockshelter. Some zoomorphic figures produced by scraping can be seen at various sites, for example, in the corridors that give access to the "windows" of the Cerca Grande Massif, or at the Lapa do Sumidouro, over alignments of dots in the upper level.

Fig. 15.7 Fine pectiform incisions in the Lapa da Pia. Municipality of Prudente de Morais, Minas Gerais (Photo: Ézio Rubbioli)

Paintings Attributed to Agriculturalists

In the more recent pictorial levels of the Lapa Vermelha I/II and Rei do Mato Cave, there are representations of what appear to be roots and tubers (cassava and peanuts?) probably cultivated and therefore attributable to horticulturalists. A stone ax with a handle and a semilunar blade typical of the Sapucaí ceramic culture was painted on the ceiling of the Lapa de Caetano. There are also representations of what appear to be semilunar blades in the Lapa da Pia, Lapa do Ballet, and Lapa da Pedra Grande.

Paintings and Incisions in Historical Times

The habit of using rock walls has not disappeared altogether. Poems, political messages, or messages of love scratched out with knife points are quite common in some of the rockshelters, such as Criciúma, Vargem da Pedra, Data, Buraco da Julieta, and Capão das Éguas. Signs inscribed on the walls associated to religious activities are constantly observed, such as those at the Lapa do Asterisco and

Experiência da Jaguara. In this last rockshelter, signs of Afro-Brazilian religion can be seen, apparently made with the blood of animals. At the Lapa da Macumba (near to the Lapinha Cave), there is a large chamber where there is a series of tiny stools made from the trees trunks and set in a circle. There are regularly spaced soot marks on the side walls. In the Lapa de Faustina, a line of small devil masks surprises the visitor.

The Rupestrian Figurations in Perspective: Heritages Accepted or Refused

Everything goes to show that each painter who added his own figures to those already on the wall took into consideration and interpreted the figures left by his ancestors, creating new rhythms in the equilibrium of the themes, colors, and styles of representations (Baeta et al. 1992; Prous and Seda 1987).

In the Lapa de Cerca Grande, for example, where figures were added to older ones already on the wall, a fence seems to have been drawn around one of the originally depicted animals. There are situations where similar figures to the existing ones have been added as a way of highlighting them, or a new item has been introduced in the scene. At that same site, fingers and horns have been applied in thin black crayon lines over anthropomorphic figures depicted in the panel known as Rochedo dos Índios. There are other examples of reworking anatomical details like legs, horns, and tails or even of repainting or refreshing faded figures (Vargem da Pedra). In contrast with that kind of restoration of existing sets of representations (ceiling of the Lapa do Ballet), other cases were detected of painting over, scraping, or peeling off, which showed how the painters had tried to erase the earlier paintings. This was also found at the Lapa do Ballet. The same behavior seems to have occurred at Cerca Grande I, where a pictorial moment in the Planalto Tradition was partly removed to make way for a new style of the same tradition. In other situations, layers of red paint were applied, probably to enhance the contrast with the new graphisms executed in a different color, mostly yellow. An example of this can be found at the Lapa do Sumidouro. There, white spots of thick paint were added to alignments with sequences of red and yellow dots, giving greater visibility to the wall. Some zoomorphic figures were outlined in other colors to enhance their visibility.

There are other situations, however, in which the more recent authors chose to make their representations on peripheral surfaces or fallen blocks of stone, possibly so as not to interfere with the older central sets of representations. That is the case with one of the engraved figures at the Lapa da Samambaia. Apart from that, there are rockshelters where the only graphisms to be found belong to a single stylistic unit such as that at the Lapa do Piquenique. That may be a sign of the author searching for a previously unused location, or one with no older figures or different kinds of figurations.

Surviving Remains of the Art of the Prehistoric Past

The rock art we see today on the rock walls is the result of a series of interventions, not only of the authors of the successive graphisms preserved on the rock surfaces but also of natural and anthropic agents in the course of time. In some shelters, parts of the wall, both great and small, have fallen or split off from the surface, and their graphisms have been lost forever. Invasions of insects and arachnids (Cerca Grande and Vargem Formosa), the growth of lichens, or the formation of calcite crystals (Lapa Vermelha I/II, Vargem da Pedra, and Serra dos Ossos) have destroyed parts of many sets of rupestrian graphisms. At the Abrigo do Trevo, the painted surface of a corridor ceiling, on which the figure of a fish was discernible, was so crumbly that it has probably dropped off naturally since the last visit by archaeologists. Some sites still have layers of patina in which there are vestiges of paint, suggesting that, at least on the wall where they are found, there were graphisms in the past. That is the case in the Sobrado Cave and in the Campinho, Acácio, and Trevo rockshelters. In the Lapa da Pia, half of the representation of a large cervid was washed away by natural flows, and only the upper front part of the animal remains. Plant-based paints and stains like annatto may have been used, and they must have disappeared over the centuries and millennia. Thick pigments like clays or white calcite do not adhere well to the cave walls and readily peel off, so that the researcher is led to underestimate the use of such coloring material. Indeed, in the case of some graphisms, what remains is only the negative of the representation, especially in the case of the Vargem Formosa, Ballet, and Samambaia. The marks left by smoke from recent fires have damaged the figurations on the lower walls at the Vargem da Pedra site. Unfortunately, deliberate depredations in the form of graffiti have also jeopardized the integrity of many archaeological figurations and sites, albeit they could be considered to be, in a way, the most recent rupestrian registration of a more recent or even contemporary history.

Conclusions: A Framework Far from the Final Artwork

The graphic expressions on rock were not the only form of prehistoric visual expression, but they are the form that has survived the passage of time until our days better than any other has, and they offer us some degree of contact with the prehistoric imaginary. From that fact stems the importance of studying it, however difficult it may be, to insert those manifestations in the chronological schemes obtained by the excavations. In the context of the Lagoa Santa region, it is the rock art, far more than the lithic, pottery, or bone industries, that enable us to evaluate the diversity of those past cultures.

Although it cannot be said that the preliminary stage of registration is over (there are most certainly other sites waiting to be discovered, and the older records still need to be verified and completed), there is a very representative collection of regional graphic manifestations available to the archaeologists of today. The greatest difficulty, which still needs to be overcome, is the lack of greater and more coherent chronological information, such as the one we have managed to obtain for the Peruaçu river valley (in the north of the Minas Gerais state) (Prous and Baeta 2001). In that light, the chronological-stylistic framework delineated by one of us (AP) for Lagoa Santa in 1980 presents obvious limitations and does not allow for the insertion of a considerable proportion of the occurrences.

Even so, it has proved possible to verify the great opposition, especially in the more ancient periods, between the graphisms of the plateau and mountain ranges of the center of Minas Gerais, including the Lagoa Santa region (Baeta 2011), on the one hand, and those of the flatlands that accompany both sides of the São Francisco River, on the other.

In the Zona da Mata region of Minas Gerais and in the middle course of the Doce River (Baeta 1998a), there are art manifestations completely different from those in Lagoa Santa or Serra do Espinhaço, and they are probably evidence of different ideologies. That is verified when comparisons are made of the regional themes and again of the topographic choices and scenographic constructions. Even in the central regions, there are minor regional differences (differentiated facies) within the Planalto Tradition. After its disappearance, certain more recent pictorial manifestations seem to be limited to the karstic region (e.g., the lines of anthropomorphic figures that typify the Ballet style). Others, in turn, can be found in the Serra do Espinhaço, like the so-called seals. The so-called Bonecões (big dolls), that is, big anthropomorphic figures with a very different morphology attributed to the Agreste Tradition, are found isolated or in pairs it sites at the foot of the Serra do Cipó and in the Diamantina region.

Obviously, the study of prehistoric graphisms is not limited merely to situating the figures in time and inserting them in stylistic categories. It is necessary to analyze the distribution of the sites in the landscape, differentiate them from one another, and characterize the main centers and the satellite locations for each of the proposed stylistic units. We need to find criteria to suggest the existence of territorial limits or ideological barriers – which most certainly must have changed in the course of (pre-) history. As for the (no doubt multiple) meanings the paintings and engravings had for their authors, we believe that there is no way at present that we can expect to discover them. That should not be a reason for frustration. After all, we know that in our own present-day society, there is no single universal comprehension of what we call contemporary "art."

References

Alvim, C. d. F. (1964). Nível mental e personalidade dos índios pré-colombianos do Vale do Rio das Velhas. *Revista da Universidade de Minas Gerais, Belo Horizonte, 14*, 131–161.

Anthonioz, S., Clément, G., Monzon, S. (1974a). Les oeuvres rupestres du massif de Caetano, région de Lagoa Santa, Minas Gerais, Brésil. Paris, *Musée de L'Homme, Institut d'Ethnologie, Archives et Documents, Micro-Édition*, R 74 039 176.

Anthonioz, S., Colombel, P., Monzon, S. (1974b). Les oeuvres rupestres Du massif Cerca Grande, région de Lagoa Santa, Minas Gerais, Brésil. Paris, *Musée de L'Homme, Institut d'Ethnologie, Archives et Documents, Micro-Édition*, R 74 039 116.

Anthonioz, S., & Monzon, S. (1977). Les Représentacions sexuelles dans l'art rupestre brésilien. *Objets et Mondes, Paris, 17*(1), 31–38.

Anthonioz, S., Monzon, S., Colombel, P. (1978). Les peintures rupestres de Cerca Grande, Minas Gerais. *Cahiers d'Archéologie d'Amérique du Sud*, Paris, *6*.

Ávila, B., Padberg-Drenkpohl, J. H., & Vidal, N. (1939). *Relatório dos trabalhos realizados nas cavernas e grutas do Valle do Rio das Velhas, Estado de Minas Gerais*. Manuscrito. Rio de Janeiro: Museu Nacional da UFRJ.

Baeta, A. (1989). *Levantamento e Documentação de Grafismos Rupestres no Carste de Lagoa Santa (Relatório)*. Belo Horizonte: Setor de Arqueologia MHN/UFMG- FAPEMIG.

Baeta, A. (1991). Arte Rupestre do Centro Mineiro: a região arqueológica de Lagoa Santa. *Revista de História, Universidade Federal de Ouro Preto, 2*(1), 196–201.

Baeta, A. (1992). La Arte Rupestre del Centro de Minas Gerais, Brasil. *Boletin de la Sociedad de Investigación de Arte Rupestre de Bolívia-SIARB, La Paz, 6*, 41–53.

Baeta, A. (1998a). *Memória Indígena no Médio Vale do rio Doce, Arte Rupestre e Identidade Krenak*. Master's Degree Dissertation, Universidade Federal de Minas Gerais.

Baeta, A. M. (1998b). *Salvamento Arqueológico da Lapa do Arco* (Relatório Técnico). Matozinhos.

Baeta, A. (2011). *Os Grafismos Rupestres e suas Unidades Estilísticas no Carste de Lagoa Santa e Serra do Cipó, MG*. Tese de Doutorado, Universidade de São Paulo.

Baeta, A. (2013a). Lapa do Ballet, muito mais que um ballet. *IN SITU, CAALE, Lagoa Santa* 3.

Baeta, A. (2013b). Vargem da Pedra – grafismos rupestres, lugares e proteção de um Monumento Natural. *IN SITU, CAALE, Lagoa Santa* 4.

Baeta, A. M., Silva, M. M. C., Prous, A. (1992). Organização do Espaço Pictural nos sítios rupestres da Região de Lagoa Santa, MG. *Anais 3o Congresso da Associação Brasileira dos Estudos do Quaternário* (pp. 417–430). Belo Horizonte, Universidade Federal de Minas Gerais.

Baeta, A., Piló, H., Nishimoto, L., Martins, H., & Casimiro, R. (2013). *Projeto Pesu Rupestre-Acervo Documentação Visual e Análise Estilística*. Lagoa Santa: Parque Estadual do Sumidouro-PESU/IEF, Artefactto Consultoria/Orange Editorial/Construir.

Baeta, A. & Prous, A. (1991). El arte rupestre del Centro de Minas Gerais, Brasil. 3° *Simpósio Internacional de Arte Rupestre*, Santa Cruz de la Sierra, SIARB *1*, 33–34.

Baeta, A. M. & Prous, A. (1992/93). Análise de Conjunto da Arte Rupestre de Santana do Riacho. *Arquivos do Museu de História Natural da UFMG, 13/14*, 333–353.

Brito, M. E., Baeta, A.M., Prous, A. (1992/93). O Levantamento Geral dos Grafismos Rupestres de Santana do Riacho. *Arquivos do Museu de História Natural da UFMG, 13/14*, 193–240.

CETEC. (1982). *Arte rupestre no Estado de Minas Gerais. A região de Lagoa Santa*. Belo Horizonte: Fundação Centro Tecnológico de Minas Gerais/CEssTEC, Série Publicações Técnicas.

Colombel, O., & Orloff, N. (1981). Importância e significado da arte rupestre. *Exposição Pré-História Brasileira. Aspectos da Arte Parietal*. Belo Horizonte: UFMG/USP.

David, H., & Moura, V. (2002). *Projeto de Conservação e adequação à visitação da Gruta do Ballet*. Belo Horizonte: Dezembro.

DNPM (1939). *Grutas de Minas Gerais* (p. 278). DNPM.

Jorge, M., Prous, A., & Ribeiro, L. (2007). *Brasil Rupestre/Brazil on Rock*. Curitiba: Zencrane Livros.

Laming-Emperaire, A. (1962). *La Signification de l'arte rupestre paleolithique*. Paris: Ed. Picard.

Laming-Emperaire, A., Prous, A., Moraes, A., Beltrão, M. da C. (1974). Grottes e Abris de la région de Lagoa Santa. *Cahiers d'Archéologie d'Amérique du Sud, 1*.

Lanari, C. H. (1909). A Lapa do Caetano. *Anais da Escola de Minas de Ouro Preto, 11*, 15–35.

Lommel, A. (1978 [1966]). A arte pré-histórica e primitiva. *Coleção O Mundo da Arte*, Rio de Janeiro, Britannica do Brasil, *1*, 176.

Mattos, A. (1940). Material lítico, cerâmico e inscrições da Lapa vermelha em Minas Gerais *Congresso Sulriograndense de História e Geografia*, Porto Alegre, vol. 3 [não consultado].
de Moura, M. T. T. (1994). O Carste da Região de Prudente de Moraes – MG. *Boletim Paulista de Geografia, 73*, 125–148.
Neves, W., & Piló, L. B. (2008). *O Povo de Luzia- em busca dos primeiros americanos*. São Paulo: Ed. Globo.
Neves, W., Araujo, A. G. M., Bernardo, D. V., Kipnis, R., & Feathers, J. K. (2012). Rock art at the Pleistocene/Holocene boundary. *Plos One, 7*(2), e322228.
Padberg-Drenpohl, J. (1926). *Relatório de duas excursões à região calcária de Lagoa Santa em 1926*. Manuscrito. Rio de Janeiro: Museu Nacional.
Padberg-Drenkpohl, J. (1929). *Relatório de excursão de 1929 à região calcárea de Lagoa Santa*. Rio de Janeiro: Manuscrito, Museu Nacional.
Pallestrini, L. (1969). *Pinturas Rupestres Brasileiras*. São Paulo: Editora Paestum.
Penna, J.P. 1964 Pinturas e Gravuras Rupestres de Minas Gerais. Origens do Homem Americano, São Paulo, Instituto de Pré-História da Universidade de São Paulo., 2° Encontro Intelectuais.2 419–421.
Prous, A. (1977). Missão de Estudo da Arte Rupestre de Lagoa Santa. *Arquivos do Museu de História Natural da UFMG, 2*, 51–65.
Prous, A. (1985/86). Direções de pesquisa na análise da arte rupestre: exemplos de análises punctuais. *Arquivos do Museu de História Natural UFMG, 10*, 196–224.
Prous. (1989). Las tentativas de datación de la arte rupestre. *Boletin de la SIARB, 3*, 19–29.
Prous, A. (1994). L'Art rupestre du Brésil. *Bulletin de la Société Préhistorique de l'Ariège, 49*, 77–144.
Prous, A. (2002). P. W. Lund e a arqueologia brasileira. *O Carste, 14*, 50–51.
Prous, A. (2007). *Arte Pré-Histórica do Brasil*. Belo Horizonte: C/Arte.
Prous, A. (2012). Le plus ancien art rupestre du Brési central: état de la question. In J. Clottes (Ed.), *Préhistoire* (pp. 719–734). Paris: Art et Société.
Prous, A., & Baeta, A. (1992/93). Elementos de Cronologia, Descrição de Atributos e Tipologia. *Arquivos do Museu de História Natural UFMG, 12/13*, 241–332.
Prous, A., & Baeta, A. M. (2001). Arte Rupestre do Rio Peruaçu. *O Carste, 13*, 36–42.
Prous, A., Lana, A. L., & Paula, F. (1980). Estilística e cronologia na arte rupestre de Minas Gerais. *Pesquisas, Série Antropologia, 31*, 121–146.
Prous, A. & Paula, F. (1979/80). L'art rupestre dans les régions explorées par Lund (Centre de Minas Gerais, Brésil). *Arquivos do Museu de História Natural da UFMG, 3*, 311–335.
Prous, A., & Seda, P. (1987). Arte Rupestre: cronologia, tradições e metodologia na arte rupestre do sudeste. *Boletim do Instituto de Arqueologia Brasileira, Série Catálogo, 3*, 177–181.
Prous, A., Baeta, A., & Rubbioli, E. (2003). *O Patrimônio Arqueológico da Região de Matozinhos – Conhecer para proteger*. Belo Horizonte: Grupo Votorantim.
Ribeiro, P. C., Carvalho, H. R., Cardoso, F., Prous, A., Barros, H. L. de, & Acosta-Avalos, D. (2007). Arqueomagnetismo aplicado ao estudo das pinturas rupestres da Serra do Cipó-MG. In A. P. de Loures Oliveira (Org.), *Arqueologia e Patrimônio de Minas Gerais* (pp. 57–71). Editar, Juiz de Fora.
Rubinger, M. M. (1956/1959). *Anotações de Campo datadas de 1956 e 1959*. Manuscrito.
Rubinger, M. M. (1979). *Pintura Rupestre-Algo Mais do que Arte Pré-Histórica*. Belo Horizonte: Ed. Interlivros.
Siqueira, A. S., Mota, J. F., & Prous, A. (1989a). Altamira: um sítio homogêneo da Tradição Planalto. *Anais da 4a Reunião Científica da Sociedade de Arqueologia Brasileira Dédalo, Universidade de São Paulo, 1*, 287–296.
Siqueira, A., Mota, J. F., & Prous, A. (1989b). L'art rupestre de La Pedra Pintada de Cocais, Minas Gerais, Brésil. *Musée de l'Homme, Institut de Ethnologie, Archives et documents, Micro-Edition*, Paris, R 89 039 534.
Soares, S. M. (1977). Arte Rupestre-significado e abordagem metodológica. *Nheengatu, 3/4*, 39–46.

Walter, H. (1948). *A pré-história da região de Lagoa Santa (Minas Gerais)*. Belo Horizonte.
Walter, H. (1958). *Arqueologia da Região de Lagoa Santa, Minas Gerais; índios pré-colombianos dos abrigos-rochedos*. Rio de Janeiro: Ed. Sedegra.
Walter, H. V., Cathoud, A., & Mattos, A. (1937). The Confins man. A contribution to the study of early man in South America. In G. G. MacCurdy *Early man* (pp. 341–348). Philadelphia: J. B. Lippincott.

Filmography

"Arqueologia no Brasil", filme em 35 mm produced by the Cinema Department of the Escola de Belas Artes da UFMG in 1977 funded by Embrafilme (text and script: A. Prous & Paulo Junqueira).
Programa Bem Cultural (Pré-História em Minas Gerais) 5 Seções, Rede Minas, Fapemig\Fundação TV Minas Cultural e Educativa, Belo Horizonte, 2012. (https://www.youtube.com/watch?v=wHI9V2poNt8)

Chapter 16
Lithic Technology in Lagoa Santa in the Early Holocene

Lucas de Melo Reis Bueno and Andrei Isnardis

Abstract In this chapter, we delineate a brief history of the interpretations about lithic technological variability in Lagoa Santa region during the early Holocene. We begin presenting some considerations about Lagoa Santa lithic industry made by studies undertaken during the nineteenth and twentieth centuries. After that, we present the general characteristics of two lithic collections that we have studied and that were made on the course of the twentieth and twenty-first centuries. We discuss the interpretations formulated regarding the composition and significance of the lithic industry in the Lagoa Santa region, especially in the aspect of its role as an indicator of time and space. The central issue concerns the proposals of artefact homogeneity and variability and their implications for understanding the process of occupation of the Lagoa Santa karst. Based on these data, we propose, even though the lithic technology of Lagoa Santa continues to be largely undefined, that the image of a simple, homogenous lithic industry of little interest is no longer sustainable. What we see is a technological organization that integrates different areas: the circulation of raw material, the combination of different techniques in exploiting those raw materials and the production of artefacts with diversified uses and meanings, constructing a mosaic of choices that define the specificity of that industry.

Introduction

Ever since the early nineteenth century, the Lagoa Santa region has played and continues to play an essential role in the formation of Brazilian archaeology as the focus of intense debate in national and international spheres. This book portrays the

L.M.R. Bueno (✉)
History Department at the Federal University of Santa Catarina,
Campus Universitário, Bairro Trindade, Florianópolis, SC CEP: 88040-970, Brazil
e-mail: lucasreisbueno@gmail.com

A. Isnardis
Anthropology and Archaeology Department and researcher of the Archaeology Sector at the Museum of Natural History and Botanical Gardens of the Federal University of Minas Gerais, Rua Gustavo da Silveira, 1.035, Santa Inês, Belo Horizonte, MG, Brazil
e-mail: isnardis@gmail.com

P. Da-Gloria et al. (eds.), *Archaeological and Paleontological Research in Lagoa Santa*, DOI 10.1007/978-3-319-57466-0_16

trajectory of that role and highlights the main points of interest and discussion. Quantitatively, the lithic remains constitute the majority of the archaeological remains collected by various researchers who have investigated the region, but, with few exceptions (see Beltrão 1974; Hurt and Blasi 1969; Walter 1958), they have received little or no attention from those scholars.

In this chapter, we delineate a brief history of considerations made, concerning the Lagoa Santa lithic industry, by studies undertaken in the course of the nineteenth and twentieth centuries, followed by a presentation of the general characteristics of two collections, namely, (1) the collection of lithic remains formed in the course of the project "Origins and Microevolution of Man in the Americas: a paleoanthropological approach" and (2) the collection of lithic remains stored in the National Museum of the Federal University of Rio de Janeiro (UFRJ). Lastly, we will discuss the interpretations formulated regarding the composition and significance of the lithic industry in the Lagoa Santa region, especially in the aspect of its role as an indicator of time and space. The central issue concerns the proposals of artefact homogeneity and variability and their implications for understanding the process of occupation of the Lagoa Santa karst.

History of the Research

For almost 200 years now, the Lagoa Santa region has been the object of archaeological research, but the lithic industry there has never been a top priority for any of the researchers. In this chapter, we present a brief synthesis of the production of several authors with the intention of highlighting the character that most of them seem to attribute to the lithic material of that region in the aspect of its technological variability.

In various works, we can readily recognize a set of characteristics attributed to the lithic collections from the excavated sites, which are almost exclusively rock-shelters. The first characteristic is the abundant presence of quartz flakes and other quartz fragments at all the sites. The second is the presence of polished stone axe blades, which present different degrees of polishing and or previous knapping. The third aspect is the notable absence of standardized artefacts that can easily be recognized because the secondary alterations produced in the flakes and other surfaces are slight and not formalized. The fourth, also shared by various authors, is the absence of great variation among the different stratigraphic layers, obviously only commented on by those authors that took the stratigraphy into account.

In the pioneering research that Peter Lund undertook in the nineteenth century and Padberg-Drenkpohl in the early decades of the twentieth century, the lithic production is not the object of any discussion at all. That is mainly due to the fact that their energies were directed at what they considered to be most notable and scientifically relevant in that regional context, namely, the possible coexistence of humans with extinct Pleistocene fauna species, as proposed by Lund and discussed by Padberg-Drenkpohl. A second reason for the lack of interest in the lithic artefacts could be associated to their lack of formal standardization, as we believe that if there had been any sophisticated artefacts with an evident morphological standard or that appealed to the aesthetic sensibility of the researchers, then those artefacts might have been the object of important considerations (especially in regard to solving the question of the possible coexistence and predation of the Pleistocene megafauna).

Unlike the two pioneer researchers, the amateur archaeologists who composed the Minas Gerais Academy of Sciences from 1930 to 1950 did conduct analyses of the lithic material and put forward their own interpretative proposals (Hurt and Blasi 1969; Mattos 1938; Walter 1958). H. V. Walter actually proposed a typology for the flaked and polished artefacts and attempted to organize them in an evolutionary scheme, an effort which Prous identified as being obviously inspired by European schemes for establishing prehistoric periods (Prous 2013). What was essentially lacking in Walter's work to support such proposals was information on the stratigraphic contexts of the artefacts (Hurt and Blasi 1969; Prous 2013).

The collaborative activities of Wesley Hurt and Oldemar Blasi marked the beginning of professional archaeological research in the Lagoa Santa region. Their research treated the lithic industries as being important, even though they were not actually the main focus of the investigations. Those authors constructed a typology for the sets of artefacts found in the area, and, combining it with contextual information, they proposed the existence of a Cerca Grande complex, embracing the totality of the various categories of remains (among them the lithic material) and corresponding to the pre-pottery occupations of the karst sites.

Within the span of that overarching complex, however, the authors did not manage to clearly distinguish periods or sub-sets because they could not detect any notable changes related to the stratigraphy of the sites. Furthermore, even though there were outstanding spatial variations, it proved impossible to understand them other than as being, perhaps, the result of sample biases or possible casual factors.

Their typology was based on inferences regarding the function of the pieces. As an example, they referred to "an amphibolite axe with a wide rounded base", "a large long axe made from an amphibolite pebble", "borer", "cutters", "end scraper", "asymmetrical knives" and so on. What is most noticeable in Hurt and Blasi's work is their interpretation of the huge numbers of quartz pieces they found. Despite the typology, their emphasis was always on those pieces that they recognized as artefacts, while the residues of their production were never the object of any in-depth analysis. Here is what Hurt (1960: 583) had to say about the lithic assemblage:

> The stone industry was composed predominantly of quartz crystal flakes. Hundreds of flakes and fragments of quartz crystal flakes were found. It is improbable that all these flakes and fragment represent the spoils and rejects from tool manufacture, but seem to be the tools themselves. If they were the spoils and rejects we would expect to find an accompanying large number of deliberately formed tools. Such was not the case, for only a few artifacts have a deliberated fabricated form or retouched edge.

The extensive work carried out by the Franco-Brazilian Archaeological Mission coordinated by Annette Laming-Emperaire included test pits at various sites and the excavation of a large area at the Lapa Vermelha IV in Pedro Leopoldo

(Laming-Emperaire et al. 1974). Various stone materials were obtained in those test pits and in surface collections. The Mission made it possible to establish the basis for a new range of approaches not only to the lithic industry but also to many other aspects of the archaeological deposits (see Chap. 7 of this volume).

The need to describe the local industries, which were obviously different from the European ones, led A. Laming-Emperaire to propose a specific vocabulary to characterize the Brazilian lithic industries in an attempt to establish a common lexicon that would make it possible to compare materials from different regions and support the effort then being made to begin the technological characterization of the various artefacts assemblages. However, the initial descriptions stemming largely from impressions in the field reaffirm the aspects referred to by Hurt and Blasi. In her article, which appeared in the publication of results edited by the Mission in 1975, M. C. Beltrão (1974: 130) synthesizes her perception of the lithic materials collected at the Sumidouro, Lapa Vermelha IV, Lapinha I, Caieiras, Cerca Grande and General Carneiro sites in these words:

> In that assemblage, we can already identify the poverty of the industry, not only from the technical point of view but also from the functional angle (the absence of arrowheads and spear points is particularly noticeable); a certain diversity of types is only found in the pottery level. Most of the tools are small or medium sized and the bigger, heavier ones among them are mostly found at the surface, sometimes associated with small Tupi-Guarani flaked quartz instruments. The most used stone is opaline or hyaline quartz followed by limestone (free translation from the French).

As a result of the activities of the French Mission's researchers, a permanent archaeology nucleus was established at the Federal University of Minas Gerais coordinated by André Prous, who was a member of the mission team. The archaeological research unfolded with the Lagoa Santa karst as its main scenario, but the lithic collections were not intensely exploited (Prous 1991a). The excavations that Prous coordinated were mostly concentrated beyond the limits of the karst in the first lines of quartzite outcrops of the Serra do Espinhaço, that form the limit of the karstic plateau to the east.

The Grande Abrigo de Santana do Riacho was the top priority for excavation, and a great amount of lithic material came from it. Just like the karst material, it consisted largely of quartz flakes. In the case of the Santana do Riacho industries, systematic analyses were carried out with special attention to identifying the technology used to manufacture the pieces, and, for the first time in the central region of Minas Gerais, the totality of the lithic material was attributed due to its importance in the analyses (Prous 1991b).

Various articles appeared based on that work describing the industries associated to different moments of occupation, the stone flaking methods (Prous 1991b) and the bipolar flaking technique (Prous and Lima 1986/90), which was predominant at Santana do Riacho, and, in the years that followed, came to be identified in various other Brazilian archaeological contexts, including some sites in the Lagoa Santa karst.

At that time, however, technological studies were restricted to the Serra do Cipó assemblage, and it proved necessary to wait for years until research could be taken up at the sites in the Lagoa Santa karst by the Origins project (Neves and Piló 2008; see Chap. 9 of this volume), in the beginning of the twenty-first century, for studies with the lithic material were undertaken once more.

In 2007, F. Pugliese's dissertation for a master's degree, entitled *Os sítios de Lagoa Santa: um estudo sobre organização tecnológica de caçadores-coletores do Brasil Central* (The Lagoa Santa sites: a study of the technological organization of the hunter-gatherers of Central Brazil), took up once more systematic analyses targeting the collections from the Lapa do Santo and Lapa das Boleiras sites in which he underscores the notable lack of variation in the artefacts within and among the different sites. The author notes that eventual variations are merely quantitative and not qualitative, and he attributes that to the greater or lesser intensity in the exploitation of resources in the vicinity of each site.

Pugliese's interpretation, relating the degree of variability among the industries to the functionality of the sites, indicates an understanding whereby the sites are areas for specific activities "given that the low degree of variability of the instruments is not consonant with residential kinds of sites" (Pugliese 2007: 123). The same author (Pugliese 2007: 125) states that "the quality and stratigraphic distribution of the lithic material together with its temporal amplitude suggest that the rockshelters were places for specific activities, and that they were intensely abandoned and re-occupied during the Early Holocene".

Thus, to some extent, the results of Pugliese's work underscore Hurt and Blasi's affirmations of a low degree of variability in the artefacts, and, whatever differences do exist among the sites and stratigraphic levels of the same site, they are largely quantitative.

Lagoa Santa Lithic Assemblages: The Origins Project Activities and the National Museum Collection

During postdoctoral research entitled "Lithic technology in Central Brazil: diversification and regionalization at the beginning of the Holocene" conducted by one of us (LB), we have analyzed lithic collections coming from excavations of archaeological sites in Lagoa Santa conducted by the Origins project. A preliminary analysis was also made of the archaeological material from the same region stored in the National Museum of the Federal University of Rio de Janeiro (MN-UFRJ).

In the case of the first analysis of the former material, the assemblage is composed by lithic collections from the sites excavated by the Origins project from 2000 to 2009 together with remains coming from spatially dispersed superficial archaeological occurrences. Among the sites in rockshelters, we took particular care with the organization of the Lapa do Santo collection, which had been the object of excavations since 2001, and of the Lapa das Boleiras site, excavated from 2001 to 2003. In the case of the former collection, we have proceeded with mass analysis in part of the collection, analyzing almost 10,500 pieces. For Lapa das Boleiras assemblage, we have made a quantitative analysis with artefacts and core pieces, comparing the results with earlier work done by Francisco Pugliese in his master dissertation (Pugliese 2007).

Apart from the two sites in rockshelters mentioned above, an effort was made to organize all the available information regarding open-air sites and archaeological occurrences, consisting mainly of lithic remains. Special attention was dedicated to the sites around the Sumidouro Lake. That group of sites and occurrences was selected because of the great antiquity associated to some of them as, for example, the Coqueirinho site dated at around 10,240 BP (Bueno 2010) and the Sumidouro site with dates around 8200 BP (Araujo and Feathers 2008).

Lapa do Santo and Lapa das Boleiras: Organization and Analysis of the Collections

The Origins project worked most intensely at the Lapa do Santo archaeological site. It carried out annual excavations from 2001 to 2009, creating an immense collection of lithic remains (Neves et al. 2004, 2008).

Because of the methodology adopted for the excavation and collection of remains at the sites, we have basically two types of samples: material that was position-plotted and material that was obtained by sifting through sieves. The first type consists of remains collected individually with their exact location registered along three coordinates x, y and z. In such material is allocated a number identifying its origin at the site, which corresponds to the number used for marking and analyzing the piece. The second type consists of material retained in sieves, all of them coming from the same unit of excavation. That unit is composed hierarchically of information on the excavation unit, level and feature.

The plotted material amounts to around 4,100 pieces and includes remains of varying categories and raw materials. Such pieces were selected during the excavation process mainly based on size criteria. According to the excavation protocol applied at that site, any lithic material bigger than 1.5 cm was to be plotted. Based on an overall examination of the plotted material and the sifted material, we found that compliance with that orientation must have fluctuated somewhat during the various stages of excavation because, while there are no pieces with dimensions of less than 1.5 cm among the plotted material, the opposite is not true; there are pieces with such dimensions to be found in the sifted material.

To facilitate the analysis and make it feasible to obtain an estimate of the general situation of the lithic collection of that site, we reorganized the entire collection grouping the remains according to the area they were excavated from. Thus, all the pieces collected from the same unit were stored in the same box and organized according to stratigraphic levels and features, independently of what stage of curating they were at. Based on that organization, we estimated that the collection consists of around 40,000 lithic remains.

Regarding the analysis for this site, we decided to make a mass analysis (Ahler 1989). To conduct it, we adopted two criteria: raw material and size. During the

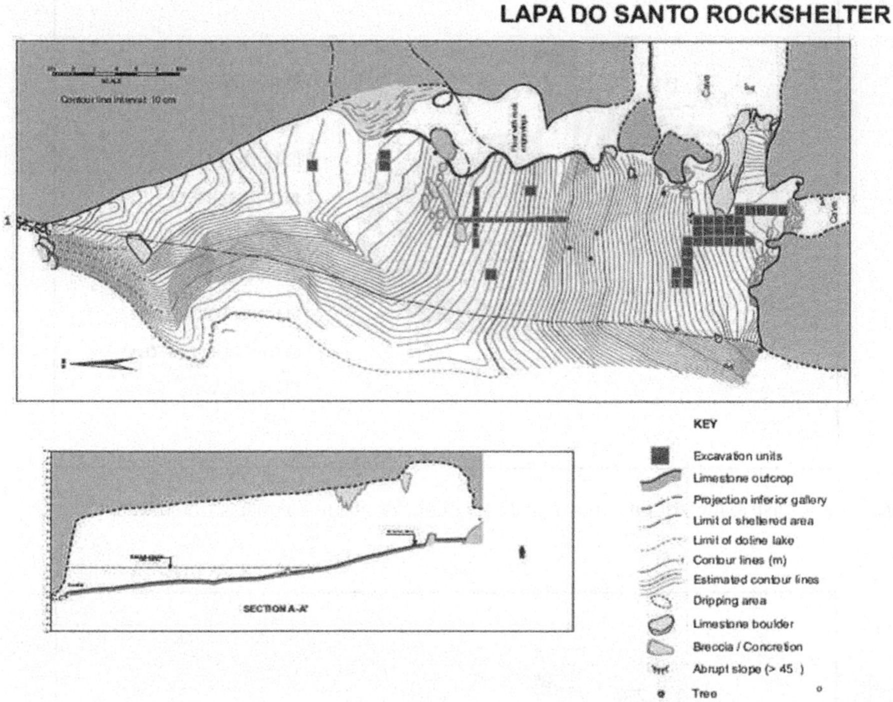

Fig. 16.1 Lapa do Santo archaeological site (Adapted from Araujo et al. 2012)

period of postdoctoral research, we analyzed 10,500 pieces using that method. First, we selected material that had been recovered from test pits opened up in the lower part of the site, because it seemed to be the part that presented greater diversity among the lithic remains. Later, we analyzed all the lithic material recovered from the F12 excavation unit (see Fig. 16.1). According to the dates obtained (Neves et al. 2008; Strauss 2010), all the remains recovered in the course of the excavations are associated to the occupation of the rockshelter during the early Holocene. The overall analysis and the descriptive statistical analysis have both been finalized for the assemblage recovered from the test pits. In the case of the F12 unit, the overall analysis has been finalized, but the statistical processing has not yet been concluded.

Thus, based on the analysis of the material from the test pits, we found the following characteristics displayed in graphic form in Figs. 16.2, 16.3, 16.4 and 16.5.

In addition to the quantitative data obtained and displayed in Figs. 16.2, 16.3, 16.4 and 16.5, a series of qualitative observations were made during the analysis. First, although we identified the presence of flint, quartzite and silicified sandstone, nevertheless, the remains made of hyaline quartz were absolutely predominant, as can be seen from Fig. 16.2. Second, we detected the existence of raw materials exogenous to the Lagoa Santa karstic region such as the quartzite plates commonly found at archaeological sites in Serra do Espinhaço, Santana do Riacho and

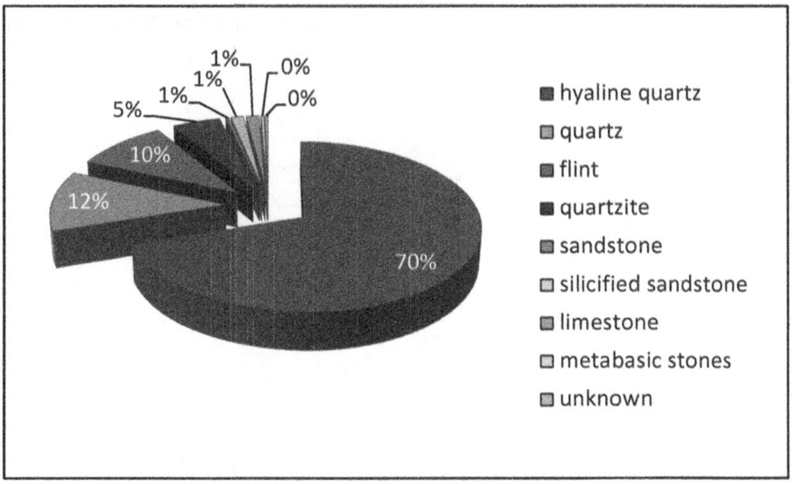

Fig. 16.2 Distribution of lithic raw materials at Lapa do Santo – remains from test pits

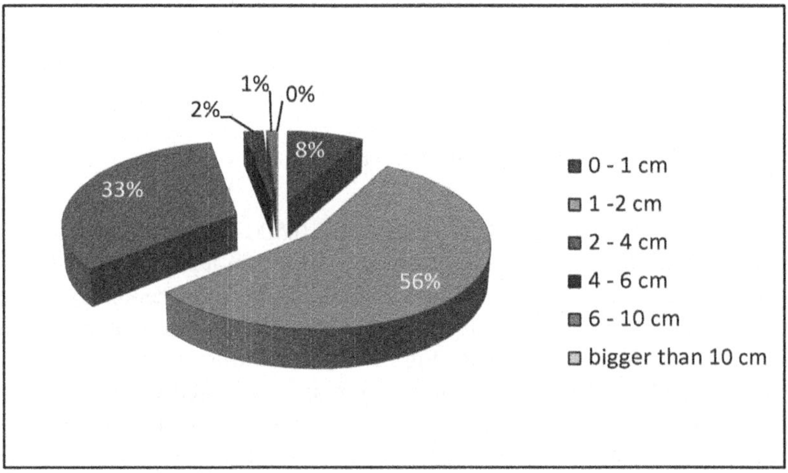

Fig. 16.3 Distribution of lithic size from Lapa do Santo – remains from test pits

Diamantina (Isnardis 2009; Prous 1991a). At the sites in the Serra do Espinhaço, those plates appear as supports for the production of artefacts with plano-convex or trapezoidal cross section and direct and subparallel or flaky reworking distributed along both edges, forming angles ranging from abrupt to semi-abrupt (Isnardis 2009). We found few examples of this kind of artefact at Lapa do Santo and all of them were fragmented, but in some stratigraphic levels, we found quartzite flakes that might have been associated to that reworking and resharpening of the cutting edges of those kinds of artefacts.

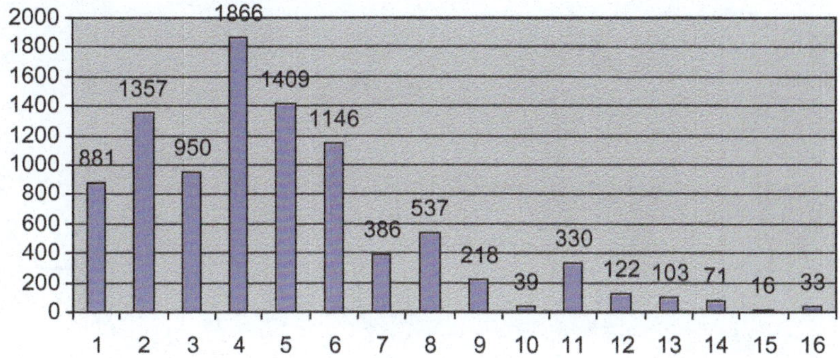

Fig. 16.4 Distribution of lithic remains through stratigraphic levels at Lapa do Santo – remains of the test pits

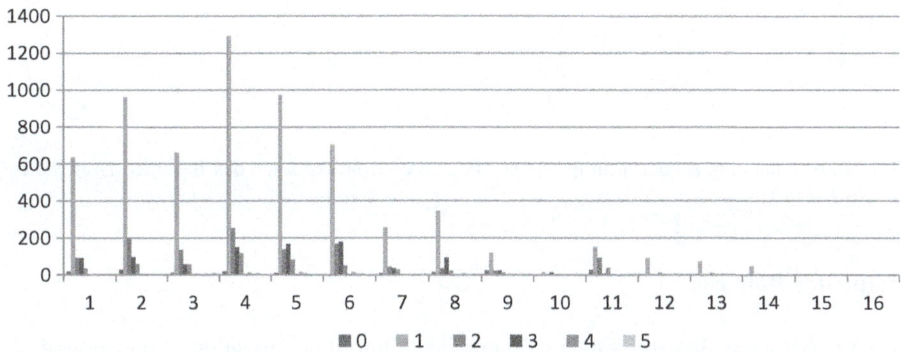

Fig. 16.5 Distribution of lithic raw materials through stratigraphic levels at Lapa do Santo – remains of the test pits

Another type of lithic remains that was found in the material from the test pits excavated at Lapa do Santo consisted of some very fine-grained, silicified sandstone flakes. This raw material is highly suitable for flaking. The flakes found were exceptional, and only a small quantity appeared, invariably associated to formal artefacts with a plano-convex cross section. They are usually small, fine, with a curved profile and a sequence of negatives on the external face whose orientation can vary from parallel to orthogonal and the base is worked. Another category of remains worth mentioning is the pyramidal quartzite cores with a pebble cortical platform and unipolar and unidirectional flaking.

Within the assemblage analyzed, there was a series of remains in flint. Two aspects caught our attention regarding that raw material: (1) the large number of fragments and remains with signs of heat-induced alterations and (2) the absence of any correlation of the fragments of that raw material with the final stages of forming artefacts. In other words, the flint, despite being scarce in the region and responding very well to knapping, did not seem to have undergone any process of appropriation different from that of the hyaline quartz.

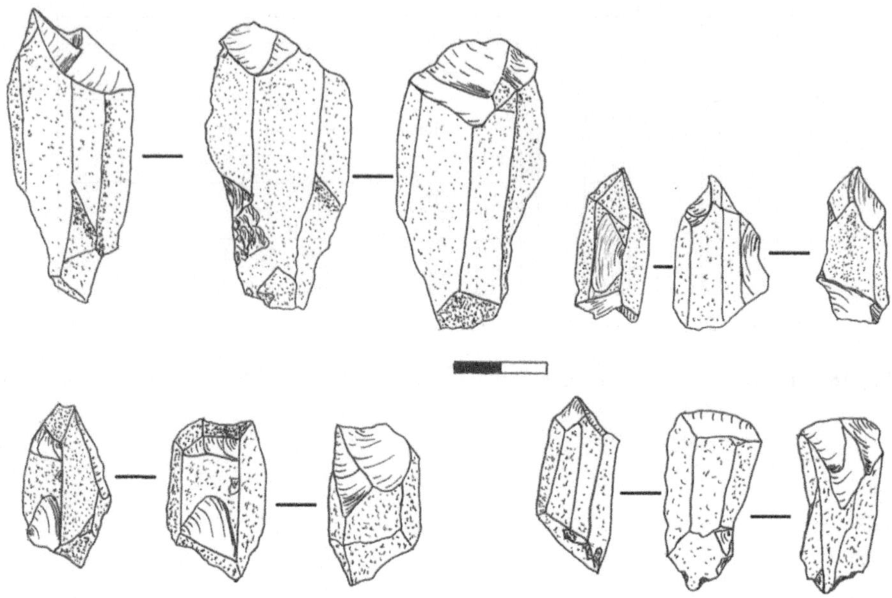

Fig. 16.6 Knapping sequence in quartz crystals cores from the Lapa das Boleiras (Drawing by Lucas Bueno)

Lapa das Boleiras

For the Lapa das Boleiras site, we first made an initial organization of the collection, locating and grouping together remains from the same collecting points and systematizing all the data obtained by Pugliese (2007) in his master's dissertation.

The data showed that 2,229 pieces had been collected from that site and that Pugliese (2007) had analyzed 1,895 (85%) of them. Considering that his analysis involved the observation of the attributes of each piece individually, we were able to extract information from his master's dissertation on the proportions of raw materials and classes of remains as well as information on their distribution among the excavated areas and the stratigraphy.

In our analysis of the material from that site, we observed qualitative aspects regarding the artefact fabrication process and the appropriation of the cores. In the specific case of quartz crystal cores, we identified the following sequence previously presented in Bueno (2012: 74; Fig. 16.6):

> These crystals first present one or a few reductions of the material by direct, unipolar percussion. Flakes are then struck off from the external, cortical surface of the platform defined at the apex of the crystal, alternating the types of flake. The first flake is totally cortical, the second flake may have a part of the body of the flake without cortex and a cortical external surface, and so on in alternating succession until the ridge being used as the percussion plane disappears, that is to say, it is exhausted. The result of that sequence, apart from the flakes it produces, is a core with a relatively flat knapping scar at one end and the

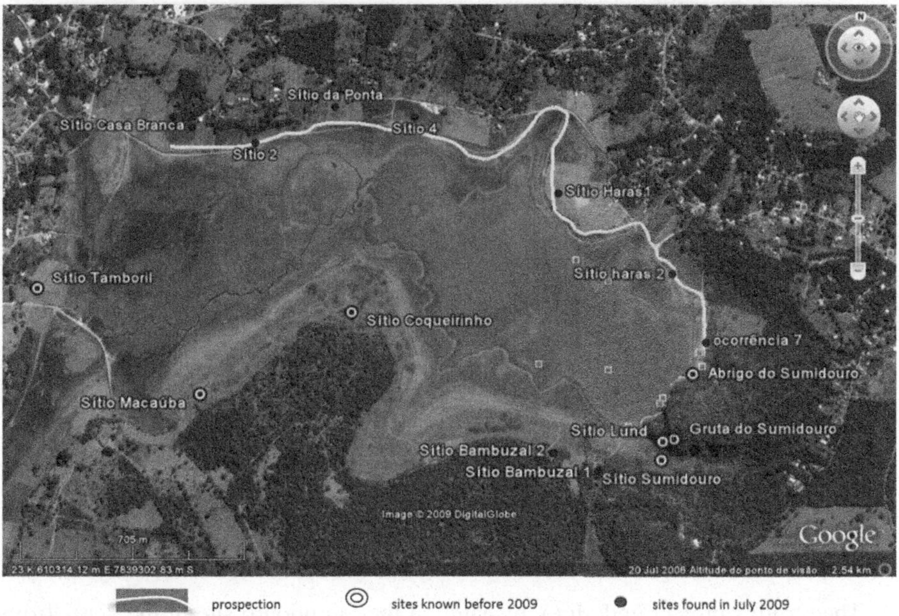

Fig. 16.7 Archaeological sites around Sumidouro Lake

root of the crystal at the other. During the process of reducing the core, it may be rotated to take advantage of the angles available on the lateral faces of the crystal, and that may also serve to reactivate a ridge that has been exhausted. These unipolar reduction strategies can also be associated to bipolar knapping strategies, whereby the force is applied in the same direction as the crystal's morphological axis, or again it may be applied in transversal direction in relation to that axis. In the latter case the form of the core, when it has been exhausted, will be somewhat globular and with the lateral cortexes still present. In some cases, such globular cores are re-used to fabricate small robust edges (Fig. 16.3). At sites like the Lapa das Boleiras and the Lapa do Santo, where there is an expressive number of cores, we can identify the association of those different strategies to the same single piece, while at the Sumidouro and Coqueirinho sites, evidence for the sequence described is obtained from the flakes there.

Open Air Sites and Occurrences Around the Sumidouro Lake

As has been mentioned, we gave priority to two of the various open-air sites around the Sumidouro Lake, namely, Sumidouro and Coqueirinho sites, where the lithic materials are chronologically situated in the early Holocene. Most of the other sites around the lake have very little lithic material, and they are associated to pottery remains (Fig. 16.7). Considering that the focus of the postdoctoral research was the occupation of the region in the early and middle Holocene, the analysis of remains from those other sites was postponed to a future moment.

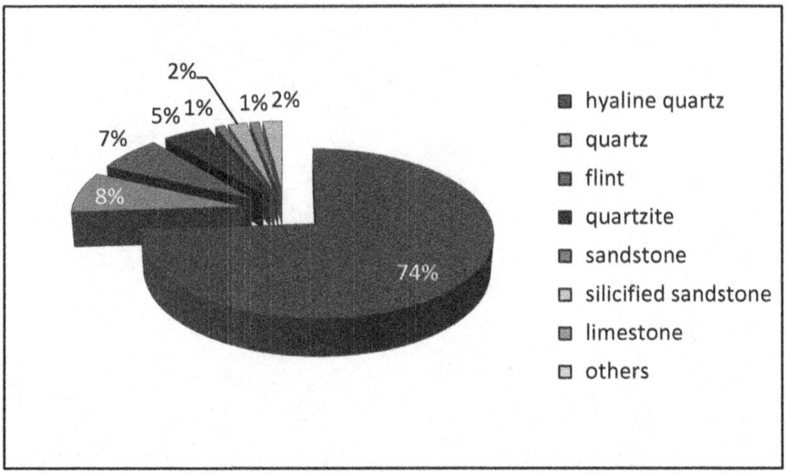

Fig. 16.8 Distribution of raw materials at Sumidouro site

The Origins project undertook research at the Sumidouro site in the years from 2003 to 2009 associated to systematic prospections in the surroundings of the Sumidouro Lake.[1] Despite the intensity of the interventions there, the collection of lithic material it produced is relatively small. Twelve test pits of 1 m^2 were made, and they were responsible for 140 lithic remains equivalent to a density of approximately 11 remains per m^2 (Bueno 2012). Their distribution in the stratigraphy and from one test pit to another did not follow any definite pattern, except perhaps for the existence of slightly greater concentration in the lower part of the site near to the Sumidouro Lake and the Lund site. In addition to the lithic remains, pottery was also found at the site, such as a small bowl apparently broken in situ as well as remains of occupation within historical times that included iron and porcelain. Despite that diversity, both the pottery and the historical materials were only found predominantly in the superficial strata, whereas the lithic material appears in the entire extension of the stratigraphy. Another material identified frequently at this site was the charcoals, many of them forming great concentrations in certain levels of the stratigraphy. There is not always a clear association between such concentrations and the other archaeological remains, and they are often present in the so-called sterile levels (Neves et al. 2008). To investigate this association, a series of dates were made for this site, both from charcoals and from sediment samples (Araujo et al. 2013; Araujo and Feathers 2008). Some of the basic data obtained from this analysis are set out in Figs. 16.8, 16.9, 16.10, 16.11 and 16.12.

Parallel to that quantitative analysis, we made a qualitative analysis of the same material. Regarding the knapping of the quartz crystals, we identified three distinct

[1] For a characterization of the history of the interventions at that site, see Neves et al. (2004, 2008) and Araujo and Feathers (2008).

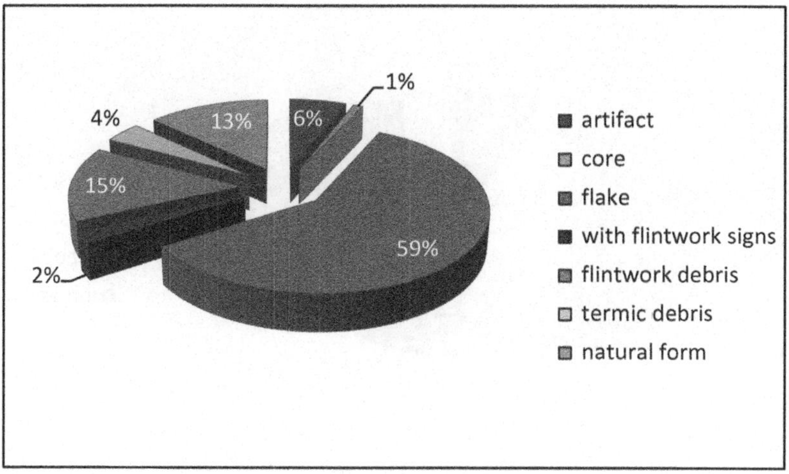

Fig. 16.9 Distribution of lithic remains categories at Sumidouro site

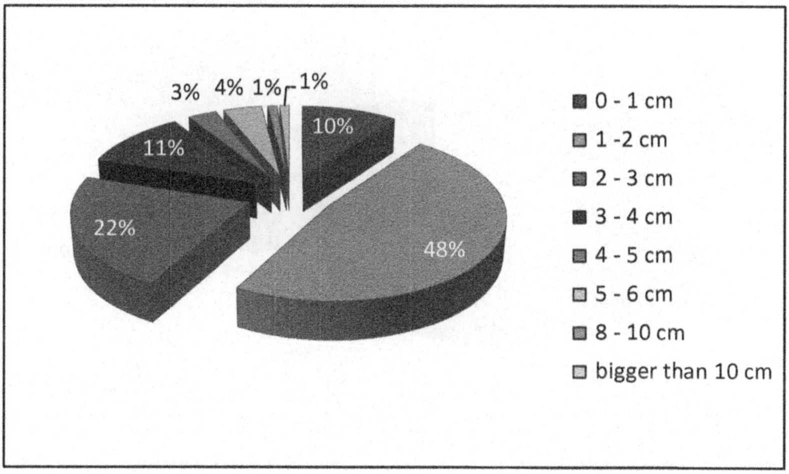

Fig. 16.10 Size of lithic remains at Sumidouro site

sequences: bipolar knapping following the longitudinal axis of the crystal, unipolar knapping of the apex of the crystal, and unipolar lateral knapping of the crystal, striking it in a transversal direction, usually on one of the flat faces at the point where three edges meet, and using as a guideline for the strike an edge that is perpendicular to the longitudinal axis of the crystal. Another aspect that must be mentioned is the small size of the flakes obtained. Many of them, as can be seen in Fig. 16.10, are less than 2 cm long. In some cases, they originated from supports that were also small, albeit there are supports in the collection that could easily provide larger flakes. The question arises here as to reason for the choice of small supports

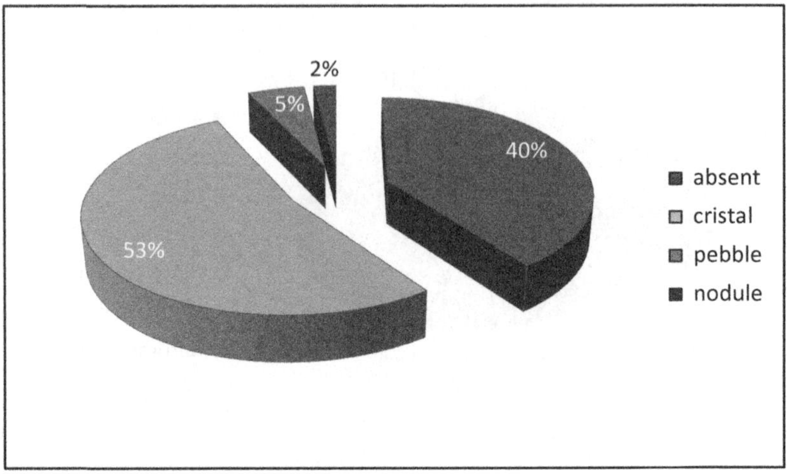

Fig. 16.11 Distribution of cortex types among lithic remains at Sumidouro site

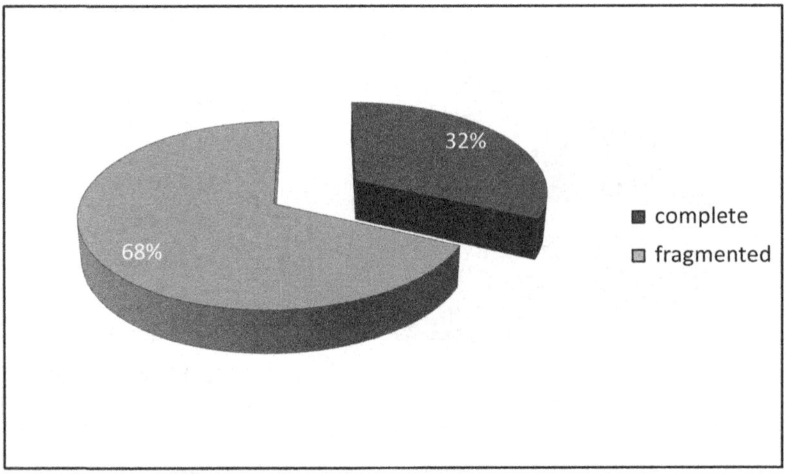

Fig. 16.12 Integrity of lithic remains at Sumidouro site

and small flakes. Considering that among the artefacts there are several that show signs of having been mounted on handles, could it be then that what we have is an industry directed towards producing compound artefacts elaborated by articulating different materials like wood, bone and stone? As we will see in the case of the Coqueirinho site, there are various elements present that favour that hypothesis.

In other archaeological collections with an expressive presence of quartz, similar characteristics to those identified above can be observed. The production of very small flakes and the intense exploitation of cores, whereby they are reduced to small cubic or sub-globular forms, have been described for the early Holocene occupations of the Bibocas II site, in Jequitaí (São Francisco River basin in the north of Minas Gerais) (Bassi 2012) and also for the Diamantina region (in the hills that divide the São Francisco and Jequitinhonha watersheds in the central-northern part of Minas Gerais) (Isnardis 2009). Bassi undertook an extensive, detailed analysis of quartz knapping methods at the Bibocas II site and showed that, contrary to what the literature suggested (Prous and Lima 1986/90), the quartz crystals can be addressed in many different ways and that multiplicity is characteristic of the Bibocas II site. The lesser diversity of methods identified for Lagoa Santa collections must therefore be the result of choices made by the stone knappers themselves and do not stem from any limitations imposed by the morphology of the quartz crystals. However, it must be taken into account that the analysis of Bassi referred to all the remains of the site collection, while in Lagoa Santa, the assemblage analysis was only a partial one, and so it is not possible to eliminate the possibility that the lesser diversity encountered is a reflection of the sample criteria.

One last feature of the Lagoa Santa industries that must be mentioned is the use of varied angles to make full use of the supports. In both the flakes and the artefacts and cores, regardless of the support involved, we have observed recurrent use of all the flat surfaces whether their angles favour knapping or not. That attitude is usually interpreted as a strategy designed to economize the raw material by exploiting it to the maximum until it can yield no further flakes. Its behaviour has been correlated with places or contexts where there is restricted access to the raw materials, either because of the distance or other accessibility restrictions. In the case of Lagoa Santa, however, none of those explanations is applicable because there is an abundance of widely dispersed crystals, making it difficult to imagine that there could have been any social control over their use. That being so, the apparent contradiction between an economizing behaviour and the ready availability of the raw material is a question entirely open to discussion.

The same contradiction between the intense and extended exploitation of the cores and the abundant presence of raw material in both regions, especially quartz, is also apparent in the cases of Diamantina and Jequitaí (Bassi 2012). In Diamantina, in particular, the quartz crystals, some of them of well over 10 cm in size, were abundant up until the twentieth century, when crystal mining began in the region. The Bibocas II site is only a short distance from a large, active quartz mine (Bassi 2012). The phenomenon appears to be recurrent, and comparative analyses among the regions, taking into account the specificities of their contexts, could be very productive for gaining a better understanding of the stone knappers' behaviour.

We used the same methodology applied to the Sumidouro site[2] material for the collection formed by lithic remains coming from the interventions at the Coqueirinho site. Although there were fewer interventions at the latter site, the collection is considerably larger and 385 remains were collected. Some of the basic data concerning the lithic material from that analysis are set out in Figs. 16.13, 16.14, 16.15, 16.16 and 16.17.[3]

[2] See Bueno (2012) for details of the methodology adopted.

[3] See Bueno (2010) for a more detailed discussion of the Coqueirinho site lithic assemblage.

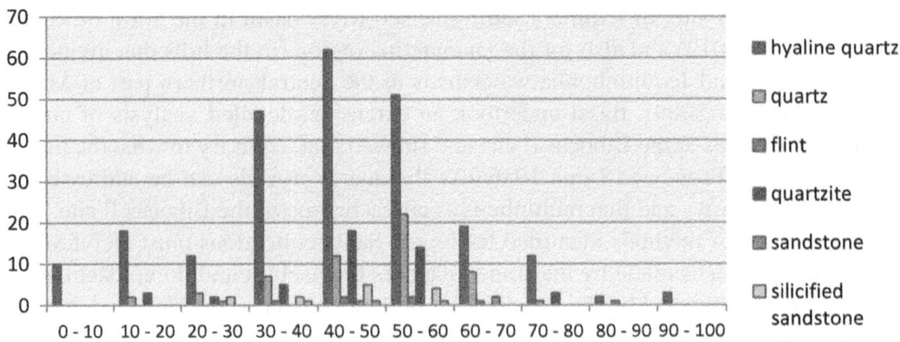

Fig. 16.13 Distribution of lithic raw materials in 10 cm levels at Coqueirinho site

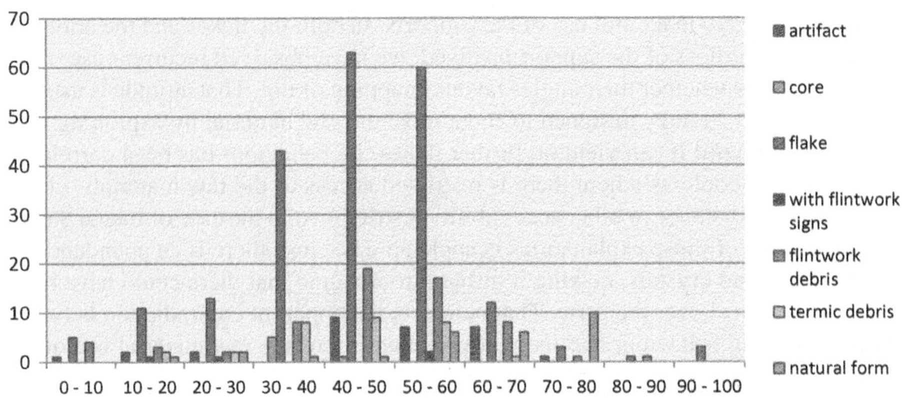

Fig. 16.14 Distribution of lithic remains categories in 10 cm levels at Coqueirinho site

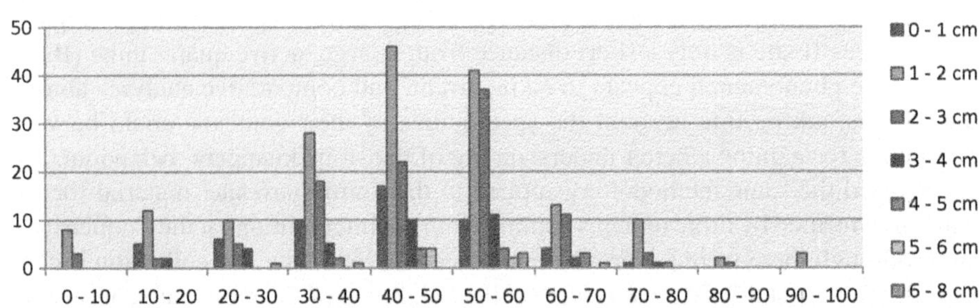

Fig. 16.15 Size of lithic remains size in 10 cm levels at Coqueirinho site

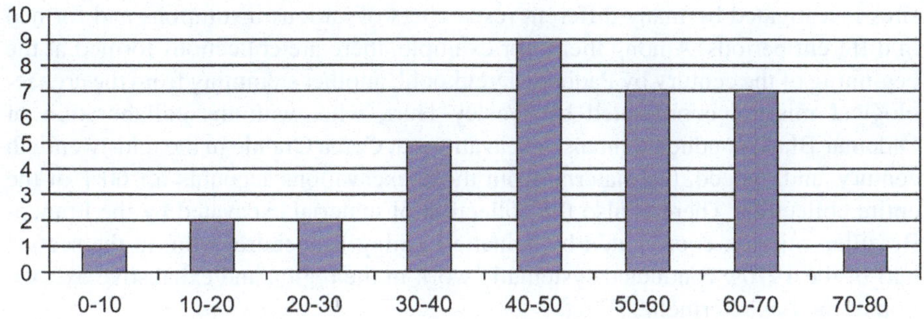

Fig. 16.16 Artefacts distribution through stratigraphic levels at Coqueirinho site

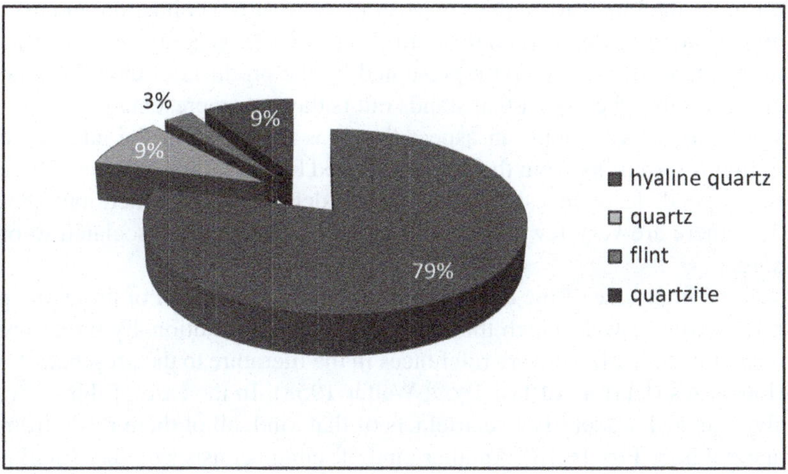

Fig. 16.17 Artefact raw materials at Coqueirinho site

The Lagoa Santa Collection in the National Museum/Federal University of Rio de Janeiro (UFRJ)

To complement the activities involving the lithic collections of the Lagoa Santa region, we made a technical visit to the National Museum in Rio de Janeiro to assess the situation and potential of the Lagoa Santa collection of lithic remains. During our stay at that institution, we were given access to the entire collection which embraces many sites in the region that were excavated during the twentieth century by different researchers, including those of foreign missions. It is worth mentioning that within that collection, there are remains originating from emblematic sites like the Lapa Vermelha IV and the Cerca Grande complex.

The lithic remains from the Lagoa Santa region in the archaeological collections of the National Museum (UFRJ) encompass specimens originating from various sites investigated by many different researchers of various institutions and formed in different periods. Among them, for example, there are collections formed at the beginning of the century by Padberg-Drenkpohl, another stemming from the archaeological mission coordinated by Wesley Hurt, who, with the collaboration of Oldemar Blasi, conducted intense excavations at Cerca Grande in the mid-twentieth century, and, indeed, the material from those excavations accounts for 60% of the entire collection. There is also the collection of material excavated by the Franco-Brazilian mission coordinated by Annette Laming-Emperaire, who, in the second half of the 1970s, conducted systematic work in the region and exhaustive excavation of the Lapa Vermelha IV site.

It was not possible to calculate the total size of the collection, but there are 20 boxes accommodating thousands of lithic remains. During the time one of us (LB) was at the institution, the natures of the contents of all the bags of material were verified in all the boxes to acquire an overall view of the collection and assess its potential for a more detailed analysis in the future. In general, there is a notable predominance of hyaline quartz represented by the products of unipolar knapping of quartz crystals. One aspect that stands out is the poor representation of any indications of bipolar knapping, an aspect that was identified in the analysis of the lithic remains stemming from the sites excavated by the Origins project. Obviously that aspect needs to be investigated in greater detail and quantified, but generally speaking, there are very few flakes or fragments that can be associated to bipolar knapping.

Given the huge size of the collection, the low representation of projectile points is also surprising. Even though this lithic industry has traditionally been classified as simple or expedient, there are references in the literature to the presence of points of various sizes (Hurt and Blasi 1969; Walter 1958). In the case of this collection, we only managed to identify five artefacts of that kind, all of them made from hyaline quartz flakes (Fig. 16.18). Another kind of remain consists of very small globular hyaline quartz cores, proof of intense utilization of that material and taking advantage of supports in various directions and angles, permitting the production of innumerable tiny flakes. This last feature is recurrent in the collections, that is, a predominance of very small remains despite the existence of much larger supports, even when the supports are actual crystals (Fig. 16.19).

Yet another category identified in the same collection is that of polished axe blades. We identified two such blades, both from the Cerca Grande 6 site (Fig. 16.20). One of them is very well polished and produced on a limestone support came from area B, test pit 1 of the units 3–4, level 4. The dating for level 3 is around 9,000 years BP (Hurt 1960). Thus, in a similar way to what we found at the Lapa do Santo and at the Lapa das Boleiras, this and perhaps one other blade found at Cerca Grande 6 (but for which we do not have a stratigraphic reference) are indicators of the knowledge and use of polishing techniques in the fabrication of lithic artefacts at least 9,000 years BP. That raises a series of questions, which only two of them we will examine here. Why was it that the practice was not disseminated and did not lead to an intensification of the production of that kind of artefact practically until 2,000 years BP, which is when we start to identify that kind of artefact more frequently in

Fig. 16.18 Projectile points in hyaline quartz. Lagoa Santa Collection. National Museum/UFRJ (Photo by Lucas Bueno)

Fig. 16.19 Knapped hyaline quartz crystal. Lagoa Santa Collection. National Museum /UFRJ (Photo by Lucas Bueno)

Fig. 16.20 Axe blades from the Cerca Grande 5 site. Lagoa Santa collection. National Museum – UFRJ (Photo by Lucas Bueno)

the archaeological record? Could the presence of this kind of artefact, usually associated to the felling and trimming of trees, be an indicator of forest management?[4]

The assemblage of remains we have described would be more than sufficient to indicate the potential of that collection, even considering the scarcity of primary information. Even so, there is yet another kind of remains that boosts its potential even further. We identified a series of artefacts produced from flint flakes, some of which are plano-convex (Fig. 16.21). This category is unprecedented in the collections studied so far and introduces new elements for discussion regarding the organization of the lithic technology associated to the occupation of rockshelters in the region. For this particular category of remains, the stratigraphic data indicate an association with layers dated at least 9,000 years BP.

Putting all those elements together, what we can discern for the Lagoa Santa region is a scenario of intense occupation, in which both standardization and variability occur. If we only take a quick look, all the sites seem to present the same general aspect, namely, thousands of flakes of hyaline quartz produced by unipolar knapping of quartz crystals and used in their crude state or with very little secondary modification. However, in the midst of that scenario of apparent homogeneity, certain elements stand out that are important insofar as they enable a characterization of greater diversity. They are the polished artefacts found in very ancient strata (a relatively rare occurrence for the older lithic industries in Brazil); the presence of formal, bifacial artefacts (including projectile points) and formal unifacial artefacts; and the presence of raw materials from outside the karst area. Another extraordinary aspect is the sheer magnitude of the collections with many of the sites generating samples of thousands of pieces. That means the region was intensely occupied with

[4] It should be remembered that polished blade artifacts may have had other uses apart from felling trees. They could be used to produce sculpted items (canoes) or to extract materials like honey from hollows in trees.

Fig. 16.21 Flint artefacts from the Cerca Grande 6 site. Lagoa Santa collection. National Museum – UFRJ (Photo by Lucas Bueno)

variation in the activities and possibly in the roles played by the sites themselves, which were articulated in much wider system of occupation that most certainly involved areas beyond the karst itself. The composition of the overall set of artefacts shows the combined use of different strategies for obtaining, circulating, and producing lithic artefacts, articulating techniques such as unipolar and bipolar knapping and polishing as well as the production of unifacial and bifacial artefacts. Underlying an apparent homogeneity, there is a significant diversity that probably corresponds to the application of differentiated strategies and performances, and in that articulation lies the key to understanding the choices that guided the production of this lithic assemblage.

A Simple Industry?

As we pointed out in the brief historical outline at the beginning of this chapter, in most of the published works consulted, the same aspects of the Lagoa Santa lithic assemblages are repeated, namely, (1) the remains are predominantly flakes and fragments of hyaline quartz resulting from knapping, (2) in some sites polished axe blades appear, (3) formal artefacts are practically absent, and (4) most of the sites indicate the absence of significant modifications in lithic technology in the entire extension of the stratigraphy.

As we have pointed out in the first section of this chapter, that description and the interpretations stemming from it give an impression of synchronic and diachronic homogeneity in the Lagoa Santa lithic industry. However, we have clearly shown, in the characterization of the lithic material from a few of the sites (Lapa do Santo, Lapa das Boleiras, Cerca Grande, Coqueirinho, Sumidouro), the existence of diversity, at least in the composition of the artefacts assemblage. There are polished artefacts that date back to the early Holocene; there is a production of formal bifacial and unifacial artefacts; there is the use of materials that are exogenous to the karst; sometimes, on the same core, there are signs that a combination of unipolar and bipolar knapping techniques was used; and, lastly, there are significant differences among the compositions of the artefacts assemblage produced from site to site (among the sites analyzed). This last aspect can be seen, for example, between the assemblages from Lapa do Santo and the Lapa das Boleiras (as discussed by Araujo and Pugliese 2010).

While all those points are important and deserve to be highlighted (see Bueno 2012 for a detailed discussion of each one), the question of raw material deserves special mention. There seems to be a consensus among the researchers that the most abundant raw material in the collections, and the one that is most readily accessible locally, is quartz in its hyaline or milky forms, and its main sources are in the vicinity of the town of Pedro Leopoldo in outcrops on the crystalline basement, or in the form of pebbles along the terraces of the Rio das Velhas river (Prous et al. 1998). We do not have a systematic inventory available for the other raw materials, but the information that is available indicate the following distribution: (1) amphibolite and other resistant rocks were probably collected from the Ribeirão da Mata stream to the north of Pedro Leopoldo; (2) compact hematite was brought in from the iron-ore bearing region of central Minas Gerais (*Quadrilátero Ferrífero*); (3) sillimanite was obtained from Conceição do Mato Dentro or from the upper Jequitinhonha river; (4) quartzite plates would have come from the Serra do Espinhaço (or even from Santana do Riacho itself); (5) the quartzite pebbles could be from the terraces of the Rio das Velhas River; (6) the jasper and the chalcedony are of uncertain origin but are probably from more than 60 km away (Araujo and Feathers 2008; Prous 1991a; Prous et al. 1998); and (7) the silexites, whose closest known source up to this moment is to the southwest in the region of the upper course of the São Francisco River (Koole 2008).

Another important point stemming from the general characterization set out in this chapter concerns the size of the lithic collections from each site. It is important to understand that the numbers are directly related to the intensity of the interventions and the dimensions of the sites, but, generally speaking, we can say that there is an enormous quantity of lithic remains that were produced during the occupations of the karst and that there are tens of thousands of remains in each one of the rockshelters that were occupied. To that we can add two other important considerations that characterize the technological system in that area: (1) there is an important association between those lithic remains and the human burials, and (2) there is no evidence of any direct spatial association between the remains of human activity and the fossils of megamammals that occupied the region in that period (Hubbe et al. 2013).

According to the delineated characterization, we can say that the Lagoa Santa lithic industry, when contrasted with the expectations stemming from theoretical aspects, presents a most unusual situation. While, on the one hand, there are elements that corroborate the idea of a very simple technology, given the use of local materials with little secondary transformation of the supports, on the other hand, there is a production of formal artefacts, use of exogenous raw materials, intense utilization of the local raw material, and, even though it was locally available, we could say that there existed both a supply and a "stockpiling" of raw material in the rockshelters, where there is evidence of continuous reoccupation. Moreover, the way the main material in the local environment, hyaline quartz, was appropriated is indicative of a strategy to make the very most of the material, with cores being knapped down to the limit using a combination of different techniques and showing the existence of refined knowledge of the nature of the raw material, how to handle it, and how to obtain the maximum possible production of small-sized flakes. In other words, despite its abundance and availability, the hyaline quartz was intensely exploited, a combination of characteristics that has often been presented as "antagonistic" in the research into the organization of lithic technology (Andrefsky 1994; Nelson 1991; Shott 1996). To gain an understanding of that configuration, we propose that the combination should be understood in a contextual framework. This proposal is also intended to reinforce the idea that concepts such as simplicity or expediency and curated have been used on many occasions much more to describe than to explain (Shott 1996). Those concepts should operate heuristically to organize and arrange the research and our questions and not as closed, uniform concepts. Each local lithic assemblage should be reflected on and analyzed in the light of those concepts, so that we can understand their specificities and singularities, and on that basis, we will be able to discuss their meaning and their potential and achieve a comprehension of the technological variability in a systemic perspective.

In that sense, the question, or questions, we select for discussion involve that combination of intensive exploitation of the quartz and its local abundance and accessibility, the immense quantity of remains, the spatial relation to human burials and at the same time the association with the megafauna and the choice of deliberately producing small flakes and artefacts.

Another important point to bring into the discussion, and one we have spoken about in the chapter, involves the presence of partially polished axe blades in strata associated to a very ancient period of occupation of the region. What we wish to underscore here is not their aforementioned utilization but instead their production and functioning. Regardless of their specific use, these artefacts were produced with handles, which is evidence of the complementariness and knowledge of the use of wood and stone, at the very least, in the manufacture of certain artefacts. We believe that this point can help us to explain the preference for small supports and low intensity of secondary modifications when the option of larger supports was readily available. As one of us has argued on a previous occasion (Bueno 2012: 81):

> our hypothesis is that the industry was marked by an overarching artifact assemblage in which compound artifacts played an important role. Small artifacts with a sharp edge, low degree of reworking, and signs of indirect handle show that they must have been used to compose artifacts mounted on wooden handles. The fact of the artifacts produced being small cannot be attributed to any physical requirement, scarcity, or form of the raw material. The maximization of the exploitation of the hyaline quartz does not stem from availability, access, or dimensions of the raw material. The production is the result of a deliberate choice, or rather of a series of choices that led to the production of artifacts with specific designs, which in turn were influenced by the selection of a given performance.

If we think about the quantity of lithic material found at the sites and the contexts of their deposition, especially regarding the other associated types of remains, then we can propose or at least sketch out hypotheses regarding the expected performances. Even though to date there have been no microwear analyses that might identify their function or the activities that those lithic materials were associated to, recent taphonomic analyses of various skeletons from the Lapa do Santo have found signs of removal of the flesh, in the form of fine, precise cuts in certain parts of the bones (Strauss 2010). To carry out such activities, and bearing in mind the powerful symbolism that would be associated to them, it would have been necessary to use precision artefacts with an active part characterized by well-defined cutting edges and a holding part that would ensure force and precision. The composition made up of small quartz flakes with a single cutting edge and some indirect way of holding them made of wood or bone is a combination that meets that need perfectly well.

Such use of micro-artefacts to elaborate compound artefacts to be used in activities of scarification and symbolic and ritual activities was presented on several occasions in papers reporting work carried out in Australia in the twentieth century (Robertson et al. 2009: 297). However, a recent revision presented by Robertson et al. (2009) and Hiscock et al. (2011) makes that hypothesis difficult to sustain at least from the point of view of Australia as a whole, and currently it is giving way to a hypothesis whereby those micro-artefacts used to manufacture compound artefacts would be used in a variety of ways, including domestic uses as well as in ritualistic and productive activities. According to Robertson et al. (2009: 298):

> For instance, Hiscock (1994, 2002, 2006, 2008) hypothesized that prehistoric Australian foragers emphasized composite tools containing backed artefacts because of their readiness and multi-functionality, employing them for almost any task.

That could well be the case for Lagoa Santa, but, once more taking into account the local context and the associations among the various types of remains that make up the archaeological record of the sites in that area, then that hypothesis of the use of such artefacts for the preparation of bodies for burial deserves to be tested, at least for the period of the early Holocene. A good first step in that direction would be to make a more detailed and comparative analysis to identify whether the cut marks identified on human bones are compatible with the cutting edges of possible lithic artefacts and to include in that activity analyses of microwear associated to the artefacts' use.

Perhaps the most important point to be made here is how essential it is to create or at least seek for contextual explanations that might help us to understand the characteristics of each industry and the choices that guided them, that orientated them and, from the broader point of view, are linked to defined performances desired to be achieved by the people who manipulated this material, who selected the raw materials, who defined how they should be approached, what should be produced and how and for what purpose they should be used (Bueno 2007).

Defining industries as simple, complex, curated or expedient is indeed a research strategy, but, on its own, it fails to explain and does not inform us of their possible meanings. One last point worth mentioning concerns the technological variability of the lithic assemblages associated to sites in rockshelters and in open air.

Analyses and comparisons among four sites in the Lagoa Santa karst, two in rockshelters and two in open air, enabled us to verify the existence of variations in the composition of the overall artefacts assemblage at each one of those sites. They are indicative of the execution of different activities and of different spatial associations, especially regarding possible connections among sites within and outside of the karst area. According to Bueno (2012: 80):

> we can propose the hypothesis that the rockshelters played the role of agglutinating locations and references for the persistent occupation of the karst, around which a variety of activities recurrently took place, leading to the setting up of small ephemeral camps. In that way, the rockshelters would represent the focal locus for marking the occupation of the karst area and would be the first or primordial point of extra-local connection, whereas the open-air sites gravitated around the rockshelters, articulating them with the locally available resources. That model is also compatible with what has been seen earlier regarding the differential variation among the site categories. The open-air sites show greater variation among them insofar as they would have been related to carrying out specific activities designed to obtain certain resources, while the shelters would congregate a diversity of relatively regular activities.

Conclusion

In the light of all that has been set out above, we have to say that the lithic technology of Lagoa Santa continues to be largely undefined. What we can state, however, is that the image of a simple, homogenous lithic industry of little interest is no longer sustainable. Furthermore, characterizing an industry as curated or expedient, in addition to not explaining very much, should not be the objective of our research if we really want to understand the cultural choices that guided the production of the artefacts assemblage that has been analyzed.

The Lagoa Santa lithic industry seems to be a highly emblematic case of the selection of local raw material to produce artefacts that require a specific design – compound artefacts mounted on handles of wood and/or bone. Here we have put forward a hypothesis for the functionality of those artefacts elaborated on the basis of their deposition contexts. However, we must also point out the need to test that hypothesis and to keep open the possibility of working in a perspective that admits that the artefacts may have been multifunctional in the way that has been discussed in Australia (Hiscock et al. 2011; Robertson et al. 2009). Furthermore, it is important to draw attention once more to the question of raw materials, because the presence of exogenous raw materials and of artefacts produced by other operative sequences with other forms and other compositions of active and prehensile parts indicates distinct functionalities and the integration of the karst with more widespread external areas, either through circulation or contact, or the effective use of a more extensive area as we have discussed on other occasions (Bueno 2010, 2012). Thus, far from being a simple industry, what we see is a technological organization that integrates different areas, the circulation of raw material, the combination of different techniques in exploiting those raw materials and the production of artefacts with diversified uses and meanings, constructing a mosaic of choices that define the uniqueness of that industry.

Acknowledgments Our thanks go to the FAPESP for financing the postdoctoral research of L. Bueno (Process 08/53474-6) in the period from September 2008 to August 2011, to Professor Walter Neves for extending an invitation to L. Bueno to participate in the Origins project, to Professor Tânia A. Lima for granting permission to study the Lagoa Santa lithic collection preserved in the archaeological storeroom of the National Museum, and to Walter, Pedro, and Mark for their invitation to participate in this volume. We are exclusively responsible for all the information set out in this chapter.

References

Ahler, S. A. (1989). Mass analysis of flaking debris: Studying the forest rather than the tree. In D. Henry & G. Odell (Eds.), *Alternative approaches to lithic analysis* (pp. 85–118). Bourder: Westview Press.

Andrefsky, W. (1994). Raw material availability and the organization of technology. *American Antiquity, 59*, 21–34.

Araujo, A., & Feathers, J. (2008). First notice of open-air paleoamerican sites at Lagoa Santa: Some geomorphological and paleoenvironmental aspects, and implications for future research. *Current Research in the Pleistocene, 25*, 27–29.

Araujo, A., & Pugliese, F. (2010). A indústria lítica. In A. Araujo & W. Neves (Eds.), *Lapa das Boleiras. Um sítio Paleoíndio do Carste de Lagoa Santa, MG, Brasil* (pp. 79–110). São Paulo: FAPESP/ANNABLUME.

Araujo, A. G. M., Neves, W. A., & Kipnis, R. (2012). Lagoa Santa revisited: An overview of the chronology, subsistence, and material culture of Paleoindian sites in Eastern Central Brazil. *Latin American Antiquity, 23*, 533–550.

Araujo, A. G. M., Strauss, A. M., Feathers, J. K., Paisani, J. C., & Schrage, T. J. (2013). Paleoindian open-air sites in tropical settings: A case study in formation processes, dating methods, and paleoenvironmental models in central Brazil. *Geoarchaeology, 28*, 195–220.

Bassi, L.F. (2012). *Tecnologia Lítica: análise diacrônica dos níveis mais antigos do sítio arqueológico Bibocas II, Jequitaí – MG*. Master Thesis, Universidade Federal de Minas Gerais.

Beltrão, M.C. (1974). Industries Lithiques trouvées au cours de la prospection 1971. In A. Laming-Emperaire, A. Prous, A. Vilhena de Moraes, M. Beltrão (Orgs.) *Grottes et abris de la region de Lagoa Santa, Minas Gerais, Brésil*. Cahiers D'Archaeologie D'Amerique du Sud, pp. 123–133.

Bueno, L. (2007). Organização tecnológica e teoria do design: entre estratégias e características de performance. In L. Bueno & A. Isnardis (Eds.), *Das Pedras aos Homens. Tecnologia Lítica na Arqueologia Brasileira* (pp. 67–94). Belo Horizonte: Editora Argvmentvm.

Bueno, L. (2010). Tecnologia lítica, cronologia e sequência de ocupação: o estudo de um sítio a céu aberto na região de Lagoa Santa, MG. *Revista do Museu de Arqueologia e Etnologia, 20*, 91–108.

Bueno, L. (2012). Entre abrigos e lagoas: tecnologia lítica e territorialidade em Lagoa Santa (Minas Gerais, Brasil). *Revista de Arqueologia, 25*, 62–83.

Hiscock, P., Clarkson, C., & Mackay, A. (2011). Big debates over little tools: Ongoing disputes over microliths on three continents. *World Archaeology, 43*, 653–664.

Hubbe, A., Hubbe, M., & Neves, W. (2013). The Brazil Megamastofauna of the Pleistocene/Holocene transition and its relationship with the early human settlement of the continent. *Earth Science Reviews, 118*, 1–10.

Hurt, W. (1960). The cultural complexes from the Lagoa Santa Region, Brazil. *American Anthropologist, 62*, 569–585.

Hurt, W., & Blasi, O. (1969). O projeto arqueológico "Lagoa Santa", Minas Gerais, Brasil. *Arquivos do Museu Paranaense, Arqueologia, 4*, 1–60.

Isnardis, A. (2009). *Entre as Pedras – as ocupações pré-históricas recentes e os grafismos rupestres da região de Diamantina, Minas Gerais*. Ph.D. Dissertation, Universidade de São Paulo.

Koole, E. (2008). *Pré-história da província cárstica do Alto São Francisco, Minas Gerais: A indústria lítica dos caçadores-coletores arcaicos*. Master Thesis, Universidade de São Paulo.

Laming-Emperaire, A., Prous, A., Vilhena de Moraes, A., & Beltrão, M. (1974). Grottes et abris de la region de Lagoa Santa, Minas Gerais, Brésil. *Cahiers D'Archaeologie D'Amerique du Sud*.

Mattos, A. (1938). *Pré-História Brasileira*. São Paulo: Companhia Editora Nacional.

Nelson, M. (1991). The study of technological organization. In M. Schiffer (Ed.), *Archaeological method and theory* (Vol. 3, pp. 57–101). Tucson: University of Arizona Press.

Neves, W., & Piló, L. B. (2008). *O Povo de Luzia*. São Paulo: Editora Globo.

Neves, W., Kipnis, R., Araujo, A., & Piló, L. (2004). *Origens e microevolução do Homem na América: uma abordagem paleoantropológica II*. Relatório Científico, FAPESP.

Neves, W., Araujo, A., Ceccantini, G., Bueno, L., De Oliveira, P., & Kipnis, R. (2008). *Origens e microevolução do homem na América: uma abordagem paleoantropológica III*. Relatório Científico, FAPESP.

Prous, A. (1991a). *Arqueologia Brasileira*. Brasília: Ed. UnB.

Prous, A. (1991b). Santana do Riacho – Tomo I. *Arquivos do Museu de História Natural* 12.

Prous, A. (2013). As muitas arqueologias das Minas Gerais. *Revista Espinhaço, 2*, 36–54.

Prous, A., & Lima, M. A. (1986/90). A tecnologia de debitagem do quartzo no centro de Minas Gerais: lascamento bipolar. *Arquivos do Museu de História Natural da UFMG, 11*, 91–111.

Prous, A., Fogaça, E., & Ribeiro, L. (Orgs.) (1998). Patrimônio Arqueológico. APA de Lagoa Santa – MG IBAMA/CPRM, Belo Horizonte.

Pugliese, F. (2007). *Os líticos de Lagoa Santa: um estudo sobre organização tecnológica de caçadores-coletores do Brasil Central*. Master Thesis, Universidade de São Paulo.

Robertson, G., Attenbrow, V., & Hiscock, P. (2009). Multiple uses for Australian backed artefacts. *Antiquity, 83*, 296–308.

Shott. (1996). An exegesis of the curation concept. *Journal of Anthropological Research, 52*, 259–280.

Strauss, A. (2010). *As práticas mortuárias dos caçadores-coletores pré-históricos da região de Lagoa Santa (MG): um estudo de caso do sítio arqueológico "Lapa do Santo"*. Master Thesis, Universidade de São Paulo.

Walter, H. V. (1958). *Arqueologia da região de Lagoa Santa, Minas Gerais*. Rio de Janeiro: Sociedade Editora e Gráfica Ltda.

Chapter 17
Towards the Development of a Tropical Geoarchaeology: Lagoa Santa as an Emblematic Case Study

Astolfo Gomes de Mello Araujo and Luís Beethoven Piló

Abstract Geoarchaeology in tropical environments has specificities that place it within a context of differential analysis. High temperature and humidity variation, the strong role played by bioturbation, the presence of deep soil profiles, and the expressive chemical alteration of the materials are some of the features of these environments. In the Lagoa Santa region, several important geoarchaeological discussions during the Origins project have approached issues related to the tropical environments, highlighting: (1) the formation processes inside rockshelters, (2) the investigation of a major event of abandonment of the region by human groups during the Middle Holocene (the "Archaic gap"), (3) the processes of formation of open-air sites, and (4) the question of the possible coexistence between humans and extinct megafauna. Including the work of researchers from different areas of knowledge, the studies generated a great richness of data and extensive discussion, which culminated in fundamental publications for the development of geoarchaeology in tropical environments.

Introduction

Geoarchaeology as a research strategy has a long history that has been addressed by several authors (Araujo 1999; Daniel 1975; Gladfelter 1977; Hassan 1979). In Brazil, the history of the systematic application of methods and techniques from the Earth Sciences to archaeological problems is relatively short, but nevertheless it has

A.G.M. Araujo (✉)
Institute of Advanced Studies (IEA-USP), and Museum of Archaeology and Ethnology (MAE-USP), University of São Paulo, Av. Prof. Almeida Prado, 1466, Cidade Universitária, São Paulo, SP, Brazil
e-mail: astwolfo@usp.br

L.B. Piló
Researcher from the Grupo Bambuí de Pesquisas Espeleológicas, Belo Horizonte, MG, Brazil
e-mail: lbpilo7@gmail.com

P. Da-Gloria et al. (eds.), *Archaeological and Paleontological Research in Lagoa Santa*, DOI 10.1007/978-3-319-57466-0_17

produced very important results (e.g. Bitencourt 2004; Kashimoto 1992; Morais 1999; Moura 1997; Parenti 2001; Rubin et al. 2003; Wagner 2014).

Specifically in the Lagoa Santa region, several researchers have made considerations about the processes related to the formation of the commonly found sheltered archaeological sites. The best example is Peter Lund's work at the Sumidouro cave (Lund 1844), which was instrumental in recognizing the possibility of great antiquity for human beings. His conclusions were based on stratigraphic observations and can be considered the first geoarchaeological work ever done in Brazil. Other scholars also contributed to the theme, albeit with different degrees of minutiae (e.g. Hurt and Blasi 1969; Laming-Emperaire et al. 1974; Lanari 1909).

The aim of this chapter, however, is to call attention to an even more recent development of something that we will call "tropical geoarchaeology" and to show how the research carried out at Lagoa Santa between 2000 and 2009 was seminal for the development of several ideas that are changing the way we perceive not only archaeological sites per se but also the processes that are responsible for their formation.

Tropical geoarchaeology implies the recognition that the application of methods and techniques from the Earth Sciences in tropical settings has to deal with a very specific scenario that needs to take into account the high temperature and moisture variation, the strong role played by bioturbation, the presence of deep soil profiles, and the heavy chemical alteration. Therefore, several lines of reasoning that were developed in other settings (such as temperate, Mediterranean, semiarid areas, etc.) cannot be directly applied to most of the cases we deal with in tropical settings (Araujo et al. 2013). In that regard, Lagoa Santa is a good study case, since the problems addressed by us in this region had to deal with questions that are directly related to tropical environments.

A Bit of History

The history of the geoarchaeological reasoning that informed the Origins project (see Chap. 9 of this volume) can be followed along four major lines:

1. The formation processes inside the rockshelters
2. The realization that there was a major abandonment event of the area during the Middle Holocene (the "Archaic gap")
3. The formation processes operating at open-air sites
4. The question of the possible contemporaneity between humans and megafauna, which has eluded so many researchers since the nineteenth century

When our interdisciplinary team led by Walter Neves arrived at Lagoa Santa in 2000, we had a very rough idea of what we would excavate. By that, we do not mean that we are talking about the archaeological bits and pieces, such as flakes, human skeletons, animal bones, and so on, but rather the question was about the geological/pedological matrix itself. We visited several sites before choosing the ones we would work on. Our first excavation was carried out in 2001 at Lapa das Boleiras, a limestone rockshelter that was well preserved and had good potential for the recovery of human remains (one of the project's main objectives). A test pit was dug in the same year at Lapa do Santo rockshelter, and since this site produced a wealth of human skeletons, its excavation was carried out for several years (see Chap. 9 of this volume). A third sheltered site, Lapa Grande de Taquaraçu, was excavated from 2003 to 2008 (Araujo et al. 2012).

During those years, several organic samples were dated, and as the radiocarbon ages started to accumulate, we observed a very unusual trend, with the ages clustering in two sharply delimited periods: one during the Early Holocene (12,460–8,700 cal BP), and other in the Late Holocene (post-2,000 cal BP; see Araujo et al. 2005, 2006). This hiatus in the Middle Holocene archaeological record was called the "Archaic Gap". Later, it also became apparent that there was a short period of reoccupation of the area in the Middle Holocene (5,100–4,200 cal BP – see Araujo 2015), but these data do not invalidate the hypothesis that climatic shifts in the Middle Holocene were responsible for the relocation of Palaeo-Indian populations at Lagoa Santa. Those events of abandonment–reoccupation–abandonment led us to think about two possibilities: either the ages were biased because most data came from the rockshelters or the entire region was indeed depopulated. The best way to test those hypotheses was to change our focus towards open-air sites. We chose Sumidouro lake shore as a starting point, due to its large size, and conducted a systematic subsurface survey. This led us to find three open-air sites (Sumidouro, Coqueirinho, and Lund) and also represents the third contribution of the project, namely, the realization that, after 200 years of research focused in rockshelters, open-air Palaeo-Indian sites did exist in the region, that most of these sites were probably very deeply buried, and that their ages followed the same pattern observed in the rockshelters. The "Archaic gap", therefore, was not an artefact of biased ages coming from rockshelters but a real tendency of the archaeological record of the region.

Ashes to Ashes or How the Anthropogenic Nature of Sediments from Lagoa Santa Rockshelters Was Recognized

The year was 2001, and we started our first excavation season inside a rockshelter, at Lapa das Boleiras. As we proceeded, we were struck by the "powderyness" of the sediment. A greyish, powdery sediment was our constant companion throughout those 30 days of fieldwork. Our noses, hairs, socks, clothes, everything was full of that powder. We talked constantly to each other about the possible explanation for that phenomenon. We were at loss and considered the possibility that the intense burning of limestone pebbles and cobbles used inside fireplaces would turn limestone into lime. A bit of research made clear that it is impossible to produce lime without a furnace and that open-air hearths would not burn limestone in such a way. On the other hand, the sediment reacted very strongly with HCl, which means it was saturated with CaCO3, and therefore it was not CaO (lime). It was also clear that the "powder" was not related to the rockshelter ceiling, since the limestone was very pure and would be dissolved by water, forming at most speleothems and not a powdery soil. We could perceive layers of reddish soil that were washed inside the site, but the majority of the matrix was, indeed, grey. The same powdery sediment was also found in Lapa do Santo, reaching the bottom of the excavated pit at 2 m deep. As the excavation proceeded in the subsequent years, it became clear that, in fact, the site was basically a mound of greyish, powdery sediment with some occasional layers of red soil. The thickness of the greyish/whitish sequence of layers reached 4 m at Lapa do Santo.

In 2003, we started the excavation of a third site, Lapa Grande de Taquaraçu. The setting was pretty much the same: a thick, 80 cm layer of soft, greyish sediment, full of charcoal fragments and archaeological materials, intermingled with (rare) fine layers of bright red soil. This archaeologically rich grey sediment was placed directly over a sterile red sediment full of limestone blocks that had collapsed from the ceiling.

We knew from published data (e.g. Missão Franco-Brasileira de Lagoa Santa 1977) that similar greyish powdery sediments were found in other sheltered archaeological sites, but we could find no clue about its origin anywhere. Different authors gave a diverse series of hypotheses for the origin of the sediment: the result of limestone weathering (e.g. Missão Franco-Brasileira de Lagoa Santa 1977:75), calcite deposition (e.g. Kipnis 2002: 184–185; Prous 1991: 83), or lacustrine sediments (Moura 1997:118). A detailed study of these greyish sediments coming from Lapa do Boquete rockshelter, northern Minas Gerais state, did not reach a conclusion about its genesis, since the standard laboratory procedure involved a decarbonation pretreatment with HCl. As the sediment was very rich in CaCO3, almost all the matrix was dissolved during the pretreatment (Moura 1997:120). The result is that several researchers spent years digging inside these sediments without a real understanding about the genesis of the matrix.

Finally, in 2005, while reading papers about geoarchaeology in rockshelters, we found a piece of information that was vital to solving the puzzle: wood ash is composed mainly by calcium carbonate, or $CaCO_3$ (Karkanas 2001; Karkanas et al. 1999; Schiegl et al. 1996; Shahack-Gross et al. 2004; Weiner et al. 2002). The colour, the powdery aspect, and the high carbonate content of the sediments could be, therefore, related to ash deposition. This today sounds so obvious that it is hard to believe that archaeologists did not acknowledge this basic information about ash composition 10 years ago. One exception was Bryan and Gruhn (1978: 297) who recognized the anthropogenic nature of the greyish powdery sediments inside Lapa Pequena rockshelter, northern Minas Gerais state. Their observations, however, did not have a real impact in the Brazilian literature, possibly because the publication was in English.

The idea that the grey sediment was mainly anthropogenic met some resistance. Especially at Lapa do Santo, a 4 m thick ash layer was considered too much for "mobile hunter-gatherers". At Lapa Grande de Taquaraçu, a visiting South American archaeologist believed that the sediment was brought inside the site by eolian activity(!).

In 2008, we published the first in-depth analysis of the anthropogenic sediments from Lapa das Boleiras rockshelter (Araujo et al. 2008). Micromorphology of thin sections of the grey sediment showed well-preserved pseudomorphs of plant calcium oxalate crystals. The matter was settled: the sediment was composed by ash and therefore was anthropogenic. Even so, the idea of huge ash accumulations was very counterintuitive. At Lapa das Boleiras, we had a mixed contribution of ash and colluvium; heavy rains could bring soil from the upper portions of the outcrop inside the site, and this process could be seen very clearly in the stratigraphy (op. cit: 3191). However, we were still left with Lapa do Santo and Lapa Grande de Taquaraçu, two sites where, given their topography, there is no way colluvial accumulation could be considered relevant. The bulk of the sediment was ash, although soil from outside was probably carried in by humans and heavily burned, forming the bright red layers already mentioned. Of course a lot of animal and plant resources, not to mention rocks, were also brought in by humans, but we did not have any compelling explanation for the bulk of the site matrix.

At Lapa Grande de Taquaraçu, a fine-grained instrumental neutron activation analysis (INAA) of the sediments showed three main, well-marked input sources (Tudela 2013): the whitish-grey sediment is indeed ash; the (rare) bright red lenses are burned colluvium; and a third component, small clay pellets found in the bottom of the stratigraphy, is related to episodic river overflows. Therefore, the bulk of the 80 cm-thick sediment is anthropogenic (see also Angeles Flores et al. 2016; Silva and Prous 2014). The anthropogenic nature of the sediment is also attested by the absolute lack of sediment accumulation after the site was abandoned. Charcoal from two archaeological facies excavated in the upper layers, a few centimetres apart, provided ages of 9,000 cal BP[1] and 1,100 cal BP. When we take into consideration that the onset of the human occupation was dated at 11,500 cal BP, this means that 80 cm of sediment (ash) were accumulated during the first 2500 years, and only a few centimetres were accumulated in the following 8000 years. The conclusion is that the absence of people results in the absence of ash production and sediment accumulation.

At Lapa do Santo, the same reasoning applies. There are small lenses of what appears to be burned soil, especially in the upper portions of the stratigraphy, that we interpret as brought in by humans and very similar to the ones we found at Taquaraçu (the same lenses that were analysed by Villagran et al. 2017). However, the majority of the archaeological matrix is ash, plus a fair amount of lithics, bones, and plant remains. A 4 m deep stratigraphy obtained during the 2001–2009 excavations shows a profile that is almost completely composed of grey sediments (see Neves et al. 2012). It is also possible that some contribution from lacustrine sediments occur in the lower portion of the stratigraphy, given the nearby presence of a doline that can be flooded episodically.

The meaning of this massive ash accumulation at Lapa do Santo has to be addressed and taken seriously. We do not have to invoke natural sediment accumulation in tortuous ways to explain something that is mainly anthropogenic. We should

[1] All radiocarbon ages were calibrated using the program CalPal-2007 (Weninger et al. 2016).

start to think of Lagoa Santa Palaeo-Indians as more sedentary than our common wisdom supposes. Lapa do Santo could be a particularly important site given its large area and strong anthropogenic activity, but several other rockshelters also show large amounts of anthropogenic sediments, such as Cerca Grande VI, Lapa do Porco Preto, Lapa da Samambaia, Abrigo da Mata da Cauaia, and many others. Are they signalling a stable, sedentary population? Independent data coming from oral health and paleopathological analyses seem to point in the same direction (Da-Gloria 2012; Da-Gloria and Larsen 2014): Lagoa Santa Palaeo-Indians show signs of population aggregation, as marked by unspecific infections, low individual mobility, inferred through external shape of the femur, and high prevalence of caries, suggesting a strong reliance on plants and not so much on meat. Once we start opening our minds to these lines of evidence, the amount of ash accumulation at the rockshelters can be accepted with less prejudice.

The "Archaic Gap", or Looking at the Familiar and Finding the Unexpected

The second major contribution of the Origins project in terms of geoarchaeological reasoning was the realization that there was a major trend of human displacement/depopulation during the Middle Holocene in large areas of Central Brazil (Araujo et al. 2003, 2005, 2006). The data amassed by several researchers was already available: hundreds of radiocarbon ages covering the Brazilian territory (e.g. Etchevarne 2000; Martin 1997; Noelli 2000; Schmitz et al. 1989) showed a decrease in the number of dates during the Middle Holocene for some portions of the country, whereas other areas did not show this trend. However, as far as we know, the only attempt to put some of this data together in a graphic way, going beyond the presentation of tables, was done by Kipnis (2002: 133), who showed the distribution of radiocarbon ages for Central Brazil. At the same time, paleoenvironmental records started to be more common. Early archaeological work (i.e. from the 1970s to the 1990s) could only rely on considerations about paleoenvironments put forward by geomorphologists, based on the observation of stone lines and other characteristics of slope deposits (e.g. Ab'Sáber 1980; Bigarella et al. 1965; Penteado 1969). This approach, albeit useful and important, suffered from some shortcomings. The use of continental stratigraphic records (i.e. soils, river terraces, dunes) for paleoenvironmental interpretations is always tricky because these records can be very fragmentary due to erosion and redeposition and also because they can show imprints of several climatic, geomorphic, and biotic events, some of them erasing completely the previous ones. In the Brazilian case, the lack of chronology was another problem, since radiocarbon or luminescence dating was rarely applied, resulting in the recognition of very broad time intervals based on relative dating. One could only talk about "Pleistocenic features" or "Holocene layers". In the 1990s, this scenario started to change with the publication of studies dealing with other paleoenvironmental proxies, such as pollen (De Oliveira 1992; Ledru 1993; Ledru et al. 1996), charcoals (Scheel-Ybert et al. 1995), and soil organic matter (Martinelli et al. 1996). In the early 2000s, speleothem records started to become available (Auler and Smart 2001), adding new data and both supporting and confronting other records. This academic scenario paved the road for new insights about the relationship between humans and environment during the Holocene.

From the Particular to the General

Our first exposure to the phenomenon we later called “Archaic Gap” was related to the ages of the Lagoa Santa skeletons which clustered into two neat extremes: either they were very old, from the Early Holocene, or they were very young, from the last 2000 years (Araujo et al. 2012). When we started to look at the ages we were accumulating during the excavations, the same pattern emerged: we had several radiocarbon and luminescence ages from the Early Holocene, several from the Late Holocene, and no ages between 8,300 cal BP and 2,000 cal BP. This led us to consider several hypotheses, some of them related to our biases (are we dating the right spots?), others related to natural biases (were the Middle Holocene layers eroded?), and still others related to biases imprinted by the Palaeo-Indians themselves (were they doing all their activities and burying their dead outside of the rockshelters during the Middle Holocene?). Of course, we also thought that a climatic explanation was in order and that a period of climatic dryness could have a major impact on the rockshelter’s attractiveness, since they were located in areas where all drainage is subterranean. In the wet season, the dolines that occur close to most rockshelters turn into lakes, and in the dry seasons, these lakes mostly disappear. An extended dry period would mean absolute lack of water close to the rockshelters. Therefore, we took measures to try to unravel the problem posed by the “Archaic Gap”. At Lapa do Santo, we started to date charcoal from a portion of the site where the stratigraphy was well preserved and seemed to contain Middle Holocene layers (“M” units). At the same time, we started the excavation at Lapa Grande de Taquaraçu that was right beside a permanent river and, therefore, not subject to the dryness issue. Regarding the open-air sites, we started the above-mentioned survey along the Sumidouro lakeshore. In a parallel way, we made a compilation of the literature regarding paleoenvironmental studies carried out in different parts of the country. The results of these actions culminated in a scenario that was, again, very counterintuitive.

First, our efforts in dating several layers from the “M” units at Lapa do Santo showed some Middle Holocene ages, but these ages were not evenly distributed. They clustered again, all falling inside a very short interval between 5,100 cal BP and 4,200 cal BP. There was, indeed, a huge gap in the occupation of the site, between 8,300 cal BP and 5,100 cal BP, followed by the short 900-year interval of reoccupation and another abandonment (Araujo 2015). This event of reoccupation suggests that the Palaeo-Indians moved the focus of their occupation elsewhere while maintaining Lagoa Santa inside their traditional territorial memory.

The excavations at Lapa Grande de Taquaraçu, located beside the Taquaraçu river, mirrored the same scenario present at Lapa das Boleiras and Lapa do Santo. The Palaeo-Indian occupation at the site started around 11,500 cal BP and finished at 9,000 cal BP (Araujo et al. 2012; Tudela 2013). No ages from the Middle Holocene were found at this site. So, the hypothesis that people left Lapa das Boleiras and Lapa do Santo during the Middle Holocene due to a local event related to drought was not robust, since a site right beside a permanent river was also abandoned. We were dealing with something larger.

The archaeological survey along the shores of Sumidouro Lake identified three open-air sites: Sumidouro, Coqueirinho, and Lund. This last site is related to a late hunter-gatherer occupation (2,260 and 2,230 cal BP). The other two sites were dated inside the Palaeo-Indian period. Coqueirinho provided an older age of 12,400 cal BP and some Middle Holocene ages (Bueno 2010), but the relationship between the charcoal and the archaeological materials is blurred by the shallowness of the site (70 cm deep) and the heavy bioturbation and erosion/redeposition processes related to the lake shore. Sumidouro, on the other hand, has a much deeper stratigraphy (up to 2.10 m), and we could clearly perceive that Middle Holocene charcoal was never associated with archaeological materials (Araujo et al. 2013). In short, the data we gathered from open-air Palaeo-Indian sites did not support the idea that people lived outside of the rockshelters during the Middle Holocene.

When we started to analyse the archaeological age database for the Brazilian territory, we were struck by a clear pattern, with a frequency curve showing a depression in the number of Middle Holocene ages coming from Central Brazil. This pattern did not appear in the Southeastern and Southern states (Araujo et al. 2005) but was perceived in other areas of South America (Araujo et al. 2006). Following the reasoning that the more people you have in the landscape, the higher the frequency of archaeological sites being produced (Rick 1987), we started to think seriously about the role of climate in this scenario.

The paleoenvironmental data coming from diverse sources and spreading throughout the Brazilian territory also provided relevant information, and perhaps the most important was the realization that different research teams reached different interpretations about what happened during the Middle Holocene in Central Brazil. Some saw the Middle Holocene as a period where dry climates were prevalent (Barberi et al. 2000; Salgado-Laboriau et al. 1997), while others saw exactly the opposite (Behling 2002; Ledru 1993; Ledru et al. 1996). Given this scenario of academic divergence, our data supported the idea that the climate was probably dryer, and, in this regard, we saw humans as an environmental proxy, perhaps not extremely fine-tuned to climatic fluctuations but nevertheless far from being immune to them (Araujo et al. 2005). Another important scientific contribution is related to the fact that the idea of the Middle Holocene as a "climatic optimum", where higher temperature was equated with higher precipitation, and so much used as basis for archaeological reasoning (e.g. Schmitz 1987; Kipnis 2002), is not plausible anymore. New data about the Middle Holocene became available, and it seems that the scenario was more complex (and more interesting) than we first thought. It is fairly possible that the abandonment of vast areas in Central Brazil was not related to dryness but to climatic instability (Araujo 2014; Racza et al. 2013). Few years of drought followed by few years of high precipitation can be very hard to detect in the pollen record but nevertheless can have a strong impact on humans.

Open-Air Sites, Slope Processes, and Paleoenvironments: Archaeology as a Data Provider

The third contribution of the Origins project is related to a theme that is of upmost importance, namely, the ability that an archaeological excavation has to provide relevant data to other disciplines, breaking down the "archaeology as method borrower" circle and starting the "archaeology as data provider" stance. This was possible because some data can only be provided by means of a careful excavation and by the presence of archaeological materials. Indeed, there is a huge difference between a layer of pebbles (or a stone line) and a layer of flaked stones. A layer of pebbles can be related to a myriad of events (Hiruma 2007) culminating in its deposition in a given spot; the pebbles can start their history as quartz veins inside a metamorphic rock, later subject to weathering, entering a river system, being deposited inside a channel, becoming a river terrace, being eroded afterwards, entering a soil system, being subject to heavy bioturbation, and forming a stone line. Alternatively, the soil's finer particles could be eroded or deflated, and the coarser pebbles could represent a paleosurface, or a lag deposit, later subject to a new cycle of soil accretion, to be buried again. The exact history of a stone line is, therefore, subject to all sorts of variables and all sorts of interpretations. Is it marking a paleosurface, or is it the remnant of heavy bioturbation? Was it simply inherited from a sedimentary rock?

On the other hand, a line of flaked stones is more commonly related to a single event, that is, the presence of people flaking at a given spot, and represents a paleosurface. Even in the worst-case scenario, when the archaeological materials are redeposited, we can say with some certainty that they came from elsewhere and, again, it is related to a paleosurface. In the case of bioturbation, we can even say how strong it was, given the vertical distribution of the pieces. Most important of all, when we find refitting pieces, as two halves of the same artefact, we have the most accurate proof that the pieces are in the place they were deposited. On the other hand, there are few instances in the geological record where you can confidently say that a set of pebbles are related to a single genetic event in a single spot, marking a paleosurface.

All these considerations were advanced to set the stage for the understanding of the context at the open-air Sumidouro site (Araujo and Feathers 2008; Araujo et al. 2013). Sumidouro showed layers of flaked stone inside a deep soil profile, and these layers helped the understanding of several issues related to pedology, geomorphology, and geochemistry.

The site is located near a lake shore, in the lower portion of a 380 m, 12% slope. This long slope has a very gentle topography, interrupted only by scattered termite mounds. Soils in the region are very deep (10 m on average; Piló 1998), overlying the limestone bedrock. The most striking feature of the soil at the site was that it showed an upper reddish layer (colours 5YR 5/6, 7.5YR 4/4, and 7.5YR 5/6) giving way to a yellowish lower soil (colours 10YR 6/6, 10YR 5/8, and 10YR 4/6). Since we were dealing with a deep tropical soil, subject to heavy bioturbation and dynamic slope processes, we did not expect a good fit between radiocarbon and single-grain optically stimulated luminescence (OSL) data and not even a good fit between the vertical positioning of pieces and their original placement. Based on the vast literature about tropical soils, bioturbation, and slope processes (Araujo 1995; Cahen and Moyersons, 1977; Boulet et al. 1995; Thomas 1994, 2004), we had several working hypotheses in mind:

I – Hypotheses regarding slope processes and site formation processes:

(a) The presence of archaeological materials in discrete levels reflects different events of human occupation, interbedded with coluvial accretion, by means of sheetwash processes, in a sub-horizontal manner.
(b) There are no discrete archaeological "layers". Soil creeping processes (Clarke et al. 1999) are responsible for the burial and, in some cases, concentration of archaeological materials. The coluvial accretion occurs not by sub-horizontal deposition but in a convolute manner, mixing materials from different occupation events, charcoal from different ages, etc.

II – Hypotheses regarding chronology and dating methods:

If Hypothesis Ia is true, charcoal fragments are good indicators of soil ages and, therefore, artefact deposition events.

If Hypothesis Ib is true, there will be no direct relationship between the soil mass, artefacts, and charcoal fragments. Direct dating of burnt artefacts (stone or ceramics) will be necessary. Luminescence dating of quartz grains might be more reliable than charcoal.

In the beginning of the excavation, we were sure that hypothesis Ib would be true: soil creeping processes and bioturbation would obliterate any behaviourally meaningful feature. This stance started to change when we perceived that both charcoal particles and archaeological materials were placed in discreet levels, instead of being randomly dispersed across the profile. While some pedologists would look at that soil profile and see in situ weathering of the limestone, we combined stratigraphy, soils, micromorphology, geochemistry, archaeological data, OSL, and radiocarbon ages and reached the conclusion that the site was formed by sheetwash, or the steady accumulation of very fine layers of soil, and that the role of soil creep and bioturbation was much less important than previously thought. Moreover, we also saw a relationship between the red and yellow horizons, paleoenvironments, and chronology: archaeological materials were always inside the red horizon, regardless of its thickness. No artefacts were found inside the yellow horizon. Again, some pedologists would think that these horizons (red = rich in hematite; yellow = reach in goethite) are related to water circulation. Archaeological data tends to support a view that sees them as related to a paleoenvironmental shift and with a chronological meaning (Araujo et al. 2013).

Once the soil profiles were understood as the product of coluvial accretion and not in situ weathering, a new panorama was open. We could devise, by means of micromorphology, at least three horizons marking paleosurface stability and also correlate the rates of coluvial accretion with theoretical expectations regarding the influence of paleoenvironments on the slope processes (*op. cit.*). Again, archaeological work provided stimulating data that could be of benefit for pedology and geomorphology.

Palaeo-Indians and Large Mammals in Brazil: Putting an End to a 200-Year-Long Controversy

The Origins project also made a decisive contribution towards resolving an issue that was the driving force behind practically all the research undertaken in the Lagoa Santa region since the nineteenth century: the coexistence of man with extinct large mammal species. A broad discussion of the archaeological, paleontological, and chronological evidence can be found in Prous (2002) and Hubbe et al. (2013).

The Lagoa Santa debate began with the research carried out by Peter Lund. In 1843, he had the opportunity to carry out a systematic excavation of the Gruta do Sumidouro (see Chaps. 2 and 3 of this volume). Lund exhumed human bones from every type of sediment except red clay. At least 30 human skeletons with various biological ages were exhumed (Lund 1844).

After analysing the yellow clay with black patches in it, Lund declared emphatically that: "It was in that mixture of extinct and extant species that the enigmatic remains of a horse and a man appeared, all in the same state of transformation and in such a way as to leave no doubt of the coexistence of those beings whose remains were buried together" (Lund 1844).

At the time, Lund's proposal received very little support (Quatrefages 1881; Reinhardt 1888) and much criticism (Lacerda and Peixoto 1876; Lütken 1883; Kate 1885; Hansen 1888; Rivet 1908; Hrdlička 1912). The main point questioned was the occurrence of periodic flooding of the cave due to the existence of a swallet, which would most certainly have churned up the sediment strata, on several occasions, together with the bones deposited in them.

At the beginning of the twentieth century, excavations in the region were begun again. At the Lapa Mortuária de Confins, H. V. Walter excavated below a stalagmitic crust incomplete remains of a human skeleton together with fragments of a horse's (*Equus*) cranium in addition to the teeth and part of the femur of a small mastodon (*Haplomastodon*) (Walter 1948). The excavation technique used and the lack of any more detailed stratigraphic control made that find somewhat questionable (Prous 2002).

At the Lapa Vermelha IV, the bones of a terrestrial sloth (*Catonyx*) were dated indirectly by dating charcoals found at the same sedimentary level at 10,895 ± 266 cal BP, but it was not possible to establish a connection between those animal remains and the hearth vestiges or the lithics instruments (Laming-Emperaire et al.

1974; Prous 2002). At a deeper stratigraphic level, however, a human skeleton was found, given the name of Luzia, and dated as from 12,914 to 13,404 cal BP (Neves et al. 1999; Prous 2002).

In the 1970s, A. Bryan of the University of Alberta (Canada), during an analysis of the paleontological material discovered by H. V. Walter, which was deposited at the Natural History Museum and Botanical Gardens of the Federal University of Minas Gerais (Museu de História Natural e Jardim Botânico – MHNJB-UFMG), identifed cut marks on the iliac bone of a mastodon (*Haplomastodon*) which were attributed to human intervention. It was proposed that the humans would have tried to disarticulate the leg bones to produce blades or make instruments (Prous 2002). The specimen was examined by paleoanthropologists and paleontologists, but they failed to arrive at a consensus as to whether it had been used by humans.

Making the most of the effects of the climate and hydrological conditions of 2001 and 2002 on the cave water dynamics (lowering of the water table and drying out of the cave), the Lapa do Sumidouro was targeted by the Origins project in an attempt to reopen the discussion of Lund's main interpretations of the sedimentary, fossil, taphonomic, and chronological records (Piló et al. 2005).

^{230}Th/^{234}U datings of stalagmitic crusts made it possible to identify a deposit of breccias with clasts with ages of over 238 ky. A red facies revealed a maximum age of 11.4 ka. At least two erosive phases were confirmed for the cave, one with a maximum age of 238 ky and the other with a maximum of 11 ky. The dates also made it possible to register two chemical sedimentation episodes in the cave: the older one with around 240 ka and the more recent one from 11 to 8 ky. Chemical and clastic deposition and erosion episodes occurred at various moments in the Pleistocene and at the beginning of the Holocene.

Yellow and red mud facies would have originated from the mantle of phyllite alteration located over the epikarst zone but not in contact with the surface of the karst and therefore sterile from the paleontological point of view. Two sedimentary facies in the Lapa do Sumidouro indicated possible connectivity with the surface of the karst relief. Regarding the taphonomic aspect, it proved possible to interpret that the human bodies had been introduced via the southern entrance to the cave. Most of the animals of the extinct fauna would have been introduced through the swallet (northern entrance).

At least four different sedimentary facies were identified, not being derivations originating from the red sediment under the action of the water table as Lund (1844) proposed. The chemical data obtained also indicated that the red and the yellow sediments are quite distinct and not synchronic and that the yellow sediments with black stains are older than the red sediments and were introduced after the erosive phases.

The ^{14}C datings of organic material found adhering to the human bones indicated a minimum calibrated age of 8 ky for the human skeletal remains found in the cave. The ^{14}C and ^{230}Th/^{234}U datings of sample 789 also established a minimum calibrated age of 8 ky for the subjacent sediments (yellowish clay with shells) with human mixed with extinct fauna bones, given that, except for the red clay, Lund found human and extinct fauna remains in all the sediments. That would seem to confirm his idea of the great antiquity of man's presence in the Lagoa Santa region. However, the data obtained actually reinforce the idea that bones of different ages may well have been removed from their primary deposition locations and mixed up with the transformed clays, thereby generating reworked and chronologically distinct fossil-bearing deposits (Piló et al. 2005).

Analysis of stratigraphic columns in other Lagoa Santa karst caves, like the Lapa do Baú, Gruta dos Macacos, Gruta da Caianga, Lapa Vermelha IV, and Gruta Cuvieri, shows that there is no single stratigraphic sequence that is typical for the fossiliferous deposits in the region. The fossiliferous facies are largely made up of matrix-supported breccias. Other facies identified as fossiliferous were muddy sand and sandy mud. The fossiliferous sediments were mainly generated by gravitational processes of the types alluvial fan and mud runoff.

Of more limited occurrence, sedimentary facies may hold articulated skeletons or in positions very close to their anatomical positions, such as those registered for the Gruta Cuvieri (Neves e Piló 2008; Hubbe et al. 2011). Such skeletons are mainly associated to the accidental falls of live animals into the depths of the caves and their posterior burial by sediments. ^{14}C and ^{230}Th/^{234} U datings in the caves clearly demonstrate the great temporal variability among the bone remains, with ages ranging from the Middle Pleistocene to the Pleistocene/Holocene transition (Auler et al. 2006).

The oldest indirect dating obtained for human remains was acquired for a skeleton (Luzia) exhumed from the Lapa Vermelha IV. Carbon datings for charcoal fragments indicated the deposition of human bones as having been somewhere between 12,914 and 13,404 cal BP (Prous and Fogaça 1999; Prous 2002). The dispersal of some of the bones cast some doubt as to the exact provenience of the skeleton, and evidence of bioturbation of the archaeological strata extended that doubt to the accuracy of the ages proposed for the material.

Studies conducted by the Origins project collaborators involving OSL datings together with granulometric, chemical, and micromorphological analyses of the sediments were used to assess stratigraphic integrity, especially the degree of mixture of sediments. The sediments nearest to where Luzia's cranium was identified generated OSL datings in the range from 12.7 to 16.0 ky and so in no way refuted the original age estimation (Feathers et al. 2010). However, W. Neves (personal communication) questions those ages because the reference that should have been used was the original position of the skeleton, 1 m above the position of the cranium.

More directly, a fragment of bone (ZMUC 4725) exhumed by Lund from the Lapa do Braga was dated to from 11,260 to 11,110 cal BP using ^{14}C AMS. Excavations at the Lapa das Boleiras also made it possible to date the first levels (BL-K12-N1) of occupation (fireplaces) of that site at 12,620–12,480 and 12,380–11,240 cal BP. According to Araújo et al. (2012), despite the evidence of humans (Luzia) crossing the region precociously, occupation in Lagoa Santa only began around 11,200 to 12,300 cal BP.

To expand knowledge of chronological aspects, the Origins project team obtained samples of Lagoa Santa remains of extinct fauna and of humans from the MHNJB-UFMG and from the Zoological Museum of the University of Copenhagen (ZMUC) for ^{14}C-AMS dating. The fragment from the MHNJB-UFMG was of the rib of a terrestrial sloth (*Catonyx*) that was discovered in the Gruta Cuvieri in 1976. That was found to have an age of 11,570–11,250 cal BP. The skeleton of the sloth was found with the bones anatomically articulated and with no presence of any human vestiges.

During a technical visit to the ZMUC, the Origins project team selected 44 fossil fragments of various extinct Pleistocene fauna belonging to the Lund Collection. It only proved possible to obtain reliable datings for six of the selected fragments (Neves and Piló 2003). The rest of the samples were devoid of collagen. One of the dates deserves special mention. It was a fragment of a saber-toothed tiger (*Smilodon*) dated at 11,080–10,160 cal BP. The dated material, which was collected by Lund during his paleontological research in the region's caves, was also not associated to human remains.

Thus, the very oldest human vestiges and the dates obtained for the regional megafauna indicated the coexistence of man with at least two genera of the megamammal fauna: *Catonyx* and *Smilodon* (Neves and Piló, 200; Piló and Neves 2003; Hubbe et al. 2009); Araújo et al. 2013; Hubbe et al. 2013).

What relations may have existed between prehistoric man and the great mammals now extinct is not very clear. Up to the moment, despite the evidence of their coexistence, there is no archaeological record in the region of any evidence whatsoever that the first humans to occupy the Lagoa Santa karst made use of the large extinct mammal for food or as a source of raw materials (Kipnis 2002; Neves and Piló 2003; Araujo et al. 2012).

Conclusions

The Origins project addressed important issues regarding the site formation processes, both at sheltered and open-air sites, the relationship between humans and extinct fauna in the Early Holocene, and the role of paleoenvironments in the reorganization of Palaeo-Indian demography. This effort culminated in the publication of several papers, books, and book chapters (see Chap. 9 of this volume), and it was only possible because researchers from different areas joined together and discussed at length these issues. Contrary to a common practice in archaeological projects, where publications tend to show sections or chapters written by different professionals without a common ground or a real synthesis, we managed to join several lines of evidence and data from different knowledge areas. We can say that we went beyond the common practice of technique borrowing and started to produce data that can be considered of relevance to other fields of knowledge. It is also important to stress that tropical settings pose problems and demand answers that must consider processes that are enhanced by high temperatures, moisture, and biological activity. Sometimes the realization of the strength of these processes leads to a somewhat "pessimistic" view of the archaeological record, seeing it as subject of very aggressive environment, where nothing but stones can survive outside of the rockshelters. We were able to show evidence of a wealth of data that can be fruitfully obtained, once we begin to realize that "geoarchaeological commonsense" is really not enough in tropical settings.

Acknowledgements We would like to thank Walter Neves, for bringing all of us together.

References

Ab'Sáber, A.N. (1980). Paleo-climas e paleo-ecologia. In P.I. Schmitz, A.S., Barbosa M.B., & Ribeiro (Eds.), *Anuário de Divulgação Científica. Temas de Arqueologia Brasileira* (pp. 33–51, vol 1(5)). Goiânia: Universidade Católica de Goiás.

Angeles Flores, R., Moreno de Sousa, J., Araujo, A. G. M., & Ceccantini, G. (2016). Before Lagoa Santa: Micro-remain and technological analysis in a lithic artifact from the Itaparica industry. *Journal of Lithic Studies, 3*(1), 6–29.

Araujo, A. G. M. (1995). Peças que sobem, peças que descem e o fim de Pompéia: Algumas observações sobre a natureza flexível do registro arqueológico. *Revista do Museu de Arqueologia e Etnologia, 5*, 3–25.

Araujo, A. G. M. (1999). As geociências e suas implicações em teoria e métodos arqueológicos. Revista do Museu de Arqueologia e Etnologia, Anais da I Reunião Internacional de Teoria Arqueológica na America do Sul. *Suplemento, 3*, 35–45.

Araujo, A. G. M. (2014). Paleoenvironments and Paleoindians in Eastern South America. In D. Stanford & A. Stenger (Eds.), *Pre-Clovis in the Americas, International Science conference proceedings* (1sted ed., pp. 221–261). Washington, DC: Smithsonian Institution.

Araujo, A. G. M. (2015). On vastness and variability: Cultural transmission, historicity, and the paleoindian record in Eastern South America. *Anais da Academia Brasileira de Ciências, 87*, 1239–1258.

Araujo, A. G. M., & Feathers, J. K. (2008). First notice of open-air Paleoamerican sites at Lagoa Santa: Some geomorphological and paleoenvironmental aspects, and implications for future research. *Current Research in the Pleistocene, 25*, 27–29.

Araujo, A. G. M., Neves, W. A., & Piló, L. B. (2003). *Eventos de seca no Holoceno e suas implicações no povoamento pré-histórico do Brasil Central. Anais do IX Congresso da Associação Brasileira de Estudos do Quaternário*. Recife: Pernambuco, Brazil, ABEQUA, CD-ROM.

Araujo, A. G. M., Neves, W. A., Piló, L. B., & Atui, J. P. V. (2005). Holocene dryness and human occupation in Brazil during the "Archaic Gap". *Quaternary Research, 64*, 298–307.

Araujo, A. G. M., Pilo, L. B., Neves, W. A., & Atui, J. P. V. (2006). Human occupation and paleoenvironments in South America: Expanding the notion of an "Archaic Gap". *Revista do Museu de Arqueologia e Etnologia, 15-16*, 3–35.

Araujo, A. G. M., Feathers, J. K., Arroyo-Kalin, M., & Tizuka, M. M. (2008). Lapa das Boleiras rockshelter: Stratigraphy and formation processes at a Paleoamerican site in Central Brazil. *Journal of Archaeological Science, 35*, 3186–3202.

Araujo, A. G. M., Neves, W. A., & Kipnis, R. (2012). Lagoa Santa revisited: An overview of the chronology, subsistence, and material culture of paleoindian sites in eastern Central Brazil. *Latin American Antiquity, 23*, 533–550.

Araujo, A. G. M., Strauss, A. M., Feathers, J. K., Paisani, J. C., & Schrage, T. J. (2013). Paleoindian open-air sites in tropical settings: A case study in formation processes, dating methods, and paleoenvironmental models in central Brazil. *Geoarchaeology, 28*, 195–220.

Auler, A. S., & Smart, P. L. (2001). Late Quaternary paleoclimate in semiarid northeastern Brazil from U-series dating of travertine and water-table speleothems. *Quaternary Research, 55*, 159–167.

Auler, A. S., Piló, L. B., Smart, P. L., Wang, X., Hoffmann, D., Richards, D. A., Edwards, R. L., Neves, W. A., & Cheng, H. (2006). U-series dating and taphonomy of Quaternary vertebrates from Brazilian caves. *Palaeogeography Palaeoclimatology Palaeoecology, 240*, 508–522.

Barberi, M., Salgado-Laboriau, M. L., & Suguio, K. (2000). Paleovegetation and paleoclimate of "Vereda de Águas Emendadas", central Brazil. *Journal of South American Earth Sciences, 13*, 241–254.

Behling, H. (2002). South and southeast Brazilian grasslands during late Quaternary times: A synthesis. *Paleogeography, Palaeoclimatology, Palaeoecology, 177*, 19–27.

Bigarella, J. J., Mousinho, M. R., & Silva, J. D. (1965). Pediplanos, pedimentos e seus depósitos correlativos no Brasil. *Boletim Paranaense de Geografia, 16*, 153–197.

Bitencourt, A. L. V. (2004). A Formação e evolução estratigráfica do abrigo do Morro Furado: processos interativos entre a morfogênese cárstica e a ocupação pré-histórica. *Canindé, 4*, 55–73.

Boulet, R., Pessenda, L. C. R., Telles, E. C., & Melfi, A. (1995). Une evaluation de la vitesse de l'accumulation superficielle de matiére par la faune du sol à partir de la datation des charbons et de l'humine du sol. Exemple des latosols des versants du lac Campestre, Salitre, Minas Gerais, Brésil. Comptes Rendus de la Academie de Sciences de Paris. *Geosciences de Surface (Pédologie), 320*, 287–294.

Bryan, A., & Gruhn, R. (1978). Results of a test excavation at Lapa Pequena, MG, Brazil. *Arquivos do Museu de História Natural da UFMG, 3*, 261–332.

Bueno, L. (2010). Tecnologia lítica, cronologia e sequencia de ocupação: o estudo de um sítio a céu aberto na região de Lagoa Santa, MG. *Revista do Museu de Arqueologia e Etnologia, 20*, 91–107.

Cahen, D., & Moeyersons, J. (1977). Sub-surface movements of stone artifacts and their implications for the prehistory of Central Africa. *Nature, 266*, 812–815.

Clarke, M. F., Williams, M. A. J., & Stokes, T. (1999). Soil creep: Problems raised by a 23 year study in Australia. *Earth Surface Processes and Landforms, 24*, 151–175.

Da-Gloria P. (2012). Health and lifestyle in the paleoamericans: Early holocene biocultural adaptation at Lagoa Santa, Central Brazil. Ph.D. dissertation, The Ohio State University.

Da-Gloria, P., & Larsen, C. S. (2014). Oral health of the Paleoamericans of Lagoa Santa, Central Brazil. *American Journal of Physical Anthropology, 154*, 11–26.

Daniel, G. (1975). *A hundred and fifty years of archaeology* (2nd ed.). London: Duckworth. 410pp.

De Oliveira, P.E. (1992). A palynological record of late quaternary vegetational and climatic change in Southeastern Brazil. Ph.D. Dissertation, Ohio State University, Columbus.

Etchevarne, C. (2000). A ocupação humana do Nordeste brasileiro antes da colonização portuguesa. *Revista USP, 44*, 112–141.

Feathers, J., Kipnis, R., Piló, L. B., Arroyo-Kalin, M., & Coblentz, D. (2010). How old Luiza? Luminescence dating and stratigraphic integrity at Lapa Vermelha, Lagoa Santa, Brazil. *Geoarchaeology, 25*, 395–436.

Gladfelter, B. G. (1977). Geoarchaeology: The geomorphologist and archaeology. *American Antiquity, 42*, 519–538.

Hansen, S. (1888). Lagoa Santa Racen. En anthropologisk Undersögelse af Jordfundne.

Hassan, F. (1979). Geoarchaeology: The geologist and archaeology. *American Antiquity, 44*, 267–270.

Hiruma, S. T. (2007). Revisão dos conhecimentos sobre o significado das "linhas de seixos". *Revista do Instituto Geológico, 27*(28), 53–64.

Hrdlička, A. (1912). Early man in South America. VII. The skeletal remains of early man in South America. Smithsonian Institution. *Bureau of American Ethnology, 52*, 153–392.

Hubbe, A., Hubbe, M., & Neves, W. A. (2009). New late-Pleistocene dates for the extinct megafauna of Lagoa Santa. *Brazil. Curr. Res. Pleistocene, 26*, 154–156.

Hubbe, A., Haddad-Martim, P., Hubbe, M., Mayer, E., Strauss, A., Auler, A., Piló, L.B., & Neves, W.A. (2011) Identification and importance of critical depositional gaps in pitfall cave environments: The fossiliferous deposit of Cuvieri Cave, Eastern Brazil. *Paleogeography, Paleoclimatology, Paleoecology 321*, 66–78.

Hubbe, A., Hubbe, M., & Neves, W. A. (2013). The Brazilian megamastofauna of the Pleistocene/Holocene and its relationship with the early human settlement of the continent. *Earth-Science Review, 118*, 1–10.

Hurt, W. R., & Blasi, O. (1969). O Projeto Arqueológico "Lagoa Santa" – Minas Gerais, Brasil (Nota Final). *Arquivos do Museu Paranaense, Nova Série, 4*, 1–62.

Karkanas, P. (2001). Site formation processes in Theopetra Cave: A record of climatic change during the late Pleistocene and early Holocene in Thessaly, Greece. *Geoarchaeology, 16*, 373–399.

Karkanas, P., Kyparissi-Apostolika, N., Bar-Yosef, O., & Weiner, S. (1999). Mineral assemblages in Theopetra, Greece: A framework for understanding diagenesis in a prehistoric cave. *Journal of Archaeological Science, 26*, 1171–1180.

Kashimoto, E.M. (1992). *Geoarqueologia no Baixo Paranapanema: Uma Perspectiva Geográfica de Estabelecimentos Humanos Pré-Históricos*. Master Thesis, Universidade de São Paulo, São Paulo.

Kate, H. (1885). Sur les crânes de Lagoa Santa. *Bull. Soc. d`Anthr. de Paris, 8*, 240–244.

Kipnis, R. (2002). Foraging societies of Eastern Central of Brazil: an evolutionary ecological study of subsistence strategies during the terminal Pleistocene and early/middle Holocene. 620p. Ph.D Dissertation University of Michigan.

Lacerda, J. B., & Peixoto, R. (1876). Contribuição para o Estudo Anthropológico das Raças Indígenas do Brasil. *Archivos do Museu Nacional, 1*, 47–76.

Laming-Emperaire, A., Prous, A., Moraes, A., & Beltrão, M. (1974). Grottes et abris de la region de Lagoa Santa. Cahiers d' Archéologie d'Amérique du Sud 1, 195p.

Lanari, C. U. (1909). Ossadas humanas fosseis encontradas numa caverna calcarea nas vizinhanças do Mocambo. *Annaes da Escola de Minas de Ouro Preto, 11*, 15–35.

Ledru, M. P. (1993). Late Quaternary environmental and climatic changes in Central Brazil. *Quaternary Research, 39*, 90–98.

Ledru, M. B., Braga, P. I., Soubies, F., Fournier, M., Martin, L., Suguio, K., & Turcq, B. (1996). The last 50,000 years in the Neotropics (Southern Brazil): Evolution of vegetation and climate. *Palaeogeography, Palaeoclimatology, Palaeoecology, 123*, 239–257.

Lund, P. W. (1844). Notícia sobre ossadas humanas fósseis achadas numa caverna do Brasil. In C. de Paula Couto (Ed.), 1950*Memórias sobre a paleontologia brasileira* (pp. 465–488). Rio de Janeiro: Instituto Nacional do Livro.

Lütken, C. F. (1883). *Exposition de quelques-uns des cranes et des autres ossements humains de Minas-Geraés dans le Brésil central découverts et déterrés par le feu Professeur P.W.Lund* (p. 40). Copenhagen: C.R. Congrès international des Americanistes.

Martin, G. (1997). *Pré-História do Nordeste do Brasil*. Recife: Editora Universitaria da Universidade Federal de Pernambuco.

Martinelli, I. A., Pessenda, L. C. R., Espinoza, E., Camargo, P. B., Telles, F. C., Cerri, C. C., Victoria, R. L., Aravena, R., Richey, J., & Trumbore, S. (1996). Carbon-13 variation with depth in soils of Brazil and climate change during the Quaternary. *Oecologia, 106*, 376–381.

Missão Franco-Brasileira de Lagoa Santa. (1977). Relatório de prospecções realizadas no Município de Montalvânia, MG. *Arquivos do Museu de História Natural da UFMG, 2*, 67–118.

Morais, J. L. (1999). A Arqueologia e o fator geo. *Revista do Museu de Arqueologia e Etnologia, 9*, 3–22.

Moura, M.T. (1997). A Evolução do Sítio Arqueológico Lapa do Boquete na Paisagem Cárstica do Vale do Rio Peruaçú: Januária (MG). Masters Thesis, Universidade de São Paulo, São Paulo.

Neves, W. A., & Piló, L. B. (2003). Solving Lund's dilemma: New AMS dates confirm that humans and megafauna coexisted at Lagoa Santa. *Curr. Res. Pleistocene, 20*, 57–60.

Neves, W.A., & Piló, L.B. (2008). O Povo de Luzia: em Busca dos Primeiros Americanos. Editora Globo. 334p.

Neves, W. A., Powell, J. F., & Ozolins, E. G. (1999). Extra-continental morphological affinities of Lapa Vermelha IV, hominid I: A multivariate analysis with progressive numbers of variables. *Homo, 50*, 263–282.

Neves, W. A., Araujo, A. G., Bernardo, D. V., Kipnis, R., & Feathers, J. K. (2012). Rock art at the Pleistocene/Holocene boundary in eastern South America. *PloS one, 7*(2), e32228.

Noelli, F. S. (2000). A ocupação humana na Região Sul do Brasil: arqueologia, debates e perspectivas - 1872 - 2000. *Revista USP, 44*, 218–269.

Parenti, F. (2001). *Le Gisement Quaternaire de Pedra Furada (Piaui, Brésil): Stratigraphie, Chronologie, Évolution Culturelle*. Paris: Editions Recherche sur les Civilisations. 322 pp.

Penteado, M. M. (1969). Novas informações a respeito dos pavimentos detríticos ("stone lines"). *Notícia Geomorfológica, Campinas, 9*(17), 15–41.

Piló L. B. (1998). Morfologia Cárstica e Materiais Constituintes: Dinâmica e Evolução da Depressão Poligonal Macacos-Baú-Carste de Lagoa Santa, Minas Gerais. Ph.D Dissertation, Universidade de São Paulo, São Paulo.

Piló, L.P., & Neves, W.A. (2003). Novas datações 14C (AMS) confirmam a tese da coexistência do homem com a megafauna plesitocênica da região cárstica de Lagoa Santa, MG. Anais do XXVII Congresso Brasileiro de Espeleologia–SBE 100–104.

Piló, L. B., Auler, A. S., Neves, W. A., Wang, X., Cheng, H., & Edwards, R. L. (2005). Geochronology, sediment provenance, and fossil emplacement at Sumidouro cave, a classic late pleistocene/early holocene paleoanthropological site in eastern Brazil. *Geoarchaeology, 20*, 751–764.

Prous, A. (1991). Fouilles de l'Abri du Boquete, Minas Gerais, Brésil. *Journal de la Société des Américanistes, 77*, 77–109.

Prous, A. (2002). O homem pré-histórico e a megamastofauna pleistocênica no Brasil – uma revisão bibliográfica. *O Carste, 14*(1), 52–59.

Prous, A., & Fogaça, E. (1999). Archaeology of the pleistocene-holocene boundary in Brasil. *Quaternary International, 53-54*, 21–41.

Quatrefages, A. (1881). L'homme fossile de Lagoa Santa en Brésil et ses descendants actuels [The Lagoa Santa fossil Man in Brazil and its present descendants]. *Comptes Rendus Academie Sciences Paris, 93*, 882–884.

Racza, M. F., De Oliveira, P. E., Bush, M., & McMichael, C. H. (2013). Two paleoecological histories spanning the period of human settlement in southeastern Brazil. *Journal of Quaternary Science, 28*, 144–151.

Reinhardt, J. T. (1888). De brasilianske Knoglehuler og de i dem forekommende Dyrelevninger. *E Museo Lundii (I), 4*, 1–56.

Rick, J. W. (1987). Dates as data: An examination of the Peruvian preceramic radiocarbon record. *American Antiquity, 52*, 55–73.

Rivet, P. (1908). La race de Lagoa Santa chez les populations précolombiennes de l'Equateur. *Bull. et Mem. Soc. d'Anthr. Paris, 5me sér.*, IX, fasc, 2, 209–274.

Rubin, J. C. R., Silva, R. T., & Barberi, M. (2003). Arqueologia e a dinâmica da paisagem: pesquisa arqueológica em ambientes fluviais. *Revista Habitus, 1*, 297–316.

Salgado-Laboriau, M. L., Casseti, V., Ferraz-Vicentini, K. R., Martin, L., Soubies, F., Suguio, K., & Turcq, B. (1997). Late Quaternary vegetational and climatic changes in cerrado and palm swamp from central Brazil. *Palaeogeography, Palaeoclimatology, Palaeoecology, 128*, 215–226.

Scheel-Ybert, R., Vernet, J.-L., Wengler, L., & Fournier, M. (1995). Carvões do solo em São Pedro, Estado de São Paulo, Brasil: datação, notas sobre o paleoambiente no Quaternário Recente, condições de depósito e origem do fogo e proposta de estudos antracológicos. *Anais do Congresso da Associação Brasileira de Estudos do Quaternário, Niterói, RJ, 5*, 169–175.

Schiegl, S., Goldberg, P., Bar-Yosef, O., & Weiner, S. (1996). Ash deposits in hayonim and kebara caves, Israel: Macroscopic, microscopic and mineralogical observations, and their archaeological implications. *Journal of Archaeological Science, 23*, 763–781.

Schmitz, P. I. (1987). Prehistoric hunters and gatherers of Brazil. *Journal of World Prehistory, 1*, 53–126.

Schmitz, P.I., Barbosa, A.S., Jacobus, A., & Ribeiro, M.B. (1989). Arqueologia nos Cerrados do Brasil Central. Pesquisas-Antropologia. Instituto Anchietano de Pesquisas, São Leopoldo.

Shahack-Gross, R., Marshall, F., Ryan, K., & Weiner, S. (2004). Reconstruction of spatial organization in abandoned Maasai settlements: Implications for site structure in the Pastoral Neolithic of East Africa. *Journal of Archaeological Science, 31*, 1395–1411.

Silva, L. V., & Prous, A. (2014). O papel dos resíduos de combustão na formação dos estratos sedimentares na Lapa do Niactor: o elemento antrópico como agente dominante na sedimentação em abrigos. *Arquivos do Museu de História Natural da UFMG, 23*, 107–141.

Thomas, M. F. (1994). *Geomorphology in the tropics – A study of weathering and denudation in low latitudes*. New York: Willey.

Thomas, M. F. (2004). Landscape sensitivity to rapid environmental change – A quaternary perspective with examples from tropical areas. *Catena, 55*, 107–124.

Tudela, D.R.G. (2013). Caracterização Físico-Química de Sedimentos do Sítio Arqueológico Lapa Grande de Taquaraçu, MG. Masters Thesis, Universidade de São Paulo, São Paulo.

Villagran, X. S., Strauss, A., Miller, C., Ligouis, B., & Oliveira, R. (2017). Buried in ashes: Site formation processes at Lapa do Santo rockshelter, east-central Brazil. *Journal of Archaeological Science, 77*, 10–34.

Wagner, G. P. (2014). The origins of the Brazilian sambaquis (shellmounds): From a historical perspective. *Cadernos do LEPAARQ XI, 21*, 209–220.

Walter, H. V. (1948). *The pre-history of the Lagoa Santa region*. Belo Horizonte: Velloso & Cia. Ltda. 165p.

Weiner, S., Goldberg, P., & Bar-Yosef, O. (2002). Three-dimensional distribution of minerals in the sediments of Hayonim Cave, Israel: Diagenetic processes and archaeological implications. *Journal of Archaeological Science, 29*, 1289–1308.

Weninger, B., Jöris, O., Danzeglocke, U. (2016). CalPal-2007.Cologne radiocarbon calibration, palaeoclimate research package. http://www.calpal.de/. Accessed 30 Oct 2016.

Index

P. Da-Gloria et al. (eds.), *Archaeological and Paleontological Research in Lagoa Santa*, DOI 10.1007/978-3-319-57466-0

G

M

The manufacturer's authorised representative in the EU is Springer Nature Customer Service Centre GmbH, Europaplatz 3, 69115 Heidelberg, Germany. If you have any concerns regarding our products, please contact ProductSafety@springernature.com

Printed and bound by CPI Group (UK) Ltd, Croydon, CR0 4YY
22/09/2026
02221052-0002